Lecture Notes in Artificial Intelligence   8249

Subseries of Lecture Notes in Computer Science

Lecture Notes in Artificial Intelligence 8249

Subseries of Lecture Notes in Computer Science

LNAI Series Editors

LNAI Founding Series Editor

Matteo Baldoni   Cristina Baroglio
Guido Boella   Roberto Micalizio (Eds.)

AI*IA 2013:
Advances in
Artificial Intelligence

XIIIth International Conference
of the Italian Association for Artificial Intelligence
Turin, Italy, December 4-6, 2013
Proceedings

 Springer

Volume Editors

Matteo Baldoni
Cristina Baroglio
Guido Boella
Roberto Micalizio
Università degli Studi di Torino
Dipartimento di Informatica
via Pessinetto 12
10149 Torino, Italy
E-mail:{baldoni, baroglio, guido, micalizio}@di.unito.it

ISSN 0302-9743   e-ISSN 1611-3349
ISBN 978-3-319-03523-9   e-ISBN 978-3-319-03524-6
DOI 10.1007/978-3-319-03524-6
Springer Cham Heidelberg New York Dordrecht London

Library of Congress Control Number: 2013952452

CR Subject Classification (1998): I.2, F.1, F.4.1, I.2.6, H.2.8, I.5, H.3-5

LNCS Sublibrary: SL 7 – Artificial Intelligence

Typesetting: Camera-ready by author, data conversion by Scientific Publishing Services, Chennai, India

Printed on acid-free paper

Springer is part of Springer Science+Business Media (www.springer.com)

*This volume is dedicated to Leonardo Lesmo,
colleague, friend, mentor to so many of us.*

Preface

AI*IA 2013 was the thirteenth in a series of international conferences on Advances in Artificial Intelligence, held bi-annually in Italy by the Italian Association for Artificial Intelligence (AI*IA). The first edition of the conference was held in Trento in 1989, then in Palermo 1991, Turin 1993, Florence 1995, Rome 1997, Bologna 1999, Bari 2001, Pisa 2003, Milan 2005, Rome 2007, Reggio Emilia 2009, Palermo 2011.

The 2013 edition of the conference was special as it celebrated the 25th anniversary of AI*IA, founded in February 1988. Turin was honored to host again this event in this very special year and so to host also the celebrations. For this important occasion, the conference format was re-designed so as to better accomplish the mission of a conference, i.e. creating a meeting place that favors the exchange of ideas.

A special day was also organized on December, 4th, 2013. The former and current presidents of the association (Luigia Carlucci Aiello, Oliviero Stock, Pietro Torasso, Roberto Serra, Marco Gori, Marco Schaerf, and Paola Mello), provoked by the journalist and scientific popularizer Andrea Vico, enlivened a panel discussion on their respective chairing stints, the challenges, the priorities, and the achieved results as well as current directions and the future of Artificial Intelligence in Italy.

In the late afternoon, the movie Blade Runner was projected at Cinema Massimo with the aim of attracting a broad public towards Artificial Intelligence research, by relying on an artistic rather than a technical language. Many times research in Artificial Intelligence gave precious cues to cinema productions looking for visions about the future. Blade Runner is an iconic example and for this reason it can, better than other movies, convey an interest towards AI. The projection was complemented by a discussion about the intersections of research and artistic productions, lead by prof. Lorenza Saitta, from the University of Piemonte Orientale Amedeo Avogadro, and prof. Roy Menarini, from the University of Udine. This special event was organized by Alessandro Mazzei and Vincenzo Lombardo with the help of the Department of Humanities, and of the Cinema Museum of Turin.

In order to foster discussion and to facilitate idea exchange, community creation, and collaboration, the conference became, in its 2013 edition, a "social" conference. In the time between acceptance and presentation, each paper had its own public web space and a forum, where authors moderated discussions on the topics of their research. Presentations at the conference, thus, had a deeper impact: by attending the conference, researchers had the chance to tighten the relationships their papers started on the web, by means of vis-à-vis discussions. On this occasion, they had the chance to profit from reserved tables, where they met a motivated and interested audience, composed of other researchers, stakeholders from the industrial world, and students. Attendees had the possibility to book directly with authors time slots for discussion, through the web, or to roam from table to table at the conference, browsing through discussions.

AI*IA 2013 received 86 submissions from 23 countries. Each submission was reviewed by at least 3 Program Committee members. The Committee decided to accept 45 papers. They are grouped in this volume in eight areas: knowledge representation and reasoning, machine learning, natural language processing, planning, distributed artificial intelligence: robotics and multi-agent systems, recommender systems and semantic web, temporal reasoning and reasoning under uncertainty, artificial intelligence applications. AI*IA 2013 was honored to host also two invited speakers: prof. Giuseppe Attardi, from the University of Pisa, and prof. Rafael H. Bordini, from the Pontificia Universidade Católica do Rio Grande do Sul.

We would like to thank all the authors for their participation and the members of the Program Committee for their excellent work during the reviewing phase. Moreover, we would like to thank all the members of the Steering Committee of AI*IA for their advice, a special thanks to Fabrizio Riguzzi, and Sara Manzoni also for their support in the organization. Finally, we are very grateful to prof. Paola Mello for her personal involvement and constant encouragement all along the months which led to AI*IA 2013.

This event was sponsored by AI*IA, by the Department of Computer Science of the University of Torino, and by the Artificial Intelligence Journal, and it obtained the "patrocinio" by Regione Piemonte. The submission, revision, and proceeding preparation phases were supported by EasyChair.

September 2013

Matteo Baldoni
Cristina Baroglio
Guido Boella
Roberto Micalizio

Organization

Program and Organizer Chairs

Matteo Baldoni	University of Torino, Italy
Cristina Baroglio	University of Torino, Italy
Guido Boella	University of Torino, Italy

Organizing Committee

Matteo Baldoni	University of Torino, Italy
Cristina Baroglio	University of Torino, Italy
Guido Boella	University of Torino, Italy
Federico Capuzzimati	University of Torino, Italy
Vincenzo Lombardo	University of Torino, Italy
Alessandro Mazzei	University of Torino, Italy
Roberto Micalizio	University of Torino, Italy

Program Committee

Giuliano Armano	University of Cagliari, Italy
Matteo Baldoni	University of Torino, Italy
Cristina Baroglio	University of Torino, Italy
Roberto Basili	University of Roma Tor Vergata, Italy
Federico Bergenti	University of Parma, Italy
Stefano Bistarelli	University of Perugia, Italy
Guido Boella	University of Torino, Italy
Luciana Bordoni	ENEA, Italy
Marco Botta	University of Torino, Italy
Stefano Cagnoni	University of Parma, Italy
Diego Calvanese	Free University of Bozen-Bolzano, Italy
Antonio Camurri	University of Genova, Italy
Amedeo Cappelli	ISTI-CNR, Italy
Luigia Carlucci Aiello	University of Roma "Sapienza", Italy
Amedeo Cesta	CNR, Italy
Antonio Chella	University of Palermo, Italy
Luca Console	University of Torino, Italy
Rosaria Conte	ICST-CNR, Italy
Gabriella Cortellessa	ISTC-CNR, Italy
Mehdi Dastani	Utrecht University, The Netherlands
Giuseppe De Giacomo	University of Roma "Sapienza", Italy

Francesco M. Donini	University of Tuscia, Italy
Floriana Esposito	University of Bari, Italy
Patrick Gallinari	University of Paris 6, France
Mauro Gaspari	University of Bologna, Italy
Nicola Gatti	Politecnico di Milano, Italy
Alfonso Emilio Gerevini	University of Brescia, Italy
Laura Giordano	University of Piemonte Orientale, Italy
Marco Gori	University of Siena, Italy
Nicola Guarino	CNR, Italy
Evelina Lamma	University of Ferrara, Italy
Nicola Leone	University of Calabria, Italy
Leonardo Lesmo	University of Torino, Italy
Bernardo Magnini	FBK, Italy
Donato Malerba	University of Bari, Italy
Alberto Martelli	University of Torino, Italy
Paola Mello	University of Bologna, Italy
Emanuele Menegatti	University of Padova, Italy
Stefania Montani	University of Piemonte Orientale, Italy
Alessandro Moschitti	University of Trento, Italy
Daniele Nardi	University of Roma "Sapienza", Italy
Angelo Oddi	ISTC-CNR, Italy
Andrea Omicini	University of Bologna, Italy
Roberto Pirrone	University of Palermo, Italy
Piero Poccianti	Consorzio Operativo Gruppo MPS, Italy
Daniele P. Radicioni	University of Torino, Italy
Fabrizio Riguzzi	University of Ferrara, Italy
Andrea Roli	University of Bologna, Italy
Francesca Rossi	University of Padova, Italy
Fabio Sartori	University of Milano, Italy
Marco Schaerf	University of Roma "Sapienza", Italy
Giovanni Semeraro	University of Bari, Italy
Rosario Sorbello	University of Palermo, Italy
Oliviero Stock	FBK, Italy
Armando Tacchella	University of Genova, Italy
Pietro Torasso	University of Torino, Italy
Eloisa Vargiu	Barcelona Digital, Spain
Marco Villani	University of Modena and Reggio Emilia, Italy
Giuseppe Vizzari	University of Milano-Bicocca, Italy

Additional Reviewers

Alviano, Mario	Baioletti, Marco	Bellandi, Andrea
Antonello, Mauro	Basile, Pierpaolo	Bellodi, Elena
Armano, Giuliano	Basso, Filippo	Benotto, Giulia

Biba, Marenglen
Bloisi, Domenico
Bonanni, Taigo Maria
Botoeva, Elena
Bragaglia, Stefano
Callaway, Charles
Cannella, Vincenzo
Chesani, Federico
Codetta Raiteri, Daniele
Colucci, Simona
D'Amato, Claudia
Damiano, Rossana
De Gemmis, Marco
Di Mauro, Nicola
Esposito, Roberto
Fanizzi, Nicola
Gavanelli, Marco
Gemignani, Guglielmo
Gena, Cristina
Gentilini, Raffaella
Ghidoni, Stefano
Giovannetti, Emiliano
Giuliani, Alessandro

Gori, Marco
Grasso, Giovanni
Greco, Gianluigi
Guerini, Marco
Iaquinta, Leo
Levorato, Riccardo
Lieto, Antonio
Lops, Pasquale
Manna, Marco
Mazzei, Alessandro
Mesejo, Pablo
Michieletto, Stefano
Mordonini, Monica
Musto, Cataldo
Narducci, Fedelucio
Nashed, Youssef S. G.
Not, Elena
Orlandini, Andrea
Patrizi, Fabio
Patti, Viviana
Peano, Andrea
Pensa, Ruggero G.
Perri, Simona

Pini, Maria Silvia
Pipitone, Arianna
Previtali, Fabio
Rossi, Fabio
Saitta, Lorenza
Santini, Francesco
Santoso, Ario
Scala, Enrico
Schifanella, Claudio
Serina, Ivan
Simkus, Mantas
Tamponi, Emanuele
Testerink, Bas
Theseider Dupré,
 Daniele
Torroni, Paolo
Ugolotti, Roberto
van der Torre, Leon
Vassos, Stavros
Venable, Kristen Brent
Vieu, Laure
Zancanaro, Massimo
Zanzotto, Fabio Massimo

Sponsoring Institutions

AI*IA 2013 was partially finded by the Artificial Intelligence Journal, by the Computer Science Department of the University of Turin, and by the Italian Association for Artificial Intelligence. AI*IA 2013 received the "patrocinio" of Regione Piemonte.

Table of Contents

Knowledge Representation and Reasoning

Machine Learning

Natural Language Processing

Planning

Distributed AI: Robotics and MAS

Recommender Systems and Semantic Web

Temporal Reasoning and Reasoning under Uncertainty

AI Applications

Comparing Alternative Solutions
for Unfounded Set Propagation in ASP

Mario Alviano, Carmine Dodaro, and Francesco Ricca

Department of Mathematics and Computer Science,
University of Calabria, 87036 Rende, Italy
{alviano,dodaro,ricca}@mat.unical.it

Abstract. Answer Set Programming (ASP) is a logic programming language
for nonmonotonic reasoning. Propositional ASP programs are usually evaluated
by DPLL algorithms combining unit propagation with operators that are specific
of ASP. Among them, unfounded set propagation is used for handling recursive
programs by many ASP solvers. This paper reports a comparison of two available solutions for unfounded set propagation, the one adopted in DLV and that
based on source pointers. The paper also discusses the impact of splitting the
input program in components according to head-to-body dependencies. Both solutions and variants have been implemented in the same solver, namely WASP.
An advantage in properly splitting the program in components is highlighted by
an experiment on a selection of problems taken from the 3rd ASP Competition.
In this experiment the algorithm based on source pointers performs better.

1 Introduction

Answer Set Programming (ASP) [1] is a declarative programming paradigm proposed
in the area of non-monotonic reasoning and logic programming. The approach to problem solving in ASP is based on the idea to represent a given computational problem by
an ASP program whose answer sets correspond to solutions, and then use a solver to
find them [2]. In its general form, allowing disjunction in rule heads and nonmonotonic
negation in rule bodies, ASP can be used to solve all problems in the second level of the
polynomial hierarchy [3]. The combination of a comparatively high expressive power
with the availability of robust and effective implementations [4] makes ASP a powerful
tool for developing advanced applications. Nonetheless, complex applications are often
demanding in terms of performance, and thus the development of faster systems is still
an interesting and challenging research topic.

The typical workflow followed for evaluating ASP programs is composed of two
steps, namely instantiation (or grounding) and model generation (or answer set computation). Instantiation consists of producing a propositional ASP program equivalent to
the input one. Model generation amounts to computing the answer sets of the propositional produced by the instantiator. In both steps non-trivial evaluation techniques are
combined together for obtaining a performant and robust system.

In this paper we focus on the model generation task, which is implemented by a
module often called ASP solver. ASP solvers are similar to SAT solvers, since both

M. Baldoni et al. (Eds.): AI*IA 2013, LNAI 8249, pp. 1–12, 2013.
© Springer International Publishing Switzerland 2013

are based on the Davis-Putnam-Logemann-Loveland (DPLL) backtracking search algorithm [5]. The role of Boolean Constraint Propagation in SAT-solvers (based only on *unit propagation* inference rule) is taken in ASP solvers by a procedure combining several inference rules. Those rules combine an extension of the well-founded operator for disjunctive programs with a number of techniques based on ASP program properties (see, e.g., [6]). Among them, unfounded set propagation is successfully used for handling recursive programs by many ASP solvers.

This paper reports a comparison of two known solutions for unfounded set propagation, the one adopted by DLV [7] and that based on source pointers [8]. These techniques were originally introduced and implemented about ten years ago within solvers based on look-ahead techniques (i.e., Smodels [8] and DLV [7]). It may be useful to verify and compare their behavior within a modern ASP solver featuring clause learning [9], backjumping [10], restarts [11], and look-back heuristics [12]. Moreover, it is interesting to study in this setting the impact of localizing the computation of unfounded sets by splitting the input program in recursive components.

In order to compare on a fair ground the above-mentioned unfounded set propagation techniques, they were implemented in the same solver, namely WASP [13].[1] The performance of the corresponding variants of WASP has been measured in an experiment where unfounded set propagation plays a crucial role. Benchmarks were taken from the 3rd ASP Competition suite [4].

In the following, after briefly introducing ASP and the model generation algorithm, we illustrate in detail unfounded set propagation; then, we describe the experiment and analyze the results. Finally, we discuss related work and draw the conclusion.

2   Preliminaries

In this section, we first introduce the syntax and the semantics of ASP. Then, we sketch an algorithm for answer set computation.

2.1   Syntax and Semantics

Let $\mathcal{A}$ be a countable set of propositional atoms. A *literal* is either an atom (a positive literal), or an atom preceded by the *negation as failure* symbol not (a negative literal). A *program* is a finite set of rules of the following form:

$$p_1 \vee \cdots \vee p_n \; :\text{-} \; q_1, \ldots, q_j, \text{ not } q_{j+1}, \ldots, \text{ not } q_m. \tag{1}$$

where $p_1, \ldots, p_n, q_1, \ldots, q_m$ are atoms and $n \geq 0$, $m \geq j \geq 0$. The disjunction $p_1 \vee \cdots \vee p_n$ is called head, and the conjunction $q_1, \ldots, q_j, \text{ not } q_{j+1}, \ldots, \text{ not } q_m$ is referred to as body. For a rule r of the form (1), the following notation is also used: $H(r)$ denotes the set of head atoms; $B(r)$ denotes the set of body literals; $B^+(r)$ and $B^-(r)$ denote the set of atoms appearing in positive and negative body literals, respectively. In the following, the complement of a literal ℓ is denoted $\bar{\ell}$, i.e., $\bar{a} = \text{not } a$ and $\overline{\text{not } a} = a$

---

[1] WASP can be downloaded at http://www.mat.unical.it/wasp.

Algorithm 1. Compute Answer Set

Input : An interpretation I for a program Π
Output: An answer set for Π or *Incoherent*

1 **begin**
2   **while** Propagate(I) **do**
3   **if** I *is total* **then**
4   **if** CheckModel(I) **then return** I;
5   **break**; // goto 12
6   ℓ := ChooseUndefinedLiteral();
7   I' := ComputeAnswerSet($I \cup \{\ell\}$);
8   **if** $I' \neq$ *Incoherent* **then**
9   **return** I';
10   **if** *there are violated learned constraints* **then**
11   **return** *Incoherent*;
12   AnalyzeConflictAndLearnConstraints(I);
13   **return** *Incoherent*;

---

for an atom a. This notation extends to sets of literals, i.e., $\overline{L} := \{\overline{\ell} \mid \ell \in L\}$ for a set of literals L.

An *interpretation* I is a set of literals, i.e., $I \subseteq \mathcal{A} \cup \overline{\mathcal{A}}$. Intuitively, literals in I are true, literals whose complements are in I are false, and all other literals are undefined. I is total if there are no undefined literals, and I is inconsistent if there is $a \in \mathcal{A}$ such that $\{a, \text{not } a\} \subseteq I$. An interpretation I satisfies a rule r if $H(r) \cap I \neq \emptyset$ whenever $B(r) \subseteq I$. On the contrary, I violates r if $\overline{H(r)} \cup B(r) \subseteq I$. Given an interpretation I for Π, a positive literal ℓ is *supported* if there exist a rule $r \in \Pi$ such that: $a \in H(r)$, $B(r) \subseteq I$ and $H(r) \setminus \{\ell\} \cap I = \emptyset$. In this case r is said to be a *supporting rule* for ℓ in I. A *model* of a program Π is a total consistent interpretation satisfying all rules of Π. The reduct [1] Π^M of Π w.r.t. M is obtained by deleting from Π each rule r such that $B^-(r) \cap I \neq \emptyset$, and then by removing all the negative literals from the remaining rules. A model M is an answer set for Π if Π^M has no model N s.t. $N^+ \subset M^+$, where N^+ and M^+ are the sets of positive literals in N and M, respectively. If M is an answer set of a program Π, then all positive literals in M are supported.

2.2 Answer Set Computation

An answer set of a given propositional program Π can be computed by Algorithm 1, which is similar to the Davis-Putnam-Logemann-Loveland procedure in SAT solvers. Initially, interpretation I is empty and extended by function Propagate, which usually includes several deterministic inference rules for pruning the search space. If an inconsistency (or conflict) is detected, function Propagate returns false. In this case, the inconsistency is analyzed and a constraint is learned (line 12). If no inconsistency is detected, function Propagate returns true and the algorithm checks whether I is total and stable (we refer to [14] for a description of the stability check). In case there are still undefined atoms, a heuristic criterion is used to chose one, say ℓ, and the computation proceeds recursively on $I \cup \{\ell\}$. If the recursive call returns an answer set, the computation ends returning it. Otherwise, if an inconsistency arises, chosen literals are unrolled until consistency of I is restored (backjumping), and the computation resumes by propagating the consequences of the constraints learned by the conflict analysis.

In Algorithm 1, an important role is played by the inference rules implemented by function Propagate, which may benefit of search space pruning operators specific of ASP. In particular, in addition to *unit propagation*, the inference rule implemented by SAT solvers, ASP solvers may rely on *support propagation* and *unfounded set propagation*. A brief description of these inference rules is provided below, and a more in-depth analysis of the unfounded set propagation is reported in the next section.

Unit Propagation. An undefined literal ℓ is inferred by unit propagation if there is a rule r that can be satisfied only by ℓ, i.e., r is such that $\bar{\ell} \in \overline{H(r)} \cup B(r)$ and $(\overline{H(r)} \cup B(r)) \setminus \{\bar{\ell}\} \subseteq I$.

Support Propagation. Answer sets are supported models: for each atom a in an answer set M there is a rule r such that $B(r) \subseteq M$ and $H(r) \cap M = \{a\}$; such a r is referred to as a *supporting rule* of a. Support propagation applies either when an atom a has no candidate supporting rules, or when a true atom has exactly one candidate supporting rule r. In the first case, atom a is derived false, while in the second case all undefined atoms in $H(r)$ are derived false and all literals in $B(r)$ are inferred true.

Unfounded Set Propagation. The answer sets of a program Π constitute an antichain of the partial ordered set defined by the subset-containment relation over the set of all interpretations for Π. Stated differently, any pair of distinct answer sets of Π is incomparable with respect to subset-containment. It turns out that supportedness is not sufficient for characterizing answer sets in general, and supported models not being answer sets are referred to as *unfounded*. Unfounded set propagation aims at detecting unfounded atoms, which are then deterministically inferred as false.

3 Unfounded Set Detection and Propagation

In this section we report a brief description of two algorithms for unfounded set detection. One was introduced by Calimeri et al. [15] and implemented in the system DLV, the other was proposed by Simons et al. [8] and implemented in the system Smodels. The algorithms are presented for receiving in input a single component of the input program, a notion also introduced in this section, but they can process the full program as well. We first provide a formal definition of unfounded set.

Definition 1 (Unfounded Set). *A set X of atoms is an unfounded set if for each r such that $H(r) \cap X \neq \emptyset$ at least one of the following conditions is satisfied:*

1. $B(r) \cap \overline{I} \neq \emptyset$;
2. $B^+(r) \cap X \neq \emptyset$;
3. $I \cap H(r) \setminus X \neq \emptyset$.

Intuitively, the first condition checks whether rule r is already satisfied by I because of a false body literal, while the last two conditions ensure that r can be satisfied by deriving all atoms in X as false.

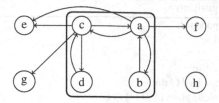

Fig. 1. Dependency graph of the program Π_1

3.1 Splitting the Program in Components

Let Π be a program. The *dependency graph* of Π, denoted $DG_\Pi = (N, A)$, is a directed graph in which (i) each atom of Π is a node in N and (ii) there is an arc in A directed from a node a to a node b whenever there is a rule r in Π such that a and b appear in $H(r)$ and $B^+(r)$, respectively. (Note that negative literals cause no arc in DG_Π. We also assume that any rule r such that $H(r) \cap B^+(r) \neq \emptyset$ is removed from Π in a previous step.) A strongly connected component (or simply component) C of DG_Π is a maximal subset of N such that each node in C is connected by a path to all other nodes in C. A component C containing more than one atom is *recursive*, or *cyclic*; it is *head-cycle free* (HCF for short) if each rule r in Π has at most one head atom in C, that is, each $r \in \Pi$ is such that $|H(r) \cap C| \leq 1$. The rules of Π can be assigned to one or more components. More specifically, a rule r is assigned to a component C if $H(r) \cap C \neq \emptyset$. Moreover, r is said to be an *external rule* of C if $B^+(r) \cap C = \emptyset$; otherwise, r is an *internal rule* of C.

Example 1. Consider the following program Π_1:

$$r_0 : a :\!- b. \quad r_2 : c :\!- d. \quad r_4 : a :\!- c, e. \quad r_6 : a :\!- f. \quad r_8 : e \vee h.$$
$$r_1 : b :\!- a. \quad r_3 : d :\!- c. \quad r_5 : c :\!- a, e. \quad r_7 : c :\!- g. \quad r_9 : f \vee g.$$

The dependency graph DG_{Π_1} of Π is reported in Fig. 1, where we also represented the unique recursive component $\{a, b, c, d\}$. The component is also HCF; its external rules are r_6 and r_7, while rules r_0–r_5 are internal.

3.2 DLV Algorithm

Algorithm 2 reports the unfounded set propagation implemented in the system DLV[15]. The algorithm is run for a cyclic, HCF component C after the satisfaction of any rule of C. In a nutshell, the algorithm computes the set of founded atoms in a given component C by first determining atoms with some external supporting rule (lines 3–6). Then, these atoms are used for deriving other founded atoms (lines 7–11). More specifically, a founded atom a derives a new founded atom a' if some internal rule of C having a in the positive body and a' in the head is such that r can still support a' and all recursive body atoms have been derived to be founded. At the end of this process, all atoms of C that have not been derived as founded form an unfounded set, which is the output of the algorithm (line 12).

Algorithm 2. DLV Algorithm

   Input   : An HCF component C
   Output: An unfounded set
1 **begin**
2 | $FoundedAtoms := \emptyset$;
3 | **foreach** *external rule r of C* **do**
4 | | Let a be the unique atom in $H(r) \cap C$;
5 | | **if** *r can support a* **then**
6 | | | $FoundedAtoms := FoundedAtoms \cup \{a\}$;

7 | **foreach** *atom* $a \in FoundedAtoms$ **do**
8 | | **foreach** *internal rule r of C s.t.* $a \in B^+(r)$ **do**
9 | | | Let a' be the unique atom in $H(r) \cap C$;
10 | | | **if** *r can support a'* **and** $B^+(r) \cap C \subseteq FoundedAtoms$ **then**
11 | | | | $FoundedAtoms := FoundedAtoms \cup \{a'\}$;

12 | **return** $C \setminus FoundedAtoms$;

Example 2. Consider Example 1. Suppose that literal *not e* is chosen by procedure ChooseUndefinedLiteral (line 6 in Alg. 1). Rules r_4 and r_5 become satisfied because their bodies are false, and Alg. 2 is run. External rules are considered first: after processing r_6, atom a is added to $FoundedAtoms$, and after processing r_7, atom c is added to $FoundedAtoms$ (note that the external atom g is still undefined). At this point, new founded atoms may be derived by processing internal rules in which founded atoms appear. In this case, from atom a and rule r_0, atom b is added to $FoundedAtoms$. Then, from atom c and rule r_3, atom d is added to $FoundedAtoms$. Finally, atoms b and d are processed but no new founded atoms are derived. The algorithm thus ends returning an empty unfounded set.

Suppose now that procedure ChooseUndefinedLiteral picks literal *not f*, thus satisfying rule r_6. Alg. 2 is run again. Rule r_6 is satisfied and hence skipped, while r_7 causes the insertion of atom c in $FoundedAtoms$ (again, note that the external atom g is still undefined). Rules in which c occurs as a positive body literal are then processed: atom d is added to $FoundedAtoms$ because of rule r_3, while r_4 causes no insertion. Also processing atom d causes no insertion. The algorithm then terminates returning the unfounded set $\{a, b\}$.

3.3 Source Pointers Algorithm

Algorithm 3 implements unfounded set propagation by means of source pointers. In a nutshell, the algorithm associates each atom in a given component with a rule modeling founded support. Such rules are referred to as source pointers. The algorithm is run after the satisfaction of any source pointer. Its input is a component C and a set of atoms S whose source pointers have been invalidated during the previous step of propagation. Atoms of C whose source pointers contain a positive body literal in S also require the computation of a new source pointer; they are thus added to S,

---

Algorithm 3. Source Pointers Algorithm

Input : An HCF component C and a set $S \subseteq C$ of atoms that lost source pointers
Output: An unfounded set

1 **begin**
2   **foreach** $a \in S$ **do**
3   **foreach** *internal rule r of C s.t.* $a \in B^+(r)$ **do**
4   Let a' be the unique atom in $H(r) \cap C$;
5   **if** *source_pointer*$(a') = r$ **then**
6                 $S := S \cup \{a'\}$;   *source_pointer*$(a') := NULL$;

7   **foreach** $a \in S$ **do**
8   **if** *FindSourcePointer(a, C, S)* **then**
9             $S := S \setminus \{a\}$; $Q := \{a\}$;
10   **foreach** $a' \in Q$ **do**
11   **foreach** *internal rule r of C s.t.* $a' \in B^+(r)$ **do**
12   Let a'' be the unique atom in $H(r) \cap C$;
13   **if** $a'' \in S$ **and** *r can support* a'' **and** $B^+(r) \cap S = \emptyset$ **then**
14                        $S := S \setminus \{a''\}$;   *source_pointer*$(a'') := r$;
15                        $Q := Q \cup \{a''\}$;

16   **return** S;

---

and the process is repeated (lines 2–6). Then, new source pointers are computed by means of function FindSourcePointer, which looks for a supporting rule whose positive body literals already have source pointers. During this process, each atom assigned to a source pointer is removed from S, which may result in new source pointer assignments (lines 7–15).

Example 3. Consider Example 1, and let us assume the following source pointers: *source_pointer*$(a) := r_6$, *source_pointer*$(b) := r_1$, *source_pointer*$(c) := r_7$, *source_pointer*$(d) := r_3$. Suppose that literal *not e* is chosen by procedure Choose-UndefinedLiteral (line 6 in Alg.1). Rules r_4 and r_5 become satisfied due to a false body literal, but Alg. 3 is not run because neither r_4 nor r_5 is a source pointer. Suppose now that procedure ChooseUndefinedLiteral picks literal *not f*, thus satisfying rule r_6, the source pointer of atom a. Alg. 3 is then run for $S := \{a\}$. First, source pointers whose bodies are not disjoint from S have to be unset. In this case, rule r_1 contains atom a in its positive body; atom b is then added to S and its source pointer unset. At this point, the algorithm looks for new source pointers. Concerning atom a, the external rules r_4 and r_6 cannot be source pointers because their bodies are false, while the internal rule r_0 cannot be a source pointer because its positive body contains an atom belonging to S. Similarly, rule r_1 cannot be a source pointer of b because $B^+(r_1) \cap S \neq \emptyset$. Finally, the algorithm terminates returning the unfounded set $S = \{a, b\}$.

Function. FindSourcePointer(Atom a, Component C, SetOfAtoms S)

```
 1  begin
 2      foreach external rule r of C s.t. a ∈ H(r) do
 3          if r can support a then
 4              source_pointer(a) := r;
 5              return TRUE;

 6      foreach internal rule r of C s.t. a ∈ H(r) do
 7          if r can support a and B⁺(r) ∩ S = ∅ then
 8              source_pointer(a) := r;
 9              return TRUE;

10      return FALSE;
```

4 Experiments

In this section we first report a few details on the benchmark problems considered in our experiment. Then, we analyze the performance of the algorithms for unfounded set detection presented in the previous section.

4.1 Benchmark Problems

We consider a selection of benchmark problems in which unfounded set propagation can be applied effectively for pruning the search space. These problems were used in the System Track of the 3rd ASP Competition [4]. In particular, we consider all the five problems belonging to the NP category of the competition featuring instances with at least one cyclic and HCF component, namely *Sokoban Decision*, *Knigth Tour*, *Labyrinth*, *Numberlink* and *Maze Generation*. We ran 50 instances for each of these problems, with the exception of Knight Tour for which only 10 instances were present in the suite of the 3rd Competition. A description of the considered problems follows.

SokobanDecision. Sokoban is a game puzzle developed by the Japanese company Thinking Rabbit, Inc. in 1982. Each instance of the problem models a board divided in cells, where some cells are occupied by walls, and other cells contains boxes. Each box occupies exactly one location, and each location contains exactly one box. The sokoban (i.e., the warehouseman) has the goal of moving the boxes distributed in the board from their initial location into in some specific cells marked as storage space. He can walk on free locations and push single boxes onto unoccupied floor locations. A box can be pushed in one direction as an atomic action.

KnightTour. This problem, used since the first ASP competition, consists in finding a tour of the chessboard for a knight piece that starts at any square, travels all squares, and comes back to the origin, following the knight move rules of chess.

Labyrinth. This is a variation of the Ravensburger's Labyrinth game where one has to guide an avatar through a dynamically changing labyrinth to certain fields. The labyrinth can be modified by pushing rows and columns, so that the avatar can reach the goal field (which changes its location when its location is pushed) from its starting field in less than a given number of allowed pushes. The following rules apply to the considered variant of the problem: (i) pushing out a field of the labyrinth causes it to reappear on the other end of the pushed row or column, respectively; (ii) the avatar can move to any field reachable via a path in which every (non-terminal) field has a connection to its successor, and vice versa; (iii) there is a unique starting and a unique goal field in the labyrinth. A solution is represented by pushes of the labyrinth's rows and columns needed to guide the avatar to its goal.

Numberlink. Numberlink is a type of logic puzzle involving finding paths to connect numbers in a grid. All the matching numbers on the grid have to be paired with single continuous lines that never cross over each other. The problem has been generalized so that the grid can be seen as an undirected graph. The goal is to find a list of paths connecting two terminal vertices for each connection such that no vertex is included in more than two paths.

MazeGeneration. This problem consists of generating a maze in an $m \times n$ grid by placing some walls in the available cells. In a valid solution each cell is either left empty or contains a wall, two distinct cells on the edges are indicated as entrance and exit, and the following conditions are satisfied: (i) each cell on the edge of the grid is a wall, except entrance and exit that are empty; (ii) there is no 2×2 square of empty cells or walls; (iii) if two walls are on a diagonal of a 2×2 square, then not both of their common neighbors are empty; (iv) no wall is completely surrounded by empty cells; (v) there is a path from the entrance to every empty cell.

4.2 Compared Methods and Hardware Settings

We implemented in WASP the two unfounded set propagation algorithms described in Section 3. Moreover, for each of those algorithms we implemented two variants, one in which the problem is split in components as described in Section 3.1, and the other in which the algorithms are applied to the entire program. Thus, we obtained the following four variants of WASP:

- $wasp^{DLV}$: WASP configured with the unfounded set propagation of DLV;
- $wasp^{DLV}_{split}$: WASP with DLV-like propagation and program split in components;
- $wasp^{SP}$: WASP configured with the algorithm based on source pointers;
- $wasp^{SP}_{split}$: WASP with source pointers and program split in components.

Concerning the hardware settings, the experiment was run on a four core Intel Xeon CPU X3430 2.4 GHz, with 4 GB of physical RAM and PAE enabled, running Linux Debian Lenny (32bit). Only one of the four processors was enabled and time and memory limits were set to 600 seconds and 3 GiB (1 GiB = 2^{30} bytes), respectively. Execution times were measured with the Benchmark Tool Run (http://fmv.jku.at/run/).

4.3 Result

The result of the experiment is reported in Table 1. The table features a column for each executable, and a row for each problem. For each pair executable-problem the table reports the number of solved instances and the average execution time. Totals are reported in the last row of the table.

In our experiment, the algorithms based on source pointers perform better than those implemented in DLV. In fact, $wasp_{split}^{SP}$ solved 18 instances more than $wasp_{split}^{DLV}$, and $wasp^{SP}$ solved 29 instances more than $wasp^{DLV}$. Looking in detail, it can be observed that the source pointer algorithm is preferable in all domains, but the differences are more significant for SokobanDecision and MazeGeneration. Indeed, $wasp_{split}^{SP}$ solved 4 instances of SokobanDecision more than $wasp_{split}^{DLV}$, spending around half of the execution time on the average in this domain. The difference between $wasp_{split}^{SP}$ and $wasp_{split}^{DLV}$ is even more marked in MazeGeneration. In fact, in this case $wasp_{split}^{SP}$ (resp. $wasp^{SP}$) solved 11 more instances than $wasp_{split}^{DLV}$ (resp. $wasp^{DLV}$). The different behaviors can be explained by a downside of the DLV-like algorithms. In fact, these algorithms have to be executed after any potential supporting rule for some atom in a cyclic HCF component is lost; on the other hand, the source pointer algorithms are only run when some source pointer is lost. This difference between the two algorithms seems to affect performances significantly.

Concerning the effect of splitting the programs in components, the results reported in Table 1 outline that it is in general advantageous to divide the input program to localize unfounded set computation. In fact, both of the algorithms $wasp_{split}^{SP}$ and $wasp_{split}^{DLV}$ outperform the corresponding 1-component versions. In particular, $wasp_{split}^{SP}$ solved (in total) 3 instances more than $wasp^{SP}$, and $wasp_{split}^{DLV}$ solved 14 more instances than $wasp^{DLV}$. Looking in detail, splitting the program pays of in both Labyrinth and Numberlink, since the instances of these problems feature a dependency graph with several cyclic components. Conversely, splitting in components introduces a little overhead (i.e., a slowdown of around 2%) for KnightTour and MazeGeneration. This behavior can be explained by the fact that the instances of these two problems feature a dependency graph with only one cyclic component. Now consider the case of SokobanDecision, where the instances feature a dependency graph composed of one big cyclic component and some cyclic components containing few atoms. Here $wasp_{split}^{DLV}$ outperforms $wasp^{DLV}$, but $wasp_{split}^{SP}$ in total is slower than $wasp^{SP}$. This can be explained by a heuristic factor. In fact, by analyzing all instances, $wasp_{split}^{SP}$ is faster than $wasp^{SP}$ in 15 instances, while it is slower in just 8 instances. To have an evidence of our conjecture, we repeated the experiment for Sokoban by forcing the same heuristic choices in

Table 1. Comparison between DLV algorithm and source pointers

	$wasp_{split}^{DLV}$		$wasp^{DLV}$		$wasp_{split}^{SP}$		$wasp^{SP}$	
	Solved	Time (s)	Solved	Time (s)	Solved	Time (s)	Solved	Time (s)
10-SokobanDecision	18	4578.71	16	3478.27	22	2739.45	22	2605.49
12-KnightTour	7	99.43	7	99.36	8	272.01	8	266.84
17-Labyrinth	29	3784.78	19	2879.38	30	3729.56	28	2773.71
20-Numberlink	32	1946.98	30	3199.03	33	2012.69	32	1982.96
33-MazeGeneration	39	3947.03	39	3852.47	50	1346.29	50	1322.99
Total	125	14356.93	111	13508.51	143	10100.00	140	8951.99

both $wasp_{split}^{SP}$ and $wasp^{SP}$. The result was that $wasp_{split}^{SP}$ solved 21 instances in 991.59 seconds and $wasp^{SP}$ solved 21 instances in 1034.96 seconds. Further analysis on the solver's statistics revealed that the main advantages of splitting the program in components are due to two factors: the computation is limited to cases for which it is useful; and learning in case of conflict produces smaller and more informative constraints.

5 Related Work

The task of computing answer sets of a propositional program is related to SAT solving. Indeed, modern ASP solvers adopt several techniques inspired by the ones that were originally used for SAT solving, like DPLL search algorithm [5], clause learning [9], backjumping [10], restarts [11], and conflict-driven heuristics [12]. Nonetheless, one of the main differences between SAT solvers and an ASP solver concerns the propagation step. An ASP solver adds to unit propagation (the main inference rule of SAT solvers), a combination of several inference rules. Among these, the inference based on the computation of unfounded sets (that has no correspondence in SAT solving) represents a crucial components for obtaining an efficient ASP solver. The two algorithms for unfounded set propagation considered in this paper were introduced by the developers of DLV [7] and Smodels/GnT [8,16].[2] In particular, Smodels uses source pointers. Unfounded set propagation is implemented in Cmodels3 [17] by learning *loop formulas*, which were introduced in [18]. A loop formula is a SAT formula which basically models an unfounded set. The solver clasp uses also a variant of the source pointers algorithm [19], where an unfounded set is computed which is not necessarily the greatest unfounded set as is done by both DLV and Smodels.

6 Conclusion

In this paper we reported a comparison between two solutions for unfounded set propagation, namely source pointers and DLV-like unfounded set propagation. Moreover, we also analyzed the effect of splitting the input program in components according to head-to-body dependencies for both alternatives. We implemented both solutions and variants in the ASP solver WASP, and ran it on a selection of benchmarks taken from the 3rd ASP Competition. In this experiment the algorithm based on source pointers performs better. Moreover, the advantages of splitting the program in components is also highlighted by the results.

References

1. Gelfond, M., Lifschitz, V.: Classical Negation in Logic Programs and Disjunctive Databases. New Generation Computing 9, 365–385 (1991)
2. Lifschitz, V.: Answer Set Planning. In: Proceedings of the 16th International Conference on Logic Programming (ICLP 1999), Las Cruces, New Mexico, USA, pp. 23–37. The MIT Press (1999)

---

[2] GnT extends the solver Smodels in order to deal with disjunctive programs.

3. Eiter, T., Gottlob, G., Mannila, H.: Disjunctive Datalog. ACM Transactions on Database Systems 22, 364–418 (1997)
4. Calimeri, F., Ianni, G., Ricca, F., Alviano, M., Bria, A., Catalano, G., Cozza, S., Faber, W., Febbraro, O., Leone, N., Manna, M., Martello, A., Panetta, C., Perri, S., Reale, K., Santoro, M.C., Sirianni, M., Terracina, G., Veltri, P.: The Third Answer Set Programming Competition: Preliminary Report of the System Competition Track. In: Delgrande, J.P., Faber, W. (eds.) LPNMR 2011. LNCS, vol. 6645, pp. 388–403. Springer, Heidelberg (2011)
5. Davis, M., Logemann, G., Loveland, D.: A Machine Program for Theorem Proving. Communications of the ACM 5, 394–397 (1962)
6. Faber, W., Leone, N., Pfeifer, G.: Pushing Goal Derivation in DLP Computations. In: Gelfond, M., Leone, N., Pfeifer, G. (eds.) LPNMR 1999. LNCS (LNAI), vol. 1730, pp. 177–191. Springer, Heidelberg (1999)
7. Leone, N., Pfeifer, G., Faber, W., Eiter, T., Gottlob, G., Perri, S., Scarcello, F.: The DLV System for Knowledge Representation and Reasoning. ACM Transactions on Computational Logic 7, 499–562 (2006)
8. Simons, P., Niemelä, I., Soininen, T.: Extending and Implementing the Stable Model Semantics. Artificial Intelligence 138, 181–234 (2002)
9. Zhang, L., Madigan, C.F., Moskewicz, M.W., Malik, S.: Efficient Conflict Driven Learning in Boolean Satisfiability Solver. In: Proceedings of the International Conference on Computer-Aided Design (ICCAD 2001), pp. 279–285 (2001)
10. Gaschnig, J.: Performance measurement and analysis of certain search algorithms. PhD thesis, Carnegie Mellon University, Pittsburgh, PA, USA, Technical Report CMU-CS-79-124 (1979)
11. Gomes, C.P., Selman, B., Kautz, H.A.: Boosting Combinatorial Search Through Randomization. In: Proceedings of AAAI/IAAI 1998, pp. 431–437. AAAI Press (1998)
12. Moskewicz, M.W., Madigan, C.F., Zhao, Y., Zhang, L., Malik, S.: Chaff: Engineering an Efficient SAT Solver. In: Proceedings of the 38th Design Automation Conference, DAC 2001, Las Vegas, NV, USA, pp. 530–535. ACM (2001)
13. Alviano, M., Dodaro, C., Faber, W., Leone, N., Ricca, F.: WASP: A native ASP solver based on constraint learning. In: Cabalar, P., Son, T.C. (eds.) LPNMR 2013. LNCS, vol. 8148, pp. 54–66. Springer, Heidelberg (2013)
14. Koch, C., Leone, N., Pfeifer, G.: Enhancing Disjunctive Logic Programming Systems by SAT Checkers. Artificial Intelligence 15, 177–212 (2003)
15. Calimeri, F., Faber, W., Leone, N., Pfeifer, G.: Pruning Operators for Disjunctive Logic Programming Systems. Fundamenta Informaticae 71, 183–214 (2006)
16. Janhunen, T., Niemelä, I.: GNT — A solver for disjunctive logic programs. In: Lifschitz, V., Niemelä, I. (eds.) LPNMR 2004. LNCS (LNAI), vol. 2923, pp. 331–335. Springer, Heidelberg (2003)
17. Lierler, Y.: CMODELS – SAT-Based Disjunctive Answer Set Solver. In: Baral, C., Greco, G., Leone, N., Terracina, G. (eds.) LPNMR 2005. LNCS (LNAI), vol. 3662, pp. 447–451. Springer, Heidelberg (2005)
18. Lin, F., Zhao, Y.: ASSAT: computing answer sets of a logic program by SAT solvers. Artificial Intelligence 157, 115–137 (2004)
19. Anger, C., Gebser, M., Schaub, T.: Approaching the core of unfounded sets. In: Proceedings of the International Workshop on Nonmonotonic Reasoning, pp. 58–66 (2006)

Mind in Degrees
The Quantitative Dimension of Mental Attitudes

Cristiano Castelfranchi

ISTC-CNR
Goal-Oriented Agents Lab (GOAL)
http://www.istc.cnr.it/group/goal

Abstract. Beliefs and Goals have a crucial "quantitative" dimension or "value/strength": the *"doxastic value"* of beliefs, and the *"motivational value"* of Goals. These "values" and "degrees" play a fundamental role in the cognitive processing: believing or not, preferring and choosing, suffering, expecting, trusting, feeling.

Keywords: Cognitive Architecture, Beliefs, Goals, Decision making, Goal dynamics, BDI.

Premise

The "attitudes" impinging on and characterizing our mental representations ("propositions"), like - in particular - Beliefs and Goals, *qualify* the use or function of such representations in the mental architecture and processing, but they also have a crucial "quantitative" dimension or "value", which plays a fundamental role in these processes: the *"doxastic value"* of (candidate) beliefs (Bs) and the *"motivational value"* of Goals (Gs).

Does this strength just consist in the degree of "activation" of those representations in our brain? Not at all; it is not the level of activation, it is the importance of that B or G in mental processing.

We will not present here a formal treatment of these dimensions and of their interaction, nor a real architecture able to integrate beliefs and goals dynamics (a still far objective); we just want to make clear the importance of the problem, several challenging and open issues and some hints for future AI research.[1]

1   Beliefs

Beliefs have a "degree" of subjective certainty, a "strength"; we can be more or less "sure" about something we believe. More precisely, we do not really "believe" that P if that "candidate" belief does not reach a given threshold of credibility; we do not

---

[1] I would like to thank the anonymous referees of the conference; they gave me crucial suggestions for the improvement of the work.

M. Baldoni et al. (Eds.): AI*IA 2013, LNAI 8249, pp. 13–24, 2013.

believe that P if we just wonder if, or doubt that P[2]. But also in the domain of beliefs proper (predictions, memories, opinions, conclusions, ...) there is a difference between something I'm absolutely certain about, that I "know", and something that I'm fairly sure that it is so, that I'm convinced that, and so on.

1.1   The Functions of the B "Value/Strength"

This "value/strength" of B plays very important roles and produces interesting behavioral and cognitive effects, apart from the "decision" to believe or not that P.

This "value/strength" of B allows for possible contradictions. It is not contradictory per se to "believe" in broad sense both that P and Not P; it is contradictory (thus to be avoided or solved) to be certain that P and to be certain (at the same time) that Not P. But it is not contradictory to believe/think that "probably/quite surely John is in his office but it might be that he is at lunch". A candidate B to become a "Belief" (not just a possible B, a data) has to win, prevail against competitors, incompatible candidate Bs. If they are not there, seemingly non problem[3] . But they might be explicitly there: we cannot come to believe that P, even if we 75% "believe", have evidence that P, since we might have also in mind 75% NotP. We have a "contradiction" in mind (an epistemic "conflict"); we have to solve the contradiction: before that, we cannot believe neither P or NotP. Thus the *absolute* "degree/strength" of the (candidate) B is not enough: does it win against competitors? Exactly like for Goals (see below).

This "value/strength" of B determines the difference between a mere anticipatory hypothesis or possibility and a true "expectation", with very different feelings and commitment.

It determines the "degree" or "strength" of surprise reactions, as well as the intensity of the possible disappointment or relief. The stronger the B the more intense the surprise if B is not validated by the evidence. The more certain I was that P, the more disappointed or relieved I will be (but of course also the G value should be taken into account: the more important the goal the more intense the relief or the disappointment (see 2).

It determines the degree of our "trust" in somebody Y or something. Because trust is made of evaluations and expectations, that is of Bs, the degree of trust (and its being sufficient or insufficient for relying on Y) directly depends on the certainty of such Bs.[4]

It determines the resistance to change a given opinion: the stronger the belief, the more convinced I am, the more I will resist to change my mind (also because depends on the "integration" of that belief, and revise it would involve the revision of several

---

[2] Under the 50% of "probability" or with too much ignorance and uncertainty.

[3] Actually a competitor is always there: (Bel P) - with a given degree of certainty - implies the (Bel NotP) with its complenentary degree of certainty. In a simplified model.

[4] Combined with the degree of the ascribed "qualities": I can be *more* or *less* sure about a *low* or *high* competence of that medical doctor. [2]

integrated and supporting Bs; a demanding and costly adjustment). This has consequences also for the argumentation moves.

It determines <u>a feeling of (un)certainty</u> (and possibly anxiety) and thus the need for additional evidence, or conversely the stopping of our search for further information.

It determines the <u>degree of other Bs</u> (see 1.2).

It determines the <u>impact of the belief on goal processing and the decision making</u>. A strong B not only exerts a stronger impact than a weak one; it also contributes to the "instrumental value" of a given piece of knowledge, as an instrument for goal achievement; the more certain a B, the more precious it is, in that it is more reliable for the action (see 3).

1.2   The Bases of the Strength of a Given B

B's strength derives from (and it is supported by) its origin: the "source" of that B.

Is B derived from something I have perceived? Or from somebody else's communication? Or is it an inference of mine, a reasoning conclusion? And are there converging and reciprocally confirming sources? or are they contradictory?

<u>Two basic principles</u> determine the strength of a candidate B:

- The more reliable, trustworthy (competent, well informed, and honest) the source (and the transmission of the information) the more credible the belief. *The doxastic value of B is a function of the credibility of its source.* That's why we preserve a good memory of belief sources: we need to "readjust" the credibility of a given source when we discover that its information was wrong; otherwise, we confirm that it was valid.

This is complicated by the fact that some sources (communication sources) can also pass their own subjective degree of certainty. For example, X (a very reliable source, 90%) says that it is rather sure (55%) that P; or, Y (a not very reliable source, 55%) says that it is really sure (90%) that P; should the final result about the strength of B:P be the same? I don't think so. And how to combine the two dimensions? Like that: 90 x 55? I don't think so. (Also current "Subjective Logics" do not solve these problems; for example, [12]).

- *The many the convergent sources, the many the supporting pieces of evidence, the stronger a B.* Catching the degree/strength of Bs by probability might create some problem we will discuss, especially in relation to source convergence. Will the convergence and confirmation of several reliable sources, increase or decrease the value of the B? It should be increased.

In such a way, *the degree of a B depends on the degree of others Bs*; not only of the meta-Bs about the source (reputation of the source, kind of source ("I saw it with my own eyes!"), ...) but also - for example - Bs premises of an *inferential process*: how sure am I that John is not in his office if I am sure (I know) that he has left Roma for a conference in Paris? Or if I'm not sure, but it might be, I do not remember well? In this case, .. let's see, let try to call him: I act to acquire information, in order to become sure.

2   Goals

Goals have a "degree" of subjective importance: not all goals are equally important; that's why we choose among them, and "prefer" those with the greater perceived value, the best expected outcome. I will argue against "hedonism"; claiming that the value or utility of a G is not the experienced or expected "pleasure" that we might derive from it.

However, in order a G be chosen and in case pursued as our "intention", the "degree/strength" of the G (its motivational force, desirability, ...) is necessary but not *sufficient*. What matters is not just its absolute but its comparative value. Although something is very important, binding, useful, attractive (repulsive) to us, this is not sufficient for pursuing it (or for avoiding); it has to win against alternatives, to be "preferred", to have not just "a lot of value" but "more value than..." At least its value has to be greater than the expected costs and thus the rival Gs of "not doing".

2.1   The Crucial Functions and Effects of Goal Value

First of all - as we just said - the possibility to decide, to make a goal prevail over the other competitors.

That's why a crucial strategy for persuading somebody Y to choose and to do something is to add value to it, to add prizes and rewards to the goal to be chosen (or harms and sanctions to the alternatives), or just to induce Y to take into account additional positive outcomes or negative ones. By changing Y's Bs we change the value of those goals, and thus the choice of Y. That's also why Gs (if not stronger enough for winning against competitors or against the foreseen costs) have to find "alliances", additional positive outcomes (that is, additional satisfied Gs) in order to prevail, to be "heavy enough": a matter of quantity.

That's why in order to stabilize our long term purposive behavior, our intentions and projects, we increase the value of an already chosen G, after the "decision" (commitment) and after the investments.

The value of a G determines the intensity of the suffering caused by its frustration (the more important the G, or the many the frustrated (unrealized) Gs, the stronger the suffering), and of disappointment, relief, or joy [13].

The value of a G also influences our esteem of the competence or skills for pursuing and achieving it, or of the person endowed with such skills, thus impacting on the comparative social hierarchy, as well as on such emotions as pride or shame.

Socially, the value of a G determines the degree of Dependence of agent X on another Y, and - vice versa - the strength of the power of Y on X. The more important the G the more dependent we are. If I depend on Y for sometime not really important for me, I can renounce to it, but if I depend on Y (and without alternative) for some goal of mine very important, indispensable for me, Y has a great power-over me.

2.2   The Origin and Bases of G's Value

There are *two fundamental kinds of value*: the "reason-based value" and the felt one, the "affective-based value".

In fact some Gs cannot be "felt" (intentions, purposes, objectives, projects, ...) whereas we can feel desires, needs, impulses, ...

- For the first kind, the value of a G is established on the basis of the *means-end hierarchical structure*, and the foreseen Pros & Cons, that is, advantages and disadvantages, costs, harms, ..[5] . The value of G1 derives from the sum of the values of the positive outcomes, that is, of superordinate Gs that one expect to be achieved (Pros), minus the sum of the expected negative outcomes: the possibly frustrated Gs (Cons).

However, it is clear that this presupposes some given value of the "top/terminal" goals, of the "ends" or "motives". The final motives and their value depend on the subject's age, gender, personality, personal experience and learning, cultural values, etc.

- For the felt Gs, the principle is completely different: *the more intense the current actual or imagined sensation* (for example, the disturbing stimulus, when one feels a "need" for something, or the imagined pleasure when one "foretastes" some pleasant experience; or the associated and evoked "somatic markers") the more pushing and stronger the G. Let's be a bit more analytic on desires and felt needs.

Felt Goals and Their Embodied Value
Desires. A desire in strict sense is *necessarily "felt"* (implying sensations), whereas (as already remarked) not all goals are "felt"; even not all the motivating goals are necessarily affectively charged and pleasant and attractive (at least in principle, in a general theory of purposive behavior). Desires in this strict sense are goals that when/if realized give *pleasure*. "Desiring" (in Italian: *star desiderando*) strictly speaking means *anticipating in one's imagination* the realization of the goal-state. The goal is represented in a sensory-motor code, thus implying the subject's experience of some sensations or feelings associated with the goal's realization. The subject is actually *imagining* these sensations and experiencing some hallucination, some anticipatory pleasure; and this is why we can "intensely" wish something. This can even imply not only the activation of somatic markers (the central neural trace of previous somatic experiences), but the actual activation of bodily reactions, like salivation, erection, etc.

Needs. The sensations implied by felt needs – differently from desires – are negative, unpleasant, disagreeable, unbearable, activating avoidance-goals, in a "prevention focus". Also for this reason we experience them as 'necessities', 'forcing' us to do or not to do something.

Moreover, we conceptualize and conceive a "need" (either felt or not) as the *lack* of something, which is conceived not just as useful but as necessary; it is a *necessary* means, as the only possible means for achieving our goal: not only if I have O (what I need for G)[6] I can realize G, but if I do not have O I cannot and will not realize G. This gives a sense of necessity, of no choice, to the general notion of "need", which – in felt needs – is reinforced by the unpleasant sensation.

---

[5] That is, other higher or lateral realized **Gs** or frustrated **Gs** with their value.
[6] Notice that (differently from "desires") "needs" are intrinsically "instrumental" goals. I need something *for* something else: either for achieving a practical goal (practical need: I need the key for opening the door), or for stopping or preventing an unpleasant sensation.

Thus we can schematically characterize a need of agent x to have or do q in order to achieve p as follows: *(Need x, q, for p)*, where q is either to possess a resource y (x needs y for p), or to do an action for achieving p; and p is the goal relative to which x needs q (which becomes my sub-goal, my means). In order for an instrumental relationship between means and ends to count as a need, four <u>belief conditions</u> must hold:

- *Necessity belief*: if not q, not p (without what I need, what I want cannot be)
- *Lack belief*: currently not q (I miss what I need – otherwise I would not be needing it)
- *Uneasiness belief*: lacking q gives me some disturbance/suffering/trouble (this often is experienced as a feeling, a bodily signal)
- *Attribution belief*: my uneasiness is because not q and not p (it is my state of need that makes me squirm).

Notice how the "anatomy" of a given motivational representation, of a kind of G, is constituted by Bs.

If x needs q, she is not only motivated by the goal that p, but also by the goal to avoid the uneasiness/suffering/trouble associated with that need. [7]

Here is the body, not the "reason", that dictates the *value* of the G! The body not only may "activate" that G, but gives it a pushing preference, value: the more *intense* the sensation the more important, unescapable, the G.

This "intrusion" of body, of proprioception, in cognitive representation management may apply even to Bs: some beliefs, as well as their degrees of certainty, are based on certain emotions and feelings and their intensity; the process is not the standard one: B: "There is a danger" ==> Fear! ; but the other way around: Fear! ==> B: "There is some danger!". A feeling of fear (activated by conditioned learning, by unconscious thoughts, by somatic markers evocation) induces an explicit belief about the external conditions.

2.3   Comparing the Uncomparable

It remains a serious <u>architectural</u> problem *how to model the process combining these two heterogeneous kinds of values in our decisions*; how we are able to choose while "evaluating" at the same time the reason-based utility and importance of a given alternative and its affective evocations.

So our mind is able to calculate, to take into account, in just one and the same decision setting, two completely heterogeneous kinds of goal "value":

i.   the "reasoned" value, based on means-end reasoning and the calculation of expected pros and cons of a given move, which are belief-based and can be changed through argumentation (Pascal's "reasons of the Reason");

ii. the felt, affective value (Pascal's "reasons of the heart" that Reason cannot understand).[8]

---

[7] On the complementarity of felt desires and felt needs see [2].

[8] Pascal's claim *"The heart has its reasons, of which reason knows nothing"* precisely means that those bases and forces for choice ("reasons") are not "arguments", are not "arguable" through reasoning on expected utilities and probabilities, on the instrumental pros and cons of the outcomes.

We not only have two independent, parallel, and competitive systems for regulating our behavior, for making a given goal prevail: one is unconscious, automatic, fast, evocation based, affective, etc; the other is based on reasoning and deliberation, slow, etc. (like in "dual system" theories, nowadays very popular). These systems strictly interact with each other; more precisely, the affective, evocative system enters the space of deliberation, introduces new dimensions on goals and beliefs, and alters the process and result of our "reasoned" decisions.[9]

Either of these kinds of value can prevail over the other: "Although I feel disgusted just at the idea of doing such-and-such, I have decided to do, because it is my duty"; "Although I know that it is bad and dangerous for me, and morally wrong, it is so attractive and pleasant for me that I cannot resist".

Thus affects do not just compete with our "deliberations", they enter our decision process and alter the *basic principles* of deliberation. Goal theory is deeply affected by affects, and this is something that has to be systematically clarified. Only a computational model and architecture might clarify this complex interaction not just competition.

2.4   Against the Hedonistic Version of the Motivational "Value"

It is wrong that we pursue our goals because - when realized - they give us "pleasure"; the greater the expected pleasure the stronger the motivating power of the goal, its attractive force; the worst the expected pain, trouble, the stronger the avoidance force. In this view there would be just one final goal, true motivation: maximizing our hedonic experience.

Pleasure is not "the" goal of our activity, and the same holds for feeling pleasure (or avoiding feeling pain). "Pleasure" – as a specific and qualitative subjective experience, sensation (not as an empty tautological label for goal satisfaction) – normally is not a goal for us: it is not what we intend to realize/achieve while acting, what moves us for performing that behavior. Of course, feeling pleasure or avoiding pain *might* become real goals and intentionally drive our actions: that is basically the mindset of the true hedonist, who acts for pleasure and not for whatever practical consequence his/her action accomplish. But typically looking for pleasure and avoiding pain are not a unique final goal of ours (a monarchic view of mind and motivation): rather, they act as "signals" for learning, and they help us learning, among other things, how to generate and evaluate goals.

Those hedonistic philosophies that identify pleasure with motivation, and relate our goal-oriented activity to pleasure motivation, should address the following, evident objections:

i.   As a matter of fact, several goals even when achieved do not give us a pleasure experience at all; they are just practical results (like to put the tablecloth on the table). And this applies not only to merely "instrumental" goals, just means, but even to our motives that might be just (often unpleasant) duties.

---

[9] Slovic's idea of "affect heuristics" just partially captures this aspect and solves the problem of its modeling [4].

ii.   If pleasure is so necessary for goal pursuit and motivated activity, why it is not necessary at all in cybernetic models of goal-directed activity and purposive systems? How is it possible to have a clearly finalistic and anticipation-driven mechanism, open to "success" and "failure", without any "pleasure"? In other terms, *what is the real function and nature of pleasure in a goal-directed system?* Moreover, pleasure seems to be present in nature (both phylogenetically and ontogenetically) well before mentally goal-directed actions. This also suggests that the function of pleasure has to be different; it does not seem to play the role of a goal.

In my view, pleasure is more related to the notion of "reward", of "reinforcement" and learning. Pleasure as an internal reward plays two fundamental roles: it attaches some value to some achieved state, which is important when the system can have more than one of such states, possibly in competition with each other; it signals that a given outcome (perhaps accidental) "deserves" to be pursued, is good, has to become a goal (that state, not the pleasure per se). In this view, pleasure is a signal and a learning device for goal creation/discovery and for evaluation. It seems very useful in a system endowed with a "generative" goal mechanism, and which needs different kinds of evaluation, more or less intuitive, fast, based on experience or on biological/inherited "preferences", and not just on reasoning (with its limits, biases, and slowness).

Moreover, as Seneca has explained, even if we will feel a pleasure and if we anticipate that pleasure this doesn't necessarily means that we do that action *for* experiencing that pleasure, "motivated" by pleasure. "Motivating" expectations are a sub-set of positive expectations. [3] [5].

3   Is There Any Relation between the *Force* of a Belief and the *Force* of a Goal?

Sure: the Cognitive Architecture is integrated and based on such a "commerce" between Bs & Gs.

The value of a goal can determine not only the utility of a B, how "precious" is that information; but it can even contribute to the subjective *certainty* of B, because we have an unconscious tendency or bias to believe what is pleasant (see for ex. [11]) and G-congruent; and we resist to remove Bs that are G-congruent.

And vice versa, the certainty of a B contributes to the evaluation of a G. What matters in fact in a choice is not only the value of the possible outcomes but also their estimated probability or plausibility, that is their doxastic certainty. Which is the value of choice G1 when I believe (expect) that *for sure* the Director will be happy of that, or that perhaps - I hope - the Director will be happy?

It is important for understanding cognitive architecture and our decision-making process to realize that: *we do not really choose among Gs, but more precisely among "expectations" or proto-expectations: Gs with related Beliefs (forecasts).*

The Role of Beliefs in Supporting Goal Processing

Let's remind our model of how G-processing is strictly (step by step) determined by Bs, and which are the B families that are relevant for arriving to formulating an Intention and then to act.

We know very little about the vicissitudes of goals in our mind, whereas - in psychology - we know everything, and in a step by step fashion, about the vicissitudes of knowledge in our mind. Except for some models of the decision processes, what do we know about goal activation and origins? Of goal selection

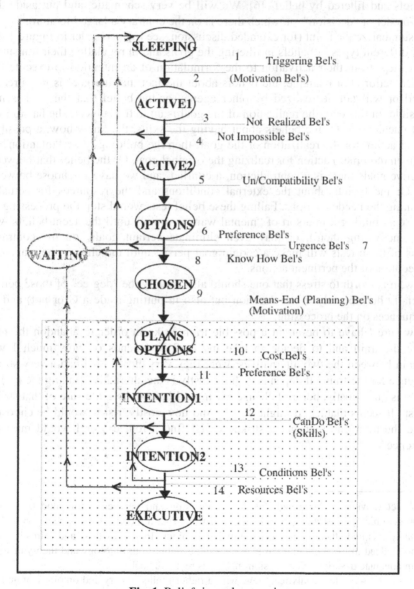

Fig. 1. Beliefs in goal processing

prior to making any choice (e.g., due to a perceived conflict)? Of goal "suspensions" (waiting for re-activation) or complete abandon? Of the evaluation of our skills, competence, know-how and possible plans (different from "planning"); or our self-confidence? Of the evaluation of the external conditions for a successful execution of the intended action? And so on.

The path of goals in our mind, from their triggering to their active pursuit (the execution of the identified instrumental act), is rather complex, with several "phases" and outcomes. We will illustrate the main phases and how they are step by step based on beliefs and filtered by beliefs [6]. We will be very schematic, and put aside the theory/model of "decision", on which there is on the contrary a large literature.[10]

To summarize the point (for extended discussion, see [6]), consider in Figure 1 the role of different types of beliefs in filtering the goals and in regulating their transition step by step, from their activation to the formulation of an intention to execute the specified action. For example, the beliefs about the fact that the goal is not already realized or will not be realized by other agents, or the belief that the goal is not impossible. In this case the realization of the goal is up to the subject; she has to find ways to achieve it. Or the beliefs about having the skills, the know-how, a possible plan and action for the realization of the goal; then the building of an "intention" *to do* a given (complex) action for realizing the original goal. Or the belief that between two active goals there is a contradiction, a conflict, and we have to choose between them. Or the beliefs about the external conditions and the resources for actually performing the needed actions. Failing these belief tests would stop the processing of a goal (e.g., putting it in a sort of "mental waiting room", until the agent beliefs will allow reactivating them), or may even eliminate certain goals; on the contrary, success of these tests will make a certain goal persist until the choice, the planning, the execution of the pertinent actions.[11]

Now, it is worth to stress that one should also specify the "degree" of those beliefs (especially but not only in the decision but also in putting aside a G or not) and its consequences on the processing.

How sure I have to be about a possible train strike in order to abandon the idea to take the train and to decide to take my car? It depends on how much I was uncertain between the two alternatives; were they very close? Or I had a very strong preference for the train? It depends on how costly is the alternative (taking the car). If to me it is quite costly, before I change my mind I have to be quite sure of that strike, not just "It seems that..." "It might be ...". And how costly, risky, would be choosing to take the train and then get a strike? Can I find a remedy, or I would miss the conference?

---

[10] This fact (having sophisticated theories of decision and yet very little models of goal dynamics) nicely illustrates a paradoxical situation. That a theory of goals be a prerequisite for a principled theory of decision-making might look obvious and already well acknowledged in the literature, but it is not so: the state of our ontology and theory of goals for an adequate description of decision-making is really dismal.

[11] For a similar – but less analytical view, with a rather limited theory and ontology of goals – see also the well-known "theory of Reasoned Action" by Fishbein and Ajzen [8].

So, the certainty of the B is a crucial dimension crossed with the other dimensions of the decision; and this might induce me to search for confirmation, reduce uncertainty, etc.

After the Decision

After our choice there is a problem of "commitment" and persistence in goal pursuit; a very crucial property for stabilizing our conduct, giving it coherence, and not wasting our efforts by continuously revising our objectives and plans for new stimuli and opportunities. Persistence in our choices is due both to the value of the G and to the dynamics of the Bs.

In our view one of the effect of a choice and investment is precisely that *the value of the chosen goal is increased*, and this makes it more dominant and stable, resistant to competitors. Celebrated Festinger' theory of "cognitive dissonance" and after-choice effect [7], can also be view in this perspective. Notice that it can be considered not only as an effect on the G value but as a feedback of the choice on the "opinions" and evaluations, that is, on the Bs supporting and consonant with the choice. They become comparatively stronger and thus the choice less revisable.

Moreover, the more the allocated resources for that G the more stable the preference (the famous "sunk costs" bias [10] due - in our view - also to that increased value).

In such condition, not necessarily the weakening of some original B will change the course of the action; reasonably we even resist more to revise the Bs congruent with our decision, and in order to abandon them we might impose a higher credibility threshold.

In sum, after the decision and the investment both the value of the winning G and the value of the supporting Bs and their dynamic properties are changed (even "irrationally" from the economic point of view), mainly in order to be stable in pursuing long term objectives.

4   Cognitive Architecture and Quantities

This quantitative aspect of the cognitive representation, architecture, and processes is not enough taken into account in current AI and "Agents" models, and in the formal logics for mental states or intentional action.

In the traditional approach there was no quantification and degrees (for example, in the classic BDI logics and models[12]); also when a representation of those "forces" is introduced the preferred approaches are still insufficient.

On the doxastic side, the "subjective probability" is quite powerful and interesting, but doesn't solve very relevant problems.

---

[12] See for general references: https://en.wikipedia.org/wiki/
Belief-desire-intention_software_model
There are important recent approaches in this direction; for example Dastani and Lorini [9].

On the motivational side, the simplifying but reductive view of "utility" and its "maximization" borrowed from Economics, and the resistance (like in Game Theory) to *explicitly* modeling the *specific* and internal "goals" of the actors, with their expectation and value are a problem.

Especially because,

- the origin and bases of those subjective values is quite obscure;
- there is not a systematic understanding and an architectural modeling of these bases and of the complex and intertwuinned functions of those quantitative dimensions; for example, how beliefs and their weight determine the processing and selection of goals and even their value.

The right characterization and formalization of these dimensions, their integration in a complex cognitive "architecture" for a goal-directed agency, and their implementation, are a long term objective, requiring exchanges among different competences in cognitive science, philosophy, logics, AI.

References

1. Broersen, J., Dastani, M., Hulstijn, J., van der Torre, L.: Goal generation in the BOID architecture. Cognitive Science Quarterly 2 (2002)
2. Castelfranchi, C., Falcone, R.: Trust Theory. A Socio-Cognitive and Computational Model. Wiley, UK (2010)
3. Castelfranchi, C.: Goals, the true center of cognition. In: Paglieri, F., Tummolini, L., Falcone, R., Miceli, M. (eds.) The Goals of Cognition. College Publications, London (2012)
4. Slovic, P., Finucane, M.L., Peters, E., MacGregor, D.G.: The affect heuristics. In: Gilovich, T., Griffin, D., Kahneman, D. (eds.) Heuristics and Biases: The Psychology of Intuitive Judgment, pp. 397–420. Cambridge University Press, Cambridge (2002)
5. Lorini, E., Marzo, F., Castelfranchi, C.: A cognitive model of altruistic mind. In: Kokinov, B. (ed.) Advances in Cognitive Economics, pp. 282–293. NBU Press, Sofia (2005)
6. Castelfranchi, C., Paglieri, F.: The role of beliefs in goal dynamics: prolegomena to a constructive theory of intentions. Synthese 155(2), 237–263 (2007)
7. Festinger, L.: Cognitive dissonance. Scientific American 207(4), 93–107 (1962)
8. Fishbein, M., Ajzen, I.: Belief, attitude, intention, and behavior: An introduction to theory and research. Addison-Wesley, Reading (1975)
9. Dastani, M.M., Lorini, E.: A logic of emotions: from appraisal to coping. In: Eleventh International Conference on Autonomous Agents and Multiagent Systems (AAMAS 2012), pp. 1133–1140 (2012)
10. Knox, R., Inkster, J.: Postdecision dissonance at post time. Journal of Personality and Social Psychology 8(4), 319–323 (1968)
11. Kunda, Z.: The case for motivated reasoning. Psychological Bulletin 108(3), 480–498 (1990)
12. Jøsang, A., McAnally, D.: Multiplication and Comultiplication of Beliefs. International Journal of Approximate Reasoning 38(1), 19–51 (2004)
13. Miceli, M., Castelfranchi, C.: Basic principles of psychic suffering: A preliminary account. Theory & Psychology 7, 769–798 (1997)

Towards an Ontology-Based Framework to Generate Diagnostic Decision Support Systems

Giuseppe Cicala, Marco Oreggia, and Armando Tacchella

Dipartimento di Informatica, Bioingegneria, Robotica e Ingegneria dei Sistemi (DIBRIS)
Università degli Studi di Genova, Via Opera Pia, 13 – 16145 Genova – Italia
{giuseppe.cicala,marco.oreggia}@unige.it,
armando.tacchella@unige.it

Abstract. The task of a Diagnostic Decision Support System (DDSS) is to deduce the health status of a physical system. We propose a framework to generate DDSS software based on formal descriptions of the application domain and the diagnostic rules. The key idea is to describe systems and related data with a *domain ontology*, and to describe diagnostic computations with an *actor-based model*. Implementation-specific code is automatically generated from such descriptions, while the structure of the DDSS is invariant across applications. Considering an artificial scalable domain related to the diagnosis of air conditioning systems, we present a preliminary experimental comparison between a hand-made DDSS, and one generated with a prototype of our framework.

1 Introduction

Diagnostic Decision Support Systems (DDSSs) support humans in the deduction of useful information about the health status of some observed physical system. Witnessing their importance in applications, there is a substantial scientific literature about DDSSs, both in the area of clinical decision support systems — see, e.g., [12] — and in the area of automation and control systems — see, e.g., [3]. From a practical point of view, the availability of digital sensors, reliable and high-capacity networks and powerful processing units, makes automated diagnosis applicable to an increasing number of systems. However, data and diagnostic rules remain domain-dependent, and the implementation of a DDSS requires the development of substantial portions of ad-hoc software which can hardly be recycled. Indeed, while most of the existing literature about DDSS focuses on improving the performances of automated diagnosis in some domain of interest, to the best of our knowledge there is no contribution in the way of generating customized DDSS from high-level specifications.

Our research aims to fill this gap by developing a framework to generate customized DDSSs using a model-based design (MBD) approach. As mentioned in [7], MBD methodologies are gaining acceptance in industry to obtain system implementations from system designs specified using graphical languages endowed with formal syntax and semantics. The key advantages of MBD methodologies are composition of the system design in a intuitive way, precise simulation of the system behavior and, when possible, static analysis of correctness. Accordingly, our DDSS generator — that we call ONDA for "ONtology-based Diagnostic Application generator" — outputs a DDSS given formal descriptions of (i) the application domain, and (ii) the diagnostic rules to be computed.

M. Baldoni et al. (Eds.): AI*IA 2013, LNAI 8249, pp. 25–36, 2013.
© Springer International Publishing Switzerland 2013

Domain and rules descriptions are supplied to ONDA using two "classical" AI formalisms, namely ontologies — in the sense of [5], i.e., "formal and explicit specification of conceptualizations" — and actor-based models — as introduced in [1] with the extensions found in [7]. Our choice of ontologies is motivated by their increasing popularity outside the AI community — mainly due to Semantic Web applications — and the added flexibility that they provide over traditional relational data models like, e.g., the ability to cope with taxonomies and part-whole relationships, and the ability to handle heterogeneous attributes. Since in our setting it is expected that large quantities of data should be handled to provide meaningful input to the DDSS, the choice of the ontology language should be restricted to those designed for tractable reasoning like, e.g., the DL-Lite family introduced in [2]. The choice of actor-based models is motivated by their support for heterogeneous modeling, i.e., a situation wherein different parts of a system have inherently distinct properties, and therefore require different types of models. For instance, in a car it is natural to capture dynamics using a continuous-time model, whereas computerized controllers like anti-lock braking are more naturally described with discrete-time models. We believe that DDSS are no exception to this pattern, since they are required to monitor and diagnose the behavior of heterogeneous systems, and they are themselves a composition of physical processes and computational elements.

In the remainder of this section, we give a brief introduction to ontologies and actor-based models. Then, in Section 2 we present the design of ONDA and the main components to generate DDSSs. Finally, In Section 3 we show an example of a practical application to monitoring and diagnosis of HVAC (Heating, Ventilation and Air Conditioning) systems in households referred to as *HVAC case study* in the following. In this scenario, we compare the performances of a DDSS generated by ONDA with those of a hand-made DDSS implementing the same diagnostic rules. We conclude the paper in Section 4 with some final remarks and an outline of our future research agenda.

1.1 Ontologies

Ontology-based data access (OBDA) relies on the concept of *knowledge base*, i.e., a pair $\mathcal{K} = \langle \mathcal{T}, \mathcal{A} \rangle$ where $\mathcal{T}$ is the *terminological box* (*Tbox* for short) specifying the intensional knowledge, i.e., known classes of data and relations among them, and $\mathcal{A}$ is the *assertional box* (*Abox* for short) specifying the extensional knowledge, i.e., factual data and their classification. Filling the Abox with known facts structured according to the Tbox is a process known as *ontology population*. One of the mainstream languages for defining knowledge bases is OWL 2 (Web Ontology Language Ver. 2) described in [8]. Since OWL 2 is a World Wide Web Consortium's recommendation, it is supported by several ontology-related tools. However, the logical underpinning of OWL 2 is the description logic $\mathcal{SROIQ}$ whose decision problem is 2NExpTime-complete according to [6]. This makes the use of the full expressive power of OWL 2 prohibitive for an application like the one we are considering.[1]

---

[1] Complexity of reasoning within OWL 2 semantics is detailed at http://www.w3.org/TR/owl2-profiles/#Computational_Properties. It is worth noticing that in full OWL 2 the computational complexity of conjunctive query answering is yet to be determined, but it is unlikely to be low.

To retain most of the practical advantages of OWL 2, but to improve on its applicability, Motik et al. introduced *OWL 2 profiles* – see [8]. Formally, an OWL 2 profile is a sub-language of OWL 2 featuring limitations on the available language constructs and their usage. In particular, the OWL 2 QL profile is described in the official W3C's recommendation as *"[the sub-language of OWL 2] aimed at applications that use very large volumes of instance data, and where query answering is the most important reasoning task."*. Given our application domain, OWL 2 QL is more appealing than both OWL 2 and other profiles, because it guarantees that conjunctive query answering and the consistency of the ontology can be evaluated efficiently.

OWL 2 QL logic underpinning is given by *DL-Lite$_\mathcal{R}$*, one of the members of the *DL-Lite* family [2]. A detailed description of *DL-Lite$_\mathcal{R}$* can be found in [2]. The most important feature of OWL 2 QL in our context is that, using the mapping techniques introduced in [10], we can keep the terminological view to reason about data, while storing the Abox elements as records in a relational database. Formally, given a knowledge base $\mathcal{K} = \langle \mathcal{T}, \mathcal{A} \rangle$, we can build a database with a set of relations (tables) $R_\mathcal{K}$ such that the query $\mathcal{K} \models \alpha$ can be translated to a relational algebra expression over $R_\mathcal{K}$ returning the same result set. The choice of OWL 2 QL guarantees that the mapping from $\mathcal{K}$ to $R_\mathcal{K}$ is feasible, and the translation of ontology-based queries to SQL queries will yield polynomially bounded expressions. In this way, we can take the best of the two approaches, i.e., use ontologies to define the conceptual view of the domain, and databases to store actual data and connect to the other performances-critical elements of the generated DDSS, like data I/O and processing components.

1.2 Actor-Based Models

In the following, we summarize some notations and definitions from [7] which are required to understand the design of diagnostic processing rules in the HVAC case study using PTOLEMY II actor-based models. We assume that if S is a set of variables that take values in some universe $\mathcal{U}$, a *valuation* over S is a function $x : S \to \mathcal{U}$ that assigns to each variable $v \in S$ some value $x(v) \in \mathcal{U}$. The set of all assignments over S is denoted by $\hat{S}$. If $x \in \hat{S}$, $v \in S$ and $\alpha \in \mathcal{U}$, then $\{x \mid v \mapsto \alpha\}$ denotes the new valuation x' obtained from x by setting v to α and leaving other variables unchanged. *Timers* are a special type of variables that take values in $\mathbb{R}_+$, i.e., non-negative real numbers. We denote with $\mathbb{R}_+^\infty$ the set $\mathbb{R}_+ \cup \{\infty\}$, where ∞ denotes positive infinity. Finally, we define $\perp \in \mathcal{U}$ and **absent** $\in \mathcal{U}$ to denote "unknown" value or "absence" of a signal at a particular point in time, respectively.

An *actor* is a tuple $A = (I, O, S, s_o, F, P, D, T)$ where I is a set of *input variables*, O is a set of *output variables*, S is a set of *state variables*, and $s_o \in \hat{S}$ is a valuation over S representing the *initial state*; F is the *fire function*, defined as $F : \hat{S} \times \hat{I} \to \hat{O}$, that produces output based on input and current state; P is the *postfire function* defined as $P : \hat{S} \times \hat{I} \to \hat{S}$ that updates the state based on the same information of the fire function; D is a *deadline function* defined as $D : \hat{S} \times \hat{I} \to \mathbb{R}_+^\infty$ and T is a *time-update function* defined as $D : \hat{S} \times \hat{I} \times \mathbb{R}_+ \to \hat{S}$. It is assumed that F, P, D, T, are total functions, and I, O, and S are pair-wise disjoint. In the following, the terms *input*, *output* and *state* refer to valuations over I, O and S, respectively.

Every actor A defines a set of behaviors whose model is inspired from the semantic models of timed or hybrid automata. A *timed behavior of* A is a sequence

$$s_0 \xrightarrow{x_0/y_o} s_0' \xrightarrow{x_0'/d_0} s_1 \xrightarrow{x_1/y_1} s_1' \xrightarrow{x_1'/d_1} s_2 \xrightarrow{x_2/y_2} s_2' \xrightarrow{x_2'/d_2} \dots$$

where for all $i \in \mathbb{N}$, $s_i, s_i' \in \hat{S}$, $d_i \in \mathbb{R}_+$, $x_i \in \hat{I}$, $y_i \in \hat{O}$ and

$$y_i = F(s_i, x_i) \quad s_i' = P(s_i, x_i) \quad d_i \leq D(s_i', x_i') \quad s_{i+1} = T(s_i', x_i', d_i).$$

Intuitively, if A is in state s_i at some time $t \in \mathbb{R}_+$ and the environment provides input x_i to A, then A instantaneously produces output y_i using the fire function F, and moves to state s_i' using the postfire function P. The environment then proposes to advance time, but it does so "respecting" any restriction on the amount of time that may elapse. A "declares" such restrictions by returning a deadline $D(s_i', x_i')$. Next, the environment chooses to advance time by some concrete delay $d_i \in R_+$, making sure that d_i does not violate the deadline provided by A. Finally, the environment notifies A that it advanced time by d_i, and A updates its state to s_{i+1} accordingly, using the time-delay function T. Notice that the actor A does not have any explicit notion of time, although it can track it using special variables (e.g., timers), and it can impose constraints on the advancement of time using deadlines.

PTOLEMY II uses a graphical syntax, with hierarchy being the fundamental modularity mechanism. This means that a model is essentially a tree of sub-models. The leaves of the tree correspond to *atomic* actors, available in the Ptolemy library of predefined actors or written in Java by users. The internal nodes of the tree correspond to *composite* actors, which are formed by composing other actors into an actor diagram. Next, for lack of space, we briefly sketch the main elements of PTOLEMY II to understand the HVAC case study. For more detailed presentation of actors, diagrams and domains we refer to [7] and to the documentation of PTOLEMY II.

- *Continuous Integrator*. This is an atomic actor parametrized by some initial value $\alpha \in \mathcal{U}$ and, conceptually, it works by integrating the incoming signal. Integrators are the building block to embed continuous dynamics in PTOLEMY II diagrams.
- *Extended State Machines (ESM)*. An ESM defines an atomic actor with a given set input and output ports. Internally, ESMs may evolve in a finite number of discrete locations (states). ESMs can be used in a hierarchical fashion to define *modal models*, i.e., a set of actors $M = \{A_c, A_1, \dots, A_n\}$ such that
  - A_c is an ESM actor, called the *controller* of the modal model. A_c must have exactly n locations denoted $l_1, \dots, l_n$ where location l_i is refined into actor A_i.
  - All actors in M have the same sets of input and output variables.
  - All S_i are pair-wise disjoint.
  If the actors $A_1, \dots, A_n$ are subject to a continuous time evolution, then modal model correspond to the definition of hybrid automata.
- *Block diagram*. Given n actors $A_i = (I_i, O_i, S_i, s_{0,i}, F_i, P_i, D_i, T_i)$, such that all I_i, O_i and S_i are pair-wise disjoint, a block diagram is a composite actor $H = \{A_1, \dots, A_n\}$. The input of H is $I = \bigcup_{i=1}^{n} I_i \setminus V$, the output of H is $O = V \setminus \bigcup_{i=1}^{n} I_i$ where $V = \bigcup_{i=1}^{n} O_i$. In other words, the set of input variables I of the composite actor is the set of all input variables of actors in H that are not connected

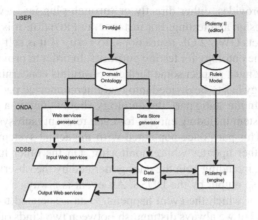

Fig. 1. Functional architecture and work-flow of the current ONDA prototype. Boxes represent applications, cylinders represent data storage, and slanted boxes represent I/O web services. Solid arrows indicate data flow among applications and storage elements, dashed arrows indicate generation of code.

to an output variable, i.e., that are not in V. Similarly, the set of output variables O is the set of all output variables V minus those that are connected to some input variable.

– *Continuous director*. Directors in PTOLEMY II can be viewed as *composition operators*: they take as input an actor diagram and return a new actor as output. The returned actor is a composite actor, but obeys the same interface as atomic actors and is therefore indistinguishable from the latter. Intuitively, composing a continuous director with a block diagram containing integrators, enables the simulation of the (hybrid) dynamic specified by the block diagram — see [7] for details.

2 ONDA Framework

In Figure 1, we show the current functional architecture and work-flow of ONDA. The picture represents three phases of the ONDA work-flow. In the USER phase, the domain ontology and the rules model are designed by the user. In the ONDA phase, the system reads and analyzes both the domain ontology and the rules model. The output of this phase is the code shown in the DDSS phase. In this phase, input web services receive data from the observed physical system and record them in the data store. The rule engine retrieves data from the data store, feeds the diagnostic rules and records their results, if any, in the data store. Output web services can then be invoked to query the data store about the results of the diagnostic rules. In the remainder of this Section, we describe each phase in detail.

In the USER phase, the user is required to provide an ontology of the observed physical system which must be written using OWL 2 QL language. While this can be accomplished in several ways, we suggest the tool PROTÉGÉ [4] because it is robust, easy to use, and it provides, either directly or through plug-ins, several add-ons that facilitate ontology design and testing. For instance, in PROTÉGÉ it is possible to check if the ontology respects OWL 2 QL restrictions, to verify if it is self-consistent, and to populate and query the ontology for testing purposes. In order to provide useful input to ONDA, the ontology must respect some further constraints concerning its contents. In particular, the ontology has to be divided into two interconnected parts, namely a *static* and a *dynamic* part. In the static part, the ontology should contain a description of the observed physical system including entities for each relevant (sub)system and relationships among them. This part, once populated with the actual systems to be observed, does not require further updates while monitoring. On the other hand, the dynamic part describes *events*, including both the ones generated by the observed system and its components, and those output by the DDSS. An event is always associated to a timestamp, i.e., the time at which the event happens. Data associated to events can be of heterogeneous types, but we always distinguish between two kinds of events, i.e., those *incoming* to the DDSS from the observed system, and those *outgoing* from the DDSS. This distinction is fundamental, because ONDA must know which events have to be associated with input and output web services, respectively. Furthermore, both events should be associated with the event generators, i.e., the elements of the static part which generate events or influence the generation of a diagnostic event. The rules model must be a sound actor diagram generated by PTOLEMY II which describes the processing to be applied to incoming data in order to generated diagnostic events. The only additional requirement on the rules model is that the set of external inputs of the diagram must coincide with the incoming events described in the ontology. Analogously, the set of external outputs of the diagram must coincide with the outgoing events.

The **ONDA** phase contains the actual DDSS generation system which consists of two modules in the current implementation, namely the *Data Store generator* and the *Web services generator*. The main flow of ONDA can be informally described as follows:

1. Given the domain ontology, a data store is generated to record data and events. The data store is a relational database which is obtained by mapping the domain ontology to suitable tables.
2. The web services generator creates services whose interface asks for incoming events of the correct type (input web services) and services which can be queried to obtain diagnostic events (output web services).
3. Currently, our working prototype uses PTOLEMY II internal engine to run the rules model as if they were code run on top of an interpreter. This solution is straightforward to implement, but has the disadvantage of being potentially too slow for real-world applications (see Section 3).

As mentioned in the Introduction, the creation of a relational database from the ontology, i.e., the Tbox $\mathcal{T}$, allows efficient storage of the corresponding Abox $\mathcal{A}$. The knowledge base $\mathcal{K} = \langle \mathcal{T}, \mathcal{A} \rangle$ can still be queried seamlessly, e.g., by using the mapping techniques described in [10]. The algorithm used by ONDA to encode an OWL 2 QL ontology into the structure of a relational database is shown in Figure 2. The main procedure is CREATEDATASTORE — Figure 2 top-left — whose tasks are the following:

procedure CREATEDATASTORE(*owlFile*)
   onto ← READOWL(*owlFile*)
   dataMap ← GETDATATYPEPROPERTIES(*onto*)
   relMap ← GETOBJECTPROPERTIES(*onto*)
   dbGraph ← VISITONTOLOGY(*onto*, *dataMap*, *relMap*)
   constrList ← FIXRELATIONSHIPS(*dbGraph*, *relMap*)
   dbStore ← CREATEDB()
   for all Node $n \in dbGraph$ **do**
      CREATETABLE(*dbStore*, n)
   end for
   for all Constraint $c \in constrList$ **do**
      CREATEFOREIGNKEYCONSTRAINT(*dbStore*, c)
   end for
end procedure

procedure FIXRELATIONSHIPS(*g*, *relMap*)
   result ← new List()
   for all Relationship $r \in relMap$ **do**
      if the cardinality of r is '1 to n' **then**
         result.add(new Constraint(*r*))
      else
         M ← new Table(*r*.getName())
         dbGraph.addNode(new Node(*M*))
         dbGraph.addEdge(*r*.getDomain(), *M*)
         dbGraph.addEdge(*r*.getRange(), *M*)
         result.add(new Constraint(*r*.getDomain(), *M*))
         result.add(new Constraint(*r*.getRange(), *M*))
      end if
   end for
end procedure

procedure VISITONTOLOGY(*onto*, *d*, *r*)
   g ← new Graph()
   VISITONTOLOGYREC(*onto*, *onto*.getThing(), *g*, *d*, *r*)
   return *g*
end procedure

procedure VISITONTOLOGYREC(*onto*, *c*, *g*, *d*, *r*)
   for all Concept $s \in c$.getSubConcepts() **do**
      father ← NIL
      if $c \neq onto$.getThing() **then**
         father ← *g*.getTable(*c*)
      end if
      T ← *g*.getTable(*s*)
      if T is NIL **then**
         T ← new Table(*s*)
         for all Attribute $a \in d$.getDataAttribute(*s*) **do**
            T.addAttribute(*a*)
         end for
         g.addNode(new Node(*T*))
         if *father* is not NIL **then**
            r.add(new Relationship(*T*, *father*, '1 to n'))
            g.addEdge(*T*, *father*)
         end if
         VISITONTOLOGY(*onto*, *s*, *g*, *d*, *r*)
      else
         if *father* is not NIL **then**
            r.add(new Relationship(*T*, *father*, '1 to n'))
            g.addEdge(*T*, *father*)
         end if
      end if
   end for
end procedure

Fig. 2. Main algorithms of the Data Store Generator component

- Read the ontology model from *owlFile* into the internal representation *onto*.
- Parse *onto* and extract the map *dataMap* between concepts and datatype properties — function GETDATATYPEPROPERTIES — and the map *relMap* between concepts and object properties (roles) — function GETOBJECTPROPERTIES.
- Visit the ontology by traversing the concept hierarchy with the function VISITONTOLOGY and create the graph *dbGraph* containing part of the relational model corresponding to *onto*.
- Finalize the relational model by considering all the relationships with the function FIXRELATIONSHIPS.
- Translate the relational model into a database, considering all the nodes of *dbGraph* and building corresponding tables — function CREATETABLE — as well as all the constraints and adding corresponding foreign key constraints — function CREATE-FOREIGNKEYCONSTRAINT.

In more detail, VISITONTOLOGY and its sister procedure VISITONTOLOGYREC — Figure 2 right — perform a visit of the subconcept hierarchy contained in *onto* to create a corresponding graph stored in *dbGraph*. Since the subconcept hierarchy forms, by definition, a directed acyclic graph, a simplified implementation of depth-first search visit is sufficient to explore *onto* exhaustively. Inside VISITONTOLOGYREC a new table T — and a corresponding node in the graph *g* — is created for each concept contained in *onto*. Furthermore, all the datatype properties corresponding to the concept of T are retrieved from the map d and added to T. These will become attributes of the entity corresponding to the concept in the final relational database. Notice that a one-to-many relationship corresponding to the inheritance relation is added to r, the set of relationships extracted considering object properties in *onto*. As long as d is implemented with a constant-time access structure, the running time of VISITONTOLOGY is linear in the size of *onto*. The procedure FIXRELATIONSHIPS — Figure 2 bottom left — finalizes the relational model by taking into account both the relationships added by VISITONTOLOGY and those initially present in *onto* as object properties (roles). Indeed, the latter are considered as many-to-many relationships by default, and a support table M named after the property is added to represent them and connect the related entities. On the other hand, one-to-many relationships, i.e., those coming from the subconcept hierarchy, are handled by adding foreign key constraints.

In the DDSS phase, we have the customized DDSS running in a loop wherein (i) data is acquired from the observed system and stored in the internal database, (ii) the rules engine process data and generates diagnostic events which are recorded on the database, and (iii) diagnostic data is served to end-user application. The details of the data acquisition on the observed system are not of concern to the DDSS generated by ONDA, because it is the responsibility of the observed system control logic to implement the data acquisition part. This choice effectively isolates the physical details of data acquisition from the rest of the DDSS. Similarly, the generated DDSS is not concerned with the details of displaying and representing diagnostic data, because these data are made available through output web-services and it is responsibility of the user applications to read such data and present them in a meaningful way.

3   Case Study: HVAC Diagnosis

Given their widespread use, monitoring and diagnosing malfunctioning Heating, Ventilation and Air Conditioning (HVAC) systems can be a route to conserve energy, avoid rapid deterioration of machinery, and reduce waste. Indeed, HVAC system are a classic topic in both model-based diagnostics — see, e.g., [11] – and data driven diagnostics — see, e.g., [9]. In our application example, we consider a minimal setting, wherein only two events are made available to the DDSS, namely on/off status of the HVAC unit, and the internal room temperature. The HVAC unit is assumed to be a warmer, i.e., a unit capable of increasing room temperature only. The warmer operates conventionally between two temperature thresholds, a minimum one defining the lowest acceptable temperature, and a maximum one defining the highest acceptable temperature in the house. When the room temperature drops below the minimum one, the warmer switches on, and it stays on until the room temperature reaches the maximum threshold. The energy consumed by the warmer is increased whenever heat dissipation of the house increases. This can be due to a temporary insulation fault condition, e.g., a window left inadvertently open. For the sake of simplicity, we do not consider here other causes of anomalous energy consumption.

In Figure 3, we show the ontology describing our HVAC domain. The main concepts in the static part of the domain are House and EventGenerator. They are related by

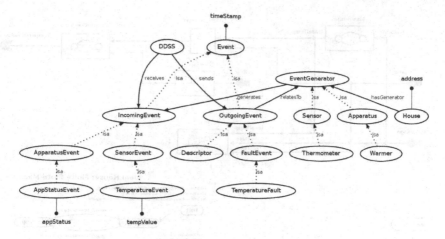

Fig. 3. Domain ontology for HVAC monitoring. Concepts are represented by ovals, concept inclusions (*is-a* relationships) are denoted by dashed arrows, roles are denoted by solid arrows, and attributes are denoted by dots attached to classes

hasGenerator, stating that every house has — possibly several — event generators attached to it. EventGenerator is the comprehensive class of elements that can generate diagnostic-relevant information. Event generators specialize into Apparatus and Sensor sub-concepts, where Sensor currently encompasses only Thermometer, i.e., the conceptualization of a temperature-reading device, and Apparatus can only be of type Warmer. The main concepts in the dynamic part of the ontology are DDSS which receives instances of IncomingEvent and sends instances of OutgoingEvent. Notice that IncomingEvent instances are connected to EventGenerator instances by the role generates, denoting that all incoming events, i.e., data from the observed system, are generated by some event generator, i.e., some field sensor. Also every OutgoingEvent instance, i.e., every diagnostic event, relatesTo some instance of EventGenerator. This is because the end user must be able to reconstruct which event generator(s) provided information that caused diagnostic rules to fire a given diagnostic event. Currently OutgoingEvent specializes to FaultEvent and Descriptor. In our case study, the only diagnostic event is an instance of TemperatureFault, which is itself a sub-concept of FaultEvent.

In Figure 4, we show some snapshots of the rules model. In the current prototype, this model is executed by PTOLEMY II engine using ad-hoc actors — not shown in Figure 4 — that connect to the data store to fetch and record events. The key element of the rules model is the Ideal House actor shown in Figure 4 (*iii*). This actor simulates the behavior of an environment in which a warmer alternates between two states: a heating state, wherein the warmer is switched on and it contributes to heat up the environment, and a cooling state, wherein the warmer is switched off and the environment cools down because of the natural heat dispersion. The dynamics of temperature in each state is simulated with a simple linear differential equation of the kind $\dot{T} = -K_c T + K_w$, where T is the temperature, $\dot{T}$ is its time-derivative, K_c is the heat-dispersion coefficient of the walls, and K_w is the warmer heating rate. This equation is

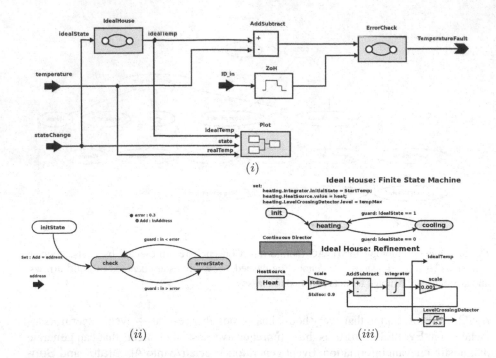

Fig. 4. Rules model for HVAC monitoring: (*i*) overall rules model (a block diagram), (*ii*) fault checking actor (a finite state machine). (*iii*) ideal house actor (a modal model whose refinement block diagrams are subject to a continuous director). In the pictures, solid right-pointing arrows denote incoming events acquired from the data store, and funnel-shaped arrows denote outgoing events, i.e., faults.

implemented by the block diagram shown Figure 4 (*iii*), where K_w corresponds to the variable heatSource. Overall, Ideal House defines a modal model, i.e., an hybrid automaton, whose switching behavior is not autonomous, but it is controlled by the signal idealState which in turn uses the on/off signal acquired from the real house apparatus. In this way, Ideal House provides a reference temperatures which, once compared with the readings received from the real household, enables the actor ErrorCheck to detect faults.

In Figure 4 (*i*) we show the overall rules model. In this diagram, we can see the Ideal House reference temperature compared with the actual temperature received from a given household. The difference between the two temperatures is fed to the actor ErrorCheck.[2] In Figure 4 (*ii*) we can see that ErrorCheck is a simple (finite) state machine whose task is to check the difference in signals to produce an error estimate — called in in the diagram. This estimate is compared with an error threshold — called error in the diagram — and there is a transition from check to error whenever the threshold is exceeded. In this way, a very simple model-based diagnosis of the real

---

[2] The actor Plot is used only for debugging purposes to track signals from Ideal House and the real household.

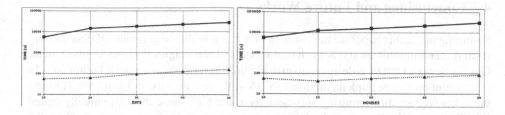

Fig. 5. Experimental results about the HVAC case study comparing the performances of the DDDS generated by ONDA (squares) and those of a manually coded DDSS (triangles). The plot on the left is about 10 monitored houses simulated for an increasing number of days. The plot on the right is about monitoring for 10 days an increasing number of houses. In both plots, the vertical axis corresponds to CPU time in seconds on a logarithmic scale.

household can be accomplished. In spite of its simplicity, it is important to notice that this model has all the elements of more complex model-based diagnosis systems, and that PTOLEMY II enables elegant implementation of much more sophisticated diagnostic rules.

In Figure 5, we show the experimental results about a DDSS generated by ONDA using the domain ontology of Figure 3 and the rules model of Figure 4, The experiment is about comparing the total running time of the ONDA-generated DDSS with respect to a hand made DDSS computing the same diagnostic rules. We consider two setups. In the first one — leftmost plot in Figure 5 — we assume that 10 houses are monitored across an increasing amount of time, i.e., from 10 to 50 days. In the second one — rightmost plot in Figure 5 — we assume that during a 10 days time span, an increasing number of houses is monitored, i.e., from 10 to 50 houses. In both cases, real houses are simulated using PTOLEMY II and a model similar to the ideal house shown in Figure 4 (i), with the addition of stochastic perturbations to model noisy sensors. Since input/output of data would be the same in both ONDA-generated DDSS and in the manually-coded one, we compare only the total time taken to retrieve data from the data store and to compute all the faults. As we can see from the plots in Figure 5, there is indeed a substantial gap between the performances of a manually-coded DDSS and the one generated by our current ONDA prototype. This is due to several factors, and the main one is the numerical method used by PTOLEMY II to solve ODEs which is triggered by our rules model shown in Figure 4. On the other hand, a manually coded DDSS can avoid such a general-purpose mechanism and perform a simple — but hard-coded — numerical simulation of the ideal house. However, we notice that the overall time spent by the ONDA-generated DDSS to process data is well below the time span over which data is generated. For instance, data generated by 50 houses over a time span of 10 days is processed in less than 9 hours. This means that ONDA-generated DDSS could be used in practice, and provide meaningful results while the observed systems are running. Moreover, if we let $m(t)$ represent the run time of the manually-coded DDSS, and $o(t)$ the run time of the ONDA-generated DDSS with 10 houses and increasing time t, we observe that there is no statistically significant trend in the ratio $o(t)/m(t)$ — established with a Mann-Kendall trend test on the ratio.

4 Conclusions and Future Works

Summing up, in this paper we have shown that it is possible to generate DDSSs starting from domain ontologies and actor-based models of the diagnostic rules to implement. Our framework ONDA is designed to provide this functionality in a push-button way. In our current prototype implementation, ONDA still relies heavily on PTOLEMY II to run the rules engine, thus requiring substantially more CPU time than a manually-coded DDSS on our HVAC case study. However, even in its present prototypical stage, the system would be usable in practice to diagnose small-to-medium scale systems with acceptable performances. In our future works, we will seek to improve on the efficiency of the rule engine, e.g., by generating more efficient code independently of the PTOLEMY II platform, and to experiment on significant industrial case studies related to monitoring and diagnosis.

References

1. Agha, G.A.: Actors: A Model of Concurrent Computation in Distributed Systems. PhD thesis, University of Michigan (1985)
2. Calvanese, D., De Giacomo, G., Lembo, D., Lenzerini, M., Rosati, R.: *DL-Lite*: Tractable Description Logics for Ontologies. In: Proceedings of the 20th National Conference on Artificial Intelligence (AAAI 2005), vol. 20, pp. 602–607 (2005)
3. Eom, S., Kim, E.: A survey of decision support system applications (1995–2001). Journal of the Operational Research Society 57(11), 1264–1278 (2005)
4. Gennari, J.H., Musen, M.A., Fergerson, R.W., Grosso, W.E., Crubézy, M., Eriksson, H., Noy, N.F., Tu, S.W.: The Evolution of Protégé: An Environment for Knowledge-Based Systems Development. International Journal of Human-Computer Studies 58(1), 89–123 (2003)
5. Gruber, T.R.: Toward principles for the design of ontologies used for knowledge sharing. International Journal of Human Computer Studies 43(5), 907–928 (1995)
6. Kazakov, Y.: $\mathcal{RIQ}$ and $\mathcal{SROIQ}$ are Harder than $\mathcal{SHOIQ}$. In: Description Logics (2008)
7. Lee, E.A., Tripakis, S., Stergiou, C., Shaver, C.: A Modular Formal Semantics for Ptolemy. Technical report, University of California at Berkley — Dept. of Electrical Engineering and Computer Science (2011)
8. Motik, B., Patel-Schneider, P.F., Parsia, B., Bock, C., Fokoue, A., Haase, P., Hoekstra, R., Horrocks, I., Ruttenberg, A., Sattler, U., et al.: OWL 2 Web Ontology Language: Structural Specification and Functional-Style Syntax. W3C Recommendation 27 (2009)
9. Namburu, S.M., Azam, M.S., Luo, J., Choi, K., Pattipati, K.R.: Data-driven modeling, fault diagnosis and optimal sensor selection for HVAC chillers. IEEE Transactions on Automation Science and Engineering 4(3), 469–473 (2007)
10. Rodrıguez-Muro, M., Calvanese, D.: Quest, an OWL 2 QL Reasoner for Ontology-based Data Access. In: OWLED 2012 (2012)
11. Sampath, M., Sengupta, R., Lafortune, S., Sinnamohideen, K., Teneketzis, D.C.: Failure diagnosis using discrete-event models. IEEE Transactions on Control Systems Technology 4(2), 105–124 (1996)
12. Sim, I., Gorman, P., Greenes, R.A., Haynes, B.R., Kaplan, B., Lehmann, H., Tang, P.C.: Clinical decision support systems for the practice of evidence-based medicine. Journal of the American Medical Informatics Association 8(6), 527–534 (2001)

Automated Reasoning in Metabolic Networks with Inhibition

Robert Demolombe, Luis Fariñas del Cerro, and Naji Obeid*

Université de Toulouse and CNRS, IRIT, Toulouse, France
robert.demolombe@orange.fr, {luis.farinas,naji.obeid}@irit.fr

Abstract. The use of artificial intelligence to represent and reason about metabolic networks has been widely investigated due to the complexity of their imbrication. Its main goal is to determine the catalytic role of genomes and their interference in the process. This paper presents a logical model for metabolic pathways capable of describing both positive and negative reactions (activations and inhibitions) based on a fragment of first order logic. We also present a translation procedure that aims to transform first order formulas into quantifier free formulas, creating an efficient automated deduction method allowing us to predict results by deduction and infer reactions and proteins states by abductive reasoning.

Keywords: Metabolic pathways, logical model, inhibition, automated reasoning.

1 Introduction

Cells in general and human body cells in particular incorporate a large series of intracellular and extracellular signalings, notably protein activations and inhibitions, that specify how they should carry out their functions. Networks formed by such biochemical reactions, often referred as *pathways*, are at the center of a cell's existence and they range from simple and chain reactions and counter reactions to simple and multiple regulations and auto regulations, that can be formed by actions defined in Figure 1. Cancer, for example, can appear as a result of a pathology in the cell's pathway, thus, the study of signalization events appears to be an important factor in biological, pharmaceutical and medical researches [14,11,7]. However, the complexity of the imbrication of such processes makes the use of a physical model as a representation seem complicated.

In the last couple of decades, scientists that used artificial intelligence to model cell pathways [10,9,16,17,6,21,15] faced many problems especially because information about biological networks contained in knowledge bases is generally incomplete and sometimes uncertain and contradictory. To deal with such issues, abduction [3] as theory completion [12] is used to revise the state of existing nodes and add new nodes and arcs to express new observations. Languages that were used to model such networks had usually limited expressivity, were specific

---

* LNCSR Scholar.

M. Baldoni et al. (Eds.): AI*IA 2013, LNAI 8249, pp. 37–47, 2013.
© Springer International Publishing Switzerland 2013

to special pathways or were limited to general basic functionalities. We, in this work, present a fragment of first order logic [19] capable of representing node states and actions in term of positive and negative relation between said nodes. Then an efficient proof theory for these fragments is proposed. This method can be extended to define an abduction procedure which has been implemented in SOLAR [13], an automated deduction system for consequence finding.

For queries about the graph that contains negative actions, it is assumed that we have a complete representation of the graph. The consequence is that the negation is evaluated according to its definition in classical logic instead of some non-monotonic logic. This approach guarantees a clear meaning of answers. Since the completion of the graph is formalized a la Reiter we used the equality predicate. It is well known that equality leads to very expensive automated deductions. This problem has been resolved by replacing completed predicates by their extensions where these predicates are used to restrict the domain of quantified variables. The result of this translation is formulated without variables where consequences can be derived very fast. This is one of the main contributions of this paper.

The rest of this paper is organized as follows. Section 2 presents a basic language and proof theory capable of describing general pathways, and shows their possible extensions to address specific and real life examples. Section 3 defines a translation procedure capable of eliminating first order variables and equality predicates and shows how it can be applied to derive new axiomatic that can be used in the automated deduction process in SOLAR. Section 4 provide some case studies, and finally section 5 gives a summary and discusses future works.

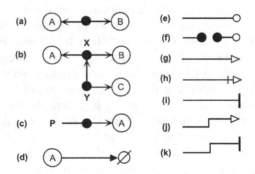

Fig. 1. Symbol definitions and map conventions

(a) Proteins A and B can bind to each other. The node placed on the line represents the A:B complex. (b) Multimolecular complexes: x is A:B and y is(A:B):C. (c) Covalent modification of protein A. (d) Degradation of protein A. (e) Enzymatic stimulation of a reaction. (f) Enzymatic stimulation in transcription. (g) General symbol for stimulation. (h) A bar behind the arrowhead signifies necessity. (i) General symbol for inhibition. (j) Shorthand symbol for transcriptional activation. (k) Shorthand symbol for transcriptional inhibition.

2 Logical Model

In this section we will present a basic language capable of modeling some basic positive and negative interaction between two or more proteins in some pathway. We will first focus on the stimulation and inhibition actions, points (g) and (i) of Figure 1, and then show how this language can be modified to express the different other actions described in the same figure.

2.1 Formal Language

Let's consider a fragment of first order logic with some basic predicates, boolean connectives ($\wedge$) and, ($\vee$) or, ($\neg$) negation, ($\rightarrow$) implication, ($\leftrightarrow$) equivalence, ($\exists$) existential and ($\forall$) universal quantifiers, and ($=$) equality.

The basic state predicates are:

- $A(x)$: with intended meaning that the protein x is *Active*.
- $I(x)$: with intended meaning that the protein x is *Inhibited*.

Having the basic state axiom $\forall x \neg(A(x) \wedge I(x))$ which indicates that a certain protein x can never be in both *Active* and *Inhibited* states at the same time.

An interaction between two or more different proteins is expressed by a predicate of the form $Action(protein_1, ..., protein_n)$. In our case we are interested by the simple *Activation* and *Inhibition* actions that are defined by the following predicates:

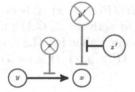   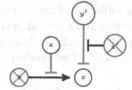

Fig. 2. Activation   **Fig. 3.** Inhibition

- $CAP(y, x)$: CAP or the *Capacity of Activation* expresses that the protein y has the capacity to activate the protein x.
- $CICAP(z, y, x)$: $CICAP$ or the *Capacity to Inhibit the Capacity of Activation* expresses that the protein z has the capacity to inhibit the capacity of the activation of x by y.
- $CIP(y', x)$: CIP or the *Capacity to Inhibit a Protein* expresses that the protein y' has the capacity to inhibit the protein x.
- $CICIP(z', y', x)$: $CICIP$ or the *Capacity to Inhibit the Capacity of Inhibition of a Protein* expresses that the protein z' has the capacity to inhibit the capacity of inhibition of x by y'.

In the next section we will define the needed axioms that will be used to model the *Activation* and *Inhibition* actions.

2.2   Action Axioms

Given the fact that a node can acquire the state active or inhibited depending on different followed pathways, one of the issues answered by abduction is to know which set of proteins is required to be active of inhibited for our target protein be active or inhibited.

Axiomatic of Activation is of the following form:

$$\forall x (\exists y (A(y) \wedge CAP(y,x) \wedge \forall z (CICAP(z,y,x) \to \neg A(z))) \wedge$$
$$\forall y' (CIP(y',x) \to (\neg A(y') \vee \exists z' (CICIP(z',y',x) \wedge A(z')))) \to A(x)) \ . \tag{1}$$

A protein x is active if there exists at least one *active* protein y that has the capacity to activate x, $CAP(y,x)$, **and** for every protein z that has the capacity to inhibit the capacity of activation of x by y, $CICAP(z,y,x)$, z is *not active*. **And** for every protein y' that has the capacity to inhibit x, $CIP(y',x)$, y' is *not active*, **or** there exist at least one *active* protein z' that has the capacity to inhibit the capacity of inhibition of x by y', $CICIP(z',y',x)$. (Figure 2)

Axiomatic of Inhibition is of the following form:

$$\forall x (\exists y' (A(y') \wedge CIP(y',x) \wedge \forall z' (CICIP(z',y',x) \to \neg A(z'))) \wedge$$
$$\forall y (CAP(y,x) \to (\neg A(y) \vee \exists z (CICAP(z,y,x) \wedge A(z)))) \to I(x)) \ . \tag{2}$$

A protein x is inhibited if there exists at least one *active* protein y' that has the capacity to inhibit x, $CIP(y',x)$, **and** for every protein z' that has the capacity to inhibit the capacity of inhibition of x by y', $CICIP(z',y',x)$, z' is *not active*. **And** for every protein y that has the capacity to activate x, $CAP(y,x)$, y is *not active*, **or** there exist at least one *active* protein z that has the capacity to inhibit the capacity of activation of x by y, $CICAP(z,y,x)$. (Figure 3)

2.3   Extension with New States and Actions

The basic language defined in 2.1 and 2.2 can be easily extended to express different and more precise node statuses and actions. For example the action of phosphorylation can be defined by the following predicates:

- $CP(z,y,x)$: CP or the *Capacity of Phosphorylation* expresses that the protein z has the capacity to phosphorylate the protein y on a certain site, knowing that x is the result of said phosphorylation.
- $CICP(t,z,y,x)$: $CICP$ or the *Capacity to Inhibit the Capacity of Phosphorylation* expresses that the protein t has the capacity to inhibit the capacity of the phosphorylation of y by z leading to x.

We can now define the new phosphorylation axiom as:

$$\forall x (\exists y1, y2 (A(y1) \wedge A(y2) \wedge CP(y1,y2,x) \wedge \forall z (CICP(z,y1,y2,x) \to \neg A(z))) \wedge$$
$$\forall y' (CIP(y',x) \to (\neg A(y') \vee \exists z' (CICIP(z',y',x) \wedge A(z')))) \to A(x)) \ .$$

$Auto-phosphorylation, Dephosphorylation, Binding, Dissociation$ etc. actions and some of the newly discovered ones such as $Methylation$ and $Ubiquitination$ [7] can formalized in a similar fashion.

3   Automated Deduction Method

In this section we define a fragment of first order logic with constants and equality, and without functions. The properties of this fragment allow us to define a procedure capable of eliminating the quantifiers in this fragment, in other words to transform the first order formulas in formulas without variables, in order to obtain an efficient automated deduction procedure with these fragments.

In the following we define a special case of $Evaluable$ formulas [4] and $Domain$ $Independent$ formulas [22] called $Restricted$ formulas, which are also different from $Guarded$ formulas [1].

Definition 1. $Restricted$ $formulas$ are $formulas$ $without$ $free$ $variables$ $defined$ by the $following$ $grammar:$

$$\varphi ::= \forall \overline{x}(P(\overline{x}, \overline{c}) \to \varphi) | \exists \overline{x}(P(\overline{x}, \overline{c}) \wedge \varphi) | \psi \ . \tag{3}$$

Where $\overline{x}$ and $\overline{c}$ represent $x_1, ..., x_n$ and $c_1, ..., c_m$ respectively, and ψ is a quantifier free formula (i.e. ψ can only appear in the scope of a restricted formula). In the following the atomic formula $P(\overline{x}, \overline{c})$ will be referenced as a $Domain$ formula.

Examples of this kind of formulas are:
$\forall x(P(x) \to Q(x))$.
$\forall x(P(x) \to \exists y(Q(y) \wedge R(x, y)))$.

Definition 2. A $completion$ $formula$ is a $formula$ of the $following$ $form:$

$$\forall x_1, ..., x_n \ (P(x_1, ..., x_n, c_1, ..., c_p) \leftrightarrow ((x_1 = a_{1_1} \wedge ... \wedge x_n = a_{1_n}) \vee ... \vee$$
$$(x_1 = a_{m_1} \wedge ... \wedge x_n = a_{m_n}))) \ . \tag{4}$$

Where P is a predicate symbol of arity $n + p$, and a_i are constants.

Completion formulas are similar to the completion axioms defined by Reiter in [18] where the implication is substituted by an equivalence.

Definition 3. $Given$ a $restricted$ $formula$ φ and a set of $completion$ for the $predicates$ in φ $noted$ $C(\varphi)$, we say $that$ $C(\varphi)$ $saturates$ φ, if and $only$ if, for $each$ $domain$ $formula$ in φ, $there$ is a $unique$ $completion$ $formula$ in C.

Definition 4. $Given$ a $domain$ $formula$ φ, we $define$ the $domain$ of the $variables$ of φ, $denoted$ $D(\mathcal{V}(\varphi), C(\varphi))$, as $follows:$

if φ is of the form $P(x_1, ..., x_n, c_1, ..., c_p)$, and $C(\varphi)$ of the form:

$$\forall x_1, ..., x_m (P(x_1, ..., x_m, c_1, ..., c_l) \leftrightarrow ((x_1 = a_{1_1} \wedge ... \wedge x_m = a_{1_m}) \vee ... \vee$$
$$(x_1 = a_{q_1} \wedge ... \wedge x_m = a_{q_m})))$$

where $n \leq m$ and $l \leq p$.

$$\text{then } D(\mathcal{V}(\varphi), C(\varphi)) = \{< a_{1_1}, ..., a_{1_n} >, ..., < a_{q_1}, ..., a_{q_n} >\} . \tag{5}$$

Quantification Elimination Procedure

Let φ be a restricted formula of the following forms: $\forall \overline{x}(\varphi_1(\overline{x}) \to \varphi_2(\overline{x}))$ or $\exists \overline{x}(\varphi_1(\overline{x}) \wedge \varphi_2(\overline{x}))$, let $C(\varphi_1(\overline{x}))$ a set of completion formulas for φ_1, then we define recursively a translation $T(\varphi, C(\varphi))$, allowing to replace universal (existential) quantifiers by conjunction (disjunction) of formulas where quantified variables are substituted by constants as follows:

– if $D(\mathcal{V}(\varphi_1), C(\varphi_1)) = \{< \overline{c_1} >, ..., < \overline{c_n} >\}$ with $n > 0$:

$$T(\forall \overline{x}(\varphi_1(\overline{x}) \to \varphi_2(\overline{x})), C(\varphi)) = T(\varphi_2(\overline{c_1}), C(\varphi_2(\overline{c_1}))) \wedge ... \wedge T(\varphi_2(\overline{c_n}), C(\varphi_2(\overline{c_n}))) .$$
$$\tag{6}$$

$$T(\exists \overline{x}(\varphi_1(\overline{x}) \wedge \varphi_2(\overline{x})), C(\varphi)) = T(\varphi_2(\overline{c_1}), C(\varphi_2(\overline{c_1}))) \vee ... \vee T(\varphi_2(\overline{c_n}), C(\varphi_2(\overline{c_n}))) .$$
$$\tag{7}$$

– if $D(\mathcal{V}(\varphi_1), C(\varphi_1)) = \varnothing$:

$$T(\forall \overline{x} \ (\varphi_1(\overline{x}) \to \varphi_2(\overline{x})) , \ C(\varphi)) = True . \tag{8}$$

$$T(\exists \overline{x} \ (\varphi_1(\overline{x}) \wedge \varphi_2(\overline{x})) , \ C(\varphi)) = False . \tag{9}$$

Note 1. It is worth nothing that in this translation process each quantified formula is replaced in the sub formulas by constants. The consequence is that if a sub formula of a restricted formula is of the form $\forall \overline{x}(\varphi_1(\overline{x}) \to \varphi_2(\overline{x}, \overline{y}))$ or $\exists \overline{x}(\varphi_1(\overline{x}) \wedge \varphi_2(\overline{x}, \overline{y}))$ where the quantifiers $\forall \overline{x}$ or $\exists \overline{x}$ are substituted by their domain values, the variables in $\overline{y}$ must have been already substituted by its corresponding constants.

Then in the theory $\mathcal{T}$ in which we have the axioms of equality and axioms of the form $\neg(a = b)$ for each constant a and b representing different objects, which are called unique name axioms by Reiter in [18], we have the following main theorem and its corresponding lemmas:

Lemma 1. *Let F be a restricted formula of the form $F : \exists \overline{x}(\varphi(\overline{x}) \wedge \psi(\overline{x}))$ where ψ is a domain formula. There exists a translation $T(F, C(\varphi))$ for any saturated completion set $C(\varphi)$ where $D(\mathcal{V}(\varphi), C(\varphi)) \neq \varnothing$.*

Proof. The proof is constructed by induction on the number of instances of $\mathcal{V}(\varphi)$ contained in $D(\mathcal{V}(\varphi), C(\varphi))$.

Lemma 2. *Let G be a restricted formula of the form $G : \forall \overline{x}(\varphi(\overline{x}) \to \psi(\overline{x}))$ where ψ is a domain formula. There exists a translation $T(F, C(\varphi))$ for any saturated completion set $C(\varphi)$ where $D(\mathcal{V}(\varphi), C(\varphi)) \neq \varnothing$.*

Proof. The proof is constructed by induction on the number of instances of $\mathcal{V}(\varphi)$ contained in $D(\mathcal{V}(\varphi), C(\varphi))$.

Theorem 1. *Let φ be a restricted formula, and $C(\varphi)$ a completion set of formulas of the domain formulas of φ, then:*

$$\mathcal{T}, \; C(\varphi) \vdash \varphi \leftrightarrow T(\varphi, C(\varphi)) \; . \tag{10}$$

Proof. The proof consists of applying induction on the number of domain formulas in a restricted formula using Lemmas 1 and 2 to prove that the theorem holds for any number domain formulas.

We will now present an example of translation from first order logic formulas composed of action and state axioms to variable free formulas:

Example 1.

Let's consider the case where a protein b has the capacity to activate another protein a, and that two other proteins c_1 and c_2 have the capacity to inhibit the capacity of activation of a by b. This proposition can be expressed by the following completion axioms:

- $\forall y(CAP(y, a) \leftrightarrow y = b)$: Expresses that b is the only protein that has the capacity to activate a.
- $\forall z(CICAP(z, b, a) \leftrightarrow z = c_1 \lor z = c_2)$: Expresses that c_1 and c_2 are the only proteins that have the capacity to inhibit the capacity of activation of a by b.

Using the activation axiom defined in section 2 and the translation procedure, we can deduce:

$$A(b) \land \neg A(c_1) \land \neg A(c_2) \land$$
$$\forall y'(CIP(y', x) \to (\neg A(y') \lor \exists z'(CICIP(z', y', x) \land A(z')))) \to A(a) \; . \tag{11}$$

Let's also consider that a protein d has the capacity to inhibit the protein a and that there is no proteins capable of inhibiting the capacity of inhibition of a by d. This proposition can be expressed by the following completion axioms:

- $\forall y(CIP(y, a) \leftrightarrow y = d)$: Expresses that d is the only protein that has the capacity to inhibit a.
- $\forall z(CICIP(z, d, a) \leftrightarrow false)$: Expresses that there are no proteins capable of inhibiting the capacity of inhibition of a by d.

Using the previous activation axiom and these completion axioms we can deduce:

$$A(b) \wedge \neg A(c_1) \wedge \neg A(c_2) \wedge \neg A(d) \rightarrow A(a) \ . \tag{12}$$

Which means that the protein a is active if the protein b is active and the proteins c_1, c_2, d are not active.

4 Queries and Results

From what we defined in sections 2 and 3, the resulting translated axioms are of the following type *conditions* $\rightarrow$ *results*, and can be chained together to create a series of reactions forming our pathway. Then questions of two different types can be answered using *deduction* or *abduction* reasoning.

Questions answered by deduction request all entities that satisfy a given property. A question can be of the following form: *what is the state (active or inhibited) of the proteins that result from the reactions formed by proteins in some knowledge base.*

And questions answered by abduction looks for minimal assumptions that must be added to the knowledge base to derive that a certain fact is true. A question can be of the following form: *what are the proteins and their respective states (active or inhibited) that should be present in order to derive that a certain protein is active or inhibited.*

Both types of questions can be addressed in SOLAR (SOL for Advanced Reasoning) [13] a first-order clausal consequence finding system based on SOL (Skip Ordered Linear) tableau calculus [8,20].

In the following we are going to show an example, based on figure 4, demonstrating abduction type queries where three coherent pathways have been found [11]

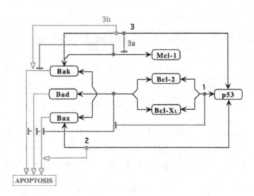

Fig. 4. Mitochondrial apoptosis induced by p53 independently of transcription

Following section 2.3 we can define new predicates to suit the needs of the pathway, as the *Capacity of Binding* $CB(z, y, x)$ and the *Capacity to Inhibit the Capacity of Binding* $CICB(t, z, y, x)$. These new predicates can be used to model the binding between p53 and Bak using the predicate $CP(p53, bak, p53\_bak)$ where p53_bak is the complex formed by such binding.

With these new predicates, new axioms can be defined that would enrich the descriptive capacities of the old ones, as seen in 2.3. Then the translation procedure applied to these axioms and to the completion axioms can be of the following form:

1. $A(p53) \land A(bak) \rightarrow A(bak\_p53)$
   bak_p53 is the result of the binding between p53 and Bak.
2. $A(bak\_p53) \rightarrow I(bak\_mcl)$
   bak_mcl is the result of binding between Bak and Mcl-1.
3. $A(bak\_p53) \land \neg A(b\_complex) \land \neg A(bak\_mcl) \rightarrow A(apoptosis)$
   b_complex is result of the binding between Bcl-2, Bcl-XL, Bak, Bad, and Bax.
4. $A(bak) \land \neg A(b\_complex) \land \neg A(bak\_mcl) \rightarrow A(apoptosis)$
5. $A(p53) \land A(bcl) \rightarrow A(p53\_bb\_complex)$
   bcl represents Bcl-2 and Bcl-XL.
   p53_bb_complex is the result of binding between p53, Bcl-2 and Bcl-XL.
6. $A(p53\_bb\_complex) \rightarrow I(b\_complex)$
7. $A(bax) \land \neg A(b\_complex) \rightarrow A(apoptosis)$
8. $A(p53) \land A(bax) \land \neg A(b\_complex) \rightarrow A(apoptosis)$
9. $A(bad) \land \neg A(b\_complex) \rightarrow A(apoptosis)$

If we want to know what are the proteins and their respective states that should be present in order to derive that the cell reached apoptosis, the answer is given by applying abduction over the previous set of compiled clauses. In the set of consequences returned by SOLAR we can find the following:

- $A(p53) \land A(bcl) \land A(bak)$: is a plausible answer, because p53 can bind to Bcl giving the *p53_bb_complex*, which can in return inhibit the *b_complex* that is responsible of inhibiting the capacity of Bak to activate the cell's apoptosis. That is why it is sufficient to for this case to have p53, Bcl, and Bak in an active state to reach apoptosis.
- Another interpretation of the previous answer is that p53 can also bind to Bak giving the *bak_p53* protein, which can in return inhibit the *bak_mcl* responsible of inhibiting the capacity of Bak to activate the cell's apoptosis. *bak_p53* can also stimulate Bak to reach apoptosis. Without forgetting that *p53_bb_complex* should be inhibiting *b_complex*.

5   Conclusion

A new language has been defined in this paper capable of modeling both positive and negative causal effects between proteins in a metabolic pathway. We showed how this basic language can be extended to include more specific actions that describes different relations between proteins. These extensions are important in this context, because there is always the possibility that new types of actions are discovered through biological experiments. We later showed how the axioms defined in such languages can be compiled against background knowledge, in order to form a new quantifier free axioms that could be used in either deduction or abduction reasoning. Although the first order axioms can be also well used to answer queries by deduction or abduction methods, the main advantage of translated axioms is their low computation time needed in order to derive consequences.

Future works can focus on extending the language used to define domain formulas, introducing for example the notion of time as in [2]. Trying to get as precise as possible in describing such pathways can help biologists discover contradictory informations and guide them during experiments knowing how huge the cells metabolic networks have become. One of the extensions that can also be introduced is the notion of *Aboutness* [5] that can limit and focus search results to what seems relevant to a single or a group of entities (proteins).

References

1. Andréka, H., Németi, I., van Benthem, J.: Modal languages and bounded fragments of predicate logic. Journal of Philosophical Logic 27(3), 217–274 (1998)
2. Chetcuti-Sperandio, N., Fariñas del Cerro, L.: A mixed decision method for duration calculus. J. Log. Comput. 10(6), 877–895 (2000)
3. Demolombe, R., Fariñas del Cerro, L.: An Inference Rule for Hypothesis Generation. In: Proc. of International Joint Conference on Artificial Intelligence, Sydney (1991)
4. Demolombe, R.: Syntactical characterization of a subset of domain-independent formulas. J. ACM 39(1), 71–94 (1992)
5. Demolombe, R., Fariñas del Cerro, L.: Information about a given entity: From semantics towards automated deduction. J. Log. Comput. 20(6), 1231–1250 (2010)
6. Erwig, M., Walkingshaw, E.: Causal reasoning with neuron diagrams. In: Proceedings of the 2010 IEEE Symposium on Visual Languages and Human-Centric Computing, VLHCC 2010, pp. 101–108. IEEE Computer Society, Washington, DC (2010)
7. Glorian, V., Maillot, G., Poles, S., Iacovoni, J.S., Favre, G., Vagner, S.: Hur-dependent loading of mirna risc to the mrna encoding the ras-related small gtpase rhob controls its translation during uv-induced apoptosis. Cell Death Differ 18(11), 1692–1701 (2011)
8. Inoue, K.: Linear resolution for consequence finding. Artificial Intelligence 56(2-3), 301–353 (1992)
9. Inoue, K., Doncescu, A., Nabeshima, H.: Hypothesizing about causal networks with positive and negative effects by meta-level abduction. In: Frasconi, P., Lisi, F.A. (eds.) ILP 2010. LNCS, vol. 6489, pp. 114–129. Springer, Heidelberg (2011)
10. Inoue, K., Doncescu, A., Nabeshima, H.: Completing causal networks by meta-level abduction. Machine Learning 91(2), 239–277 (2013)
11. Kohn, K.W., Pommier, Y.: Molecular interaction map of the p53 and mdm2 logic elements, which control the off-on swith of p53 response to dna damage. Biochem. Biophys. Res. Commun. 331(3), 816–827 (2005)

12. Muggleton, S., Bryant, C.H.: Theory completion using inverse entailment. In: Cussens, J., Frisch, A.M. (eds.) ILP 2000. LNCS (LNAI), vol. 1866, pp. 130–146. Springer, Heidelberg (2000)
13. Nabeshima, H., Iwanuma, K., Inoue, K., Ray, O.: Solar: An automated deduction system for consequence finding. AI Commun. 23(2-3), 183–203 (2010)
14. Pommier, Y., Sordet, O., Rao, V.A., Zhang, H., Kohn, K.W.: Targeting chk2 kinase: molecular interaction maps and therapeutic rationale. Curr. Pharm. Des. 11(22), 2855–2872 (2005)
15. Ray, O.: Automated abduction in scientific discovery. In: Magnani, L., Li, P. (eds.) Model-Based Reasoning in Science and Medicine. SCI, vol. 64, pp. 103–116. Springer, Heidelberg (2007)
16. Ray, O., Whelan, K., King, R.: Logic-based steady-state analysis and revision of metabolic networks with inhibition. In: Proceedings of the 2010 International Conference on Complex, Intelligent and Software Intensive Systems, CISIS 2010, pp. 661–666. IEEE Computer Society, Washington, DC (2010)
17. Reiser, P.G.K., King, R.D., Kell, D.B., Muggleton, S.H., Bryant, C.H., Oliver, S.G.: Developing a logical model of yeast metabolism. Electronic Transactions in Artificial Intelligence 5, 233–244 (2001)
18. Reiter, R.: On closed world data bases. In: Readings in Nonmonotonic Reasoning, pp. 300–310. Morgan Kaufmann Publishers Inc., San Francisco (1987)
19. Shoenfield, J.R.: Mathematical logic. Addison-Wesley series in logic. Addison-Wesley Pub. Co. (1967)
20. Siegel, P.: Representation et utilisation de la connaissance en calcul propositionnel. Thèse d'État, Université d'Aix-Marseille II, Luminy, France (1987)
21. Tamaddoni-Nezhad, A., Kakas, A.C., Muggleton, S., Pazos, F.: Modelling inhibition in metabolic pathways through abduction and induction. In: Camacho, R., King, R., Srinivasan, A. (eds.) ILP 2004. LNCS (LNAI), vol. 3194, pp. 305–322. Springer, Heidelberg (2004)
22. Ullman, J.D.: Principles of database systems. Computer software engineering series. Computer Science Press (1980)

Multicriteria Decision Making Based on Qualitative Assessments and Relational Belief

Amel Ennaceur[1], Zied Elouedi[1], and Eric Lefevre[2]

[1] LARODEC, University of Tunis, Institut Supérieur de Gestion, Tunisia
amel_naceur@yahoo.fr, zied.elouedi@gmx.fr
[2] LGI2A, Univ. Lille Nord of France, UArtois EA 3926, France
eric.lefevre@univ-artois.fr

Abstract. This paper investigates a multi-criteria decision making method in an uncertain environment, where the uncertainty is represented using the belief function framework. Indeed, we suggest a novel methodology that tackles the challenge of introducing uncertainty in both the criterion and the alternative levels. On the one hand and in order to judge the criteria weights, our proposed approach suggests to use preference relations to elicitate the decision maker assessments. Therefore, the Analytic Hierarchy Process with qualitative belief function framework is adopted to get adequate numeric representation. On the other hand, our model assumes that the evaluation of each alternative with respect to each criterion may be imperfect and it can be represented by a basic belief assignment. That is why, a new aggregation procedure that is able to rank alternatives is introduced.

1 Introduction

In real life decision making, the decision maker is faced with many situations in which he has to make a decision between different alternatives. However, the most preferable one is not always easily selected. Thus, he often needs to make judgments about alternatives that are evaluated on the basis of different criteria [16]. In this context, the problem is called a multi-criteria decision making (MCDM) problem.

Within this MCDM framework, a large number of methods has been proposed. On the one hand, the outranking approach introduced by Roy, where some methods like Electre and Promethee are developed [4]. On the other hand, the value and utility theory approaches mainly started by Keeney and Raiffa [8], and then implemented in a number of methods. However, classical methods applying both multi-attribute utility theory and outranking model do not take into account imperfection in their parameters. To cope with this problem, several approaches have been developed. The idea was to combine theories managing uncertainty or imprecision, such as probability theory, belief function theory and fuzzy set theory, with MCDM methods [2], [9].

In this context, belief function theory has shown its efficiency. In fact, there are several MCDM approaches which have been developed such as DS/AHP approach [2], belief Analytic Hierarchy Process (AHP) method [5], [7], etc.

M. Baldoni et al. (Eds.): AI*IA 2013, LNAI 8249, pp. 48–59, 2013.

In spite of many advantages of the presented approaches, they have a crucial limitation. First, they still treat criteria weights and alternatives performances in the same way. By nature, the criteria weights are relative to each other. Therefore, it is reasonable to elicit their importance by pair-wise comparison. However, the evaluation of an alternative on each criterion should be independent of each other. Second, the relative importance of criteria can be provided by the expert based on some previous experience or obtained by some elaborated approach. As a result, uncertain, imprecise and subjective data are usually present which make the decision-making process complex and challenging. Therefore, we consider MCDM problem for which expert estimates the evaluation of alternatives according to ordinal criteria. The information provided can be uncertain and/or imprecise.

Based on these reasons, we suggest a new methodology for MCDM problems based on belief function theory. In fact, our model integrates one of the most important weight calculation procedures with a new ranking process of alternatives in an uncertain environment. In the proposed methodology, AHP with its qualitative belief functions extension is applied to evaluate criteria and weighting them in the presence of uncertainty. Then, in the second step, a fusion procedure is proposed to aggregate the alternatives priorities and the criteria weights.

In this paper, section 2 and 3 describe an overview of the basic concepts of respectively the belief function theory and the qualitative belief function methods. Then, in the main body of the paper, we present our new approach namely belief MCDM method. Finally, our method will be illustrated by an example.

2 Belief Function Theory

2.1 Basic Concepts

Let Θ be the frame of discernment representing a finite set of elementary hypotheses related to a problem domain. We denote by 2^Θ the set of all the subsets of Θ [12].

The impact of a piece of evidence on the different subsets of the frame of discernment Θ is represented by the so-called basic belief assignment (bba), denoted by m [12]:

$$\sum_{A \subseteq \Theta} m(A) = 1. \tag{1}$$

For each $A \subseteq \Theta$, the value $m(A)$, named a basic belief mass (bbm), represents the portion of belief committed exactly to the event A. The events having positive bbm's are called focal elements. Let $\mathcal{F}(m) \subseteq 2^\Theta$ be the set of focal elements of the bba m.

Associated with m is the belief function (bel) is defined for $A \subseteq \Theta$ and $A \neq \emptyset$ as:

$$bel(A) = \sum_{\emptyset \neq B \subseteq A} m(B) \text{ and } bel(\emptyset) = 0. \tag{2}$$

The degree of belief $bel(A)$ given to a subset A of the frame Θ is defined as the sum of all the basic belief masses given to subsets that support A without supporting its negation.

2.2 Combination

In the Transferable Belief Model (TBM), one interpretation of the belief function theory [13], the basic belief assignments induced from distinct pieces of evidence can be combined using the conjunctive rule [13]:

$$(m_1 \bigcirc\!\!\!\!\!\!\wedge\, m_2)(A) = \sum_{B,C \subseteq \Theta, B \cap C = A} m_1(B)m_2(C), \quad \forall A \subseteq \Theta. \tag{3}$$

$m_1 \bigcirc\!\!\!\!\!\!\wedge\, m_2$ is the bba representing the combined impact of two pieces of evidence.

2.3 Discounting

The technique of discounting allows to take into consideration the reliability of the information source that generates the bba m. Let $\beta = 1 - \alpha$ be the degree of reliability ($\alpha \in [0,1]$) assigned to a particular belief function. If the source is not fully reliable, the bba it generates is "discounted" into a new less informative bba denoted m^α:

$$m^\alpha(A) = (1 - \alpha)m(A), \forall A \subset \Theta \tag{4}$$

$$m^\alpha(\Theta) = \alpha + (1 - \alpha)m(\Theta) \tag{5}$$

2.4 Uncertainty Measures

In the case of the belief function framework, the bba is defined on an extension of the powerset: 2^Θ and not only on Θ. In the powerset, each element is not equivalent in terms of precision. Indeed, $\theta_i \subset \Theta$ ($i \in \{1, 2\}$) is more precise than $\theta_1 \cup \theta_2 \subseteq \Theta$. In order to try to quantify this imprecision, different uncertainty measures (UM) have been defined, such as the composite measures introduced by Pal et al. [10] such as:

$$H(m) = \sum_{A \in \mathcal{F}(m)} m(A) \log_2\left(\frac{|A|}{m(A)}\right). \tag{6}$$

The interesting feature of $H(m)$ is that it has a unique maximum.

2.5 Decision Making

The TBM considers that holding beliefs and making decision are distinct processes. Hence, it proposes a two level model: (1) The credal level where beliefs are entertained and represented by belief functions. (2) The pignistic level where beliefs are used to make decisions and represented by probability functions called the pignistic probabilities, denoted $BetP$ [14]:

$$BetP(A) = \sum_{B \subseteq \Theta} \frac{|A \cap B|}{|B|} \frac{m(B)}{(1 - m(\emptyset))}, \forall A \in \Theta \tag{7}$$

3   Qualitative Belief Function Method

The problem of eliciting qualitatively expert opinions and generating basic belief assignments have been addressed by many researchers [1] [15] [6]. In this subsection, we present the approach of Ben Yaghlane et al. [1], since in next section we will use this method to elicitate the expert preferences. This method is chosen since it handles the issue of inconsistency in the pair-wise comparisons. Also, the originality of this method is its ability to generate quantitative information from qualitative preferences only.

Giving two alternatives, an expert can usually express which of the propositions is more likely to be true, thus they used two binary preference relations: the preference and the indifference relations, defined as follows:

$$a \succ b \Leftrightarrow bel(a) - bel(b) \geq \varepsilon \tag{8}$$

$$a \sim b \Leftrightarrow |bel(a) - bel(b)| \leq \varepsilon \tag{9}$$

ε is considered to be the smallest gap that the expert may discern between the degrees of belief in two propositions a and b. Note that ε is a constant specified by the expert before beginning the optimization process.

Then, a mono-objective technique was used to solve such constrained optimization problem:

$$Max_m UM(m)$$
$$s.t.$$
$$bel(a) - bel(b) \geq \varepsilon \quad (a \text{ is prefered to } b)$$
$$bel(a) - bel(b) \leq \varepsilon \quad (a \text{ is indifferent to } b) \tag{10}$$
$$bel(a) - bel(b) \geq -\varepsilon \quad (a \text{ is indifferent to } b)$$
$$\sum_{a \in \mathcal{F}(m)} m(a) = 1, m(a) \geq 0, \forall a \subseteq \Theta; m(\emptyset) = 0,$$

where the first, second and third constraints are derived from the previous equations. The last constraint ensures that the total amount of masses allocated to the focal elements of the bba is equal to one, also it specifies that masses are non negative and imposes that the bba to be generated must be normalized.

A crucial step that is needed before beginning the task of generating belief functions, is the identification of the candidate alternatives.

4   MCDM Method Based on Qualitative Assessments and Relational Belief

The new framework called Belief MCDM method mixes a multi-criteria decision making method inspired by the Analytic Hierarchy Process (AHP) and belief function theory. The originality of our approach is to apply the qualitative AHP to compute the importance of criteria and to replace the aggregation step by a new fusion process. Its main aim is to take into account both imprecision and uncertain assessments. In other words, the assumption was made that the performances of alternatives are provided on the form of bba while the weights of criteria are introduced using qualitative assessment and preference relations.

4.1   Assigning Criteria Weight via Qualitative AHP

In most multi-criteria methods, a numerical value is assigned to each criterion expressing its relative importance. This reflects the corresponding criterion weight. In fact, there are many elicitation techniques, but the AHP has some advantages. One of the most important advantage of the AHP attributes to its pair-wise comparison scheme.

In fact, the AHP method is a decision-making technique developed by Saaty [11]. This method elicits preferences through pair-wise comparisons which are constructed from decision maker's answers. Indeed, the expert can use both objective information about the elements as well as subjective opinions about the elements' relative meaning and importance. The responses to the pair-wise comparison question use a nine-point scale [11], which translates the preferences of a decision maker into numbers. An eigenvector method is applied to solve the reciprocal matrix for determining the criteria importance and alternative performance. The simple additive weighting method is used to calculate the utility for each alternative across all criteria.

However, standard AHP do not handle the problem of uncertainty. Therefore, in the proposed methodology, AHP with its belief extension, namely qualitative AHP, is applied to obtain more decisive judgments by prioritizing the evaluation criteria and weighting them in the presence of imperfection.

Step 1: Let Ω be a set of criteria where $\Omega = \{c_1, \ldots, c_m\}$. In this first stage, qualitative AHP computations are used for forming a pair-wise comparison matrix in order to determine the criteria weights using preferences relations only. Thus to express his preferences, the decision maker has only to express his opinions qualitatively, based on knowledge and experience that he provides in response to a given question rather than direct quantitative information. Therefore, he only selects the related linguistic variable using preference modeling instead of using a nine-point scale. It is illustrated in Table 1.

Table 1. Preferences relation matrix

	c_1	c_2	$\ldots$	c_m
c_1	-	P_{12}	$\ldots$	P_{1m}
c_2	-	-	$\ldots$	P_{2m}
$\ldots$	-	-	-	$\ldots$
c_m	-	-	-	-

In this table, P_{ij} is a preference assessment. It may be:

1. a strict preference relation $\succ$ iff $(c_i \succ c_j) \wedge \neg(c_j \succ c_i)$
2. an indifference relation $\sim$ iff $(c_i \succ c_j) \wedge (c_j \succ c_i)$
3. an unknown relation (no relation is given).

Under this approach, the expert is not constrained to quantify the degree of preferences and to fill all pair-wise comparisons matrix. He is able to express his preferences freely.

Step 2: Once the pair-wise comparison matrix is complete, our objective is then to compute the importance of each criterion. In fact, within our model, we propose to transform these preference relations into numerical values using the belief function framework. By adopting our approach, we try to closely imitate the expert reasoning without adding any additional information. Therefore, we suggest to apply Ben Yaghlane et al. approach [1] to convert the preferences relations into constraints of an optimization problem whose resolution, according to some uncertainty measures (UM) such as H (Equation 6), allows the generation of the least informative or the most uncertain belief functions. Indeed, we assume that the criterion weight is then described by a basic belief assignment and it is denoted by m^Ω. It can then be determined by the resolution of an optimization problem as defined is the previous section (Equation 10).

Furthermore, the proposed method addresses the problem of inconsistency. In fact, if the preference relations are consistent, then the optimization problem is feasible. Otherwise, no solutions will be found.

Finally, to obtain a relative importance of each criterion, we propose to transform the obtained bba m^Ω into pignistic probabilities:

$$BetP^\Omega(c_i) = \omega_i, \forall i = 1, \ldots, m \tag{11}$$

4.2 Aggregation of the Assessments with Respect to All Criteria for Each Alternative

Step 3: In our proposed method the evaluation of each alternative to each criterion is introduced as a basic belief assignment (bba). This comes from the fact that in most cases the input data cannot be defined within a reasonable degree of accuracy. In Table 2, a belief decision matrix representing all the alternatives' performances with respect to each criterion is introduced, where m_j^i represents the belief assessment of each alternative a_j with respect to each criterion c_i.

Table 2. Belief decision matrix

Alternatives	Criteria			
	c_1	c_2	...	c_m
a_1	m_1^1	m_1^2	...	m_1^m
...				
a_n				

Under this approach, we consider MCDM problem, where alternatives are evaluated with regard to ordinal criteria. Since it is generally thought that ordinal criteria are difficult to assess directly and the decision maker is required to give an accurate evaluation of the performances of all the alternatives on the given criteria, which is usually inaccurate, unreliable or even unavailable, especially in an uncertain environment. For the sake of simplicity, we assume that all criteria have the same assessment grades. In line with this assumption, many methods were defined such as the evidential reasoning approach and the AHP method. Let X be the set of assessment grades: $X = \{x_1, \ldots, x_h\}$.

To summarize, alternative a_j is assessed on each criterion c_i, using the same set of the ordinal assessment grades x_k which are required to be mutually exclusive and exhaustive. The ordinal assessment grades constitute our frame of discernment in the belief function theory.

Step 4: By using standard MCDM method, the performance matrix is obtained by multiplying the weighting vector by the decision matrix. Therefore, in this stage, we must update the alternative evaluations (bbas) with the importance of their respective criteria. In this context, our approach proposes to regard the criteria weight as a measure of reliability [5]. In fact, the idea is to measure most heavily the bba evaluated according to the most importance criteria and conversely for the less important ones. If we have c_i an evaluation criterion, then we get β_i its corresponding measure of reliability.

$$\beta_i = \frac{\omega_i}{\max_k \omega_k} \qquad \forall i, k = \{1, \ldots, m\} \tag{12}$$

Then, we get [5]:

$$m_j^{i, \alpha_i}(x_k) = \beta_i m_j^i(x_k), \ \forall x_k \subset X. \tag{13}$$

$$m_j^{i, \alpha_i}(X) = (1 - \beta_i) + \beta_i m_j^i(X). \tag{14}$$

where m_j^i is the relative bba for the alternative a_j, and we denote $\alpha_i = 1 - \beta_i$.

Step 5: Using Equation 14 and 14, an overall performance matrix is calculated. In this step, the main difficulty of our approach is how to combine and to compare different bbas. An intuitive definition of the strategy is to combine them using the conjunctive rule of combination, since we can assume that for each alternative performance is considered as a distinct source of evidence, and provides opinions towards the preferences of particular alternative. In addition, the obtained bba is defined on a common frame of discernment. So, the conjunctive rule may then be applied. This rule is used in order to aggregate the bbas induced by the expert for every alternative on all the criteria. The objective is to yield a combined bba that represents the performance of every alternative on the overall objective:

$$m_{a_j} = \bigcirc m_j^{i, \alpha_i}, \quad i = \{1, ..., m\} \tag{15}$$

Step 6: The results of the calculations are in the form of a bba. So each alternative is characterized by a single bba. As defined above, the main problem arises in comparing these bbas. In the present work, we consider first belief dominance (FBD) [3] in order to compare evaluations expressed by belief functions. By using this concept, we obtain a preference relation between each pair of alternatives. First, we start by computing the ascending belief function noted $\overrightarrow{bel}_i$ induced by m_{a_i} and associating to the evaluation of alternative a_i which is defined such as: $\overrightarrow{bel}_i(A_k) = \sum_{C \subseteq A_k} m_{a_i}(C)$ for all $A_k \in \overrightarrow{S}(X)$, where k is the number of assessments grades, $A_k = \{x_1, \ldots, x_k\}$, and $\overrightarrow{S}(X)$ denote the set of $\{A_1, \ldots, A_l\}$.

Then, the descending belief function noted $\overleftarrow{bel}_i$ induced by m_{a_i} and associating to the evaluation of alternative a_i is defined such as: $\overleftarrow{bel}_i(B_k) = \sum_{C \subseteq B_k} m_{a_i}(C)$ for all $B_k \in \overleftarrow{S}(X)$, where $B_k = \{x_k, \ldots, x_1\}$.

Thus, a bba m_{a_i} is said to dominate a bba m_{a_j} if and only if the following conditions are satisfied simultaneously:

- For all $A_k \in \overrightarrow{S}(X)$ $\overrightarrow{bel}_i(A_k) \leq \overrightarrow{bel}_j(A_k)$
- For all $B_k \in \overleftarrow{S}(X)$ $\overleftarrow{bel}_i(B_k) \geq \overleftarrow{bel}_j(B_k)$

The first condition means that there is greater belief mass of $\overrightarrow{bel}_j(A_k)$ than that of $\overrightarrow{bel}_i(A_k)$. On the contrary, the second condition means that there is greater belief mass of $\overleftarrow{bel}_i(B_k)$ than that of $\overleftarrow{bel}_j(B_k)$. In the case where the two conditions are not verified simultaneously, m_i does not dominate m_j ($\overline{FBD}$). Furthermore, it permits establishing four partial preference situations:

- if $m_i \ FBD \ m_j$ and $m_j \ FBD \ m_i$, then m_i is indifferent from m_j.
- if $m_i \ FBD \ m_j$ and $m_j \ \overline{FBD} \ m_i$, then m_i is strictly preferred to m_j.
- if $m_i \ \overline{FBD} \ m_j$ and $m_j \ FBD \ m_i$, then m_j is strictly preferred to m_i.
- if $m_i \ \overline{FBD} \ m_j$ and $m_j \ \overline{FBD} \ m_i$, then m_i and m_j are incomparable.

Once these relations are determined, the decision maker can then identify the subsets of best alternatives.

To summarize, Figure 1 shows the decision maker's process of the proposed approach.

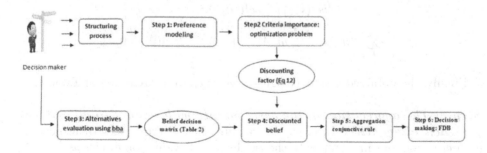

Fig. 1. Decision making process

5   Illustrative Example

In this section, we apply our proposed MCDM method to deal with a relatively simple decision problem of purchasing a car. The problem involves three criteria: Ω = {Comfort (c_1), Style (c_2), Fuel efficiency (c_3)}, and three selected alternatives: Θ = {Peugeot (a_1), Renault (a_2), Ford (a_3)}.

Step 1: The determination of the criteria importance: Along with our new MCDM method, a judgment matrix based on the pair-wise comparison process using preference modeling is defined in Table 3. As mentioned above, the decision maker was asked to indicate his level of preference between the selected criteria.

Table 3. Preference relation matrix for criterion level

Criteria	c_1	c_2	c_3
c_1	-	$\succ$	$\succ$
c_2	-	-	$\sim$
c_3	-	-	-

From Table 3, we remark that the decision maker has identified his preferences qualitatively. He identifies that $\{c_1\}$ is evaluated to be more important than $\{c_3\}$ and $\{c_1\}$ is evaluated to be more preferred than $\{c_2\}$.

Now, for deriving the weights of criteria, we apply our presented model. Therefore, we must transform these qualitative assessments into an optimization problem (Equation 10), then we solve the obtained system to compute the criteria importance. We assume that $\varepsilon = 0.01$ and the uncertainty measures is H since it has a unique maximum as defined in Equation 6. We obtain then the following optimization problem example:

$$Max_m H(m) = -m(\{c_1\}) * log_2(1/m(\{c_1\})) - m(\{c_2\})log_2(1/m(\{c_2\}))$$
$$-m(\{c_3\}) * log_2(1/m(\{c_3\})) - m(\Omega) * log_2(3/m(\Omega));$$
$$s.t.$$
$$bel(\{c_1\}) - bel(\{c_2\}) \geq \varepsilon$$
$$bel(\{c_1\}) - bel(\{c_3\}) \geq \varepsilon$$
$$bel(\{c_2\}) - bel(\{c_3\}) \leq \varepsilon \qquad (16)$$
$$bel(\{c_2\}) - bel(\{c_3\}) \geq -\varepsilon$$
$$\sum_{c_i \in \mathcal{F}(m)} m(c_i) = 1, m(A) \geq 0, \forall A \subseteq \Omega; m(\emptyset) = 0,$$

Finally, the obtained results (weighting vector) are represented in Table 4.

Step 2: The aggregation of the assessments with respect to all criteria for each alternative: In order to evaluate each alternative with respect to each criterion, our first step is to choose the set of evaluation grades as:

X= {poor, indifferent, average, good, excellent}.

Table 4. The weights assigned to the subset of criteria

Criteria	c_1	c_2	c_3	Ω
m^Ω	0.238	0.208	0.208	0.346
$BetP^\Omega$	0.352	0.324	0.324	
β_i	1	0.92	0.92	

In fact, the same set of evaluation grades is used for the three qualitative criteria. We propose to consider three alternatives. The evaluation of each alternative with respect to each criterion is given by a bba (see Table 5). For instance, to evaluate the alternative a_1, the expert hesitates between the fourth and the fifth assessment grades. He is sure that a_1 is either good or excellent without being able to refine his judgment.

Table 5. The bbas assigned to the alternatives performances

Criteria	c_1	c_2	c_3
a_1	$m_1^1(\{x_4\}) = 0.3$ $m_1^1(\{x_4, x_5\}) = 0.7$	$m_1^2(\{x_1\}) = 0.4$ $m_1^2(X){=}0.6$	$m_1^3(\{x_4\}) = 1$
a_2	$m_2^1(\{x_2, x_3\}) = 0.6$ $m_2^1(\{x_3, x_4\}) = 0.4$	$m_2^2(\{x_5\}) = 0.7$ $m_2^2(X) = 0.3$	$m_2^3(\{x_4, x_3\}) = 0.4$ $m_2^3(\{x_3\}) = 0.3$ $m_2^3(X) = 0.3$
a_3	$m_3^1(\{x_3\}) = 1$	$m_3^2(\{x_1, x_2\}) = 0.2$ $m_3^2(\{x_2, x_3\}) = 0.8$	$m_3^3(\{x_4, x_5\}) = 0.5$ $m_3^3(X) = 0.5$

By applying equation 14 and 14 using the discounting technique, the overall performance matrix was calculated and the decision matrix was determined in Table 6.

Table 6. The bbas assigned to the alternatives performances after discounting

Criteria	c_1	c_2	c_3
a_1	$m_1^{\alpha_1}(\{x_4\}) = 0.3$ $m_1^{\alpha_1}(\{x_4, x_5\}) = 0.7$	$m_1^{\alpha_2}(\{x_1\}) = 0.368$ $m_1^{\alpha_2}(X){=}0.632$	$m_1^{\alpha_3}(\{x_4\}) = 0.92$ $m_1^{\alpha_3}(X) = 0.08$
a_2	$m_2^{\alpha_1}(\{x_2, x_3\}) = 0.6$ $m_2^{\alpha_1}(\{x_3, x_4\}) = 0.4$	$m_2^{\alpha_2}(\{x_5\}) = 0.644$ $m_2^{\alpha_2}(X) = 0.356$	$m_2^{\alpha_3}(\{x_4, x_3\}) = 0.368$ $m_2^{\alpha_3}(\{x_3\}) = 0.276$ $m_2^{\alpha_3}(X) = 0.356$
a_3	$m_3^{\alpha_1}(\{x_3\}) = 1$	$m_3^{\alpha_2}(\{x_1, x_2\}) = 0.184$ $m_3^{\alpha_2}(\{x_2, x_3\}) = 0.736$ $m_3^{\alpha_2}(X) = 0.08$	$m_3^{\alpha_3}(\{x_4, x_5\}) = 0.46$ $m_3^{\alpha_3}(X) = 0.54$

In order to determine the overall performance of each alternative, the conjunctive rule of combination is used to combine the obtained bbas. Therefore, for each alternative, we propose to combine its corresponding bbas. Table 7 shows the results.

Table 7. The combined bbas

	$\emptyset$	$\{x_4\}$	$\{x_4, x_5\}$
m_{a_1}	0.368	0.5966	0.0354

	$\emptyset$	$\{x_3\}$	$\{x_2, x_3\}$	$\{x_3, x_4\}$
m_{a_2}	0.6441	0.1769	0.076	0.103

	$\emptyset$	$\{x_1, x_2\}$	$\{x_2, x_3\}$	$\{x_4, x_5\}$	X
m_{a_3}	0.5126	0.01	0.3974	0.0368	0.0432

Now and after getting a single bba for each alternative, we suggest to apply the FBD concept to compare and to rank the obtained bbas. Therefore, the ascending belief function and the descending belief function are calculated.

For instance, to calculate the ascending belief function and the descending belief function for the alternative a_1, we get:

$$a_1 = \begin{cases} \overrightarrow{bel_1^1}(\{x_1\}) = 0 \\ \overrightarrow{bel_1^1}(\{x_1, x_2\}) = 0 \\ \overrightarrow{bel_1^1}(\{x_1, x_2, x_3\}) = 0 \\ \overrightarrow{bel_1^1}(\{x_1, x_2, x_3, x_4\}) = 0.5966 \\ \overrightarrow{bel_1^1}(\{x_1, x_2, x_3, x_4, x_5\}) = 1 \end{cases} \qquad a_1 = \begin{cases} \overleftarrow{bel_1^1}(\{x_5\}) = 0 \\ \overleftarrow{bel_1^1}(\{x_4, x_5\}) = 0.632 \\ \overleftarrow{bel_1^1}(\{x_3, x_4, x_5\}) = 0.632 \\ \overleftarrow{bel_1^1}(\{x_2, x_3, x_4, x_5\}) = 0.632 \\ \overleftarrow{bel_1^1}(\{x_1, x_2, x_3, x_4, x_5\}) = 1 \end{cases}$$

Similarly, the FDB is computed for the alternatives a_2 and a_3, and finally, the preference situations between the alternatives are established. The results are given in the Table 8.

Table 8. Observed belief dominances between the alternatives

	a_1	a_2	a_3
a_1	$-$	FBD	$\overline{FBD}$
a_2	$\overline{FBD}$	$-$	$\overline{FBD}$
a_3	$\overline{FBD}$	$\overline{FBD}$	$-$

From this table, we have the alternative a_2 is outranked by alternative a_1, so it cannot be chosen. Moreover, alternative a_3 is incomparable to a_1. Then, a_1 and a_3 are the set of best alternatives according to our expert.

6   Conclusion

In this paper, a new MCDM approach in an uncertain environment was developed. Our proposed method, named belief MCDM, is based on the belief function framework. Indeed, to compute the weight of criteria, we apply a qualitative AHP approach then a new aggregating procedure is applied in order to update the alternatives performances. The approach developed is simple and comprehensible in concept, efficient in modeling human evaluation processes which makes it of general use for solving practical qualitative MCDM problems.

References

1. Ben Yaghlane, A., Denoeux, T., Mellouli, K.: Constructing belief functions from expert opinions. In: Proceedings of the 2nd International Conference on Information and Communication Technologies: from Theory to Applications (ICTTA 2006), Damascus, Syria, pp. 75–89 (2006)
2. Beynon, M., Curry, B., Morgan, P.: The Dempster-Shafer theory of evidence: An alternative approach to multicriteria decision modelling. OMEGA 28(1), 37–50 (2000)
3. Boujelben, M.A., Smet, Y.D., Frikha, A., Chabchoub, H.: A ranking model in uncertain, imprecise and multi-experts contexts: The application of evidence theory. International Journal of Approximate Reasoning 52, 1171–1194 (2011)
4. Brans, J., Vincke, P., Marechal, B.: How to select and how to rank projects: The PROMOTEE method. European Journal of Operational Research 24, 228–238 (1986)
5. Ennaceur, A., Elouedi, Z., Lefevre, E.: Handling partial preferences in the belief AHP method: Application to life cycle assessment. In: Pirrone, R., Sorbello, F. (eds.) AI*IA 2011. LNCS (LNAI), vol. 6934, pp. 395–400. Springer, Heidelberg (2011)
6. Ennaceur, A., Elouedi, Z., Lefevre, E.: Introducing incomparability in modeling qualitative belief functions. In: Torra, V., Narukawa, Y., López, B., Villaret, M. (eds.) MDAI 2012. LNCS, vol. 7647, pp. 382–393. Springer, Heidelberg (2012)
7. Ennaceur, A., Elouedi, Z., Lefevre, E.: Reasoning under uncertainty in the AHP method using the belief function theory. In: Greco, S., Bouchon-Meunier, B., Coletti, G., Fedrizzi, M., Matarazzo, B., Yager, R.R. (eds.) IPMU 2012, Part IV. CCIS, vol. 300, pp. 373–382. Springer, Heidelberg (2012)
8. Keeney, R., Raiffa, H.: Decisions with multiple objectives: Preferences and value tradeoffs. Cambridge University Press (1976)
9. Laarhoven, P.V., Pedrycz, W.: A fuzzy extension of Saaty's priority theory. Fuzzy Sets and Systems 11, 199–227 (1983)
10. Pal, N., Bezdek, J., Hemasinha, R.: Uncertainty measures for evidential reasoning I: A review. International Journal of Approximate Reasoning 7, 165–183 (1992)
11. Saaty, T.: The Analytic Hierarchy Process. McGraw-Hill, New-York (1980)
12. Shafer, G.: A Mathematical Theory of Evidence. Princeton University Press (1976)
13. Smets, P.: The combination of evidence in the Transferable Belief Model. IEEE Pattern Analysis and Machine Intelligence, 447–458 (1990)
14. Smets, P.: The application of the Transferable Belief Model to diagnostic problems. International Journal of Intelligent Systems 13, 127–158 (1998)
15. Wong, S., Lingras, P.: Representation of qualitative user preference by quantitative belief functions. IEEE Transactions on Knowledge and Data Engineering 6, 72–78 (1994)
16. Zeleny, M.: Multiple Criteria Decision Making. McGraw-Hill Book Company (1982)

PreDeLo 1.0: A Theorem Prover
for Preferential Description Logics

Laura Giordano[1], Valentina Gliozzi[2], Adam Jalal[3],
Nicola Olivetti[4], and Gian Luca Pozzato[2]

[1] DISIT, Università del Piemonte Orientale, Alessandria, Italy
laura.giordano@unipmn.it
[2] Dip. Informatica, Universitá di Torino, Italy
{valentina.gliozzi,gianluca.pozzato}@unito.it
[3] BE/CO/DA CERN, Geneva. Switzerland
adam.jalal@cern.ch
[4] Aix-Marseille Univ., CNRS, LSIS UMR 7296, France
nicola.olivetti@univ-amu.fr

Abstract. We describe PreDeLo 1.0, a theorem prover for preferential Description Logics (DLs). These are nonmonotonic extensions of standard DLs based on a typicality operator $\mathbf{T}$, which enjoys a preferential semantics. PreDeLo 1.0 is a Prolog implementation of labelled tableaux calculi for such extensions, and it is able to deal with the preferential extension of the basic DL $\mathcal{ALC}$ as well as with the preferential extension of the *lightweight* DL *DL-Lite$_{core}$*. The Prolog implementation is inspired by the "lean" methodology, whose basic idea is that each axiom or rule of the tableaux calculi is implemented by a Prolog clause of the program. Concerning $\mathcal{ALC}$, PreDeLo 1.0 considers two extensions based, respectively, on Kraus, Lehmann and Magidor's *preferential* and *rational* entailment. In this paper, we also introduce a tableaux calculus for checking entailment in the rational extension of $\mathcal{ALC}$.

1 Introduction

Nonmonotonic extensions of Description Logics (DLs) have been actively investigated since the early 90s [4,1,3,7,14,9,6]. A simple but powerful nonmonotonic extension of DLs is proposed in [14,9,13]: in this approach "typical" or "normal" properties can be directly specified by means of a "typicality" operator $\mathbf{T}$ enriching the underlying DL; the typicality operator $\mathbf{T}$ is essentially characterized by the core properties of nonmonotonic reasoning axiomatized by either *preferential logic* [15] or *rational logic* [16]. In these logics one can consistently express defeasible inclusions and exceptions such as: typical students do not pay taxes, but working students do typically pay taxes, but working students having children normally do not: $\mathbf{T}(Student) \sqsubseteq \neg TaxPayer$; $\mathbf{T}(Student \sqcap Worker) \sqsubseteq TaxPayer$; $\mathbf{T}(Student \sqcap Worker \sqcap \exists HasChild.\top) \sqsubseteq \neg TaxPayer$. In order to perform useful inferences, in [9] we have introduced a nonmonotonic extension of the logic $\mathcal{ALC}$ plus $\mathbf{T}$ based on a minimal model semantics. The resulting logic, called $\mathcal{ALC} + \mathbf{T}_{min}$, supports typicality assumptions, so that if one knows that john is a student, one can nonmonotonically assume that he is also a

M. Baldoni et al. (Eds.): AI*IA 2013, LNAI 8249, pp. 60–72, 2013.

typical student and therefore that he does not pay taxes. As an example, for a TBox specified by the inclusions above, in $\mathcal{ALC} + \mathbf{T}_{min}$ the following inferences hold: TBox $\models_{\mathcal{ALC}+\mathbf{T}_{min}} \mathbf{T}(Student \sqcap Tall) \sqsubseteq \neg TaxPayer$ (as being tall is irrelevant with respect to paying taxes); TBox $\cup \{Student(john)\} \models_{\mathcal{ALC}+\mathbf{T}_{min}} \neg TaxPayer(john)$; TBox $\cup \{Student(john), Worker(john)\} \models_{\mathcal{ALC}+\mathbf{T}_{min}} TaxPayer(john)$ (giving preference to more specific information); TBox $\cup \{Student(john), Tall(john)\} \models_{\mathcal{ALC}+\mathbf{T}_{min}} \neg TaxPayer(john)$; TBox $\cup \{\exists HasBrother. Student(john)\} \models_{\mathcal{ALC}+\mathbf{T}_{min}} \exists HasBrother. \neg TaxPayer(john)$ (minimal consequence applies to individuals not explicitly named in the ABox as well, without any ad-hoc mechanism). The approach based on the operator $\mathbf{T}$ has been extended in [10] to low complexity DLs such as *DL-Lite$_{core}$* [5], which are nonetheless well-suited for encoding large knowledge bases (KBs). It is known that query entailment in $\mathcal{ALC}+\mathbf{T}_{min}$ is in CO-NEXP^{NP}, whereas it drops to Π_2^p for *DL-Lite$_c$*$\mathbf{T}_{min}$. In [9] and [11] we have also introduced tableau calculi for deciding minimal entailment in $\mathcal{ALC} + \mathbf{T}_{min}$ and *DL-Lite$_c$*$\mathbf{T}_{min}$, respectively.

In this work we focus on proof methods and theorem proving for nonmonotonic extensions of DLs. On the one hand, we introduce a tableaux calculus also for the nonmonotonic logic $\mathcal{ALC} + \mathbf{T}_{min\mathbf{R}}$, resulting from the application of the minimal model semantics mentioned above to $\mathcal{ALC}$, when the KLM logic underlying the typicality operator $\mathbf{T}$ is the rational one $\mathbf{R}$ [16]. This calculus performs a two-phase computation: in the first phase, candidate models (complete open branches) falsifying the given query are generated, in the second phase the minimality of candidate models is checked by means of an auxiliary tableau construction. On the other hand, we describe PreDeLo 1.0, a theorem prover for preferential Description Logics $\mathcal{ALC}+\mathbf{T}_{min}$, $\mathcal{ALC}+\mathbf{T}_{min\mathbf{R}}$ and *DL-Lite$_c$*$\mathbf{T}_{min}$ implementing labelled tableaux calculi mentioned above in Prolog. PreDeLo 1.0 is inspired by the methodology introduced by the system leanT^AP [2], even if it does not fit its style in a rigorous manner. The basic idea is that each axiom or rule of the tableaux calculi is implemented by a Prolog clause of the program: the resulting code is therefore simple and compact.

As far as we know, PreDeLo 1.0 is the first theorem prover for preferential DLs. In general, the literature contains very few proof methods for nonmonotonic extensions of DLs. We provide some experimental results to show that the performances of PreDeLo 1.0 are promising. PreDeLo 1.0 also comprises a graphical interface written in Java, and it is available for free download at http://www.di.unito.it/~pozzato/predelo/.

2   The Logics $\mathcal{ALC} + \mathbf{T}_{min}$ and $\mathcal{ALC} + \mathbf{T}_{min\mathbf{R}}$

The logics $\mathcal{ALC} + \mathbf{T}_{min}$ and $\mathcal{ALC} + \mathbf{T}_{min\mathbf{R}}$ are obtained by adding to standard $\mathcal{ALC}$ the typicality operator $\mathbf{T}$ [14,13]. The intuitive idea is that $\mathbf{T}(C)$ selects the *typical* instances of a concept C. We can therefore distinguish between the properties that hold for all instances of concept C ($C \sqsubseteq D$), and those that only hold for the normal or typical instances of C ($\mathbf{T}(C) \sqsubseteq D$).

The language $\mathcal{L}$ of the logics is defined by distinguishing *concepts* and *extended concepts* as follows. Given an alphabet of concept names $\mathcal{C}$, of role names $\mathcal{R}$, and of individual constants $\mathcal{O}$, $A \in \mathcal{C}$ and $\top$ are *concepts* of $\mathcal{L}$; if $C, D \in \mathcal{L}$ and $R \in \mathcal{R}$, then $C \sqcap D, C \sqcup D, \neg C, \forall R.C, \exists R.C$ are *concepts* of $\mathcal{L}$. If C is a concept, then C and $\mathbf{T}(C)$ are *extended concepts*, and all the boolean combinations of extended concepts are extended concepts of $\mathcal{L}$. A KB is a pair (TBox,ABox). TBox contains inclusion relations (subsumptions) $C \sqsubseteq D$, where $C \in \mathcal{L}$ is an extended concept of the form either C' or $\mathbf{T}(C')$, and $D \in \mathcal{L}$ is a concept. ABox contains expressions of the form $C(a)$ and $R(a, b)$, where $C \in \mathcal{L}$ is an extended concept, $R \in \mathcal{R}$, and $a, b \in \mathcal{O}$.

In order to provide a semantics to the operator $\mathbf{T}$, we extend the definition of a model used in "standard" terminological logic $\mathcal{ALC}$. The idea is that the operator $\mathbf{T}$ is characterized by a set of postulates that are essentially a reformulation of the Kraus, Lehmann and Magidor's axioms of either *preferential logic* $\mathbf{P}$ [15] or *rational logic* $\mathbf{R}$ [16]: roughly speaking, the assertion $\mathbf{T}(C) \sqsubseteq D$ corresponds to the conditional assertion $C \mathrel{\vrule height 1.2ex depth 0pt width 0pt}\!\!\sim D$ of $\mathbf{P}$ (respectively $\mathbf{R}$). $\mathbf{T}$ has therefore all the "core" properties of nonmonotonic reasoning as it is axiomatized by $\mathbf{P}$ or $\mathbf{R}$. The idea is that there is a global preference relation among individuals, in the sense that $x < y$ means that x is "more normal" than y, and that the typical members of a concept C are the minimal elements of C with respect to this relation. In this framework, an element $x \in \Delta$ is a *typical instance* of some concept C if $x \in C^I$ and there is no C-element in Δ *more typical* than x. The typicality preference relation is partial.

Definition 1. *Given a preference relation $<$, which is an irreflexive and transitive relation over a domain Δ, for all $S \subseteq \Delta$, we define $Min_<(S) = \{x : x \in S \text{ and } \nexists y \in S \text{ s.t. } y < x\}$. We say that $<$ satisfies the* Smoothness Condition *iff for all $S \subseteq \Delta$, for all $x \in S$, either $x \in Min_<(S)$ or $\exists y \in Min_<(S)$ such that $y < x$.*

Definition 2. *A model of $\mathcal{ALC} + \mathbf{T}_{min}$ is any structure $\langle \Delta, <, I \rangle$, where: Δ is the domain; I is the extension function that maps each extended concept C to $C^I \subseteq \Delta$, and each role R to a $R^I \subseteq \Delta \times \Delta$; $<$ is an irreflexive and transitive relation over Δ satisfying the Smoothness Condition. I is defined in the usual way (as for $\mathcal{ALC}$) and, in addition, $(\mathbf{T}(C))^I = Min_<(C^I)$. A model of $\mathcal{ALC} + \mathbf{T}_{min\mathbf{R}}$ is a model of $\mathcal{ALC} + \mathbf{T}_{min}$ whose preference relation $<$ is further assumed to be modular: for all $x, y, z \in \Delta$, if $x < y$ then either $x < z$ or $z < y$.*

Given a model $\mathcal{M}$ of Definition 2, I can be extended so that it assigns to each individual a of $\mathcal{O}$ a distinct element a^I of the domain Δ (unique name assumption). We say that $\mathcal{M}$ satisfies an inclusion $C \sqsubseteq D$ if $C^I \subseteq D^I$, and that $\mathcal{M}$ satisfies $C(a)$ if $a^I \in C^I$ and $R(a, b)$ if $(a^I, b^I) \in R^I$. Moreover, $\mathcal{M}$ satisfies TBox if it satisfies all its inclusions, and $\mathcal{M}$ satisfies ABox if it satisfies all its formulas. $\mathcal{M}$ satisfies a KB (TBox,ABox), if it satisfies both its TBox and its ABox.

The semantics of the typicality operator can be specified by modal logic. The interpretation of $\mathbf{T}$ can be split into two parts: for any x of the domain Δ, $x \in (\mathbf{T}(C))^I$ just in case (i) $x \in C^I$, and (ii) there is no $y \in C^I$ such that $y < x$. Condition (ii) can be represented by means of an additional modality $\Box$, whose semantics is given by the preference relation $<$ interpreted as an accessibility relation. The interpretation of $\Box$ in $\mathcal{M}$ is as follows: $(\Box C)^I = \{x \in \Delta \mid \text{for every } y \in \Delta, \text{ if } y < x \text{ then } y \in C^I\}$. We immediately get that $x \in (\mathbf{T}(C))^I$ if and only if $x \in (C \sqcap \Box \neg C)^I$. From now on, we consider $\mathbf{T}(C)$ as an abbreviation for $C \sqcap \Box \neg C$.

Even if the typicality operator $\mathbf{T}$ itself is nonmonotonic (i.e. $\mathbf{T}(C) \sqsubseteq E$ does not imply $\mathbf{T}(C \sqcap D) \sqsubseteq E$), what is inferred from a KB can still be inferred from any KB' with KB $\subseteq$ KB'. In order to perform nonmonotonic inferences, in [9] we have strengthened the above semantics for the case of $\mathcal{ALC} + \mathbf{T}_{min}$ by restricting entailment to a class of minimal (or preferred) models. Intuitively, the idea is to restrict our consideration to models that *minimize the untypical instances of a concept*. In this work, we also extend this nonmonotonic entailment to $\mathcal{ALC} + \mathbf{T}_{min^{\mathbf{R}}}$.

Given a KB, we consider a finite set $\mathcal{L}_{\mathbf{T}}$ of concepts: these are the concepts whose untypical instances we want to minimize. We assume that the set $\mathcal{L}_{\mathbf{T}}$ contains at least all concepts C such that $\mathbf{T}(C)$ occurs in the KB or in the query F, where a *query* F is either an assertion $C(a)$ or an inclusion relation $C \sqsubseteq D$. As we have just said, $x \in C^I$ is typical for C if $x \in (\square \neg C)^I$. Minimizing the untypical instances of C therefore means to minimize the objects falsifying $\square \neg C$ for $C \in \mathcal{L}_{\mathbf{T}}$. Hence, for a given model $\mathcal{M} = \langle \Delta, <, I \rangle$, we define:

$$\mathcal{M}_{\mathcal{L}_{\mathbf{T}}}^{\square^-} = \{(x, \neg \square \neg C) \mid x \notin (\square \neg C)^I, \text{ with } x \in \Delta, C \in \mathcal{L}_{\mathbf{T}}\}.$$

Definition 3 (Preferred and minimal models). *Given a model $\mathcal{M} = \langle \Delta, <, I \rangle$ of a knowledge base* KB, *and a model $\mathcal{M}' = \langle \Delta', <', I' \rangle$ of* KB, *we say that $\mathcal{M}$ is preferred to $\mathcal{M}'$ w.r.t. $\mathcal{L}_{\mathbf{T}}$, and we write $\mathcal{M} <_{\mathcal{L}_{\mathbf{T}}} \mathcal{M}'$, if (i) $\Delta = \Delta'$, (ii) $\mathcal{M}_{\mathcal{L}_{\mathbf{T}}}^{\square^-} \subset \mathcal{M}_{\mathcal{L}_{\mathbf{T}}}'^{\square^-}$, (iii) $a^I = a^{I'}$ for all $a \in \mathcal{O}$. $\mathcal{M}$ is a minimal model for* KB *(w.r.t. $\mathcal{L}_{\mathbf{T}}$) if it is a model of* KB *and there is no other model $\mathcal{M}'$ of* KB *such that $\mathcal{M}' <_{\mathcal{L}_{\mathbf{T}}} \mathcal{M}$.*

Definition 4 (Minimal Entailment in $\mathcal{ALC} + \mathbf{T}_{min}$ (resp. $\mathcal{ALC} + \mathbf{T}_{min^{\mathbf{R}}}$)). *A query F is minimally entailed in $\mathcal{ALC} + \mathbf{T}_{min}$ (resp. $\mathcal{ALC} + \mathbf{T}_{min^{\mathbf{R}}}$) by* KB *with respect to $\mathcal{L}_{\mathbf{T}}$ if F is satisfied in all models of* KB *that are minimal with respect to $\mathcal{L}_{\mathbf{T}}$. We write* KB $\models_{\mathcal{ALC} + \mathbf{T}_{min}} F$ *(resp.* KB $\models_{\mathcal{ALC} + \mathbf{T}_{min^{\mathbf{R}}}} F$).

As an example, consider the TBox of the Introduction. We have that TBox $\cup$ {*Student* $(john)$} $\models_{\mathcal{ALC} + \mathbf{T}_{min^{\mathbf{R}}}} \neg TaxPayer(john)$, since $john^I \in (Student \sqcap \square \neg Student)^I$ for all minimal models $\mathcal{M} = \langle \Delta <, I \rangle$ of the TBox. In contrast, by the nonmonotonic character of minimal entailment, TBox $\cup$ {*Student(john), Worker(john)*} $\models_{\mathcal{ALC} + \mathbf{T}_{min^{\mathbf{R}}}} TaxPayer(john)$.

3   The Logic DL-$Lite_c \mathbf{T}_{min}$

In this section, we present the extension of the logic DL-$Lite_{core}$ [5] with the $\mathbf{T}$ operator called DL-$Lite_c \mathbf{T}_{min}$ [10]. The language of DL-$Lite_c \mathbf{T}_{min}$ is defined as follows.

Definition 5. *We consider an alphabet of concept names $\mathcal{C}$, of role names $\mathcal{R}$, and of individuals $\mathcal{O}$. Given $A \in \mathcal{C}$ and $R \in \mathcal{R}$, we define*

$$C_L := A \mid \exists S.\top \mid \mathbf{T}(A) \qquad S := R \mid R^- \qquad C_R := A \mid \neg A \mid \exists S.\top \mid \neg \exists S.\top$$

A DL-$Lite_c \mathbf{T}_{min}$ KB *is a pair (TBox, ABox). TBox contains a finite set of concept inclusions of the form $C_L \sqsubseteq C_R$. ABox contains assertions of the form $C(a)$ and $R(a, b)$, where C is a concept C_L or C_R, $R \in \mathcal{R}$, and $a, b \in \mathcal{O}$.*

A model $\mathcal{M}$ for $DL\text{-}Lite_c\mathbf{T}_{min}$ is any structure $\langle \Delta, <, I \rangle$, defined in Definition 2, where I is extended to deal with inverse roles: given $R \in \mathcal{R}$, $(R^-)^I = \{(a, b) \mid (b, a) \in R^I\}$. Furthermore, $<$ is multilinear: if $u < z$ and $v < z$, then either $u = v$ or $u < v$ or $v < u$. In [10] a Π_2^p upper bound for the complexity of $DL\text{-}Lite_c\mathbf{T}_{min}$ is provided by a small model theorem. Intuitively, what allows us to keep the size of the small model polynomial is that, for each atomic role R, a single element of the domain can be used to satisfy all occurrences of the existential $\exists R.\top$. Also, the same element of the domain can be used to satisfy all occurrences of existential concepts of the form $\exists R^-.\top$.

4   A Tableau Calculus for $\mathcal{ALC} + \mathbf{T}_{minR}$

In this section we introduce a tableau calculus $\mathcal{TAB}_{min}^{\mathcal{ALC}+\mathbf{T}R}$ for deciding whether a query F is minimally entailed from a KB in the logic $\mathcal{ALC} + \mathbf{T}_{minR}$. The calculus performs a two-phase computation: in the first phase, a tableau calculus, called $\mathcal{TAB}_{PH1}^{\mathcal{ALC}+\mathbf{T}R}$, simply verifies whether KB $\cup \{\neg F\}$ is satisfiable in a model of Definition 2, building candidate models; in the second phase another tableau calculus, called $\mathcal{TAB}_{PH2}^{\mathcal{ALC}+\mathbf{T}R}$, checks whether the candidate models found in the first phase are *minimal* models of KB, i.e. for each open branch of the first phase, $\mathcal{TAB}_{PH2}^{\mathcal{ALC}+\mathbf{T}R}$ tries to build a model of KB which is preferred to the candidate model w.r.t. Definition 3. The whole procedure is formally defined at the end of this section (Definition 6).

The calculus $\mathcal{TAB}_{min}^{\mathcal{ALC}+\mathbf{T}R}$ tries to build an open branch representing a minimal model satisfying KB $\cup \{\neg F\}$. The negation of a query $\neg F$ is defined as follows: if $F \equiv C(a)$, then $\neg F \equiv (\neg C)(a)$; if $F \equiv C \sqsubseteq D$, then $\neg F \equiv (C \sqcap \neg D)(x)$, where x does not occur in KB. $\mathcal{TAB}_{min}^{\mathcal{ALC}+\mathbf{T}R}$ makes use of labels, denoted with $x, y, z, \ldots$. Labels represent individuals either named in the ABox or implicitly expressed by existential restrictions. These labels occur in *constraints* (or *labelled* formulas), that can have the form $x \xrightarrow{R} y$ or $x : C$ or $y < x$, where x, y are labels, R is a role and C is either a concept or the negation of a concept of $\mathcal{ALC} + \mathbf{T}_{minR}$ or has the form $\square\neg D$ or $\neg\square\neg D$, where D is a concept.

Let us now analyze the two components of $\mathcal{TAB}_{min}^{\mathcal{ALC}+\mathbf{T}R}$, starting with $\mathcal{TAB}_{PH1}^{\mathcal{ALC}+\mathbf{T}R}$.

4.1   The Tableaux Calculus $\mathcal{TAB}_{PH1}^{\mathcal{ALC}+\mathbf{T}R}$

A tableau of $\mathcal{TAB}_{PH1}^{\mathcal{ALC}+\mathbf{T}R}$ is a tree whose nodes are pairs $\langle S \mid U \rangle$. S is a set of constraints, whereas U contains formulas of the form $C \sqsubseteq D^L$, representing subsumption relations $C \sqsubseteq D$ of the TBox. L is a list of labels, used in order to ensure the termination of the tableau calculus. A branch is a sequence of nodes $\langle S_1 \mid U_1 \rangle, \langle S_2 \mid U_2 \rangle, \ldots, \langle S_n \mid U_n \rangle \ldots$, where each node $\langle S_i \mid U_i \rangle$ is obtained from its immediate predecessor $\langle S_{i-1} \mid U_{i-1} \rangle$ by applying a rule of $\mathcal{TAB}_{PH1}^{\mathcal{ALC}+\mathbf{T}R}$, having $\langle S_{i-1} \mid U_{i-1} \rangle$ as the premise and $\langle S_i \mid U_i \rangle$ as one of its conclusions. A branch is closed if one of its nodes is an instance of a (Clash) axiom, otherwise it is open. A tableau is closed if all its branches are closed.

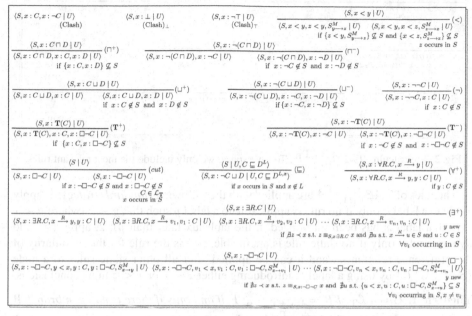

Fig. 1. The calculus $\mathcal{TAB}_{PH1}^{ALC+}\mathbf{T_R}$.

The rules of $\mathcal{TAB}_{PH1}^{ALC+\mathbf{T_R}}$, which is inspired by the calculus $\mathcal{TAB}_{PH1}^{ALC+\mathbf{T}}$ for $ALC + \mathbf{T}_{min}$ introduced in [9], are presented in Fig. 1. Rules $(\exists^+)$ and $(\Box^-)$ are called *dynamic* since they can introduce a new variable in their conclusions. The other rules are called *static*. We do not need any extra rule for the positive occurrences of $\Box$, since these are taken into account by the computation of $S_{x \to y}^M$ of $(\Box^-)$. The (cut) rule ensures that, given any concept $C \in \mathcal{L}_\mathbf{T}$, an open branch built by $\mathcal{TAB}_{PH1}^{ALC+\mathbf{T_R}}$ contains either $x : \Box\neg C$ or $x : \neg\Box\neg C$ for each label x: this is needed in order to allow $\mathcal{TAB}_{PH2}^{ALC+\mathbf{T_R}}$ to check the minimality of the model corresponding to the open branch. As mentioned above, given a node $\langle S \mid U \rangle$, each formula $C \sqsubseteq D$ in U is equipped with the list L of labels to which the rule $(\sqsubseteq)$ has already been applied. This avoids multiple applications of such rule to the same subsumption by using the same label.

In order to check the satisfiability of a KB, we build its *corresponding constraint system* $\langle S \mid U \rangle$, and we check its satisfiability. Given KB=(TBox,ABox), its *corresponding constraint system* $\langle S \mid U \rangle$ is defined as follows: $S = \{a : C \mid C(a) \in \text{ABox}\} \cup \{a \xrightarrow{R} b \mid R(a,b) \in \text{ABox}\}$; $U = \{C \sqsubseteq D^\emptyset \mid C \sqsubseteq D \in \text{TBox}\}$. KB is satisfiable if and only if its corresponding constraint system $\langle S \mid U \rangle$ is satisfiable. In order to verify the satisfiability of KB $\cup \{\neg F\}$, we use $\mathcal{TAB}_{PH1}^{ALC+\mathbf{T_R}}$ to check the satisfiability of the constraint system $\langle S \mid U \rangle$ obtained by adding the constraint corresponding to $\neg F$ to S', where $\langle S' \mid U \rangle$ is the corresponding constraint system of KB. To this purpose, the rules of the calculus $\mathcal{TAB}_{PH1}^{ALC+\mathbf{T_R}}$ are applied until either a contradiction is generated $(clash)$ or a model satisfying $\langle S \mid U \rangle$ can be obtained from the resulting constraint system.

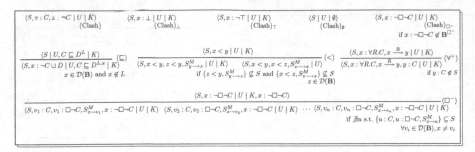

Fig. 2. The calculus $\mathcal{TAB}_{PH2}^{\mathcal{ALC}+\mathbf{T_R}}$. To save space, we only include the most relevant rules.

The rules of $\mathcal{TAB}_{PH1}^{\mathcal{ALC}+\mathbf{T_R}}$ are applied with the following *standard strategy*: 1. apply a rule to a label x only if no rule is applicable to a label y such that $y \prec x$ (where $y \prec x$ says that label x has been introduced in the tableaux later than y); 2. apply $(<)$ and dynamic rules only if no static rule is applicable. $(<)$ is the rule for the modularity of the preference relation $<$, and it can be applied after all other static rules to a node $\langle S, x < y \mid U \rangle$ by using a label z, introducing either $z < y$ or $x < z$ in its conclusions.

Theorem 1. *Given $\mathcal{L_T}$, $KB \models_{\mathcal{ALC}+\mathbf{T}_{min}\mathbf{R}} F$ if and only if there is no open branch $\mathbf{B}$ in the tableau built by $\mathcal{TAB}_{PH1}^{\mathcal{ALC}+\mathbf{T_R}}$ for the constraint system corresponding to KB $\cup \{\neg F\}$ such that the model represented by $\mathbf{B}$ is a minimal model of KB.*

Thanks to the side conditions on the application of the rules and the blocking machinery adopted by the dynamic ones, it can be shown that any tableau generated by $\mathcal{TAB}_{PH1}^{\mathcal{ALC}+\mathbf{T_R}}$ for $\langle S \mid U \rangle$ is finite; due to space limitations, proofs are confined in the accompanying technical report [8].

4.2   The Tableaux Calculus $\mathcal{TAB}_{PH2}^{\mathcal{ALC}+\mathbf{T_R}}$

Let us now introduce the calculus $\mathcal{TAB}_{PH2}^{\mathcal{ALC}+\mathbf{T_R}}$ which, similarly to the calculus $\mathcal{TAB}_{PH2}^{\mathcal{ALC}+\mathbf{T}}$ introduced in [9] for $\mathcal{ALC} + \mathbf{T}_{min}$, checks whether each open branch $\mathbf{B}$ built by $\mathcal{TAB}_{PH1}^{\mathcal{ALC}+\mathbf{T_R}}$ represents a minimal model of the KB.

Given an open branch $\mathbf{B}$ of a tableau built from $\mathcal{TAB}_{PH1}^{\mathcal{ALC}+\mathbf{T_R}}$, let $\mathcal{D}(\mathbf{B})$ be the set of labels occurring in $\mathbf{B}$. Moreover, let $\mathbf{B}^{\square^-}$ be the set of formulas $x : \neg\square\neg C$ occurring in $\mathbf{B}$, that is to say $\mathbf{B}^{\square^-} = \{x : \neg\square\neg C \mid x : \neg\square\neg C$ occurs in $\mathbf{B}\}$.

A tableau of $\mathcal{TAB}_{PH2}^{\mathcal{ALC}+\mathbf{T_R}}$ is a tree whose nodes are tuples of the form $\langle S \mid U \mid K \rangle$, where S and U are defined as in $\mathcal{TAB}_{PH1}^{\mathcal{ALC}+\mathbf{T_R}}$, whereas K contains formulas of the form $x : \neg\square\neg C$, with $C \in \mathcal{L_T}$. The basic idea of $\mathcal{TAB}_{PH2}^{\mathcal{ALC}+\mathbf{T_R}}$ is as follows. Given an open branch $\mathbf{B}$ built by $\mathcal{TAB}_{PH1}^{\mathcal{ALC}+\mathbf{T_R}}$ and corresponding to a model $\mathcal{M}^{\mathbf{B}}$ of KB $\cup \{\neg F\}$, $\mathcal{TAB}_{PH2}^{\mathcal{ALC}+\mathbf{T_R}}$ checks whether $\mathcal{M}^{\mathbf{B}}$ is a minimal model of KB by trying to build a model of KB which is preferred to $\mathcal{M}^{\mathbf{B}}$. To this purpose, it keeps track (in K) of the negated box used in $\mathbf{B}$ ($\mathbf{B}^{\square^-}$) in order to check whether it is possible to build a model of KB containing less negated box formulas.

The rules of $\mathcal{TAB}_{PH2}^{ALC+\mathbf{T_R}}$ are shown in Figure 2. The tableau built by $\mathcal{TAB}_{PH2}^{ALC+\mathbf{T_R}}$ closes if it is not possible to build a model smaller than $\mathcal{M}^{\mathbf{B}}$, it remains open otherwise. Since by Definition 3 two models can be compared only if they have the same domain, $\mathcal{TAB}_{PH2}^{ALC+\mathbf{T_R}}$ tries to build an open branch containing all the labels appearing in $\mathbf{B}$, i.e. those in $\mathcal{D}(\mathbf{B})$. To this aim, the dynamic rules use labels in $\mathcal{D}(\mathbf{B})$ instead of introducing new ones in their conclusions. The rule ($\sqsubseteq$) is applied to *all the labels of* $\mathcal{D}(\mathbf{B})$ (and not only to those appearing in the branch). The rule ($\square^-$) is applied to a node $\langle S, x : \neg\square\neg C \mid U \mid K, x : \neg\square\neg C \rangle$, that is to say when the negated box formula $x : \neg\square\neg C$ also belongs to the open branch $\mathbf{B}$. Also in this case, the rule introduces a branch on the choice of the individual $v_i \in \mathcal{D}(\mathbf{B})$ to be used in the conclusion. In case a tableau node has the form $\langle S, x : \neg\square\neg C \mid U \mid K \rangle$, and $x : \neg\square\neg C \notin \mathbf{B}^{\square^-}$, then $\mathcal{TAB}_{PH2}^{ALC+\mathbf{T_R}}$ detects a clash, called (Clash)$_{\square^-}$: this corresponds to the situation where $x : \neg\square\neg C$ does not belong to $\mathbf{B}$, while the model corresponding to the branch being built contains $x : \neg\square\neg C$, and hence is *not* preferred to the model represented by $\mathbf{B}$. The calculus $\mathcal{TAB}_{PH2}^{ALC+\mathbf{T_R}}$ also contains the clash condition (Clash)$_\emptyset$. Since each application of ($\square^-$) removes the negated box formulas $x : \neg\square\neg C$ from the set K, when K is empty all the negated boxed formulas occurring in $\mathbf{B}$ also belong to the current branch. In this case, the model built by $\mathcal{TAB}_{PH2}^{ALC+\mathbf{T_R}}$ satisfies the same set of $x : \neg\square\neg C$ (for all individuals) as $\mathbf{B}$ and, thus, it is not preferred to the one represented by $\mathbf{B}$.

Let KB be a knowledge base whose corresponding constraint system is $\langle S \mid U \rangle$. Let F be a query and let S' be the set of constraints obtained by adding to S the constraint corresponding to $\neg F$. $\mathcal{TAB}_{PH2}^{ALC+\mathbf{T_R}}$ is *sound and complete* in the following sense: an open branch $\mathbf{B}$ built by $\mathcal{TAB}_{PH1}^{ALC+\mathbf{T_R}}$ for $\langle S' \mid U \rangle$ is satisfiable in a minimal model of KB iff the tableau in $\mathcal{TAB}_{PH2}^{ALC+\mathbf{T_R}}$ for $\langle S \mid U \mid \mathbf{B}^{\square^-} \rangle$ is closed.

The termination of $\mathcal{TAB}_{PH2}^{ALC+\mathbf{T_R}}$ is ensured by the fact that dynamic rules make use of labels belonging to $\mathcal{D}(\mathbf{B})$, which is finite, rather than introducing "new" labels in the tableau. Also, it is possible to show that the problem of verifying that a branch $\mathbf{B}$ represents a minimal model for KB in $\mathcal{TAB}_{PH2}^{ALC+\mathbf{T_R}}$ is in NP in the size of $\mathbf{B}$.

The overall procedure $\mathcal{TAB}_{min}^{ALC+\mathbf{T_R}}$ is defined as follows:

Definition 6. *Let KB be a knowledge base whose corresponding constraint system is* $\langle S \mid U \rangle$. *Let F be a query and let S' be the set of constraints obtained by adding to S the constraint corresponding to $\neg F$. The calculus $\mathcal{TAB}_{min}^{ALC+\mathbf{T_R}}$ checks whether a query F can be minimally entailed from a KB by means of the following procedure:*

- *the calculus $\mathcal{TAB}_{PH1}^{ALC+\mathbf{T_R}}$ is applied to $\langle S' \mid U \rangle$;*
- *if, for each branch $\mathbf{B}$ built by $\mathcal{TAB}_{PH1}^{ALC+\mathbf{T_R}}$, either: (i) $\mathbf{B}$ is closed or (ii) the tableau built by the calculus $\mathcal{TAB}_{PH2}^{ALC+\mathbf{T_R}}$ for $\langle S \mid U \mid \mathbf{B}^{\square^-} \rangle$ is open, then the procedure says* YES *else the procedure says* NO

The following results hold (again, due to space limitations, proofs are confined in [8]):

Theorem 2. $\mathcal{TAB}_{min}^{ALC+\mathbf{T_R}}$ *is a sound and complete decision procedure for verifying if* $KB \models_{ALC+\mathbf{T}_{min}\mathbf{R}}^{\mathcal{L_T}} F$.

Theorem 3. *The problem of deciding whether* $\text{KB} \models_{\mathcal{ALC}+\mathbf{T}_{min}\mathbf{R}} F$ *is in* CO-NEXP^{NP}.

The calculus $\mathcal{TAB}_{min}^{\mathcal{ALC}+\mathbf{T}}$, introduced in [9], is similar to $\mathcal{TAB}_{min}^{\mathcal{ALC}+\mathbf{T}_\mathbf{R}}$: due to space limitations, we omit details. In [11], we have also introduced a tableau calculus $\mathcal{TAB}_{min}^{Lite_c\mathbf{T}}$ for deciding query entailment in $DL\text{-}Lite_c\mathbf{T}_{min}$; we have also shown that the problem of deciding whether $\text{KB} \models_{DL\text{-}Lite_c\mathbf{T}_{min}} F$ by means of $\mathcal{TAB}_{min}^{Lite_c\mathbf{T}}$ is in Π_2^p.

5 Design of `PreDeLo 1.0`

We present a SICStus Prolog implementation of the tableaux calculi for nonmonotonic extensions of DLs introduced in Section 4. The program, called `PreDeLo 1.0`, is inspired by the "lean" methodology of lean$T^A P$, even if it does not follow its style in a rigorous manner. The program comprises a set of clauses, each one of them implements a sequent rule or axiom of the tableau calculi. The proof search is provided for free by the mere depth-first search mechanism of Prolog, without any additional ad hoc mechanism. `PreDeLo 1.0` also comprises a Java graphical user interface interacting with the Prolog engine by means of the jasper package.

`PreDeLo 1.0` comprises two main predicates, called `prove` and `prove_phase2`, implementing, respectively, Phase 1 and Phase 2 of the tableau calculi.

Phase 1: the `prove` predicate. Concerning the first phase of the calculi, `PreDeLo 1.0` represents a tableaux node $\langle S \mid U \rangle$ with two Prolog lists: `S` and `U`. Elements of `S` are either pairs `[X, F]`, representing formulas of the form $x : F$, or triples of the form either `[X,R,Y]` or `[X,<,Y]`, representing either roles $x \xrightarrow{R} y$ or the preference relation $x < y$, respectively. Elements of `U` are pairs of the form `[[C inc D],L]`, representing $C \sqsubseteq D^L \in U$ described in Section 4.1.

As an example, the Prolog lists `S=[[x,neg ti c],[y,<,x],[y,c],[y,box neg c],[x, fe r in d],[x,r,z]]` and `U=[ [c inc (d and e), [ ] ],[top inc c, [y]] ]` represent the constraint system $\langle x : \neg\mathbf{T}(C), y < x, y : C, y : \Box\neg C, x : \forall R.D, x \xrightarrow{R} z \mid C \sqsubseteq (D \sqcap E)^\emptyset, \top \sqsubseteq C^{\{y\}}\rangle$.

The calculi $T^{\mathcal{ALC}+\mathbf{T}}$ are implemented by a top-level predicate

```
prove(+ABox,+TBox,[+X,+F],-Tree).
```

This predicate succeeds if and only if the query $x : F$ is minimally entailed from the KB represented by `TBox` and `ABox`. When the predicate succeeds, then the output term `Tree` matches a Prolog term representing the closed tableaux found by the prover that will be displayed by the Java graphical interface. For instance, in order to prove that $\neg TaxPayer(john)$ is minimally entailed from the KB whose TBox is $\{\mathbf{T}(Student) \sqsubseteq \neg TaxPayer, \mathbf{T}(Student \sqcap Worker) \sqsubseteq TaxPayer\}$ and whose ABox is $\{Student(john)\}$, one queries `PreDeLo 1.0` with

```
prove([[john,student]], [[ti student inc not taxp,[ ]], [ti
(student and worker) inc taxp, [ ]]],[john,not taxp],Tree).
```

The top-level predicate prove/4 invokes a second-level one:

prove(+S,+U,+Lt,+Labels,+ABOX,-Tree)

having 6 arguments. In detail, S corresponds to ABox enriched by the negation of the query $x : F$, whereas Lt is a list corresponding to the set of concepts $\mathcal{L}_{\mathbf{T}}$. Labels is the set of labels belonging to the current branch, whereas ABOX is used to store the initial ABox (i.e. without the negation of the query) in order to eventually invoke phase 2 on it, in order to look for minimal models of the initial KB.

Each clause of the prove/6 predicate implements an axiom or rule of the calculi $\mathcal{TAB}^{ACC+\mathbf{T}}_{PH1}$, $\mathcal{TAB}^{ACC+\mathbf{T_R}}_{PH1}$ and $\mathcal{TAB}^{Lite_c\mathbf{T}}_{PH1}$. To search a closed tableaux for $\langle S \mid U \rangle$, PreDeLo 1.0 proceeds as follows. First of all, if $\langle S \mid U \rangle$ is a clash, the goal will succeed immediately by using one of following clauses:

```
prove(S,U,_,_,_Tree):-member([X,C],S),member([X, neg C],S),
                      buildTree(S,U,Tree),!.
prove(S,U,_,_,_,Tree):-member([_,neg top],S),buildTree(S,U,Tree),!.
prove(S,U,_,_,_,Tree):-member([_,bottom],S),buildTree(S,U,Tree),!.
```

implementing (Clash), (Clash)$_\top$ and (Clash)$_\bot$, respectively. The auxiliary predicate buildTree is used in order to build the derivation used by the Java interface (we omit further details due to space limitations). If $\langle S \mid U \rangle$ is not an instance of the axioms, then the first applicable rule will be chosen, e.g. if S contains an intersection [X,C and D], then the clause implementing the ($\sqcap^+$) rule will be chosen, and PreDeLo 1.0 will be recursively invoked on its unique conclusion. PreDeLo 1.0 proceeds in a similar way for the other rules. The ordering of the clauses is such that the application of the dynamic rules is postponed as much as possible: this implements the strategy ensuring the termination of the calculi described in the previous section.

As an example, the clause implementing ($\mathbf{T}^+$) is as follows:

```
1. prove(S,U,Lt,Labels,ABOX,Tree):-member([X,ti C],S),
2.      (\+(member([X,C],S)); \+(member([X, box neg C],S))),!,
3.      prove([[X,C]|[[X, box neg C]|S]],U,Lt,Labels,ABOX,Tree1),!,
4.      buildTree(S,U,Tree1,Tree),!.
```

In line 1, the standard Prolog predicate member is used in order to find a formula of the form $x : \mathbf{T}(C)$ in the list S. In line 2, the side conditions on the applicability of such a rule are checked: the rule can be applied if either $x : C$ or $x : \Box\neg C$ do not belong to S. In line 3 PreDeLo 1.0 is recursively invoked on the unique conclusion of the rule, in which $x : C$ and $x : \Box\neg C$ are added to the list S.

The very last clause of prove is the following one:

```
prove(S,U,Lt,_,ABOX,[]) :- !,
    getNegBox(S,Bb,K),initLabels(S,Db),clearSubsumptions(U,U1),!,
    \+toplevelphase2(ABOX,U1,Lt,K,Bb,Db),!.
```

invoked when no other clauses are applicable. In this case, the branch built by the prover represents a model for the initial set of formulas, therefore the predicate toplevelphase2 is invoked in order to check whether such a model is minimal or not for the initial KB. In detail, according to the procedure of Definition 6, prove succeeds in case the calculus of phase 2 finds an open tableau, namely when the predicate `toplevelphase2` fails. Auxiliary predicates `getNegBox`, `initLabels` and `clearSubsumptions` are used to compute sets $\mathcal{D}(\mathbf{B})$, $\mathbf{B}^{\square^-}$ and K used in the calculi described in Section 4.2.

Phase 2: the `prove_phase2` predicate. Given an open branch built by phase 1, the predicate `toplevelphase2` first applies an optimization preventing useless applications of ($\sqsubseteq$), then it invokes the predicate

$$prove\_phase2\,(+S,+U,+Lt,+K,+Bb,+Db).$$

S and U contain the initial KB (without the query), whereas K, Bb and Db are Prolog lists representing K, $\mathbf{B}^{\square^-}$ and $\mathcal{D}(\mathbf{B})$ as described in Section 4.2. Lt is as for `prove/6`.

Also in this case, each clause of `prove_phase2` implements an axiom or rule of the calculi $\mathcal{TAB}_{PH2}^{\mathcal{ALC}+\mathbf{T}}$, $\mathcal{TAB}_{PH2}^{\mathcal{ALC}+\mathbf{T_R}}$ and $\mathcal{TAB}_{PH2}^{Lite_c\mathbf{T}}$. To search a closed tableaux, PreDeLo 1.0 first checks whether the current node $\langle S \mid U \mid K \rangle$ is a clash. For examples, here are the clauses implementing, respectively, clashes (Clash)$_\emptyset$ and (Clash)$_{\square^-}$:

```
prove_phase2(_,_,_,[   ],_,_):-!.
prove_phase2(S,_,_,_,Bb,_):-member([X,neg box neg C],S),
                    \+member([X,neg box neg C],Bb),!.
```

detecting, the first one, the case in which K is empty (that is to say all negated boxed formulas occurring in the open branch of phase 1 also occur in the current branch), the second one, the case in which there is a $x : \neg\square\neg C \in S$ such that $x : \neg\square\neg C$ does not occur in the open branch built by phase 1, i.e. $x : \neg\square\neg C \notin \mathbf{B}^{\square^-}$. In both cases, the model built by phase 2 is not preferred to the one built by phase 1, then the current branch of phase 2 must be closed. As for the predicate `prove/6` of phase 1, if $\langle S \mid U \mid K \rangle$ is not an instance of the axioms, then the first applicable rule will be chosen, and PreDeLo 1.0 will be recursively invoked on its conclusions. As an example, the clause implementing ($\mathbf{T}^+$) is as follows:

```
prove_phase2(S,U,Lt,K,Bb,Db) :- select([X,ti C],S,S1),
      prove_phase2([[X,C]|[[X,box neg C]|S1]],U,Lt,K,Bb,Db),!.
```

Notice that, according to the calculi $\mathcal{TAB}_{PH2}^{\mathcal{ALC}+\mathbf{T}}$, $\mathcal{TAB}_{PH2}^{\mathcal{ALC}+\mathbf{T_R}}$ and $\mathcal{TAB}_{PH2}^{Lite_c\mathbf{T}}$, the principal formula to which the rule is applied is removed from the current node: to this aim, the SICStus Prolog predicate `select` is used rather than `member`.

5.1   Performances of `PreDeLo 1.0`

The performances of PreDeLo 1.0 are promising. We have tested it by running SICStus Prolog 4.0.2 on an Apple MacBook Pro, 3.06 GHz Intel Core 2 Duo, 4GB RAM machine. We have randomly generated KBs with different sizes (from 10 to 100 ABox formulas and TBox inclusions) as well as different numbers of named individuals. Figure 3 shows the answers of PreDeLo 1.0 within a given time limit: in less than 10 seconds, it is able to answer in more than the 75% of tests. Notice that, as far as we know, it does not exist a set of acknowledged benchmarks for defeasible DLs.

$\mathcal{TAB}^{ALC+T}_{min}$

	1ms	10ms	1s	10s
yes	8%	8%	14%	14%
no	50%	54%	49%	60%
time out	42%	38%	37%	26%

$\mathcal{TAB}^{ALC+T_R}_{min}$

	1ms	10ms	1s	10s
yes	8%	9%	12%	15%
no	51%	51%	55%	56%
time out	41%	40%	33%	29%

$\mathcal{TAB}^{Lite_c T}_{min}$

	1ms	10ms	1s	10s
yes	10%	12%	15%	16%
no	57%	60%	71%	79%
time out	33%	28%	14%	5%

Fig. 3. Performances of `PreDeLo 1.0` over randomly generated KBs (ABox with 50 assertions, TBox with 50 inclusions), obtained from 15 different atomic concepts and 10 named individuals

6 Conclusions

In this work we have focused on proof methods and theorem proving for preferential extensions of DLs. In detail, we have given three different original contributions: 1. we have introduced the nonmonotonic logic $ALC + \mathbf{T}_{min}\text{R}$, extending ALC with a typicality operator $\mathbf{T}$ based on rational logic $\mathbf{R}$ plus a minimal model semantics; 2. we have introduced a labelled tableaux calculus for checking minimal entailment in $ALC + \mathbf{T}_{min}\text{R}$; we have also provided a complexity upper bound for it, namely that is in CO-NEXP^{NP} as for $ALC + \mathbf{T}_{min}$; 3. we have presented `PreDeLo 1.0`, a theorem prover implementing tableaux calculi for reasoning in all these preferential DLs.

Of course, many optimizations are possible and we intend to study them in future work. In order to improve the performances of `PreDeLo 1.0`, we are currently developing a distributed extension, whose basic idea is that the two phases of the calculus are performed by different machines: a "master" machine M executes the first phase of the tableaux calculus, whereas other computers will be devoted to perform the second phase on open branches detected by M. When M finds an open branch, it invokes the second phase of the calculus on a different "slave" machine S_1, going on performing the first phase on other branches, rather than waiting for the result of S_1. When another open branch is detected, then another machine S_2 is involved in the procedure in order to perform the second phase of the calculus on that branch. In this way, the second phase is performed simultaneously on different branches, leading to a significant increasing of the performances of the theorem prover.

Furthermore, we aim at extending the implementation of `PreDeLo 1.0` to the low complexity DLs of the $\mathcal{EL}$ family, allowing for conjunction ($\sqcap$) and existential restriction ($\exists R.C$). Despite their relatively low expressivity, they are relevant for several applications, in particular in the bio-medical domain; for instance, small extensions of $\mathcal{EL}$ can be used to formalize medical terminologies, such as GALEN, SNOMED, and the Gene Ontology used in bioinformatics. Reasoning in $\mathcal{EL}$ is tractable. An extension of $\mathcal{EL}^{\perp}$ with the typicality operator $\mathbf{T}$ has been proposed in [10], where it has also been shown that, for a specific fragment, minimal entailment is in Π^p_2.

References

1. Baader, F., Hollunder, B.: Priorities on defaults with prerequisites, and their application in treating specificity in terminological default logic. J. of Autom. Reas. 15(1), 41–68 (1995)
2. Beckert, B., Posegga, J.: leantap: Lean tableau-based deduction. JAR 15(3), 339–358 (1995)

3. Bonatti, P.A., Faella, M., Sauro, L.: Defeasible inclusions in low-complexity DLs. J. Artif. Intell. Res. (JAIR) 42, 719–764 (2011)
4. Bonatti, P.A., Lutz, C., Wolter, F.: The complexity of circumscription in DLs. J. Artif. Intell. Res. (JAIR) 35, 717–773 (2009)
5. Calvanese, D., Giacomo, G., Lembo, D., Lenzerini, M., Rosati, R.: Tractable reasoning and efficient query answering in DLs: the DL-Lite family. J. Autom. Reas. 39(3), 385–429 (2007)
6. Casini, G., Straccia, U.: Rational closure for defeasible description logics. In: Janhunen, T., Niemelä, I. (eds.) JELIA 2010. LNCS, vol. 6341, pp. 77–90. Springer, Heidelberg (2010)
7. Donini, F.M., Nardi, D., Rosati, R.: Description logics of minimal knowledge and negation as failure. ACM Trans. Comput. Log. 3(2), 177–225 (2002)
8. Giordano, L., Gliozzi, V., Olivetti, N., Pozzato, G.L.: A tableaux calculus for $\mathcal{ALC}+\mathbf{T}_{min}\mathbf{R}$, Tech. rep. (2013), http://www.di.unito.it/~pozzato/tralctrm.pdf
9. Giordano, L., Gliozzi, V., Olivetti, N., Pozzato, G.L.: A nonmonotonic Description Logic for reasoning about typicality. Artificial Intelligence 195, 165–202 (2013)
10. Giordano, L., Gliozzi, V., Olivetti, N., Pozzato, G.L.: Reasoning about typicality in low complexity DLs: the logics $\mathcal{EL}^{\perp}\mathbf{T}_{min}$ and DL-lite$_c\mathbf{T}_{min}$. In: IJCAI, pp. 894–899 (2011)
11. Giordano, L., Gliozzi, V., Olivetti, N., Pozzato, G.L.: A tableau calculus for a nonmonotonic extension of the Description Logic $DL\text{-}Lite_{core}$. In: Pirrone, R., Sorbello, F. (eds.) AI*IA 2011. LNCS, vol. 6934, pp. 164–176. Springer, Heidelberg (2011)
12. Giordano, L., Gliozzi, V., Olivetti, N., Pozzato, G.L.: A tableau calculus for a nonmonotonic extension of $\mathcal{EL}^{\perp}$. In: Brünnler, K., Metcalfe, G. (eds.) TABLEAUX 2011. LNCS, vol. 6793, pp. 180–195. Springer, Heidelberg (2011)
13. Giordano, L., Gliozzi, V., Olivetti, N., Pozzato, G.L.: Preferential vs Rational Description Logics: which one for Reasoning About Typicality? In: ECAI, pp. 1069–1070 (2010)
14. Giordano, L., Gliozzi, V., Olivetti, N., Pozzato, G.L.: $\mathcal{ALC} + \mathbf{T}_{min}$: a preferential extension of Description Logics. Fundamenta Informaticae 96, 1–32 (2009)
15. Kraus, S., Lehmann, D., Magidor, M.: What does a conditional knowledge base entail? Artificial Intelligence 55(1), 1–60 (1992)
16. Lehmann, D., Magidor, M.: Nonmonotonic reasoning, preferential models and cumulative logics. Artificial Intelligence 44(1-2), 167–207 (1990)

Automated Selection of Grounding Algorithm in Answer Set Programming

Marco Maratea[1], Luca Pulina[2], and Francesco Ricca[3]

[1] DIBRIS, Univ. degli Studi di Genova, Viale F. Causa 15, 16145 Genova, Italy
marco@dist.unige.it
[2] POLCOMING, Univ. degli Studi di Sassari, Viale Mancini 5, 07100 Sassari, Italy
lpulina@uniss.it
[3] Dip. di Matematica ed Informatica, Univ. della Calabria, Via P. Bucci, 87030 Rende, Italy
ricca@mat.unical.it

Abstract. Answer Set Programming (ASP) is a powerful language for knowledge representation and reasoning. ASP is exploited in real-world applications and is also attracting the interest of industry thanks to the availability of efficient implementations. ASP systems compute solutions relying on two modules: a *grounder* that produces, by removing variables from the rules, a ground program equivalent to the input one; and a *model generator* (or *solver*) that computes the solutions of such propositional program. In this paper we make a first step toward the exploitation of automated selection techniques to the grounding module. We rely on two well-known ASP grounders, namely the grounder of the DLV system and GRINGO and we leverage on automated classification algorithms to devise and implement an automatic procedure for selecting the "best" grounder for each problem instance. An experimental analysis, conducted on benchmarks and solvers from the 3rd ASP Competition, shows that our approach improves the evaluation performance independently from the solver associated with our grounder selector.

1 Introduction

Answer Set Programming (ASP) [6,13,14] is a powerful language for knowledge representation and reasoning that is exploited in both real-world and industrial applications (see, e.g. [15] for an overview). In order to deal with industrial-level problems, performance of ASP systems is a factor of paramount importance. Given a non-ground ASP program, ASP systems compute solutions relying on two main modules: a *grounder* that produces a propositional ASP program and a *model generator* (or *solver*) that takes as input a propositional program and returns a solution.

The recent application of automated algorithm selection techniques to ASP solving (see, e.g., [9,22]) has noticeably improved the performance of ASP systems. On the other hand, the adoption of such techniques has been confined to the model generator module, mainly because they are often obtained by importing to ASP techniques already applied to propositional satisfiability (SAT) or Quantified SAT (see, e.g, [26]). Limiting the choice to only one grounder does not allow to exploit the various optimization techniques, possibly applied to different problem class (e.g., *P* and *NP* problem classes

M. Baldoni et al. (Eds.): AI*IA 2013, LNAI 8249, pp. 73–84, 2013.
© Springer International Publishing Switzerland 2013

as classified in the 3rd ASP Competition [4]), implemented only in one grounder (e.g. Magic sets [1] in the presence of queries).

In this paper we make a first step toward the exploitation of automated selection techniques to the grounding module. We rely on two well-known ASP grounders, namely the grounder of the DLV system [20] (DLV-G in the following) and GRINGO [12,10] and we leverage on automated classification algorithms to automatically select the "best" grounder. More in details, our starting point is an experimental analysis conducted on the domains of benchmarks belonging to both P and NP classes of the 3rd ASP Competition, involving state-of-the-art ASP solvers and the aforementioned grounders. We then applied classification methods by relying on characteristics of the various encoding (*features*), with the aim of automatically select the most appropriate grounder. We then implemented a system based on these ideas and the results of our experimental analysis show that the performance of the considered solvers are boosted by the usage of the proposed system.

To sum up, the main contributions of this paper are: (i) the application of automated selection techniques to the grounding module in ASP computation, to complement a recent body of research only focused on solvers; (ii) the implementation of a system based on these techniques; and (iii) an experimental analysis of the new system, involving a large variety of benchmarks, grounders and solvers, that shows the benefits of the approach.

The paper is structured as follows. Section 2 introduces needed preliminaries about ASP. Section 3 shows the main answer set computation methods, with focus on the grounding module. Section 4 then describes the ideas we have applied to reach the above-mentioned goals. Section 5 presents implementation details of the system implemented along the line reported in the previous section, and its results. The paper ends with some conclusions in Section 6.

2   Answer Set Programming

In this section we recall Answer Set Programming syntax and semantics.

Syntax. A variable or a constant is a *term*. An *atom* is denoted $p(t_1, ..., t_n)$, where p is a *predicate* of arity n and $t_1, ..., t_n$ are terms. A *literal* is either a *positive literal* p or a *negative literal* not p, where p is an atom. A *(disjunctive) rule* r has the following form:

$$a_1 \lor \ldots \lor a_n :\!- b_1, \ldots, b_k, \text{ not } b_{k+1}, \ldots, \text{ not } b_m. \tag{1}$$

where $a_1, \ldots, a_n, b_1, \ldots, b_m$ are atoms. The disjunction $a_1 \lor \ldots \lor a_n$ is the *head* of r, while the conjunction $b_1, \ldots, b_k, \text{not } b_{k+1}, \ldots, \text{not } b_m$ is the *body* of r. A rule having precisely one head literal (i.e. $n = 1$) is called a *normal rule*. If the body is empty (i.e. $k = m = 0$), it is called a *fact* and the $:\!-$ sign is usually omitted. A rule without head literals (i.e. $n = 0$) is usually referred to as an *integrity constraint*. A rule r is *safe* if each variable appearing in r appears also in some positive body literal of r.

An *ASP program* $\mathcal{P}$ is a finite set of safe rules. A not -free (resp., $\lor$-free) program is called *positive* (resp., *normal*). A term, an atom, a literal, a rule, or a program is *ground* if no variables appear in it.

Hereafter, we denote by $H(r)$ the set $\{a_1, \ldots, a_n\}$ of the head atoms, and by $B(r)$ the set $\{b_1, \ldots, b_k, \texttt{not } b_{k+1}, \ldots, \texttt{not } b_m\}$ of the body literals. $B^+(r)$ (resp., $B^-(r)$) denotes the set of atoms occurring positively (resp., negatively) in $B(r)$. A predicate p is referred to as an *EDB* predicate if, for each rule r having in the head an atom whose name is $p \in H(r)$, r is a fact; all others predicates are referred to as *IDB* predicates. The set of facts in which *EDB* predicates occur, denoted by $EDB(\mathcal{P})$, is called *Extensional Database (EDB)*, the set of all other rules is the *Intensional Database (IDB)*.

Semantics. Let $\mathcal{P}$ be an ASP program. The *Herbrand universe* of $\mathcal{P}$, denoted as $U_{\mathcal{P}}$, is the set of all constants appearing in $\mathcal{P}$. In the case when no constant appears in $\mathcal{P}$, an arbitrary constant is added to $U_{\mathcal{P}}$. The *Herbrand base* of $\mathcal{P}$, denoted as $B_{\mathcal{P}}$, is the set of all ground atoms constructable from the predicate symbols appearing in $\mathcal{P}$ and the constants of $U_{\mathcal{P}}$. Given a rule r occurring in a program $\mathcal{P}$, a *ground instance* of r is a rule obtained from r by replacing every variable X in r by $\sigma(X)$, where σ is a substitution mapping the variables occurring in r to constants in $U_{\mathcal{P}}$. We denote by $Ground(\mathcal{P})$ the set of all the ground instances of the rules occurring in $\mathcal{P}$.

An *interpretation* for $\mathcal{P}$ is a set of ground atoms, that is, an interpretation is a subset I of $B_{\mathcal{P}}$. A ground positive literal A is true (resp., false) w.r.t. I if $A \in I$ (resp., $A \notin I$). A ground negative literal $\texttt{not } A$ is true w.r.t. I if A is false w.r.t. I; otherwise $\texttt{not } A$ is false w.r.t. I. Let r be a rule in $Ground(\mathcal{P})$. The head of r is true w.r.t. I if $H(r) \cap I \neq \emptyset$. The body of r is true w.r.t. I if all body literals of r are true w.r.t. I (i.e., $B^+(r) \subseteq I$ and $B^-(r) \cap I = \emptyset$) and otherwise the body of r is false w.r.t. I. The rule r is *satisfied* (or *true*) w.r.t. I if its head is true w.r.t. I or its body is false w.r.t. I. A *model* for $\mathcal{P}$ is an interpretation M for $\mathcal{P}$ such that every rule $r \in Ground(\mathcal{P})$ is true w.r.t. M. A model M for $\mathcal{P}$ is *minimal* if there is no model N for $\mathcal{P}$ such that N is a proper subset of M. The set of all minimal models for $\mathcal{P}$ is denoted by $\mathrm{MM}(\mathcal{P})$. In the following, the semantics of ground programs is first given, then the semantics of general programs is given in terms of the answer sets of its instantiation. Given a *ground* program $\mathcal{P}_g$ and an interpretation I, the *reduct* of $\mathcal{P}_g$ w.r.t. I is the subset $\mathcal{P}_g^I$ of $\mathcal{P}_g$ obtained by deleting from $\mathcal{P}_g$ the rules in which a body literal is false w.r.t. I.[1] Let I be an interpretation for a ground program $\mathcal{P}$. I is an *answer set* (or stable model) for $\mathcal{P}$ if $I \in \mathrm{MM}(\mathcal{P}^I)$ (i.e., I is a minimal model for the program $\mathcal{P}^I$) [7].

Queries. A program $\mathcal{P}$ can be coupled with a *query* in the form $q?$, where q is a literal. Let $\mathcal{P}$ be a program and $q?$ be a query, $q?$ is *true* iff for every answer set A of $\mathcal{P}$ it holds that $q \in A$. Basically, the semantics of queries corresponds to cautious reasoning, since a query is true if the corresponding atom is true in all answer sets of P.

3   Answer Sets Computation

In this section we overview the evaluation of ASP programs and recall the available solutions mentioning the techniques underlying the state of the art implementations.

The evaluation of ASP programs is traditionally carried out in two phases: program instantiation and model generation. As a consequence, an ASP system usually couples

---

[1] This definition, introduced in [7], is equivalent to the one of Gelfond and Lifschitz [14].

two modules: the *grounder* or *instantiator* and the *ASP model generator* or *solver*. In the following we provide a more detailed description of the instantiation, since the target of this work is improving performance of this phase.

ASP Program Instantiation. In general, an ASP program $\mathcal{P}$ contains variables, and the process of *instantiation or grounding* aims to eliminate these variables in order to generate a propositional ASP program equivalent to $\mathcal{P}$. Note that, the full theoretical instantiation $Ground(\mathcal{P})$ introduced in previous section contains all the ground rules that can be generated applying every possible substitution of variables. A modern instantiator module does not produce the full ground instantiation $Ground(\mathcal{P})$ (which is unnecessarily huge in size), but employees several techniques to produce one that is both equivalent and usually much smaller than $Ground(\mathcal{P})$. Notice that grounding is an EXPTIME-hard task, indeed in general it may produce a program that is of exponential size w.r.t. the input program. Thus, having an instantiator able to produce a comparatively small program in a reasonable time is crucial to achieve good (or even acceptable) performance in evaluating ASP programs. For instance, DLV-G generates a ground instantiation that has the same answer sets as the full one, but is much smaller in general [20].

In order to generate a small ground program equivalent to $\mathcal{P}$, a modern instantiator usually exploits some structural information of the input program. The evaluation proceeds bottom-up, starting from the information contained in the facts and evaluating the rules according to the positive body-to-head dependencies. Such dependencies can be identified by means of the *Dependency Graph* (DG) of $\mathcal{P}$. The DG of $\mathcal{P}$ is a directed graph $G(\mathcal{P}) = \langle N, E \rangle$, where N is a set of nodes and E is a set of arcs. N contains a node for each IDB predicate of $\mathcal{P}$, and E contains an arc $e = (p, q)$ if there is a rule r in $\mathcal{P}$ such that q occurs in the head of r and p occurs in a positive literal of the body of r. The graph $G(\mathcal{P})$ induces a subdivision of $\mathcal{P}$ into subprograms (also called *modules*) allowing for a modular evaluation. We say that a rule $r \in \mathcal{P}$ *defines* a predicate p if p appears in the head of r. For each strongly connected component (SCC) C of $G(\mathcal{P})$, the set of rules defining all the predicates in C is called *module* of C and is denoted by $\mathcal{P}_c$.

The DG induces a partial ordering among its SCCs which is followed during the evaluation. Basically, this order allows to perform a layered evaluation of the program; one module at time in such a way that data needed for the instantiation of a module C_i have been already generated by the instantiation of the modules preceding C_i. This way, ground instances of rules are generated using only atoms which can possibly be derived from $\mathcal{P}$, and thus avoiding the combinatorial explosion that may occur in the case of a full instantiation [19]. Modules containing recursive rules are evaluated according to fix-point techniques originally introduced in the field of deductive databases [29]. In turn, each rule in a module is processed by applying a variable to constant matching procedure, which is basically implemented as a backtracking algorithm. Actually, modern instantiators implement a backjumping technique [19]. Note that the instantiation of a rule is very similar to the evaluation of a conjunctive query, which is a process exponential both in the size of the query (number of elements in the body) and in the number of variables. An additional aspect influencing the cost of instantiating a rule is the way variables are bound (trough joins or builtin operators), as it also happens for conjunctive queries [29].

At the time of this writing, the two most prominent instantiators for ASP programs, which are capable of parsing the core language employed in the 3rd ASP competition, are DLV-G and GRINGO. These two grounders are both based on the above-mentioned techniques, but employ specific variants and heuristics which are described in the related literature [20,25,12], and that will be outlined in Section 5 when the results are analyzed.

Answer Set Solving.   The subsequent computations, which constitute the non-deterministic part of ASP programs evaluation, are then performed on the ground instantiation by an *ASP solver*. ASP solvers employ algorithms very similar to SAT solvers, i.e., specialized variants of DPLL [5] search.

There are several different approaches to ASP solving that range from native solvers (i.e., implementing ASP-specific techniques), to rewriting-based solutions (e.g., rewriting programs into SAT and calling a SAT solver). Among the ones that participated to the 3rd ASP Competition, we recall the native ASP solvers SMODELS [27], DLV [20] and CLASP [11]. SMODELS is one the first ASP systems made available; and DLV is one of the first robust implementations able to cope with disjunctive programs. Both feature look-ahead based techniques and ASP-specific search space pruning operators. CLASP is a native ASP solver relying on conflict-driven nogood learning. Among the rewriting-based ASP solvers we mention CMODELS [21], IDP [24], and the LP2SAT [16] family that resort on a translation to SAT. There are also proposals, like the LP2DIFF [17] family, rewriting ASP in difference logic and calling a Satisfiability Modulo Theories solver to compute answer sets.

4   Automated Selection of Grounding Algorithm

In our previous work [23,22], our aim was to build an efficient ASP solver on top of state-of-the-art systems, leveraging on machine learning techniques to automatically choose the "best" available solver on a per-instance basis. Our analysis focused on *ground* instances, and, to do that, we ran each non-ground instance with GRINGO – the same setting used in the 3rd ASP Competition –, letting our system ME-ASP to choose the best solver to fire. In order to extend ME-ASP to cope with non-ground instances, and considering that in [23] we report that ME-ASP was not able to cope with a number of instances due to GRINGO failures during the grounding stage, to obtain a more efficient system we investigate the application of algorithm selection techniques to the grounding phase, by relying on DLV-G and GRINGO.

GRINGO was used at the 3rd ASP Competition for all participant solvers because it features an easy numeric format (i.e. the one of LPARSE [28]), but, in general, it is not clear which grounder represents the choice that allows to reach the best possible performance.

In order to investigate this point, we design an experiment aimed to highlight the performance of a pool of state-of-the-art ASP solvers on a pool of problem instances.

Table 1. Pool of ASP problems involved in the reported experiments. Notice that "Grammar-BasedIE" and "MCSQuerying" are shorthands for the problems named "GrammarBasedInformationExtraction" and "MultiContextSystemQuerying", respectively.

Problem	Class	Problem	Class
DisjunctiveScheduling	NP	HydraulicLeaking	P
HydraulicPlanning	P	GrammarBasedIE	P
GraphColouring	NP	HanoiTower	NP
KnightTour	NP	MazeGeneration	NP
Labyrinth	NP	MCSQuerying	NP
Numberlink	NP	PackingProblem	NP
PartnerUnitsPolynomial	P	Reachability	P
SokobanDecision	NP	Solitaire	NP
StableMarriage	P	WeightAssignmentTree	NP

Concerning the solvers, we selected the pool comprised in the multi-engine solver ME-ASP, namely CLASP, CMODELS, DLV and IDP. As reported in [23], these solvers are representative of the state-of-the-art solver (SOTA), i.e., considering a problem instance, the oracle that always fares the best among available solvers.

The benchmarks considered for the experiment belong to the suite of the 3rd ASP Competition. This is a large and heterogeneous suite of benchmarks encoded in ASP-Core, which was already employed for evaluating the performance of state-of-the-art ASP solvers. That suite includes planning domains, temporal and spatial scheduling problems, combinatorial puzzles, graph problems, and a number of application domains, i.e., database, information extraction and molecular biology field [2]. In more detail, we have employed the encodings used in the System Track of the competition of all evaluated problems belonging to the categories P and NP, and all the problem instances evaluated at the competition[3]. Notice that with *instance* we refer to the complete input program (i.e., encoding+facts) to be fed to a solver for each instance of the problem to be solved. In Table 1 we report the problems involved in our experiment. We evaluated 10 instances per problem – the same ones evaluated at the System Track of the 3rd ASP Competition –, for a total amount of 180 instances.

In Table 2 we report the results of the experiment described above. All the experiments ran on a cluster of Intel Xeon E31245 PCs at 3.30 GHz equipped with 64 bit Ubuntu 12.04, granting 600 seconds of CPU time for the whole process (grounding + solving) and 2GB of memory to each system. We present Table 2 in two parts – top and bottom – organized as follows. The first column reports the problem name and it is followed by two groups of columns. Each group is labeled with the considered solver name and it is composed of two sub-groups, denoting the grounder (groups "DLV-G" and "GRINGO"). Finally, each sub-group is composed of two columns, in which we report the total amount of solved instances and the total CPU time (in seconds) spent to solve them (columns "#" and "Time", respectively).

---

[2] An exhaustive description of the benchmark problems can be found in [3].

[3] Both encodings and problem instances are available at the competition website [3].

Table 2. Results of a pool of solvers using different grounders on the instances evaluated at the 3rd ASP Competition

	CLASP				CMODELS			
	DLV-G		GRINGO		DLV-G		GRINGO	
	#	Time	#	Time	#	Time	#	Time
DisjunctiveScheduling	10	425.20	5	75.45	9	977.82	4	947.17
GrammarBasedIE	10	2323.72	10	254.33	10	2344.88	10	266.69
GraphColouring	3	20.90	3	144.55	4	423.54	4	357.90
HanoiTower	6	751.65	7	1058.87	7	314.03	7	137.11
HydraulicLeaking	10	2095.85	7	2819.15	10	2087.08	7	2850.30
HydraulicPlanning	10	883.17	10	154.30	10	878.80	10	164.57
KnightTour	7	148.76	7	82.52	6	71.19	6	18.50
Labyrinth	9	720.53	10	237.50	8	399.77	9	580.40
MazeGeneration	10	35.54	10	4.94	10	120.03	10	5.94
MCSQuerying	10	136.38	10	130.82	10	155.84	10	133.55
Numberlink	8	254.06	7	9.64	4	161.19	4	353.41
PackingProblem	9	2285.40	–	–	9	2247.19	–	–
PartnerUnitsPolynomial	8	263.70	2	63.94	8	262.93	–	–
Reachability	9	528.69	6	110.50	8	427.97	5	439.69
SokobanDecision	10	425.09	10	512.59	10	814.07	10	893.03
Solitaire	3	206.73	2	81.51	4	303.09	3	488.84
StableMarriage	1	61.47	–	–	–	–	–	–
WeightAssignmentTree	8	416.69	1	15.62	5	1100.25	–	–

	DLV				IDP			
	DLV-G		GRINGO		DLV-G		GRINGO	
	#	Time	#	Time	#	Time	#	Time
DisjunctiveScheduling	1	35.09	1	44.60	10	415.90	5	69.36
GrammarBasedIE	10	2020.14	10	280.51	10	2285.07	10	280.53
GraphColouring	–	–	–	–	3	510.15	3	531.83
HanoiTower	–	–	–	–	8	408.31	9	474.39
HydraulicLeaking	10	1936.64	7	2830.51	10	2130.24	7	2843.08
HydraulicPlanning	10	837.25	10	164.01	10	889.13	10	173.13
KnightTour	5	711.54	5	15.81	9	1002.47	10	1092.94
Labyrinth	3	71.48	3	71.36	5	49.66	6	21.92
MazeGeneration	8	629.47	8	472.77	10	37.24	10	7.16
MCSQuerying	10	31.08	10	160.17	10	138.18	6	56.58
Numberlink	4	5.50	4	10.08	8	130.06	8	80.31
PackingProblem	8	1910.86	–	–	9	2215.05	–	–
PartnerUnitsPolynomial	1	443.06	–	–	8	264.21	–	–
Reachability	10	58.89	4	69.51	5	302.53	5	1201.39
SokobanDecision	6	182.42	6	280.80	10	1306.88	9	926.54
Solitaire	4	85.50	–	–	5	216.67	5	109.06
StableMarriage	–	–	–	–	–	–	–	–
WeightAssignmentTree	10	549.09	–	–	5	113.23	1	12.07

Looking at Table 2, concerning the results of CLASP, we report that it was able to solve 141 (out of 180) instances using the DLV-G grounder, while it stops at 107 using GRINGO. In particular, looking at the table, we can see that CLASP mainly benefits from the usage of DLV-G – in terms of total amount of solved instances – in DisjunctiveScheduling, PackingProblem, PartnerUnitsPolynomial, and WeightAssignmentTree problems. Looking at the performance of CMODELS, we can see a very similar picture; notice that DLV-G +CMODELS solves 132 instances, while GRINGO +CMODELS stops at 99. While we can find the same picture looking at IDP performance (it solves 135 formulas if coupled with DLV-G, while 104 if coupled with GRINGO), the picture changes in a noticeable way if we look at the performance of DLV, because – as expected – it performs better using its native grounder (in terms of total amount of solved instances) instead of GRINGO – it solves 100 instances instead of 68. Looking at results in which a solver solves the same number of instances with both grounders, we report that in most cases the usage of GRINGO leads to lower CPU times.

To investigate this phenomenon and possibly getting advantage on this picture, we computed some problem characteristics called *features*. This is the first proposal of *non-ground* ASP features in ASP solving: our choice is to compute "simple" but meaningful features, that are cheap-to-compute and that can help to discriminate among problems and/or classes, in order to employ the "best" grounder on each instance considered.

Most of the features we extract are related to peculiarities of ASP that can lead to select the most appropriate grounders, e.g. if the program contains a query we want to choose DLV-G given it implements specialized techniques to deal with ASP programs with queries [1]. Such features are: fraction of non-ground rules, either normal or disjunctive, presence of queries, ground and partially grounded queries; maximum Strongly Connected Components (SCC) size, number of Head-Cycle Free (HCF) and non-HCF components, degree of support for non-HCF components; features indicating if the program is recursive, tight and stratified; and number of builtins. This set of ASP peculiar features is complemented with features that take into account other characteristics such as problem size, balancing measures and proximity to horn, and are the following: number of predicates, maximum body size, ratio of positive and negative literals in each body of non-ground rules, and its reciprocal, fraction of unary, binary and ternary non-ground rules, and fraction of horn rules.

In order to highlight the differences between DLV-G and GRINGO, we extract the features described above on the instances *submitted* at the 3rd ASP Competition, i.e., a pool of more than one thousand instances related to the problems listed in Table 1, from which we discarded the instances involved in the experiment of Table 2. In Figure 1 we report the boxplots related to the distribution of two features. The differences between DLV-G and GRINGO herewith reported motivates the work presented in the next section.

5   Implementation and Experiments

In this section we show the implementation of our automated grounder selector, representing a first attempt towards the exploitation of automated selection techniques to predict both grounder and solver, given an ASP non-ground instance. Our current

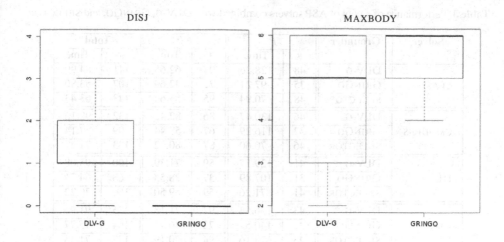

Fig. 1. Distributions of the features: number of disjunctive rules (DISJ) (left) and maximum body size (MAXBODY) (right) considering problems *submitted* to the 3rd ASP Competition for which DLV-G allows a better performance with respect to GRINGO (distribution on the left of each plot) and vice-versa (distribution of the right of each plot)

implementation is composed of two main elements, namely a feature extractor able to analyze non-ground ASP programs, and a decision making module. In particular, the latter has been implemented as an if-then-else decision list, computed with the support of the PART decision list generator [8], a classifier that returns a human readable model based on if-then-else rules.

The resulting model highlights some specific cases in which there is a clear difference in performance between two grounders. DLV-G is always chosen when one has to deal with queries. GRINGO is usually preferable for instantiating non-disjunctive and recursive encodings with many components and is preferable, in particular, when most of the rules of the encoding feature a short body. DLV-G is usually the right choice also when the encoding contains rules having large bodies (say, bodies with 4 or more literals) and the program has a simple structure (few components). The reasons behind this result can be found both in the techniques implemented in the two grounders and in some more specific implementation choices. For instance queries are processed far better by DLV-G, which exploits specific techniques like magic sets [2], while this is a feature that is not well supported in GRINGO.[4] GRINGO is a newer implementation which (we argue) was initially optimized do deal with non-disjunctive encodings, since also CLASP (the solver developed by the same team) does not support disjunction. Finally, we argue that DLV-G performs better with rules having comparatively long bodies because it features both a more sophisticated indexing technique (w.r.t. GRINGO)

---

[4] In the competition, as well as in our experiment, only ground queries are used and when calling GRINGO a straight technique is employed to handle propositional queries: q? is replaced by the constraint :-not q, concluding the q is cautiously true when the resulting program has no answer set.

Table 3. Performance of a pool of ASP solvers combined with DLV-G, GRINGO, and SELECTOR

Solver	Grounder	P		NP		Total	
		#	Time	#	Time	#	Time
CLASP	DLV-G	48	128.26	93	62.65	141	84.99
	GRINGO	35	97.21	72	32.69	107	53.80
	SELECTOR	48	70.94	95	59.64	143	63.43
CMODELS	DLV-G	46	130.47	86	82.42	132	99.16
	GRINGO	32	116.29	67	58.44	99	77.14
	SELECTOR	46	70.60	87	80.72	133	77.22
DLV	DLV-G	41	129.17	59	71.39	100	95.08
	GRINGO	31	107.89	37	28.53	68	64.71
	SELECTOR	41	71.26	59	69.50	100	70.22
IDP	DLV-G	43	136.54	92	71.13	135	91.96
	GRINGO	32	140.57	72	46.97	104	75.77
	SELECTOR	43	74.16	94	70.19	137	71.43

and effective join ordering heuristics [18]; indeed, these techniques are expected to pay off especially in these cases. The grounder selector presented in this paper is available for download at http://www.mat.unical.it/ricca/downloads/ GR-SELECTOR-AIIA.zip.

Aim of our next experiment is to test the performance of the pool of solvers introduced in Section 4 using the proposed tool as grounder (called SELECTOR in the following). Table 3 shows the results of the experiment described above on the benchmark instances evaluated at the 3rd ASP Competition. The table is structured as follows: "Solver" and "Grounder" report the solver and the grounder name, respectively; for each grounder+solver S, "P", "NP" report the number of instance solved by S ("#") and the average CPU time (in seconds) spent on such instances ("Time") related to the benchmark instances comprised in the classes P and NP, respectively.

Looking at Table 3, we can see that all the considered solvers benefit from the usage of SELECTOR as grounder. Looking at the performance of CLASP, we can see that SELECTOR+CLASP it is able to solve 2 instances more than DLV-G +CLASP, and 36 instances more than GRINGO +CLASP. The increase of performance is noticeable concerning NP instances, while if we look at the performance of P instance, we report that SELECTOR+CLASP is able to solve the same instances of DLV-G +CLASP, also if in this case the average CPU time per solved instance related to SELECTOR+CLASP is 55% of the one related to DLV-G +CLASP.

Looking now at the performance of CMODELS, we can see that the picture is very similar to the one related to CLASP. SELECTOR+CMODELS solves 34 instances more that GRINGO +CMODELS, while the gap with DLV-G +CMODELS stops to 1. Similar considerations can be reported in the case of IDP. Finally, considering the performance of DLV, we can see that the picture slightly changes. In this case, SELECTOR+DLV is never superior to DLV-G +DLV- in terms of total amount of solved instances – but we report better performance in terms of average CPU time per solved instance, i.e., SELECTOR+DLV is about 25% faster than DLV-G +DLV.

6   Conclusions

ASP systems are obtained by combining a grounder, which eliminates variables, and a solver, that computes the answer sets. It is well known that both components play a central role in the performance of the system. Algorithm selection techniques, up to now, have been applied only on the second component. In this paper we make a first step toward the exploitation of automated selection techniques for the grounding component.

In particular, we implemented a new system able to automatically select the most appropriate grounder for solving the instance at hand out of two state-of-the-art ASP instantiators. An experimental analysis, conducted on benchmarks and solvers from the 3rd ASP Competition, shows that our grounder selector improves the evaluation performance independently from the solver associated.

As far as future work is concerned we are exploring the possibility to implement a selector that is able to predict the best grounder+solver pair among a set of possible combinations.

Acknowledgments. The authors wish to thank the anonymous reviewers for their comments and suggestions to improve the quality of the paper. M. Maratea and L. Pulina are partially supported by Regione Autonoma della Sardegna and Autorità Portuale di Cagliari, L.R. 7/2007, Tender 16 2011, CRP-49656 "Metodi innovativi per il supporto alle decisioni riguardanti lottimizzazione delle attività in un terminal container".

References

1. Alviano, M., Faber, W., Greco, G., Leone, N.: Magic sets for disjunctive datalog programs. Artificial Intellegence 187, 156–192 (2012)
2. Alviano, M., Faber, W., Leone, N.: Disjunctive asp with functions: Decidable queries and effective computation. Theory and Practice of Logic Programming 10(4-6), 497–512 (2010)
3. Calimeri, F., Ianni, G., Ricca, F.: The third answer set programming system competition (2011), https://www.mat.unical.it/aspcomp2011/
4. Calimeri, F., et al.: The Third Answer Set Programming Competition: Preliminary Report of the System Competition Track. In: Delgrande, J.P., Faber, W. (eds.) LPNMR 2011. LNCS, vol. 6645, pp. 388–403. Springer, Heidelberg (2011)
5. Davis, M., Putnam, H.: A Computing Procedure for Quantification Theory. Journal of the ACM 7, 201–215 (1960)
6. Eiter, T., Gottlob, G., Mannila, H.: Disjunctive Datalog. ACM Transactions on Database Systems 22(3), 364–418 (1997)
7. Faber, W., Leone, N., Pfeifer, G.: Recursive aggregates in disjunctive logic programs: Semantics and complexity. In: Alferes, J.J., Leite, J. (eds.) JELIA 2004. LNCS (LNAI), vol. 3229, pp. 200–212. Springer, Heidelberg (2004)
8. Frank, E., Witten, I.H.: Generating accurate rule sets without global optimization. In: Proceedings of ICML 1998, p. 144. Morgan Kaufmann Pub. (1998)
9. Gebser, M., Kaminski, R., Kaufmann, B., Schaub, T., Schneider, M.T., Ziller, S.: A portfolio solver for answer set programming: Preliminary report. In: Delgrande, J.P., Faber, W. (eds.) LPNMR 2011. LNCS, vol. 6645, pp. 352–357. Springer, Heidelberg (2011)
10. Gebser, M., Kaminski, R., König, A., Schaub, T.: Advances in *gringo* series 3. In: Delgrande, J.P., Faber, W. (eds.) LPNMR 2011. LNCS, vol. 6645, pp. 345–351. Springer, Heidelberg (2011)

11. Gebser, M., Kaufmann, B., Neumann, A., Schaub, T.: Conflict-driven answer set solving. In: Twentieth International Joint Conference on Artificial Intelligence (IJCAI 2007), pp. 386–392. Morgan Kaufmann Publishers, Hyderabad (2007)
12. Gebser, M., Schaub, T., Thiele, S.: GrinGo: A New Grounder for Answer Set Programming. In: Baral, C., Brewka, G., Schlipf, J. (eds.) LPNMR 2007. LNCS (LNAI), vol. 4483, pp. 266–271. Springer, Heidelberg (2007)
13. Gelfond, M., Lifschitz, V.: The Stable Model Semantics for Logic Programming. In: Logic Programming: Proceedings Fifth Intl Conference and Symposium, pp. 1070–1080. MIT Press (1988)
14. Gelfond, M., Lifschitz, V.: Classical Negation in Logic Programs and Disjunctive Databases. New Generation Computing 9, 365–385 (1991)
15. Grasso, G., Leone, N., Manna, M., Ricca, F.: ASP at work: Spin-off and applications of the DLV system. In: Balduccini, M., Son, T.C. (eds.) Logic Programming, Knowledge Representation, and Nonmonotonic Reasoning. LNCS (LNAI), vol. 6565, pp. 432–451. Springer, Heidelberg (2011)
16. Janhunen, T.: Some (in)translatability results for normal logic programs and propositional theories. Journal of Applied Non-Classical Logics 16, 35–86 (2006)
17. Janhunen, T., Niemelä, I., Sevalnev, M.: Computing Stable Models via Reductions to Difference Logic. In: Erdem, E., Lin, F., Schaub, T. (eds.) LPNMR 2009. LNCS, vol. 5753, pp. 142–154. Springer, Heidelberg (2009)
18. Leone, N., Perri, S., Scarcello, F.: Improving ASP instantiators by join-ordering methods. In: Eiter, T., Faber, W., Truszczyński, M. (eds.) LPNMR 2001. LNCS (LNAI), vol. 2173, pp. 280–294. Springer, Heidelberg (2001)
19. Leone, N., Perri, S., Scarcello, F.: BackJumping Techniques for Rules Instantiation in the DLV System. In: Proc. of NMR 2004, pp. 258–266 (2004)
20. Leone, N., Pfeifer, G., Faber, W., Eiter, T., Gottlob, G., Perri, S., Scarcello, F.: The DLV System for Knowledge Representation and Reasoning. ACM Transactions on Computational Logic 7(3), 499–562 (2006)
21. Lierler, Y.: CMODELS – SAT-Based Disjunctive Answer Set Solver. In: Baral, C., Greco, G., Leone, N., Terracina, G. (eds.) LPNMR 2005. LNCS (LNAI), vol. 3662, pp. 447–451. Springer, Heidelberg (2005)
22. Maratea, M., Pulina, L., Ricca, F.: The multi-engine ASP solver ME-ASP. In: del Cerro, L.F., Herzig, A., Mengin, J. (eds.) JELIA 2012. LNCS, vol. 7519, pp. 484–487. Springer, Heidelberg (2012)
23. Maratea, M., Pulina, L., Ricca, F.: A Multi-Engine approach to Answer Set Programming. Theory and Practice of Logic Programming (in press)
24. Mariën, M., Wittocx, J., Denecker, M., Bruynooghe, M.: Sat(id): Satisfiability of propositional logic extended with inductive definitions. In: Kleine Büning, H., Zhao, X. (eds.) SAT 2008. LNCS, vol. 4996, pp. 211–224. Springer, Heidelberg (2008)
25. Perri, S., Scarcello, F., Catalano, G., Leone, N.: Enhancing DLV instantiator by backjumping techniques. Annals of Mathematics and Artificial Intelligence 51(2-4), 195–228 (2007)
26. Pulina, L., Tacchella, A.: A self-adaptive multi-engine solver for quantified boolean formulas. Constraints 14(1), 80–116 (2009)
27. Simons, P., Niemelä, I., Soininen, T.: Extending and Implementing the Stable Model Semantics. Artificial Intelligence 138, 181–234 (2002)
28. Syrjänen, T.: Lparse 1.0 User's Manual (2002), http://www.tcs.hut.fi/Software/smodels/lparse.ps.gz
29. Ullman, J.D.: Principles of Database and Knowledge Base Systems. Computer Science Press (1989)

Entity-from-Relationship Modelling

Claudio Masolo[1] and Alessandro Artale[2]

[1] Laboratory for Applied Ontology, ISTC-CNR, Trento, Italy
[2] KRDB Research Centre, Free University of Bozen-Bolzano, Italy

Abstract. We investigate the problem of the (re-)*identification* of new entities emerging from the presence of relationships in a conceptual schema. While the general practice is to *reify* the relation and thus individuating as many individuals as many tuples in the original relation, we propose here different identification methods based on the possibility of *grouping* the tuples according to some predefined criteria. The proposed modelling mechanism is general enough to capture in a uniform way also the case where *wholes* emerge from their *components* or *constituents*.

Keywords: conceptual modelling constructs, ontological foundations, formal semantics.

1 Introduction

In conceptual modelling and object-oriented modelling, *roles* are usually considered as *anti-rigid* and *relationally dependent* unary predicates. For instance, the predicate `Customer` is modeled as a subclass of `Person` and it is anti-rigid because persons are only contingently customers. `Customer` is relationally dependent because, to exist, customers require the existence of companies with which they trade, i.e., *customers are persons that trade with companies* as showed in Fig. 1. This model can be refined by specifying some temporal constraints. For instance, one could just assume that the `trade` relation is *mandatory*—at every time at which a person is a customer she must be trading with a company, but not necessarily the same company—or *essential*—at every time at which a person is a customer she must be trading with the *same* company [7,1].

Steimann [17] shows that expressing roles by means of *ISA* relations suffers some problems. Some of them, e.g., attribute *overriding* and *hiding*, can be solved by managing in the right way the attributes along the inheritance mechanism. Other issues are more difficult to solve, in particular the *counting problem* [8]: what do we count to answer the question *"how many passengers did Alitalia transport in 2011?"* To match the standard answer provided by airlines, assuming that the same person can fly several times with the same airline, it is not correct to just count persons. Because of these problems some approaches [11,15,17,19] refuse the idea that roles are subsumed by kinds and propose a separation between *role* and *kind* hierarchies—see [13] for a comparison with alternative solutions based on events. In this view, customers and passengers are not persons, they are *adjunct* entities that existentially depend

M. Baldoni et al. (Eds.): AI*IA 2013, LNAI 8249, pp. 85–96, 2013.

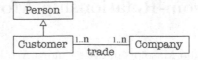

Fig. 1. A standard way to represent roles

on persons but obey different identity criteria—i.e., to identify customers or passengers it is not enough to identify persons—and the *role* a given person is *playing* is not anymore described by a simple sub-class (as in the case of Fig. 1).

The general idea that new types of entities *emerge* from the holding of a *generating* relation (e.g., `trade`, for customers)—a case of *supervenience*—seems to apply also to types that are not usually considered as *roles*. For instance, hammers exist because of a morfo-topological relation between a handle and a head, or statues exist because of an amount of matter shaped in a given way.[1] Differently from the usual conceptualization of roles, hammers and statues can survive a change in their *substrata*. For example, hammers survive the substitution of the handle while statues survive a renovation. This fact prevents the link between the emergent entities and their substrata—hammer vs. handles and heads or statues vs. amounts of matter—to be an ISA relation. We will see that also for customers—a prototypical example of role—one can consider both identification and re-identification criteria not based only on persons.

In this work we propose a general mechanism with a twofold aim: (*i*) to capture cases of emergence of entities in a unified way, and (*ii*) to formally establish the *identification* (at a given time) and *re-identification* (through time) criteria for the emergence of entities. Since these criteria are based on the grouping of the instances of the generating relation, we will make use of the relatively well known mechanism of *reification* of relations (see [9], section 10.5) to capture them (with the addition of further constraints). As a secondary effect all the emergent properties will be *rigid*—e.g., customers depend on but are not persons, statues depend on but are not amounts of matter, etc.—avoiding some of the problems briefly introduced above.

2   From Relations to Entities

The guiding idea of our approach is that, according to some *identification* criteria, new entities can *emerge* from either *many-to-many* relations or (more in general) from *n*-ary relations.

Our running example focuses on the concept of customer; artefacts, like statues and hammers, are considered in Section 4. Let us start from the schema in Fig. 2.a and the *customer database*:

$$\text{trade(luc, microsoft)}, \quad \text{trade(luc, skype)},$$
$$\text{trade(john, microsoft)}, \quad \text{trade(john, amazon)}.$$

---

[1] The notion of statue emerges from the generating *attribute* 'shape'. Attributes must be seen here as binary relations between classes and concrete values (see Section 4).

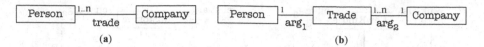

Fig. 2. A many-to-many relation and its reification

Since **Person** does not participate in a mandatory way to **trade**, one can identify customers as exactly those persons that further trade with companies:[2]

$$\text{Customer}(x) \rightarrow \text{Person}(x) \wedge \exists y.\text{trade}(x, y),$$
$$\text{trade}(x, y) \rightarrow \text{Customer}(x),$$

leading to the diagram of Fig. 1. In this case, **Customer** is ISA-related to **Person**—we call this the *ISA-identification* criterion. According to the *ISA-identification* criterion, in our example, there are *two* customers: **luc** and **john**.

Other notions of being a customer can be of interest from a modelling perspective. For instance the particular relation a given person has with a specific company can also identify a customer, e.g., **trade(john, amazon)** and **trade(john, microsoft)** can *support* the existence of *two* different customers. According to this new identification criterion, customers are not persons and, in the current example, we can count up to *four* different customers, one for each **trade**-instance, i.e., there is a one-to-one correspondence between customers and **trade**-instances. To model this kind of identification criteria, we need to refer to **trade**-instances. Following ORM practice [9], we *reify* (*objectify*)—a practice that is common also in the UML community with the use of association classes—the **trade**-relation into the class **Trade** (see Fig. 2.b) which is in one-to-one correspondence with the *extension* of the relation **trade**.[3] We remind here the first-order semantics of reification classes:

$$\text{Trade}(x) \rightarrow \exists^{=1} y_1.\text{arg}_1(x, y_1) \wedge \text{Person}(y_1) \wedge \exists^{=1} y_2.\text{arg}_2(x, y_2) \wedge \text{Company}(y_2),$$
$$\text{arg}_1(x, y_1) \wedge \text{arg}_1(x', y_1) \wedge \text{arg}_2(x, y_2) \wedge \text{arg}_1(x', y_2) \rightarrow x = x'.$$

Correspondingly, in the reified model of Fig. 2.b, we have:

$$\text{Trade}(\|\text{luc, skype}\|), \ \text{arg}_1(\|\text{luc, skype}\|, \text{luc}), \ \text{arg}_2(\|\text{luc, skype}\|, \text{skype}), \dots$$

Thus the fact that each **Trade**-instance corresponds to a customer can be made explicit—e.g., in UML we can model customer as the association class of the trade relation. More complex identification criteria can be considered. For instance, by knowing that **skype** is *owned by* **microsoft**, one can assume that all the facts that involve a single person trading with companies of the same business group support the existence of a single customer. Thus, in the previous example, we can count *three* different customers, i.e., **trade(luc, microsoft)** and **trade(luc, skype)** support one customer while **trade(john, amazon)** and

---

[2] *Relations* are denoted with small starting letters, *classes* with capital starting letters.
[3] To improve the readability, we denote the instance reifying an instance **r** of an *n*-ary relation with $\|a_1, \dots, a_n\|$ where a_i is the arg_i of **r**.

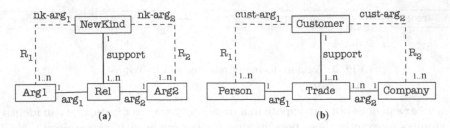

Fig. 3. The general schema for identification criteria (a) and a special case (b)

`trade(john, microsoft)` support two customers. Symmetrically, one could abstract from persons by grouping into a single customer all the `Trade`-instances that involve members of the *same family* (trading with a given company).

The general the schema in Fig. 3a. allows us to capture identification criteria by *grouping* tuples of the reified binary[4] relation, `Rel`, according to constraints specified over the different arguments of the relation. The modelling choices rely on the introduction of the primitive binary relation `support`, that individuates the instances of the reified relations that are grouped into a new entity, and on n dashed relations[5]—one for each argument of the n-ary relation `Rel`—labelled by `nk-arg`$_i$ and representing the composition of `support` with the argument `arg`$_i$ of the generating relation `Rel`, i.e.:

$$\texttt{nk-arg}_i(x,y) \triangleq \exists z.\texttt{support}(z,x) \wedge \texttt{arg}_i(z,y). \qquad (1)$$

Every dashed line is additionally labeled by a binary relation R_i used in the *indentification criterion* to establish when instances of `NewKind` can be identified:

$$\texttt{NewKind}(x_1) \wedge \texttt{NewKind}(x_2) \rightarrow \qquad (2)$$
$$(x_1 = x_2 \leftrightarrow (R_1 \ \texttt{nk-arg}_1)(x_1, x_2) \wedge \ldots \wedge (R_n \ \texttt{nk-arg}_n)(x_1, x_2))$$

where $(R_i \ \texttt{nk-arg}_i)(x_1, x_2)$ is defined as:

$$(R_i \ \texttt{nk-arg}_i)(x_1, x_2) \triangleq \forall y_1 y_2 (\texttt{nk-arg}_i(x_1, y_1) \wedge \texttt{nk-arg}_i(x_2, y_2) \rightarrow R_i(y_1, y_2)) \quad (3)$$

Starting from the general model in Fig. 3b, different notions of customer can be formalized via the relations R_i. The ISA-identification criterion is captured by $R_1(y_1, y_2) \triangleq y_1 = y_2$ and $R_2(y_1, y_2) \triangleq \top$ (where $\top$ is the constant **true**) that will force the existence of two different customers, c_1, c_2, such that:[6]

$$\texttt{support} = \{\langle \|\texttt{luc, microsoft}\|, c_1\rangle, \langle \|\texttt{luc, skype}\|, c_1\rangle,$$
$$\langle \|\texttt{john, microsoft}\|, c_2\rangle, \langle \|\texttt{john, amazon}\|, c_2\rangle\}.$$

---

[4] The extension to the case of n-ary relations is straightforward.

[5] These relations are dashed since they are always present in the model (at least implicitly) but they are required to specify an identification criterion.

[6] For compactness we write $P = \{p_1, \ldots, p_i\}$ instead of $P(p_1) \wedge \cdots \wedge P(p_i)$.

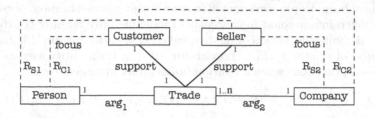

Fig. 4. Both `Customer` and `Seller` are in a one-to-one relation to `Trade`-instances

To group all the tuples with the same person trading companies *owned* by the same business group we enforce that:

$$R_1(y_1, y_2) \triangleq y_1 = y_2, \quad R_2(y_1, y_2) \triangleq y_1 = y_2 \vee \mathtt{own}(y_1, y_2) \vee \mathtt{own}(y_2, y_1) \quad (4)$$

According to (4), by extending our customer database with `own(microsoft, skype)`, we obtain three different customers, c_1, c_2, c_3, such that:

$$\mathtt{support} = \{\langle \|\mathtt{luc, microsoft}\|, c_1\rangle, \langle \|\mathtt{luc, skype}\|, c_1\rangle,$$
$$\langle \|\mathtt{john, microsoft}\|, c_2\rangle, \langle \|\mathtt{john, amazon}\|, c_3\rangle\}.$$

When R_1 is the `sibling` relation we can group along the person dimension.

We now illustrate the need for a new primitive called `focus`. Let us assume that in the identification criteria of `Customer` and `Seller` both R_1 and R_2 are the identity. Thus, both `Customer`- and `Seller`-instances are in a one-to-one correspondence with `Trade`-instances. To differentiate `Customer` from `Seller` (actually, they are disjoint classes) we rely on the intuition that one looks for persons trading with companies to identify customers while, to identify sellers, one looks for companies trading with persons. Customers existentially depends on both persons and companies but the dependence with persons is "stronger". This is supported by the fact that customers mainly inherit attributes from persons (in particular, customers are spatially co-located with persons, not with companies), while sellers from companies. We represent this asymmetry between the dashed relations, `nk-arg`$_i$, by *selecting* at least one of them as the `focus`— e.g., `Customer` focuses on `cust-arg`$_1$ while `Seller` on `sell-arg`$_2$ (see Fig. 4).

The generation mechanism can be *iterated*: new classes can be obtained by grouping entities that, in turn, have been generated. Starting from the *ground*, entities can then be layered, and *levels* can be stacked—see [12,14] for more details on the notion of level and the issues concerned with this notion.

3 Re-identification through Time

Time can be considered as any other argument, e.g., `trade` becomes a ternary relation linking persons and companies at a given time. However, time plays a fundamental role to understand how entities can evolve, what changes they survive, what *persistence* criteria they respect. While the nature and the structure of time have been deeply investigated in knowledge representantion, persistence has been taken into account in conceptual modelling only recently [10,16,3,2].

We use the following notion of rigidity, called *existential rigidity* [18], based on the primitive of *existence*—$\mathsf{E}(x,t)$ stands for "at time t, the entity x exists"—and on the assumption that when a relation holds all its arguments exist:

$$\mathsf{P}(x_1,\ldots,x_n,t) \to \forall t'(\mathsf{E}(x_1,t') \wedge \cdots \wedge \mathsf{E}(x_n,t') \to \mathsf{P}(x_1,\ldots.x_n,t')).$$

Assuming that `Customer` is a subclass of `Person` we cannot model it as rigid, indeed a person can exist without being trading with a company. On the contrary, `Customer`, as introduced in Fig. 3b (or more generally, every `NewKind` as introduced in Fig. 3a) is assumed to be rigid, i.e., customers are always customers. In our approach, as showed in Fig. 3b, `Customer` is not a subclass of `Person` and no `Person`-instances *migrate* to `Customer` while their existence depend on the supporting `Trade`-instances. When a person is trading with a company, a new entity focusing on that person comes into existence as a customer.

We make the following hypothesis: (i) time is point-based, linear, and discrete; (ii) the primitive E is in our vocabulary; (iii) all the entities, including relation-instances, exist at least at one time and can persist; (iv) the $\mathtt{arg_i}$ relations are existentially rigid, i.e., relation-instances do not change arguments. In this framework, $\mathtt{trade(luc, skype, t_1)}$ and $\mathtt{trade(luc, skype, t_2)}$ are reified as

$$\mathtt{Trade(\|luc,skype\|,t_1) \wedge arg_1(\|luc,skype\|,luc,t_1) \wedge arg_2(\|luc,skype\|,skype,t_1)}$$
$$\mathtt{Trade(\|luc,skype\|,t_2) \wedge arg_1(\|luc,skype\|,luc,t_2) \wedge arg_2(\|luc,skype\|,skype,t_2)}$$

that, by our assumption, implies that $\mathsf{E}(\|\mathtt{luc, skype}\|, \mathtt{t_1}) \wedge \mathsf{E}(\|\mathtt{luc, skype}\|, \mathtt{t_2})$.

We now analyze the special role of time in the (re-)identification of entities. Let us start by systematically adding a temporal parameter to all the predicates involved in (i) our general schema of Fig. 3a and (ii) the identification criterion (2) (including the definitions (1) and (3)). In this Section, when referring to formulas or schema axioms of Section 2 we will intend their temporal version. While $=$ and `NewKind` are assumed as rigid, this does not hold for `support` and `focus`. At a given time different `NewKind`-instances cannot share the same supporting `Rel`-instance but the same `Rel`-instance can support different `NewKind`-instances at different times and a `NewKind`-instance can also be supported by different `Rel`-instances at different times. For example, let us introduce `Customer` as in Fig. 3b and assume the *temporal customer database* where $\mathtt{own(microsoft, skype}, t)$ holds at all times t (i.e., `own` is existentially rigid) and:

$$\mathtt{Trade(\|luc,skype\|, t_1), Trade(\|luc,amazon\|, t_1), Trade(\|luc,microsoft\|, t_1),}$$
$$\mathtt{Trade(\|luc,microsoft\|, t_2), Trade(\|luc,skype\|, t_3), Trade(\|luc,amazon\|, t_3).}$$

The identification criterion (4) acts only synchronically. It enforces the groupings depicted in Table 1a where, at $\mathtt{t_1}$, $\mathtt{c_1}$ is supported by two `Trade`-instances, namely $\|\mathtt{luc, skype}\|$ and $\|\mathtt{luc, microsoft}\|$. However (4), and (2) in general, does not specify whether and how `Trade`-instances can support the same customer at different times. Customers are not necessarily instantaneous. For instance, a single customer could be supported by $\|\mathtt{luc, amazon}\|$ both at $\mathtt{t_1}$ and $\mathtt{t_3}$.

To explicitly characterize these possibilities, we introduce an *across-time* grouping criterion, called *persistence criterion*, with the following general form:

$$\text{NewKind}(x_1,t_1) \land \text{NewKind}(x_2,t_2) \land t_1 \neq t_2 \to \tag{5}$$
$$(x_1 = x_2 \leftrightarrow (\text{R}_1 \text{ nk-arg}_1)(x_1,x_2,t_1,t_2) \land \ldots \land (\text{R}_n \text{ nk-arg}_n)(x_1,x_2,t_1,t_2))$$

where, similarly to (3), $(\text{R}_n \text{ nk-arg}_n)(x_1,x_2,t_1,t_2)$ has the following general form:[7]

$$(\text{R}_i \text{ nk-arg}_i)(x_1,x_2,t_1,t_2) \triangleq \tag{6}$$
$$\forall y_1 y_2 (\text{nk-arg}_i(x_1,y_1,t_1) \land \text{nk-arg}_i(x_2,y_2,t_2) \to \text{R}_i(y_1,y_2,t_1,t_2))$$

For conciseness, in the following we will define persistence criteria only by specifying the R_i since the rest is obvious from the formulas (5) and (6).

Consider the temporal customer database together with the identification criterion (4) and the following persistence criteria (where $\perp$ is the constant **false**):

$$R_1(y_1, y_2, t_1, t_2) \triangleq \perp, \qquad R_2(y_1, y_2, t_1, t_2) \triangleq \top, \tag{7}$$

$$R_1(y_1, y_2, t_1, t_2) \triangleq y_1 = y_2, \quad R_2(y_1, y_2, t_1, t_2) \triangleq y_1 = y_2, \tag{8}$$

$$R_1(y_1, y_2, t_1, t_2) \triangleq y_1 = y_2,$$
$$R_2(y_1, y_2, t_1, t_2) \triangleq y_1 = y_2 \lor (\text{own}(y_1, y_2, t_1) \land \text{own}(y_1, y_2, t_2)) \lor$$
$$(\text{own}(y_2, y_1, t_1) \land \text{own}(y_2, y_1, t_2)), \tag{9}$$

$$R_1(y_1, y_2, t_1, t_2) \triangleq y_1 = y_2, \quad R_2(y_1, y_2, t_1, t_2) \triangleq \top. \tag{10}$$

Tables 1.a, 1.b, and 1.c depict the **Trade**-instances that support (at different times) the customers generated by adding to (4), respectively, the criteria (7), (8), and (9). The criterion (7) specifies an empty condition, therefore it enforces all the customers to be instantaneous and no additional across-time grouping is done. The criterion (8), 'same person, same company', identifies c_2 with c_5 (in Table 1.a) but it does not identify c_1 with c_3 since **skype** $\neq$ **microsoft**. Relaxing the identity of companies by requiring only that they belong to the same group, as specified in the criterion (9), we have $c_1 = c_3 = c_4$ and $c_2 = c_5$ (in Table 1.a), i.e., only two customers. However the persistence and identification criterion can (negatively) interact. For instance, the criterion (10), when applied to Table 1.a, enforces to identify both $c_1 = c_3$ and $c_2 = c_3$ that contradicts the identification criterion (4) enforcing $c_1 \neq c_2$, in its turn. In this case the two criteria are incompatible and no customers can be generated. By maintaining (10) but assuming the identification criterion where $R_1(y_1, y_2) \triangleq y_1 = y_2$ and $R_2(y_1, y_2) \triangleq \top$ ('same person'), we eliminate the incompatibility and obtain a single customer supported by all the **Trade**-instances.

To avoid **NewKind**-instances with *intermittent* existence[8] (c_2 in Table 1.b) one can substitute $t_1 \neq t_2$ with $\text{succ}(t_1, t_2)$ in (5) and adopt the *convexity* principle:

$$\text{NewKind}(x, t_1) \land \text{NewKind}(x, t_2) \land t_1 < t < t_2 \to \text{NewKind}(x, t)$$

---

[7] The fact of considering only constraints on the arguments of the supporting entities is in line with the idea that **NewKind**-instances are generated by **Rel**-instances.

[8] Note that intermittent existence makes sense, for instance, for **Trade**-instances.

Table 1. (a), (b), (c) depict the customers generated, respectively, by (7), (8), (9)

	t_1	t_2	t_3
c_1	$\|luc,microsoft\|, \|luc,skype\|$		
c_2	$\|luc,amazon\|$		
c_3		$\|luc,microsoft\|$	
c_4			$\|luc,skype\|$
c_5			$\|luc,amazon\|$

(a)

	t_1	t_2	t_3
c_1	$\|luc,microsoft\|, \|luc,skype\|$		
c_2	$\|luc,amazon\|$		$\|luc,amazon\|$
c_3		$\|luc,microsoft\|$	
c_4			$\|luc,skype\|$

(b)

	t_1	t_2	t_3
c_1	$\|luc,microsoft\|, \|luc,skype\|$	$\|luc,microsoft\|$	$\|luc,skype\|$
c_2	$\|luc,amazon\|$		$\|luc,amazon\|$

(c)

The above examples make clear the central role of time in making explicit persistence and identification criteria. However, they also make clear that a huge range of possibilities exists, and only few of them can be characterized by considering Customer as a subclass of Person.

4   (Re-)Identifying Wholes

We show how our modelling principle can be adopted also for the *emergence* of *wholes* starting from their components or substrata. In the following we provide such evidence in the case of *constitution* and *composition*.

Constitution. Consider the classical example of the statue Goliath and the lump of clay Lumpl [6]. Lumpl, but not Goliath, would survive a squeezing; Goliath, but not Lumpl, would survive the loss of some parts, e.g., a tiny piece of the finger. Lumpl already existed before the sculptor bought it. Goliath, by a continuous and complete renovation of the clay it is made of, could survive the destruction of all parts of Lumpl. The example can be generalized to other cases, e.g., collection of cells vs. organ, body vs. person, piece of metal vs. traffic sign. Accepting the previous arguments, the persistence conditions of Lump and Statue are different and Lump and Statue cannot be linked by a subclass relation. In conceptual modelling one often introduces a *constitution* relation between Statue and Lump. In our framework, the whole is identified starting from its *attributes*—to be understood as binary relations between classes and concrete values—and no more from generic relations.

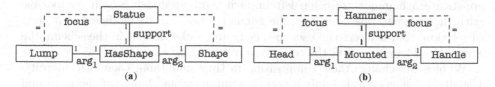

Fig. 5. Composition (a) and constitution (b)

Let us consider Fig. 5a where **Lump** has the attribute **has_shape**—reified by the class **HasShape**—associating a shape to each lump of clay (**Shape**-instances have to be understood as concrete values). At every time, a statue (e.g., Goliath) focuses on a single lump of clay (e.g., Lumpl) that has a specific shape (one of the statue-shapes), i.e., at every time, it is supported by a unique **HasShape**-instance. Statues are then entities that, to exist, require a lump of clay to have a specific (kind of) shape.[9] One can assume different persistence criteria for statues, e.g.

$$R_1(y_1, y_2, t_1, t_2) \triangleq \top, \quad R_2(y_1, y_2, t_1, t_2) \triangleq y_1 = y_2 \tag{11}$$
$$R_1(y_1, y_2, t_1, t_2) \triangleq \top, \quad R_2(y_1, y_2, t_1, t_2) \triangleq \mathsf{SShape}(y_1, t_1) \wedge \mathsf{SShape}(y_2, t_2) \tag{12}$$

A statue can change its substratum, the **Lump**-instance on which it focuses, but this substratum must have a constant shape, as specified in (11), or a tolerance could be accepted, e.g., its shape belongs to the **Shape**-subclass **SShape** as in (12). Thus, statues could have an essential shape and a mandatory substratum.

Composition. Consider the case of a *hammer* that requires a heavy metal head to be mounted at right angle at the end of a handle. While in the case of statues the link with lumps is stronger than the one with shapes, hammers seem similarly linked to both heads and handles, i.e., hammers have a *multiple focus* on both heads and handles. This is supported by the fact that the values of some attributes of an hammer are combinations of the attributes' values of both its head and its handle. For instance, the weight of an hammer is the sum of the weights of its head and its handle. This situation is modeled in Fig. 5b assuming a one-to-one relation between hammers and **Mounted**-instances.

Multiple focus represents an increment in the complexity of the *structure* of the built entities. Assuming that $I \neq \emptyset$ is the set of selected focus arguments, the following axiom holds:

$$\mathtt{focus}(x, y) \leftrightarrow \bigvee_{i \in I} \exists z.\mathtt{support}(z, x) \wedge \mathtt{arg_i}(z, y)$$

Hammers are composed by heads and handles that satisfy a *unity*-criterion and are structurally linked in a given way. In conceptual modelling, this situation is usually modelled by linking with a *part-of* relation the classes **Head** and **Handle** to **Hammer** but the conditions that the handle and the head need to satisfy to

---

[9] The case of customers that focus only on **Person** and are one-to-one related to **Trade**-instances can be understood as a particular case of constitution.

constitute a hammer are often left implicit. Our approach is still compatible with the standard one—actually, the focus relation can be modelled as a part-of relation—but it forces the user to specify and make explicit in the schema the *structure* of hammers (via the reification of the mounted relation).

Wholes can change their composition in time maintaing their own identity. Usually, hammers survive only a non-contemporaneous change of the head and the handle. Assuming hammers are non intermittent, to express the previous constraint, one needs to generalize (5) to allow for disjunctions of conjunctions of $(R_i \ \text{nk-arg}_1)$ clauses. The persistence criterion for Hammer becomes:

$$\text{Hammer}(x_1, t_1) \wedge \text{Hammer}(x_2, t_2) \wedge \text{succ}(t_1, t_2) \rightarrow$$
$$(x_1 = x_2 \leftrightarrow ((R_1 \ \text{ham-arg}_1)(x_1, x_2, t_1, t_2) \wedge (R_2 \ \text{ham-arg}_2)(x_1, x_2, t_1, t_2)) \vee$$
$$((R'_1 \ \text{ham-arg}_1)(x_1, x_2, t_1, t_2) \wedge (R'_2 \ \text{ham-arg}_2)(x_1, x_2, t_1, t_2)))$$

where the R_i should respect:

$$R_1(y_1, y_2, t_1, t_2) \triangleq y_1 = y_2, \quad R_2(y_1, y_2, t_1, t_2) \triangleq \top$$
$$R'_1(y_1, y_2, t_1, t_2) \triangleq \top, \quad\quad\quad R'_2(y_1, y_2, t_1, t_2) \triangleq y_1 = y_2.$$

5 Final Remarks and Criticisms

This paper discusses a modelling mechanism which allows to identify new entities starting from the existence of a generating relationship with the addition of explicit identification and persistence criteria. We showed that the approach is flexible enough to capture various aggregation mechanisms and to represent constitution and composition that usually require additional primitive relations.

Our approach could be criticized from an applicative perspective. It expands the domain of quantification by introducing new (kinds of) entities that are not necessary in a model based on ISA relations like the one in Fig. 1. In addition, it adds new primitives (namely focus and support), it requires complex generation patterns, and it loses the power of the inheritance mechanism through subsumption (for instance, in Fig. 3b, the attributes that Customer inherits from Person need to be explicitly specified). Why do we need to pay this price?

A first pay back is in terms of generality, systematicity, and expressivity. The general schema in Fig. 3a, together with the persistence criteria, is able to capture, in a *uniform* and *explicit* way, different notions of customer. One of them corresponds to the model in Fig. 1 but, for the other notions, plain subsumption is not enough and the domain has to be extended—as already recognized by [7,11,15,17,19]. However note that in [7] customers are still persons but with a *qua-individual*—in particular, a PersonQuaCustomer-instance—*inhering in* them. For example, the fact that John and Microsoft are trading is reified into one *relator* and two *qua-individuals* that existentially depend on the relator, i.e., john_qua_trading_with_microsoft inheres in john and is a PersonQuaCustomer-instance, while microsoft_qua_trading_with_john inheres in microsoft and is a CompanyQuaSeller-instance. Guizzardi claims that

PersonQuaCustomer-instances can be taken into account to solve the counting problem—even though customers are persons and not PersonQuaCustomer-instances. On the other hand, the introduction of qua-individuals does not allow to specify identification and persistence criteria to refine the grouping mechanism—e.g., to distinguish the case when someone trades with companies owned by the same group from the case when trading with just different companies. For instance, the fact that john is trading with two different companies gives origin to two PersonQuaCustomer-instances but only to one Customer-instance—john, in this example—independently of the relation between the companies involved. Furthermore, by considering customers as mereological sums (assuming an extensional mereology, see [4]) of a Person-instance and a PersonQuaCustomer-instance Customer is no more a subclass of Person. In addition, even assuming customers as sums of persons and qua-individuals, it is not possible to identify a single customer generated by (synchronic) facts involving different persons (e.g., family members), or to re-identify customers that change their substratum through time. Consequently, the approach in [7] cannot address the notions of composition and constitution either. In our approach, we reify the trade-instances but we do not introduce the qua-individuals. Hence, while in [7] customers are linked to persons by identity and to Trade-relators through the qua-individuals inhering in them, in our framework we need two additional primitives: (i) focus, to link customers to persons, and (ii) support, to link customers to Trade-instances with no intermediate qua-individuals.

Two additional peculiarities of our proposal are relevant from a conceptual perspective. First, the generation of the (classes of) entities relies on a *ground model* (also called *ground level* or simply *ground*) made of the set of 'basic' classes and relations (Person, Trade, and Company in the example of Fig. 4) from which all the other ones (Customer and Seller in Fig. 4) are built. Conceptual dependencies among classes are then explicit in the model and it becomes clear what are the 'basic' concepts and the way the 'complex' ones are built and layered. Second, the proposed mechanism makes explicit the way complex objects are 'generated' from (i) grounding data (facts) and (ii) identification and persistence criteria. Notice that the extensions of classes generated by non equivalent criteria are disjoint even when a one-to-one correspondence between their instances can be established. The user can then chose how two concepts are connected independently of the set-theoretical relations of their extensions. The *intension* of concepts can then be (partially) characterized without being 'subjected' to the extensional constraints typical of the ISA relation.

From an ontological perspective, some adherent to the Quinean maxim *"to be is to be the value of a variable"*, refuse the additional *ontological commitment* necessary to adopt our modelling strategy: customers are not 'real', they are just conceptual construction. However, in the perspective of conceptual modelling, a strong ontological commitment towards the class-instances is not necessary and the offered analytic power becomes more relevant. In addition, our framework can be quite easily connected to a *constructivist* approach that starts to become recognized in the philosophy of the measurement theory [5].

References

1. Artale, A., Guarino, N., Keet, C.M.: Formalising temporal constraints on part-whole relations. In: Brewka, G., Lang, J. (eds.) Proceedings of the 11th International Conference on Principles of Knowledge Representation and Reasoning (KR 2008), pp. 673–683. AAAI Press (2008)
2. Artale, A., Franconi, E.: Foundations of temporal conceptual data models. In: Borgida, A.T., Chaudhri, V.K., Giorgini, P., Yu, E.S. (eds.) Conceptual Modeling: Foundations and Applications. LNCS, vol. 5600, pp. 10–35. Springer, Heidelberg (2009)
3. Artale, A., Kontchakov, R., Ryzhikov, V., Zakharyaschev, M.: Complexity of reasoning over temporal data models. In: Parsons, J., Saeki, M., Shoval, P., Woo, C., Wand, Y. (eds.) ER 2010. LNCS, vol. 6412, pp. 174–187. Springer, Heidelberg (2010)
4. Casati, R., Varzi, A.: Parts and Places. The Structure of Spatial Representation. MIT Press, Cambridge (1999)
5. van Fraassen, B.C.: Scientific Representation: Paradoxes of Perspective. Clarendon Press, Oxford University Press (2008)
6. Gibbard, A.: Natural property rights. Nous 10, 77–88 (1976)
7. Guizzardi, G.: Ontological Foundations for Structural Conceptual Models. Phd, University of Twente (2005)
8. Gupta, A.: The Logic of Common Nouns: an investigation in quantified modal logic. Phd thesis, Yale University (1980)
9. Halpin, T., Morgan, T.: Information Modeling and Relational Databases, 2nd edn. Morgan Kaufmann (2008)
10. Jensen, C.S., Snodgrass, R.T.: Temporal data management. IEEE Transactions on Knowledge and Data Engineering 111(1), 36–44 (1999)
11. Loebe, F.: An Analysis of Roles. Toward Ontology-Based Modelling. Master's thesis, University of Leipzig (2003)
12. Masolo, C.: Levels for conceptual modeling. In: De Troyer, O., Bauzer Medeiros, C., Billen, R., Hallot, P., Simitsis, A., Van Mingroot, H. (eds.) ER 2011 Workshops. LNCS, vol. 6999, pp. 173–182. Springer, Heidelberg (2011)
13. Masolo, C., Vieu, L., Kitamura, Y., Kozaki, K., Mizoguchi, R.: The counting problem in the light of role kinds. In: Davis, E., Doherty, P., Erdem, E. (eds.) Proceedings of the AAAI Spring Symposium: Logical Formalizations of Commonsense Reasoning. AAAI Press (2011)
14. Masolo, C.: Understanding ontological levels. In: Lin, F., Sattler, U. (eds.) Proceedings of the Twelfth International Conference on the Principles of Knowledge Representation and Reasoning (KR 2010), pp. 258–268. AAAI Press (2010)
15. Mizoguchi, R., Sunagawa, E., Kozaki, K., Kitamura, Y.: A model of roles within an ontology development tool: Hozo. J. of Applied Ontology 2(2), 159–179 (2007)
16. Spaccapietra, S., Parent, C., Zimanyi, E.: Conceptual Modeling for Traditional and Spatio-Temporal Applications—The MADS Approach. Springer (2006)
17. Steimann, F.: On the representation of roles in object-oriented and conceptual modelling. Data and Knowledge Engineering 35, 83–106 (2000)
18. Welty, C., Andersen, W.: Towards ontoclean 2.0: A framework for rigidity. Journal of Applied Ontology 1(1), 107–116 (2005)
19. Wieringa, R., de Jonge, W., Spruit, P.: Using dynamic classe and role classe to model object migration. Theory and Practice of Object Systems 1(1), 31–83 (1995)

Supervised Learning and Distributional Semantic Models for Super-Sense Tagging

Pierpaolo Basile, Annalina Caputo, and Giovanni Semeraro

Dept. of Computer Science
University of Bari Aldo Moro
Via E. Orabona, 4 - 70125 Bari Italy
{pierpaolo.basile,annalina.caputo,giovanni.semeraro}@uniba.it

Abstract. Super-sense tagging is the task of annotating each word in a text with a super-sense, i.e. a general concept such as animal, food or person, coming from the general semantic taxonomy defined by the WordNet lexicographer classes. Due to the small set of involved concepts, the task is simpler than Word Sense Disambiguation, which identifies a specific meaning for each word. The small set of concepts allows machine learning algorithms to achieve good performance when coping with the problem of tagging. However, machine learning algorithms suffer from data-sparseness. This problem becomes more evident when lexical features are involved, because test data can contain words with low frequency (or completely absent) in training data. To overcome the sparseness problem, this paper proposes a supervised method for super-sense tagging which incorporates information coming from a distributional space of words built on a large corpus. Results obtained on two standard datasets, SemCor and SensEval-3, show the effectiveness of our approach.

1 Introduction

Super-sense tagging is the task of annotating each word in a text with a concept coming from the general semantic taxonomy defined by the WordNet [10] lexicographer classes called super-senses. A super-sense defines a general concept such as animal, body, person, communication, motion. An overview of WordNet super-senses for each part-of-speech (PoS) is reported in Table 1.

Super-senses are assigned to words such as nouns, verbs, adjectives and adverbs, which are the word-classes defined in WordNet. Super-sense tagging can be considered as an halfway task between Named Entity Recognition (NER) [11] and Word Sense Disambiguation (WSD) [20]. The former involves a small set of categories, for example: Person, Organization, Location, Time. The latter requires a very large set of senses with very specific meanings.

The NER task mainly deals with proper nouns, or some specific domain concepts like biological entities. The small number of classes allows to implement robust supervised algorithms that are able to obtain good performance. Conversely, the goal of a WSD algorithm consists in assigning the appropriate meaning or sense, chosen from a set of predefined possibilities, to each word occurrence. Typically, the set of meanings for each word is provided by a sense inventory like WordNet. WSD algorithms deal with

M. Baldoni et al. (Eds.): AI*IA 2013, LNAI 8249, pp. 97–108, 2013.
© Springer International Publishing Switzerland 2013

Table 1. WordNet super-senses

PoS	Super-senses
adj	all, pert (relational adjectives), ppl (participial adjectives)
adv	all
noun	Tops (nouns that appear as super senses), act, animal, artifact, attribute, body, cognition, communication, event, feeling, food, group, location, motive, object, person, phenomenon, plant, possession, process, quantity, relation, shape, state, substance, time
verb	body, change, cognition, communication, competition, consumption, contact, creation, emotion, motion, perception, possession, social, stative, weather

a large number of classes, potentially all the possible meanings of open-class words[1]. Due to the complexity at this level of granularity, WSD algorithms are not able to reach performance of NER algorithms.

Differently from named entity classes, super-senses are not strictly related to proper nouns, and provide a more abstract set of meanings than WSD, which simplifies the problem of sense disambiguation. Super-sense tagging combines the small-sized set of categories typical of NER with meanings provided by WordNet super-senses, therefore it can be considered as a simpler version of WSD, where a smaller number of meanings is involved. The small set of senses allows to use robust supervised learning algorithms trained on a hand-tagged corpus. However, supervised learning algorithms suffer from data-sparseness. This problem becomes more evident when lexical features are involved (like in super-sense task), because test data can contain words with low frequency (or completely absent) in training data.

In this paper, we propose a method for super-sense tagging that exploits distributional features in Support Vector Machines (SVM) [5]. Distributional features derive from *WordSpaces* [22] built exploiting the distributional nature of words. In such a way, we propose to remedy the data sparseness problem by exploiting the paradigmatic relations (i.e. *in absentia*) encoded in such spaces. *WordSpaces* are based on the geometrical representation of words as points in a vector space. The points representation is built directly from the text taking into account word co-occurrences in some predefined contexts. In such way, *WordSpaces* encode the paradigmatic relations between words, i.e. the relations between elements that can appear as alternative in a point. Hence, concepts with similar or related meanings are represented by points close to each other in such spaces (geometric metaphor of meaning). This property follows from the *distributional hypothesis* [12], according to which the meaning of a word is determined by its usage in the language, and that words that appear in the same *contexts* are semantically related. In this paper, we take into account two definitions of context: Wikipedia pages and Wikipedia categories. The main idea of our work is to improve the robustness of our super-sense tagging approach by extending lexical information through distributional analysis. Using distributional analysis in semantic spaces we expect that words with similar meaning are represented close to each other in *WordSpace*. We can rely on this

---

[1] http://www.ucl.ac.uk/internet-grammar/wordclas/open.htm, Open and Closed Word Classes.

property to solve the problem of data-sparseness by adding distributional information about words as features into the supervised learning strategy.

The paper is organized as follows. Section 2 provides details about the exploited distributional approach, while Section 3 describes the methodology adopted to solve super-sense tagging. Section 4 discusses the data used for test and training, while Section 5 reports the results of the evaluation. Related works are briefly analyzed in Section 6, while the last section gives some final observations.

2 WordSpaces

The method used in this work to represent words in the *WordSpace* relies on the distributional approach. This approach represents words as vectors in a high-dimensional space [23,22]. In a *WordSpace* words and concepts are represented by points in a mathematical space, and this representation is learned from text in such a way that concepts with similar or related meanings are near to one another in that space (geometric metaphor of meaning). Therefore, semantic similarity between words can be represented as proximity in that n-dimensional space. The main characteristic of the geometric metaphor of meaning is not that meanings are represented as locations in a semantic space, but rather that similarity between words can be expressed in spatial terms, as proximity in a high-dimensional space.

One of the great virtues of the distributional approach is that word spaces can be built using entirely unsupervised analysis of free text. In addition, they make very few language-specific assumptions (only tokenized text is needed). According to the *distributional hypothesis* [12], the meaning of a word is determined by the rules of its use in the context of ordinary and concrete language behaviour. This means that words are semantically similar if they share *contexts* (surrounding words). If "green" and "yellow" frequently occur in the same context, for example near the word "color", the hypothesis states that they are semantically related or similar. Co-occurrence between words is defined with respect to a context, for example a window of terms of fixed length, a document, or a Wikipedia page/category. It should be pointed out that a word is represented by a vector in a high-dimensional space. Since distributional methods are expected to handle efficiently high-dimensional vectors, a common choice is to adopt *dimensionality reduction* algorithms that allow to represent high-dimensional data in a lower-dimensional space without losing information. For example, *Latent Semantic Analysis* (LSA) [16] collects the text data in a co-occurrence matrix, which is then decomposed into smaller matrices with Singular-Value Decomposition (SVD), by capturing latent semantic structures in the text data. The main drawback of SVD is scalability.

In this paper, we adopt *Random Indexing* (RI) that, differently from LSA, targets the problem of dimensionality reduction by removing the need for the matrix decomposition or factorization. RI incrementally accumulates *context* vectors, which can be later assembled into a new space, thus it offers a novel way of conceptualizing the construction of context vectors.

RI relies on the concept of Random Projection [8]: high-dimensional vectors randomly chosen are "nearly orthogonal". This yields a result that is comparable to orthogonalization methods, such as SVD, but saving computational resources. Specifically, RI creates the *WordSpace* in two steps:

1. a *random* vector is assigned to each context. This vector is sparse, high-dimensional and ternary, which means that its elements can take values in {-1, 0, 1}. The random vector contains a small number of randomly distributed non-zero elements (seeds), and the structure of this vector follows the hypothesis behind the concept of Random Projection. We can consider the random vector as a fingerprint assigned to each context;
2. random vectors are accumulated incrementally by analyzing contexts in which terms occur. In particular, the *semantic* vector assigned to each word is the sum of the random vectors of the contexts in which the term occurs. It should be pointed out that random vectors are added by multiplying them by the term frequency.

In this work, we exploit RI to build two spaces using two different definitions of context:

WikiPages A random vector is assigned to each Wikipedia page;
WikiCategories Categories can identify more general concepts in the same way as super-senses. In this case, for each category a random vector is created.

We use SemanticVectors package[2] [26], an open-source tool, to build the two *WordSpaces*. SemanticVectors creates semantic *WordSpaces* from free natural language text using the Random Indexing technique.

Before building the *WordSpaces*, we need to index all Wikipedia pages using the last dump provided by Wikipedia foundation. During the indexing step we extract page categories using a regular expression[3] and add these as meta-data to each page. After this first indexing step, we build a second index containing a document for each category: the document is the sum of all those documents tagged with the category. That index is necessary to build the *WordSpace* that relies on Wikipedia categories as context. We use Apache Lucene[4], an open-source API, for indexing.

Finally, we run SemanticVectors tool on each index, obtaining as result the two *WordSpaces*, WikiPages and WikiCategories. Each space consists of the most frequent 150,000 terms. Table 2 reports information about *WordSpaces*, in particular: the number of contexts (pages or categories) C, the space dimensions D and the number of no-zero elements (seeds) in random vectors S. In order to reduce the space dimension we perform some text filtering operations to remove stop-words and words with low occurrences. In particular, we consider pages with more than 2,000 words and terms that occur at least in ten pages.

3   Methodology

The problem of super-sense tagging is very close to the sequence labelling one. Given a text composed by a sequence of words, the goal is to annotate each word with the correct super-sense.

---

[2] Available on-line: http://code.google.com/p/semanticvectors/
[3] Categories are defined in the page using mediawiki syntax.
[4] Lucene is available on-line: lucene.apache.org.

Table 2. WordSpaces information

WordSpace	C	D	S
WikiPages	1,138,106	1,000	10
WikiCategories	335,759	400	10

Table 3a reports a sequence of words tagged with their super-senses in bold face. Tags are in IOB2 format[5]: B for the word at the begin of the annotation, I for inside and O for outside words. IOB2 schema allows to annotate multi-word expressions. The small number of involved labels allows us to use a supervised learning strategy. The classes of our learning problem are represented by all the possible super-sense tags included between the begin, inner and outside tags. A super-sense could occur both in B and I tags. Hence, the training step requires a hand-annotated corpus in IOB2 format; details about the dataset are reported in Section 4.

We use a set of lexical/morphological and contextual features to represent each word w in the training data, in particular:

1. The word w plus contextual words: w_{-1} and w_{+1} (the first word to the left and the first word to the right of w);
2. The lemma l_w of the word w plus contextual lemmas: l_{w-1} and l_{w+1};
3. The part-of-speech (PoS) tag pos_w of the word w plus contextual PoS-tags: pos_{w-1} and pos_{w+1};
4. The super-sense assigned to the most frequent sense of the word w. The most frequent sense in WordNet is the first sense assigned to that word, for example for the word *bank* the first sense is *"sloping land (especially the slope beside a body of water)"* and its super-sense is *noun.object*;
5. The first letter of the PoS-tag, which generally identifies the word-class: noun, verb, adjective and adverb;
6. A binary feature that indicates if the word begins with an upper-case character;
7. Distributional features: words in the *WordSpace* are represented by high-dimensional vectors as described in Section 2. We use as features all the components of the word vector w.

The number of contextual words to consider in each feature was selected after a tuning step performed through K-fold validation on the training data. Table 3b shows the set of features for the word *law* taken from the example reported in Table 3a.

As learning method we adopt Support Vector Machines (SVM). In particular, we propose two systems:

1. The first system is based on SVM without distributional information (i.e. excluding the feature 7). We use an open-source tool, *YAMCHA* [15], that is a generic text chunker based on SVM adopted in several NLP tasks such as PoS tagging, named entity recognition and phrase chunking. We use this method as a baseline of our approach.

---
[5] http://www.cnts.ua.ac.be/conll2002/ner/

Table 3. Examples

(a) Example of super-sense tagging.

Implementation	B-noun.act
of	O
Georgia	B-noun.location
's	O
automobile	B-noun.artifact
title	B-noun.communication
law	*B-noun.communication*
was	O
also	B-adv.all
recommended	B-verb.communication
by	O
the	O
outgoing	B-adj.all
jury	B-noun.group
.	O

(b) The set of features for the word *law*.

feature	value
1	law
	title
	was
2	law
	title
	be
3	NN
	NN
	VBD
4	noun.communication
5	N
	N
	V
6	false
7	w

2. The second system relies on distributional information. Distributional features are numeric and cannot be represented in *YAMCHA*, which manages only discrete values. Moreover, distributional features are represented by high-dimensional vectors. For these reasons we adopt *LIBLINEAR* [9], a library for large linear classification. *LIBLINEAR* provides good results when a large number of features is involved. It uses a linear mapping instead of non-linear kernels such as polynomial kernels, adopted by *YAMCHA*. LIBLINEAR implements linear support vector machines which are very efficient on large sparse data sets.

4 Dataset and System Setup

The dataset used to perform the training step is SemCor. SemCor is a collection of 352 documents annotated with WordNet synsets[6]. In particular, SemCor consists of three parts: in the first two ones all words are tagged, while in the last one only the verbs are tagged. For training, we use only the first two sections, for a total of 186 documents and 419,982 annotated words. We transform the SemCor dataset, which is in XML-like format, in IOB2 format in order to train YAMCHA. During the transformation step we add all the features as described in Section 3. To build the second system based on LIBLINEAR we need to transform the IOB2 dataset in the LIBLINEAR data-format, which requires a line for each example (i.e. word). Each line contains the $class_{id}$, in our case the tag assigned to the word, and the list of features as pairs $\langle feature_{id} : feature_{value} \rangle$.

---

[6] We use SemCor annotated with WordNet 2.0, available on-line:
http://www.cse.unt.edu/~rada/downloads.html.

LIBLINEAR requires that each data instance (example) is represented as a vector of real numbers. For each value assumed by no-numeric feature, a $feature_{value}$ is generated. In this case, the $feature_{value}$ can assume only two values: 1 if the $feature_{id}$ occurs in the data instance, 0 otherwise.

For testing, we adopt SensEval-3 All-words dataset[7] [25]. This dataset consists of 3 documents annotated with WordNet synsets from which we retrieve the super-senses, for a total of 5,511 words. The dataset was originally created for the evaluation of WSD algorithms. We convert the dataset in IOB2 format as for the training data.

For the evaluation, in order to compare the systems output against the super-sense labels in testing data, we use a modified version of the script provided by ConLL-2000 shared task for text chunking[8]. Text chunking is a sequence labelling task in which words are annotated with phrase chunks, for example noun and verb phrases. The script provides information about precision, recall and F-measure for each super-sense type.

5 Evaluation

The evaluation has a twofold goal:

1. Proving that distributional information is able to improve the performance of a supervised learning strategy for super-sense tagging. For that reason we compare our system with *YAMCHA*, which does not include distributional features.
2. Comparing the two *WordSpaces*. We train two different models: $LIBL_{pages}$ which uses the distributional features provided by the *WordSpace* built on Wikipedia pages, and $LIBL_{cat}$ which relies on Wikipedia categories.

We used two simple systems as baselines:

random randomly chooses a super-sense from the set of possible super-senses assigned to the word;
1stsense labels words with the super-sense assigned to the most frequent sense.

The evaluation is performed on both the datasets: SemCor and SensEval-3. On Sem-Cor dataset we perform a 5-folds cross-validation. Table 4 reports the results in terms of precision (P), recall (R) and F-measure (F). The last column ($\triangle\%$) shows the improvement in F-measure values with respect to the $1^{st}sense$ baseline.

The second dataset is used for testing: first we train the systems on SemCor and then we test them on SensEval-3. Experimental results are reported in Table 5.

Results show that our methods ($LIBL_{pages}$ and $LIBL_{cat}$) are able to outperform the baselines in both the datasets. In particular, on SensEval-3 we obtain very good results: an improvement of 16.11% and 16.31% with respect to the $1^{st}sense$ baseline. This outcome underlines as our methods are able to deal with data-sparseness problem as also highlighted by improvements in recall values. Indeed, our best strategy $LIBL_{cat}$

---

[7] We use SensEval-3 annotated with WordNet 2.0, available on-line:
   http://www.cse.unt.edu/~rada/downloads.html.
[8] http://www.cnts.ua.ac.be/conll2000/chunking/.

Table 4. Results of 5-folds cross-validation on SemCor

System	P	R	F	$\triangle\%$
random	66.82	65.23	66.01	-
$1^{st}sense$	83.43	81.44	82.42	-
YAMCHA	90.99	91.08	91.04	+10,42
$LIBL_{pages}$	91.74	91.75	91.75	+11,32
$LIBL_{cat}$	91.77	91.81	**91.79**	+11,36

Table 5. Results on SensEval-3 All-words dataset

System	P	R	F	$\triangle\%$
random	47.56	60.69	53.33	-
$1^{st}sense$	61.44	78.41	68.89	-
YAMCHA	69.53	85.91	76.86	+11,56
$LIBL_{pages}$	73.37	87.92	79.99	+16,11
$LIBL_{cat}$	73.43	88.17	**80.13**	+16,31

achieves an improvement of 12,44% in terms of recall with respect to the baseline. Since SemCor and SensEval-3 are two different corpora containing documents from different domains, they are prone to data sparseness. SemCor can contain words that do not occurred in training data. For that reason, improvements are not so pronounced on SemCor, where training and testing data refer to the same domain. On the other hand, distributional features help our systems to improve their performance on SensEval-3, where training and testing data come from different domains.

We want to report also the improvement of our best system ($LIBL_{cat}$) with respect to the results reported by Ciaramita and Altun [3] on both SemCor and SensEval-3. However, it should be pointed out that we used a different training set. Our system was trained only on the first two datasets of SemCor, while in [3] authors exploited all the three parts of SemCor. On SemCor, we obtain an improvement of 19.73%, 18.14%, 18.93% with respect to precision, recall and F-measure, while on SensEval-3 the improvements are of 8.6%, 19.56%, 13.60% respectively. These figures highlight consistent improvements in recall across both datasets.

Regarding the second goal of the evaluation, the two $WordSpace$s obtain comparable results. The difference between $LIBL_{pages}$ and $LIBL_{cat}$ is not significant.

Tables 6 and 7 report fine-grained results for SemCor and SensEval-3 datasets respectively, focusing on subsets of the evaluation. In particular they take into account the five most frequent noun and verb super-senses. Both tables report the performance in terms of F-measure; the best result is reported in bold face.

In SemCor (see Table 6), methods based on distributional features are always able to achieve the best performance.

In SensEval-3 (see Table 7), the improvement with respect to nouns still remains significant, while for verbs there are two cases (verb.stative and verb.perception) where

Table 6. Results on SemCor considering the five most frequent super-senses

SemCor					
NOUNS					
Super-sense	#n	$1^{st}sense$	$YAMCHA$	$LIBL_{pages}$	$LIBL_{cat}$
noun.person	14848	68.34	97,73	**97.76**	97.68
noun.artifact	9008	86.92	87.63	89.02	**89.11**
noun.act	8146	79.95	81.05	86.18	**86.61**
noun.cognition	7132	76.65	83.63	84.45	**84.82**
noun.communication	7010	80.30	87.92	89.22	**89.29**
VERBS					
Super-sense	#n	$1^{st}sense$	$YAMCHA$	$LIBL_{pages}$	$LIBL_{cat}$
verb.stative	12394	85.28	92.88	92.95	**92.99**
verb.communication	6422	83.61	84.26	**87.93**	87.80
verb.cognition	4351	79.30	82.93	**84.34**	84.16
verb.change	4153	72.21	77.31	79.74	**79.75**
verb.motion	3866	76.73	82.40	83.33	**83.37**

our systems obtain results slightly below the $1^{st}sense$ baseline. This result can be due to the average polysemy of verbs that is twice that of nouns, resulting in a more difficult disambiguation.

Comparing differences between our *WordSpaces* systems and *YAMCHA*, we can note that in SemCor they are not so pronounced, while they become clearer in SensEval-3, corroborating our hypothesis about the data sparseness. This proofs that distributional information contributes to build a more stable learning method.

Table 7. Results on SensEval-3 All-words considering the most five frequent super-senses

SensEval-3 All-words					
NOUNS					
Super-sense	#n	$1^{st}sense$	$YAMCHA$	$LIBL_{pages}$	$LIBL_{cat}$
noun.person	145	84.74	92.96	**94.16**	93.20
noun.artifact	133	74.20	86.43	89.29	**90.17**
noun.act	97	70.16	67.58	73.10	**76.19**
noun.cognition	70	61.54	79.14	81.75	**84.44**
noun.event	61	75.21	90.09	**92.17**	**92.17**
VERBS					
Super-sense	#n	$1^{st}sense$	$YAMCHA$	$LIBL_{pages}$	$LIBL_{cat}$
verb.stative	187	**80.99**	72.82	80.84	80.65
verb.communication	90	79.29	80.00	**86.55**	86.05
verb.motion	82	71.91	76.51	77.99	**78.48**
verb.cognition	62	82.35	71.23	**83.58**	81.75
verb.perception	62	**82.64**	76.79	82.14	81.42

6   Related Work

Sequence labelling strategies are very common in natural language tasks such as named entity recognition and text chunking. However, they are not usual in Word Sense Disambiguation, where knowledge intensive methods or classification strategies are used [20]. Some approaches to the tagging strategy in WSD were proposed in [24,18,19] and rely on Hidden Markov Model. Regarding super-sense tagging, early approaches exploit lexical acquisition techniques (only nouns) [4,7], but they are mainly focused on classification rather than tagging. A first attempt to use a labelling method for supersense was proposed by [3], which adopts a perceptron-trained Hidden Markov Model. In that work, the authors evaluate the system on SemCor and SensEval-3 obtaining improvements with respect to the first-sense baseline. It is important to point out that their system relies on all the three parts of SemCor and it works only for nouns and verbs. This makes impossible a direct comparison with our system.

An adaptation of the previous work to Italian language was proposed in [21] where MultiSemCor, a parallel sense labelled corpus of SemCor, was adopted as training. Due to the lower quality of the Italian training data the system achieves a slightly lower accuracy with respect to the system working on English. However, in [1] the authors propose a new strategy based on Maximum Entropy method which improves the performance on Italian texts. The authors do not use MultiSemCor as training, but they build a new corpus starting from the Italian Syntactic-Semantic Treebank (ISST).

A variant of the proposed work evaluated on the Italian language was described in [2]. That system accounted for a bigger set of features able to take into account some specificities of the Italian language. The system was built exploiting the Italian Syntactic-Semantic Treebank (ISST) for both training and testing. Differently from SemCor, which initially comprised only terms tagged with synsets, the ISST dataset was created specially for the super-sense tagging task. The evaluation was performed in the context of EVALITA 2011, the third evaluation campaign of Natural Language Processing and Speech tools for Italian.

In [6], the authors prove the effectiveness of distributional features for supervised learning to the Semantic Role Labelling task. In particular, they propose a learning method based on SVM that uses distributional features coming from a *WordSpace* built using LSA.

Regarding the use of super-senses in other tasks, [14] exploits super-senses to build useful latent semantic features in syntactic parse re-ranking. Another attempt to use super-senses as features was proposed by [17], in which super-senses and collections are adopted as features in a Word Sense Disambiguation system called SuperSenseLearner. A very interesting approach to the use of super-senses in Information Retrieval was proposed by [13]. The authors disambiguate queries and documents using super-senses instead of WordNet synsets and obtain promising results on TREC collection when their system is combined with pseudo relevance feedback.

We believe that super-sense tagging not only can be useful in NLP tasks which rely on semantic features, but it can also be used in more complex problems such as Information Retrieval, Information Extraction and Question Answering, where large amounts of text need to be processed. In these contexts super-sense tagging could be a more effective and scalable alternative to the classical Word Sense Disambiguation method.

7 Conclusions

This paper describes a new supervised approach to perform super-sense tagging. The idea is to incorporate features that come from a *WordSpace* built adopting distributional approach. The insight is to solve the problem of data-sparseness exploiting the similarity between words in a *WordSpace*. In particular, we use Random Indexing as a method to build *WordSpace* on Wikipedia contents. We propose two *WordSpace*s: the former uses Wikipedia pages while the latter relies on Wikipedia categories. As training data we use SemCor, which is a corpus annotated with WordNet synsets, while SensEval-3 All-words is used for testing. Evaluation results prove the effectiveness and robustness of our approach. The introduction of distributional features improves both precision and recall, and results in a significant improvement in terms of F-measure.

As future work, we plan to investigate three different aspects: 1) experimenting with other learning strategies different from SVM; 2) exploiting different methods for building *WordSpace*s, such as LSA; 3) investigating super-senses tagging in real applications, for example Information Retrieval.

References

1. Attardi, G., Dei Rossi, S., Di Pietro, G., Lenci, A., Montemagni, S., Simi, M.: A Resource and Tool for Super-sense Tagging of Italian Texts. In: Proceedings of the 7th International Conference on Language Resources and Evaluation, LREC 2010 (2010)
2. Basile, P.: Super-Sense Tagging Using Support Vector Machines and Distributional Features. In: Magnini, B., Cutugno, F., Falcone, M., Pianta, E. (eds.) EVALITA 2012. LNCS, vol. 7689, pp. 176–185. Springer, Heidelberg (2012)
3. Ciaramita, M., Altun, Y.: Broad-coverage sense disambiguation and information extraction with a supersense sequence tagger. In: Proceedings of the 2006 Conference on Empirical Methods in Natural Language Processing, pp. 594–602. Association for Computational Linguistics (2006)
4. Ciaramita, M., Johnson, M.: Supersense tagging of unknown nouns in WordNet. In: Proceedings of the 2003 Conference on Empirical Methods in Natural Language Processing, vol. 10, pp. 168–175. Association for Computational Linguistics (2003)
5. Cortes, C., Vapnik, V.: Support-vector networks. Machine Learning 20(3), 273–297 (1995)
6. Croce, D., Basili, R.: Structured learning for semantic role labeling. In: Pirrone, R., Sorbello, F. (eds.) AI*IA 2011. LNCS, vol. 6934, pp. 238–249. Springer, Heidelberg (2011)
7. Curran, J.: Supersense tagging of unknown nouns using semantic similarity. In: Proceedings of the 43rd Annual Meeting on Association for Computational Linguistics, pp. 26–33. Association for Computational Linguistics (2005)
8. Dasgupta, S., Gupta, A.: An elementary proof of a theorem of Johnson and Lindenstrauss. Random Structures & Algorithms 22(1), 60–65 (2003)
9. Fan, R., Chang, K., Hsieh, C., Wang, X., Lin, C.: LIBLINEAR: A library for large linear classification. The Journal of Machine Learning Research 9, 1871–1874 (2008)
10. Fellbaum, C.: WordNet: An Electronic Lexical Database. MIT Press (1998)
11. Grishman, R., Sundheim, B.: Message Understanding Conference-6: a brief history. In: Proceedings of the 16th Conference on Computational Linguistics, COLING 1996, vol. 1, pp. 466–471. Association for Computational Linguistics, Stroudsburg (1996)
12. Harris, Z.: Mathematical Structures of Language. Interscience, New York (1968)

13. Kim, S., Seo, H., Rim, H.: Information retrieval using word senses: root sense tagging approach. In: Proceedings of the 27th Annual International ACM SIGIR Conference on Research and Development in Information Retrieval, pp. 258–265. ACM (2004)
14. Koo, T., Collins, M.: Hidden-variable models for discriminative reranking. In: Proceedings of the Conference on Human Language Technology and Empirical Methods in Natural Language Processing, pp. 507–514. Association for Computational Linguistics (2005)
15. Kudo, T., Matsumoto, Y.: Fast Methods for Kernel-Based Text Analysis. In: Proceedings of the 41st Annual Meeting of the Association for Computational Linguistics, pp. 24–31. Association for Computational Linguistics, Sapporo (2003)
16. Landauer, T.K., Dumais, S.T.: A Solution to Plato's Problem: The Latent Semantic Analysis Theory of Acquisition, Induction, and Representation of Knowledge. Psychological Review 104(2), 211–240 (1997)
17. Mihalcea, R., Csomai, A., Ciaramita, M.: Unt-yahoo: Supersenselearner: Combining senselearner with supersense and other coarse semantic features. In: Proceedings of the Fourth International Workshop on Semantic Evaluations (SemEval 2007), pp. 406–409. Association for Computational Linguistics, Prague (2007)
18. Molina, A., Pla, F., Segarra, E.: A Hidden Markov Model Approach to Word Sense Disambiguation. In: Garijo, F.J., Riquelme, J.-C., Toro, M. (eds.) IBERAMIA 2002. LNCS (LNAI), vol. 2527, pp. 655–663. Springer, Heidelberg (2002)
19. Molina, A., Pla, F., Segarra, E.: WSD System Based on Specialized Hidden Markov Model (upv-shmm-eaw). In: SENSEVAL-3/ACL 2004 (2004)
20. Navigli, R.: Word Sense Disambiguation: A survey. ACM Comput. Surv. 41, 10:1–10:69 (2009)
21. Picca, D., Gliozzo, A., Ciaramita, M.: Supersense tagger for Italian. In: Proceedings of the 6th International Conference on Language Resources and Evaluation, LREC 2008 (2008)
22. Sahlgren, M.: The Word-Space Model: Using distributional analysis to represent syntagmatic and paradigmatic relations between words in high-dimensional vector spaces. Ph.D. thesis, Stockholm: Stockholm University, Faculty of Humanities, Department of Linguistics (2006)
23. Schütze, H.: Automatic word sense discrimination. Computational Linguistics 24(1), 97–123 (1998)
24. Segond, F., Schiller, A., Grefenstette, G., Chanod, J.: An experiment in semantic tagging using hidden markov model tagging. In: ACL/EACL Workshop on Automatic Information Extraction and Building of Lexical Semantic Resources for NLP Applications, pp. 78–81 (1997)
25. Snyder, B., Palmer, M.: The English all-words task. In: Mihalcea, R., Edmonds, P. (eds.) Senseval-3: Third International Workshop on the Evaluation of Systems for the Semantic Analysis of Text, pp. 41–43. Association for Computational Linguistics, Barcelona (2004)
26. Widdows, D., Ferraro, K.: Semantic Vectors: A Scalable Open Source Package and Online Technology Management Application. In: Proceedings of the 6th International Conference on Language Resources and Evaluation, LREC 2008 (2008)

A Heuristic Approach to Handling Sequential Information in Incremental ILP

Stefano Ferilli[1,2] and Floriana Esposito[1,2]

[1] Dipartimento di Informatica – Università di Bari
{ferilli,esposito}@di.uniba.it
[2] Centro Interdipartimentale per la Logica e sue Applicazioni – Università di Bari

Abstract. When using Horn Clause Logic as a representation formalism, the use of uninterpreted predicates cannot fully account for the complexity of some domains. In particular, in Machine Learning frameworks based on Horn Clause Logic, purely syntactic generalization cannot be applied to these kinds of predicates, requiring specific problems to be addressed and tailored strategies and techniques to be introduced. Among others, outstanding examples are those of numeric, taxonomic or sequential information. This paper deals with the case of (multidimensional) sequential information.Coverage and generalization techniques are devised and presented, and their integration in an incremental ILP system is used to run experiments showing its performance.

1 Introduction

Automatic reasoning and learning in many real-world domains requires structured representations that are able to represent and handle complex relationships among the involved entities and their properties. These capabilities go beyond traditional representations where any description must fit a fixed number of atomic descriptors, such as feature vectors. First-Order Logic (*FOL* for short) formalisms overcome these limits and allow to express relationships among objects. Inductive Logic Programming (ILP) [17] is the branch of Machine Learning based on Logic Programming as a representation language. The cost for the increased expressive power provided by relationships is a worse computational complexity of the procedures exploited by FOL, due to the problem of *indeterminacy* in mapping portions of one formula onto portions of another.

Usually, predicates that make up the description language used to tackle a specific problem are defined by the knowledge engineer that is in charge of setting up the reasoning or learning task, and are handled as purely syntactic entities by the systems. However, the purely syntactic level (*à la* Herbrand) is often too limiting for an effective application of this kind of techniques to real-world problems. This work specifically focuses on sequential information, where items (objects, events, situations, etc.) in the description are related, among others, by a relationship of adjacency along some dimension (or possibly many different dimensions: time, space, etc.). In this case the transitivity of the immediate adjacency relationships cannot be overlooked, leading to a more general relationship for which a recursive definition can be cast.

M. Baldoni et al. (Eds.): AI*IA 2013, LNAI 8249, pp. 109–120, 2013.
© Springer International Publishing Switzerland 2013

The problem of handling sequential information in logic descriptions, and specifically of carrying out learning tasks with it, is not new in the literature. However, the perspective of this work is different from the mainstream research carried out so far. While most works have focused on sequential information on a single dimension [13, 14, 15, 16, 2, 11], we face the multidimensional setting, and allow complex interrelationships among any combination of events and involved objects. Thus, for instance, events in different dimensions can be related to each other, which prevents simple extension of single-dimension approaches to multiple orthogonal dimensions. Also, we go beyond strictly linear sequences requiring a total ordering relationship among events, and allow for 'parallel' events along the same dimension. Moreover, we aim at learning rules in the classical ILP fashion, but allowing preconditions to include sequential information, while other works aimed at classifying sequences [13], or at inferring predictive models for them [14, 15, 2], or at extracting frequent patterns [16, 6]. A final, fundamental peculiarity of our technique is its incrementality, which we deem as mandatory in real-world domains. Indeed, it processes examples one by one, and if needed suitably adapts the current model for each of them, without requiring the whole set of examples to be available when learning starts.

The rest of this paper is organized as follows. Section 2 introduces the representation formalism on which our proposal is based. Then, Sections 3 and 4 describe the proposed generalization and coverage techniques for representations including sequential information, respectively. Section 5 presents experiments that show the effectiveness of the proposed approach. Lastly, Section 6 concludes the paper and outlines future work directions.

2   Representation

We will deal with the case of linked Datalog [4] clauses. The choice of the description language, and of its level of granularity, is up to the experimenter, which is responsible for its consistency. From now on, we will call *events* the items (terms) on which sequential relationships can be set. Just like any other object, we allow events to have properties and relationships to other events and/or objects. We reserve the following predicates to express sequential information among events:

next/3 expresses immediate adjacency between two event: next(I_1, I_2, D) indicates that I_2 immediately follows I_1 along dimension D;

after/4 is a generalization expressing any kind of sequentiality relationship between two events: after$(I_1, I_2, D, [R_1, R_2])$ is to be intended as "I_2 follows I_1 along dimension D after at least R_1, and at most R_2, adjacency steps".

We assume that observations are always expressed in terms of immediate adjacency (next/3 relationships), while models are expressed in terms of general after/4 sequential relationships only. Also, due to lack of space, in this paper we will not stress the range-related part of the after/4 atoms, for which reason that argument is dropped in the following.

Example 1. Let us consider the following description:

$$\texttt{a}(s_1,t), \ \texttt{tl}(s_1), \ \texttt{hl}(s_1), \ \texttt{a}(s_x,t), \ \texttt{ntm}(s_x), \ \texttt{nhm}(s_x), \ \texttt{a}(s_2,u),$$
$$\texttt{tm}(s_2), \ \texttt{hm}(s_2), \ \texttt{wm}(s_2), \ \texttt{ff}(s_2), \ \texttt{gg}(s_2), \ \texttt{a}(s_3,t), \ \texttt{tm}(s_3), \ \texttt{hm}(s_3),$$
$$\texttt{a}(s_4,v), \ \texttt{tm}(s_4), \ \texttt{hh}(s_4), \ \texttt{a}(s_5,v), \ \texttt{wh}(s_5), \ \texttt{hm}(s_5), \ \texttt{c}(s_1,s_3),$$
$$\texttt{c}(s_1,s_2), \ \texttt{d}(u,v,w), \ \texttt{d}(u,t), \ \texttt{next}(s_1,s_x,default), \ \texttt{next}(s_x,s_2,default),$$
$$\texttt{next}(s_2,s_3,default), \ \texttt{next}(s_1,s_4,x), \ \texttt{next}(s_2,s_5,y)$$

It is an observation, as indicated by the presence of constants as arguments and by the use of **next** atoms to express sequential information. It involves 6 events $(s_1, s_2, s_3, s_4, s_5, s_x)$ along 3 dimensions: *default*, x and y. The immediate adjacency sequence along dimension *default* is s_1-s_x-s_2-s_3; along dimension x it is s_1-s_4 and along dimension y it is s_2-s_5.

Example 2. Let us consider the following description:

$$\texttt{a}(E_1,T), \ \texttt{tl}(E_1), \ \texttt{hl}(E_1), \ \texttt{a}(E_2,U), \ \texttt{tm}(E_2), \ \texttt{hl}(E_2), \ \texttt{ff}(E_2),$$
$$\texttt{gg}(E_2), \ \texttt{a}(E_3,T), \ \texttt{tm}(E_3), \ \texttt{hm}(E_3), \ \texttt{a}(E_4,V), \ \texttt{tl}(E_4), \ \texttt{hh}(E_4),$$
$$\texttt{wm}(E_4), \ \texttt{a}(E_5,V), \ \texttt{th}(E_5), \ \texttt{hl}(E_5), \ \texttt{c}(E_1,E_3), \ \texttt{c}(E_2,E_1),$$
$$\texttt{d}(U,V,W), \texttt{d}(V,T), \ \texttt{after}(E_1,E_2,default), \ \texttt{after}(E_2,E_3,default),$$
$$\texttt{after}(E_1,E_4,x), \ \texttt{after}(E_2,E_5,y)$$

It is a model, since terms denoting events and objects are variables and sequential information is expressed by **after** atoms. It involves 5 events $(E_1, E_2, E_3, E_4, E_5)$ along 3 dimensions: *default*, x and y. The adjacency sequence along dimension *default* is E_1-E_2-E_3; along dimension x it is E_1-E_4 and along dimension y it is E_2-E_5. Event E_1 has properties **tl** and **hl**, it has relationship **a/2** with object T and is involved in non-sequential relationship **c/2** to event E_3 and from event E_2. Objects V and T are connected by relationship **d/2**, and so on.

To handle particular domains, sequentiality is not restricted to be a linear relationship: i.e., many events may follow a given event along the same dimension. This is to be interpreted as a parallelism, or as a kind of independence, among these events. Thus, rather than being simple 'strings' of atoms, our representations induce a Directed Acyclic Graph for each dimension. For instance, this is useful when describing process models, where parallel activities may take place. Also, note that relationships can be expressed among events referring to different dimensions, which establish interactions among the dimensions and hence ensure full multidimensionality handling.

3   Generalization

Consider two clauses, C' and C'', involving sequential information according to the formalism defined in the previous section. Clearly, to be properly handled, sequential information needs to be suitably interpreted according to a background knowledge expressing transitivity of sequential relationships, and able to handle ranges for the allowed (minimum and maximum) number of intermediate steps between two events. Unfortunately, the purely logical setting ignores such a background knowledge, ensuring that two descriptions match only if they have exactly the same number of steps between two events. On the other hand, keeping the purely logical setting is desirable since it ensures general applicability of the logical representation and inference techniques.

A naive algorithm for generalizing two descriptions involving sequential predicates might be the following:

1. set an association between a subset of events in the former description and corresponding events in the latter
2. generalize the corresponding sequential relationships
3. generalize the rest of the descriptions consistently with the result of (2)

Since one is usually interested in specific generalizations fulfilling particular properties (e.g., being the least general ones), an optimization problem is cast where all possible such generalizations must be computed for identifying the best one. So, let us estimate the problem complexity. Step 1 introduces indeterminacy; step 2 is deterministic after step 1 has established an association of events; step 3 causes additional indeterminacy depending on the specific properties and relationships at hand.

The problem is that step 1 alone introduces a significant amount of indeterminacy, as shown in the following. Consider two sequences of events $S' = \langle s'_i \rangle_{i=1,\ldots,n}$ and $S'' = \langle s''_j \rangle_{j=1,\ldots,m}$, and assume (without loss of generality) that $n \leq m$. A generalization might associate any subset of events in the shorter sequence with events in the longer one. Let us assume that these associations must be injective. There are $2^n - 1$ such subsets (excluding the empty one), and specifically there are $\binom{n}{k}$ thereof of size k, for each $k = 1, \ldots, n$. Given a subset of size k, there are $\binom{m}{k}$ corresponding k-tuples of events in S'' that can be associated to them by the generalization, assuming that the original sequences are to be preserved by the generalization (i.e., that given two events s'_i, s'_j in S', with $i < j$, to be associated to events s''_h, s''_k, with $h < k$, s'_i must be associated to s''_h and s'_j must be associated to s''_k). Thus, overall there are $\sum_{k=1}^{n} \binom{n}{k} \cdot \binom{m}{k}$ possible associations to be checked. It is clearly unpractical checking all of them for selecting the best one. For instance, given $n = 7$ and $m = 10$, there are 19447 possible associations of the steps in S' with those in S''.

We propose to use a heuristic for the first step, that directly selects a single, most promising association to be exploited as a base for steps 2 and 3. Our starting point is the intuition that two events should be associated if they are similar to each other, and that the best association should obtain the highest possible overall similarity among all the possible associations. So, we need first of all a description for each event. The simplest such description (let us call it 0-level description) selects, for each event, all non-sequential literals in the description that include that event in their arguments, and possibly other non-event arguments. Then, i-level descriptions, for $i \geq 1$, are recursively obtained from $(i - 1)$-level descriptions by including all literals that have at least one argument in common with those in the $(i - 1)$-level description. Clearly, larger values of i provide more precise descriptions (and hence more effective learned models), but require more computational effort (making less efficient the learning step). Given i-level descriptions $d_i(s')$ and $d_i(s'')$ for two events, after building two corresponding clauses $E' = \text{dummy}(s')$:- $d_i(s')$ and $E'' = \text{dummy}(s'')$:- $d_i(s'')$ it is possible to apply the similarity measure $\text{fs}(E', E'')$ for Datalog clauses proposed in [8]. The similarity values are obviously affected by the choice of i.

The overall association is finally obtained using a greedy approach: the similarity of all pairs of i-level descriptions of events, one from each clause, is computed; the pairs are ranked by decreasing similarity, and the rank is scanned top-down, starting from the empty generalization and progressively extending it by adding the generalization of the descriptions of each pair whose association (involving both events and other objects) is compatible with the cumulative association of the generalization computed so far.

Example 3. Let us generalize the clauses C' and C'' having as heads, respectively, $\text{h}(E_1)$ and $\text{h}(s_1)$, and as bodies, respectively, the descriptions in Examples 2 and 1. Considering 0-level descriptions for events, we have:

- $\text{h}(E_1)$:- $\text{a}(E_1,T)$, $\text{tl}(E_1)$, $\text{hl}(E_1)$. for event E_1
- $\text{h}(s_1)$:- $\text{a}(s_1,t)$, $\text{tl}(s_1)$, $\text{hl}(s_1)$. for event s_1

...and so on for the other steps. Note that $\text{h}(s_1)$ does not contain $\text{c}(s_1, s_3)$ nor $\text{c}(s_1,s_2)$ because no other events than s_1 are allowed at the 0-level (and the same for $\text{h}(E_1)$). Since the arguments in the head must necessarily match, the base association $\{s_1/E_1\}$ is set. Generalizing the above two clauses, this also binds T to t, yielding $\{s_1/E_1, t/T\}$. Then, the similarity function applied to the remaining pairs of terms returns the following ranking (by decreasing order):

$\text{fs}(E_3, s_3) = -2.59166666666667$
$\text{fs}(E_3, s_4) = -2.46666666666667$
$\text{fs}(E_3, s_5) = -2.46666666666667$
$\text{fs}(E_3, s_2) = -2.45505952380952$
$\text{fs}(E_2, s_2) = -2.42708333333333$
$\text{fs}(E_4, s_4) = -2.41666666666667$
$\text{fs}(E_2, s_3) = -2.38214285714286$
...

Thus, E_3 is associated to s_3, which excludes all the next associations involving either of these terms. The next valid association is s_2/E_2, which again excludes all subsequent associations involving either of these terms. Proceeding in this way, the next selected associations are s_4/E_4 and s_5/E_5, after which no further valid association is found.

Once the pairs of associated events in the two clauses have been determined, the corresponding sequential predicates are generalized. First, we simplify the sequential descriptions. For each pair of events in the same description, such that no other selected event falls in between them, a single 'compound' after atom is created.

For instance, given the association $\{S'/s', S''/s''\}$, and the sequential chains $\{\text{after}(S',S,d), \text{after}(S,S'',d)\} \subseteq C'$ and $\{\text{after}(s',s_1,d), \text{after}(s_1,s_2,d), \text{after}(s_2,s'',d)\} \subseteq C''$, such that S, s_1 and s_2 were not selected by any association, these atoms would be replaced by $\text{after}(S', S'', d) \in C'$ and by $\text{after}(s', s'', d) \in C''$.

Then, these simplified atoms can be generalized.

Finally, two clauses are created, each having as a head a dummy atom having as arguments all the associated terms (events and objects) in the current partial generalization, and as a body all non-sequential literals not used in the event generalization step. These two clauses are generalized using standard (non-sequential) algorithms (e.g., the one proposed in [8]). The heads ensure that all the information concerning associated events is fixed, and the generalization is only in charge of finding the best mapping among the remaining literals. This deals with the second kind of indeterminacy of step 3 in the algorithm.

For instance, if the associations fixed in step 1 are:
$\{s_1/E_1, t/T, s_3/E_3, s_2/E_2, u/U, s_4/E_4, s_5/E_5, v/V\}$,   then the dummy heads would be $\text{dummy}(E_1, T, E_3, E_2, U, E_4, E_5, V)$ and $\text{dummy}(s_1, t, s_3, s_2, u, s_4, s_5, v)$.

4   Coverage

Another crucial issue is how to check whether a (sequential) model covers a (sequential) observation. We evaluated three possible alternatives:

1. associating the sequential information first, and then completing the coverage check with the contextual information: the advantage is that the useless cost of context coverage is avoided for event associations that do not comply with the sequences; the disadvantage is that many choice points might be available about the possible event associations
2. associating the contextual information first, and then completing the coverage with the sequential information: the advantage is that there should be fewer possible ways of covering an observation (compared to all possible event associations), among which choosing one that also fulfills the sequential constraints; the disadvantage is that event associations that are not compliant with the sequential constraints might be generated, and these wrong associations must be later identified and filtered out
3. interleaving the association of sequential and contextual information for each event (defining a suitable boundary in breadth for the contextual description) and subsequently completing the check with the contextual information not yet considered for the single events: advantages and disadvantages are a trade-off among the previous cases

Solution 2 is more immediate, because it can exploit existing (efficient) coverage procedures for non-sequential representations and simplifies the sequence check. Although this check is a potential cause of inefficiency, the amount of indeterminacy it brings should be less than for the other options. The corresponding algorithm is as follows:

1. apply preliminary coverage check to the contextual and cross-event information, obtaining a covering association A also including event bindings

2. complete the coverage check: For all sequence atoms a in the model:
   (a) identify the corresponding initial and final events in the observation using the fixed event association A
   (b) perform sequence checking according to the following algorithm; if this check fails, backtrack on (1)

As regards sequence checking, we need to find all paths between two events in the directed graph induced by the sequence, where nodes are events and edges connect events between which a sequence atom is present in the description. To simplify the complexity of this operation, we proceed as follows. Let us consider the set I of sequence atoms in the observation concerning a fixed dimension.

1. create the sequentiality graph G induced by I
2. compute the topological sort T of G (i.e., the list of nodes in I in which a node u appears after a node v if there exists a path from v to u in G)
3. Among all associations A between events in the model and events in the observation, obtained by the previous algorithm, find at least one for which the sequential part is covered, to be checked as follows:
   (a) For all sequence atoms in the model, $L = \mathtt{after(X,Y,D)}$:
       i. locate events s and t associated to X and Y, respectively, by A
       ii. extract from T the sublist $S = [s, ..., t]$: if $S = []$ (i.e., t does not follow s in T) then fail
       iii. driven by the sequence of events in S, check whether in I there exists a 'simple' chain (i.e., not involving events that have already been bound by A) of sequence atoms that lead from s to t; if there is no such a chain, then fail

5   Evaluation

We evaluated our algorithms in the Process Mining domain, where the sequential relationships among activities are fundamental to correctly capture the process model and, specifically, conditions on activity execution. Here we will refer to WoMan [10], an incremental process mining algorithm in which the proposed operators were embedded, and to its application to learning user's daily routines in a Smart Environment domain [9], for predicting his needs and comparing the actual situation with the expected one.

5.1   Process Mining and WoMan

A *process* is a sequence of *events* associated to activities performed by agents [5]. A *workflow* is a (formal) specification of how a set of tasks can be composed to result in valid processes [18]. An *activity* is the actual execution of a task. A *case* is a particular execution of activities in a specific order compliant to a given workflow, along an ordered set of *steps* (time points) [12]. Workflows can be modeled as directed graphs where nodes are associated to tasks/activities, and edges represent the potential flow of control among activities. Edges (i.e., tasks) can be labeled with pre-conditions on the state of the process, which determine whether they will be traversed (i.e., executed) or not [1]. *Process Mining* [19]) aims at inferring workflow models from examples of cases. WoMan (short for *Workflow Manager*) is an ILP Process Mining method that provides the expressiveness of FOL to describe both cases and their contextual information in a unified framework. While details of WoMan and its comparison with other process mining system (also some working in the ILP framework) are provided in [10], here we will quickly recall its representation formalism.

WoMan descriptions of cases are based on two predicates:

activity(S,T) : at step S task T is executed;
next(S',S'') : step S'' follows step S'.

where the vocabulary of activities is the (fixed and context-dependent) set of constants representing the allowed tasks, and each step is denoted by a unique identifier. Cases are expressed as conjunctions of ground atoms built on these predicates. For instance, in a hypothetical daily-routine 'morning' workflow we might have:

activity$(s_0,$wake_up$)$, next(s_0,s_1), activity$(s_1,$toilet$)$, next(s_0,s_2), activity$(s_2,$radio$)$, next(s_2,s_3), activity$(s_3,$book$)$, next(s_1,s_4), next(s_3,s_4), activity$(s_4,$wth$)$, next(s_4,s_5), activity$(s_5,$weight$)$, activity$(s_7,$tea$)$, next(s_5,s_6), activity$(s_6,$dress$)$, next(s_6,s_7), next(s_6,s_8), activity$(s_8,$tv$)$, next(s_7,s_9), next(s_8,s_9), activity$(s_9,$door$)$

This formalism provides an explicit representation of parallel executions in the task flow (e.g., 'toilet' with 'radio' and 'book') and allows to smoothly add further information and relationships concerning the steps and tasks and the context in which they appear, using domain-dependent predicates. For instance, the previous 'morning' workflow case description might be extended as follows:

early(s_0), happy(s_0), early(s_1), long_duration(s_1), early(s_2), news(s_2,n'), about(n',s), sports(s), bad(n'), upset(s_2), early(s_3), calm(s_4), early(s_4), windy(s_4), early(s_5), short_duration(s_5), short_interval(s_5,s_6), on_time(s_6), pullover(s_6), long_duration(s_6), late(s_7), late(s_8), news(s_8,n''), about(n'',s), updates(n'',n'), interesting(n''), late(s_9)

This naturally overcomes the limitation of using propositional conditions for edges in the model, as typical in the Process Mining literature. So, while learning the workflow structure for a given case, examples for learning task pre-conditions are generated as well, and provided to a learning system. For each activity(s,t) atom in the case description an example for learning pre-conditions for task t is created using the subset of atoms in the description associated to steps up to s only. Indeed, when applying the workflow model, pre-conditions for performing a task at a given time can only be checked against the events that took place up to that moment. In our example, for activity$(s_3,$book$)$ we would have:

book(s_3) :- activity$(s_0,$wake_up$)$, next(s_0,s_1), activity$(s_1,$toilet$)$, next(s_0,s_2), activity$(s_2,$radio$)$, next(s_2,s_3), activity$(s_3,$book$)$, early(s_0), happy(s_0), early(s_1), long_duration(s_1), early(s_2), news(s_2,n'), about(n',s), sports(s), bad(n'), upset(s_2), early(s_3).

For compliance with the incrementality of the proposed approach to learning the workflow structure, this learning system must be incremental as well. A suitable learner is InTheLEx [7], that is also endowed with a positive-only-learning feature [3] (useful because only examples of workflow cases actually carried out are typically available). This example, together with others, might lead to infer preconditions such as:

```
book(Y) :- activity(X,radio), news(X,N), bad(N), upset(X),
    next(X,Y), activity(Y,book), early(Y).
```
(in order to read a book, the actor must be upset because of having heard bad news on the radio, and it must be early)

It is clear that, in this setting, being able to express pre-conditions based on the whole flow of contexts and events that took place in the past is necessary to fully capture complex cases.

5.2 Smart Environment Dataset

The proposed approach was tested on a real-world dataset taken from the CASAS repository (http://ailab.wsu.edu/casas/datasets.html), concerning daily activities of people living in cities all over the world. In particular, we selected the Aruba dataset, involving an elderly person visited from time to time by her children. While in this dataset sequentiality is limited to one dimension only, we considered it a good testbed because the experimenter can better check the basic algorithm's behavior than in more complex (multidimensional) cases. The Aruba dataset reports data concerning 220 days, represented as a sequence of timestamped sensor data, some of which annotated with a label indicating the beginning or end of a meaningful activity. We obtained a set of cases by splitting this dataset into daily cases according to the following logic: a new day started at the first activity after sleeping that is not followed by a new sleeping activity after just one intermediate activity (e.g., going to the toilet). We collected a context description for each task execution in each case, and filled the case descriptions with this information. Then, 2-level descriptions (a trade-off between effectiveness and efficiency) of positive examples to learn preconditions were obtained for each task, i.e. regularities in context that were present in all executions of each task. 5976 examples were obtained.

Figure 1 (top) shows the rules learned using the proposed procedure for two sample activities: 'work' and 'wash_dishes'. A comparison with the corresponding theory learned by InTheLEx on the same examples without using sequential information (shown in the bottom part of Figure 1), suggests the following:

- both theories singled out the set of sensors that directly determine the execution of the target activities: {sensor_m018} for washing dishes, and {sensor_m026} for work, where the status must be 'on'
- preconditions learned by the proposed method use generalized temporal links (after predicate) to express the temporal sequence that underlies specific activities, while those learned by InTheLEx are limited, as expected, to specific sequences of activities connected by next predicates only

```
wash_dishes(A) :- after(B,A,default), activity(B,meal_preparation),
    after(D,B,default), activity(D,sleeping), sensor_m006(B,E),
    status_off(E), sensor_m007(B,F), status_off(F), sensor_m008(B,_),
    sensor_m018(B,G), status_on(G), sensor_m020(B,_).
work(A) :- after(B,A,default), activity(B,meal_preparation),
    sensor_m018(B,D), status_on(D), sensor_m026(A,E), status_on(E).
```

```
wash_dishes(A) :- next(B,A), next(_,B).
work(A) :- sensor_m026(A,C), status_on(C).
```

Fig. 1. Preconditions for some kinds of activities, learned using the sequence-aware technique (top) or the base InTheLEx system (bottom) on the full dataset.

Clearly, the sequence-aware preconditions are more specific, which avoids over-generalization and provides the reader with better insight in the activity. Having to learn from positive examples only, this specificity is desirable.

The ARUBA dataset does not provide for cross-relationships among contextual objects (they are all separate and independent sensors), nor across events. However, by interpreting the above rules implicit relationships can be identified: e.g., movement sensor m026 is in the study, and m018 is in the kitchen, which is coherent with the kind of activity the preconditions refer to.

Noteworthily, the fact that after so many examples the preconditions do not degenerate to be empty is encouraging and confirms that the system is able to find regularities when they actually exist. Indeed, for other activities (e.g., 'relax'), both methods returned an empty precondition. This suggests that these activities may indeed take place at any moment, independently of the sequence of events. Also, the presence of sequence atoms in the preconditions learned by the sequence-aware procedure confirms that they help to improve the quality of the outcome, compared to the plain InTheLEx outcome on the same data.

As to the quantitative evaluation, we adopted a 10-fold cross-validation approach to assess the predictive accuracy of the learned preconditions[1]. Due to the positive-examples-only setting, accuracy is determined as the number of covered test examples over the overall number of test examples. No assumption can be made for negative examples, because it might well be that the same precondition holds for different activities. So, they are not mutually exclusive and positive examples for an activity cannot be automatically considered as negative for the others. Of course, in this setting, more general rules will tend to cover more cases, and having an empty precondition for a task would trivially reach 100% accuracy on that task. Thus, for the only purpose of evaluating the system's accuracy under more stressing conditions, the learned models were also tested on non-trivial cases only, by removing all the preconditions in the model and the test examples corresponding to trivial activities (i.e., those for which empty preconditions were learned). To stress the procedure even more, the coverage

---

[1] Note that *soundness* is inappropriate here, because we are evaluating the preconditions, not the workflow model.

test was skipped during the learning phase: while saving time, this might cause useless generalizations (because the implemented generalization is actually an approximation of the theoretical least general one).

The average cross-validation accuracy was 99.86% on the whole dataset, and 98.25% on the non-trivial portion only. This shows that the system is actually able to induce significant preconditions even using a huge number of positive examples only, and without exploiting any bias to avoid overgeneralization. In the same way, filtering out trivial cases lowers accuracy, but with very limited impact. The sequence-aware preconditions for all activities in the dataset were learned in less than half a second per example on average. This ensures that our method can be applied on-line to the given environment, which is useful because the model can be continuous kept up-to-date without causing unnecessary delays in the normal activities of the system or of the involved people.

6 Conclusions

Horn clause Logic is a powerful representation language for automated learning and reasoning in domains where relations among objects must be expressed to fully capture the relevant information. Often uninterpreted predicates cannot fully account for the complexity of some domains. Among others, outstanding examples are those of numeric, taxonomic or sequential information. This paper deals with the case of (multidimensional) sequential information, where a number of points (events) in several dimensions (time, space, etc.) are connected by sequential predicates, whose transitive closure is to be implicitly assumed.

Coverage and generalization techniques are devised and presented, and their integration in an incremental ILP system is used to run experiments showing its performance in the Daily Routines domain. Future work will concern deeper empirical evaluation of the behavior of the proposed approach also in multidimensional settings, and the development of a specialization operator for non-monotonic inductive inference.

Acknowledgments. Thanks to Nicola Di Mauro for his suggestions, and Antonio Vergari for the useful discussions and for running some experiments. This work was partially funded by the Italian PON 2007-2013 project PONO2_00563_3489339 'Puglia@Service'.

References

[1] Agrawal, R., Gunopulos, D., Leymann, F.: Mining process models from workflow logs. In: Schek, H.-J., Saltor, F., Ramos, I., Alonso, G. (eds.) EDBT 1998. LNCS, vol. 1377, pp. 469–483. Springer, Heidelberg (1998)

[2] Anderson, C.R., Domingos, P., Weld, D.S.: Relational markov models and their application to adaptive web navigation. In: Proceedings of the Eighth ACM SIGKDD International Conference on Knowledge Discovery and Data Mining (KDD 2002), pp. 143–152. ACM Press (2002)

[3] Bombini, G., Di Mauro, N., Esposito, F., Ferilli, S.: Incremental learning from positive examples. In: Atti del 24-esimo Convegno Italiano di Logica Computazionale 2009, CILC 2009 (2009)

[4] Ceri, S., Gottlöb, G., Tanca, L.: Logic Programming and Databases. Springer, Heidelberg (1990)

[5] Cook, J.E., Wolf, A.L.: Discovering models of software processes from event-based data. Technical Report CU-CS-819-96, Department of Computer Science, University of Colorado (1996)

[6] Esposito, F., Di Mauro, N., Basile, T.M.A., Ferilli, S.: Multi-dimensional relational sequence mining. Fundamenta Informaticae 89(1), 23–43 (2008)

[7] Esposito, F., Semeraro, G., Fanizzi, N., Ferilli, S.: Multistrategy theory revision: Induction and abduction in inthelex. Machine Learning 38(1/2), 133–156 (2000)

[8] Ferilli, S., Basile, T.M.A., Biba, M., Di Mauro, N., Esposito, F.: A general similarity framework for horn clause logic. Fundamenta Informaticæ 90(1-2), 43–46 (2009)

[9] Ferilli, S., De Carolis, B., Redavid, D.: Logic-based incremental process mining in smart environments. In: Ali, M., Bosse, T., Hindriks, K.V., Hoogendoorn, M., Jonker, C.M., Treur, J. (eds.) IEA/AIE 2013. LNCS, vol. 7906, pp. 392–401. Springer, Heidelberg (2013)

[10] Ferilli, S., Esposito, F.: A logic framework for incremental learning of process models. Fundamenta Informaticae (to appear)

[11] Gutmann, B., Kersting, K.: TildeCRF: Conditional random fields for logical sequences. In: Fürnkranz, J., Scheffer, T., Spiliopoulou, M. (eds.) ECML 2006. LNCS (LNAI), vol. 4212, pp. 174–185. Springer, Heidelberg (2006)

[12] Herbst, J., Karagiannis, D.: An inductive approach to the acquisition and adaptation of workflow models. In: Proceedings of the IJCAI 1999 Workshop on Intelligent Workflow and Process Management: The New Frontier for AI in Business, pp. 52–57 (1999)

[13] Jacobs, N.: Relational sequence learning and user modelling. PhD thesis, K.U.Leuven, Department of Computer Science, Faculty of Science (2004)

[14] Kersting, K., De Raedt, L., Gutmann, B., Karwath, A., Landwehr, N.: Relational sequence learning. In: De Raedt, L., Frasconi, P., Kersting, K., Muggleton, S.H. (eds.) Probabilistic ILP 2007. LNCS (LNAI), vol. 4911, pp. 28–55. Springer, Heidelberg (2008)

[15] Kersting, K., Raiko, T., Kramer, S., De Raedt, L.: Towards discovering structural signatures of protein folds based on logical hidden markov models. In: Pac. Symp. Biocomput., pp. 192–203 (2003)

[16] Lee, S.D., De Raedt, L.: Constraint based mining of first order sequences in seqLog. In: Meo, R., Lanzi, P.L., Klemettinen, M. (eds.) Database Support for Data Mining Applications. LNCS (LNAI), vol. 2682, pp. 154–173. Springer, Heidelberg (2004)

[17] Muggleton, S.: Inductive logic programming. New Generation Computing 8(4), 295–318 (1991)

[18] van der Aalst, W.M.P.: The application of petri nets to workflow management. The Journal of Circuits, Systems and Computers 8, 21–66 (1998)

[19] Weijters, A.J.M.M., van der Aalst, W.M.P.: Rediscovering workflow models from event-based data. In: Proceedings of the 11th Dutch-Belgian Conference of Machine Learning (Benelearn 2001), pp. 93–100 (2001)

How Mature Is the Field of Machine Learning?

Marcello Pelillo and Teresa Scantamburlo

Università Ca' Foscari Venezia
{pelillo,scantamburlo}@dsi.unive.it

Abstract. We propose to address the question whether the fields of machine learning and pattern recognition have achieved the level of maturity in the sense suggested by Thomas Kuhn. This is inextricably tied to the notion of a paradigm, one of the cornerstones of twentieth-century philosophy of science, which however is notoriously ambiguous, and we shall argue that the answer to our inquiry does depend on the specific interpretation chosen. Here we shall focus on a "broad" interpretation of the term, which implies a profound commitment to a set of beliefs and values. Our motivating question can in fact be seen simply as an excuse to analyze the current status of the machine learning field using Kuhn's image of scientific progress, and to discuss the philosophical underpinnings of much of contemporary machine learning research.

Keywords: Machine learning, philosophical foundations, philosophy of science.

1 Introduction

According to Thomas Kuhn, the "acquisition of a paradigm and of the more esoteric type of research it permits is a sign of maturity in the development of any given scientific field" [18, p. 11]. In this paper, we propose to address the question whether the fields of machine learning and pattern recognition have achieved the level of maturity in the sense implied by the quotation above.[1]

Note that Kuhn's notion is quite different from (and indeed more profound than) the commonsensical view which maintains that "mature scientific disciplines are expected to develop experimental methodologies, comparative evaluation techniques, and theory that is based on realistic assumptions" [16, p. 112]. Under this interpretation, one would be tempted to respond with an emphatic "yes" to the question posed above. Indeed, in the last 25 years or so researchers have dramatically changed their attitude to the evaluation of new algorithms and techniques, and it seems that Langley's well-known incitement to make machine learning an "experimental science" [22] has been taken seriously by the community. Since its birth in the late 1990's, for example, the UCI ML repository keeps

---

[1] A note on terminology: although throughout the paper we consistently use the term "machine learning," we in fact refer to *both* the fields of machine learning and pattern recognition both being, as Bishop pointed out [3], simply two facets of the same field (but see the discussion in [9] for a different perspective).

M. Baldoni et al. (Eds.): AI*IA 2013, LNAI 8249, pp. 121–132, 2013.
© Springer International Publishing Switzerland 2013

growing at a fast pace and at the time of writing it contains 244 different data sets on the most disparate problems and applications. On the other hand, there is an increasing level of sophistication in the way in which the performance of the algorithms are quantitatively evaluated, and we saw an evolution from simple scalar performance measures such as the classification accuracy to more elaborated ones such as ROC curves and statistical tests. However, a deeper analysis reveals that the situation is more controversial than it appears, as there is more to science than simply experimental analysis, and the equation "scientific = experimental" is too naive to satisfactorily capture the multifaceted nature of the "scientific method" (granted that there exists one [11]).

In this paper, however, we do not intend to enter into this discussion, but would like to attack the question from a purely Kuhnian perspective according to which, as anticipated above, the notion of maturity in science is inextricably tied to the concept of a "paradigm," one of the cornerstones of twentieth-century philosophy of science. In one sense, our motivating question can in fact be interpreted simply as an excuse to analyze the current status of the machine learning and pattern recognition fields using the conceptual tools provided by Kuhn. Note that under this interpretation, there is no pretense of judging the "scientificity" of a given research area or to provide a demarcation, à la Popper, between scientific and non-scientific fields. Indeed, using Kuhn's suggestion, Aristotelian physics, which dominated the scene for over two millennia, has to be considered as mature as today's physics although of course, according to the modern interpretation of the term, we would not dream of calling it a "science." [2]

The publication of Kuhn's *Structure of Scientific Revolutions* in 1962 was a momentous event in the modern history of ideas. It provoked itself a revolution in the way we think at science whose far-reaching effects are felt in virtually all academic as well as popular circles. What made Kuhn's image particularly successful in describing the nature of scientific progress is, no doubt, his notion of a paradigm. Unfortunately, the reception of the term by the philosophical and scientific communities was controversial, and a number of difficulties persuaded Kuhn to clarify his position in his famous 1969 *Postscript* [18]. Indeed, as he himself admitted, the term was used in a vague and ambiguous way throughout the book but, besides minor stylistic variations, he identified two very different usages of the term. On the one hand, he aimed to describe some accepted examples which serve as a model for the solution of new puzzles (the "narrow" sense), whereas, on the other hand, he meant a more profound commitment to a set of beliefs and values (the "broad" sense).

In this paper, we aim to approach the question posed in the title by exploiting both interpretations of the concept, and the discussion will make it clear that the answer depends on which sense one considers. Note that Cristianini [7] has recently undertaken a study similar in spirit to ours, but he seems to have emphasized mostly the first, narrow, sense of the term. In contrast, we shall focus more on the broad interpretation. This will give us the opportunity to discuss

---

[2] See Kuhn's autobiographical fragment contained in [19] for a rehabilitation of Aristotle as a physicist.

the philosophical (often tacit) assumptions underlying much of contemporary machine learning research and to undertake a critical reflection of its current status. In particular, we will see how deep is the bond with essentialism, one of the oldest and most powerful ideas in the whole history of philosophy, and we shall maintain that the community is gently moving away from it, a phenomenon which, we shall speculate, seems to parallel the rejection of the essentialist hypothesis by modern science.

2   Machine Learning between Science and Engineering

Before undertaking our exploration we need to motivate the pertinence of Kuhn's contribution within the context of machine learning. This is required by the fact that Kuhn's investigation is essentially directed towards well-established sciences such as physics or chemistry. Thus, it would seem sensible to ask whether machine learning is an appropriate subject of study under a Kuhnian approach.

We feel the question of particular relevance in view of the fact that the context in which machine learning research has grown up is interdisciplinary in a profound way. Such a distinct character of the evolution of the field has not always been recognized by researchers so that the aspects involved are usually seen in isolation. This resulted in the common reductionist tendency to conceive the fields of machine learning and pattern recognition as either engineering or science. Some scholars, indeed, assume that, as well as providing technical solutions, the fields of pattern recognition and machine learning deal with fundamental questions pertaining to categorization, abstraction, generalization, induction, etc., and, in so doing, their contribution is in fact scientific [10,42]. In some cases, the approach of machine learning has been associated even to the scientific practice of physics [38] or, more generally, to experimental sciences [22]. Conversely, nowadays it prevails the idea that these areas are primarily engineering disciplines. For example, Pavlidis, one of the pioneers of the field, recalling Hermann Hesse's novel *Das Glassperlenspiel* claims that "the prospects are bright if we approach pattern recognition as an engineering problem and try to solve important special cases while staying away from the Glassperlenspiel. The prospects are grim if we keep looking for silver bullets that will solve "wholesale" a large range of general problems, especially if we harbor the illusion of doing things the way the human brain does"[27, p. 7]. And this idea has been more recently echoed by von Luxburg et al. [41].

This sharp opposition between science and technology stems from an oversimplified view of their mutual relationship. However, in the light of some new achievements in the philosophy of technology (see, e.g., [12]), it turns out that, granted that there are indeed important differences, at the conceptual level the boundary between the two camps is more blurred than is commonly thought, and that they stand to each other in a kind of circular, symbiotic relationship. Indeed, technology can be considered as an activity producing new knowledge on a par with ordinary science. The so called operative theories [5] in technology look like those of science and their contribution goes beyond the mere application of scientific knowledge. On the other hand, even science can be brought closer to technology when its progress is expressed in terms of "immanent goals." This idea lies at the heart of Laudan's problem-solving approach to science [23] and could well characterize much of the work in the field of machine learning.

The profound interaction between scientific and technological components is a key to understand the machine learning activity and other research areas within artificial intelligence. The history of the field, in fact, counts numerous examples of this fecund relationship. The case of neural networks is particularly significant, as their original formulation had a genuine scientific motivation, that is, the wish of studying and imitating the brain but, in the phase of their renaissance, technical matters prevailed. Indeed, with the (re)invention of the back-propagation algorithm for multi-layer neural networks and, above all, thanks to the impressive results obtained by these new models on practical problems such as zip code recognition and speech synthesis, a new wave of excitement spread across the artificial intelligence community. At that point, however, it was already clear that these models had no pretense of being biologically plausible [6]. Pavlidis nicely summed up this state of affairs by noting that "the neural networks that have been in vogue during the last 15 years may be interesting computational devices but they are not models of the brain. (Except maybe of the brains of people who make that claim sincerely)" [27, p. 2]. Bayesianism is another interesting example of the gate allowing machine learning to move from theoretical issues to more practical aims. Introduced as a theory which can characterize the strength of an agent's belief, it provided many inference algorithms with a practical machinery. On the other hand, recent advances in density estimation techniques, such as nonparametric Bayesian methods, have been successfully applied to approach a variety of cognitive processes [36].

To sum up, the recent contributions of philosophy of technology and of the philosophy of science lead us to rethink the classical dichotomy between science and technology, which is still holding in some subfields of artificial intelligence, as they appear closer than we used to think. Historical examples suggest that machine learning and pattern recognition work, indeed, as a bridge between the two and many ideas from science result in technological innovation and vice versa [30]. In reference to our discussion, this means that a contribution from philosophy of science should not be considered irrelevant for these two fields since the scientific side is as much important as the technological one. Accordingly, we do think that Kuhn's analysis is not only appropriate to the machine learning research but could also contribute to get a deeper understanding of its nature.

3   Kuhn's Notion of a Paradigm

In his *Structure* (as the book is known) [18], Kuhn provides an account of scientific development that is dramatically different from the standard idea of a steady, cumulative progress. According to him, a science traverses several discontinuities alternating "normal" and "revolutionary" phases. During normal periods the development of a science is driven by adherence to a "paradigm" whose function is to support scientists in their "puzzle-solving" activity with a number of practical and theoretical tools, including theories, values and metaphysical assumptions.

When some worrying puzzles remain unsolved (the so-called "anomalies") and the current approach loses progressively its original appeal, a discipline enters a period of crisis. At this point, the activity is characterized by "a proliferation of competing articulations, the willingness to try anything, the expression of explicit discontent, the recourse to philosophy and to debate over fundamentals" [18, p. 91]. Finally, the crisis is resolved by a scientific revolution leading to the replacement of the current paradigm by a new one. The revolution results in a paradigm shift, after which a discipline returns to a normal phase, based this time on a new accepted framework.

As we have seen before, there are two distinct uses of the notion of a paradigm. At first, Kuhn uses the term "paradigm" to refer to some concrete achievements that can work as models or examples and supply explicit rules for the solution of the remaining puzzles. In the history of science examples of this notion abound and include, e.g., Newton's mechanics and Franklin's theory of electricity, which implicitly defined the legitimate problems and methods of a research field for succeeding generations of practitioners. A second way to apply the term "paradigm" refers to a more global sense and includes, above all, concepts, theoretical principles, metaphysical assumptions, worldviews, etc.

In his *Postscript*, Kuhn introduced the idea of a broad paradigm in terms of a "disciplinary matrix", which could be seen as a theoretical and methodological framework wherein scientists conduct their research. This framework includes the basic assumptions of a discipline providing a community with the practical and theoretical indications, for instance, about how to lead investigations or what to expect from experiments. Among the elements which compose this matrix, we aim to focus on symbolic generalizations and metaphysical paradigms.

A research community could easily present formal expressions or codified terms that could live in the acceptance of all members for several years. These are what Kuhn calls "symbolic generalizations" and their function goes basically in two directions. That is, they can work as laws of nature, such as $\mathbf{F} = m\mathbf{a}$, or "elements combine in constant proportion by weight" [18, p. 183], but they can also serve to settle some fundamental definitions assigning symbols to specific meanings. Note that, according to Kuhn "all revolutions involve, among other things, the abandonment of generalizations the force of which had previously been in some part that tautologies" [18, p. 184].

A second type of component is given by the metaphysical parts of a paradigm. Metaphysical elements can be beliefs or models and incorporate tacit or implicit knowledge. In practice these components shape the general disposition and the methodological attitude of a scientist suggesting particular metaphor or worldviews. The strength of such components is that of determining what will be accepted as as an explanation and, above all, the importance of unsolved puzzles.

4   Paradigms in Machine Learning: The Broad Perspective

Forms of narrow paradigms can be easily found in machine learning research. The evolution of the field, indeed, is a story of great achievements that were able to create strong traditions around them. An obvious example is provided by neural networks which played a key role in the early as well as later developments of the field. More recent examples include, e.g., kernel methods and spectral clustering. Some of these success stories are collected in [7], which nicely describes the transition from the "knowledge-driven" to the "learning-driven" paradigm in artificial intelligence.

4.1   The Disciplinary Matrix

Here we would like instead to focus on the broad sense of the notion of a paradigm and see whether the contours of a disciplinary matrix come up through the concrete practice of the discipline within the community. Hence, with Kuhn, we could ask: "what do its members share that accounts for the relative fulness of their professional communication and the relative unanimity of their professional judgments?"[18, p. 182]. To address this issue we will consider the components presented above: the symbolic generalization and the metaphysical paradigm.

First, note that the majority of traditional machine learning techniques are centered around the notion of "feature"[8,3]. Indeed, within the field there is a widespread tendency to describe objects in terms of numerical attributes and to map them into a Euclidean (geometric) vector space so that the distances between the points reflect the observed (dis)similarities between the respective objects. This kind of representation is attractive because geometric spaces offer powerful analytical as well as computational tools that are simply not available in other representations. In fact, classical machine learning methods are tightly related to geometrical concepts and numerous powerful tools have been developed during the last few decades, starting from linear discriminant analysis in the 1920's, to perceptrons in the 1960's, to kernel machines in the 1990's.

In the light of Kuhn's perspective we could think of such a representational attitude in terms of a collection of symbolic generalizations which lead the community to take some definitions or principles for granted. Indeed, the development of the field has been accompanied by the deployment of codified terms such as "feature extraction," "feature vector," "feature space," etc., and even by a formal vocabulary which is the basis of the subsequent mathematical manipulation. As a whole, symbolic generalizations have contributed to the general acceptance of a clear idea of what categories are and how they do form, that is the conviction that a classifier groups a set of objects under the same label because of some common features.

But the content of such generalizations might be read also at the level of the metaphysical paradigm. This brings us to discuss the philosophical assumptions behind machine learning and pattern recognition research. In fact, as pointed out in [8], their very foundations can be traced back to Aristotle and his mentor Plato who were among the firsts to distinguish between an "essential property" from an "accidental property" of an object, so that the whole field can naturally be cast as the problem of finding such essential properties of a category. As Watanabe put it [42, p. 21]: "whether we like it or not, under all works of pattern recognition lies tacitly the Aristotelian view that the world consists of a discrete number of self-identical objects provided with, other than fleeting accidental properties, a number of fixed or very slowly changing attributes.

Some of these attributes, which may be called features, determine the class to which the object belongs." Accordingly, the goal of a pattern recognition algorithm is to discern the essences of a category, or to "carve the nature at its joints." In philosophy, this view takes the name of *essentialism* and has contributed to shape the puzzle-solving activity of machine learning research in such a way that it seems legitimate to speak about an essentialist paradigm.

4.2   Essentialism and Its Discontents

Essentialism has profoundly influenced most of scientific practice until the nineteenth century, even though early criticisms came earlier with the dawn of modern science and the new Galilean approach. Later, William James, deeply influenced by Darwin, went so far as to argue that "[t]here is no property ABSOLUTELY essential to any one thing. The same property which figures as the essence of a thing on one occasion becomes a very inessential feature upon another" [17, p. 959]. Nowadays, anti-essentialist positions are associated with various philosophical movements including pragmatism, existentialism, decostructionism, etc., and is also maintained in mathematics by the adherents of the structuralist movement, a view which goes back to Dedekind, Hilbert and Poincaré, whose basic tenet is that "in mathematics the primary subject-matter is not the individual mathematical objects but rather the structures in which they are arranged" [34, p. 201]. Basically, for an anti-essentialist what really matters is relations, not essences. The influential American philosopher Richard Rorty nicely sums up this "panrelationalist" view with the suggestion that there are "relations all the way down, all the way up, and all the way out in every direction: you never reach something which is not just one more nexus of relations" [35, p. 54].

During the 19th and the 20th centuries, the essentialist position was also subject to a massive assault from several quarters outside philosophy, and it became increasingly regarded as an impediment to scientific progress. Strikingly enough, this conclusion was arrived at independently in at least three different disciplines, namely physics, biology, and psychology.

In physics, anti-essentialist positions were held (among others) by Mach, Duhem, Poincaré, and in the late 1920's Bridgman, influenced by Einstein's achievements, put forcefully forward the notion of operational definitions precisely to avoid the troubles associated with attempting to define things in terms of some intrinsic essence [4]. For example, the (special) theory of relativity can be viewed as the introduction of operational definitions for simultaneity of events and of distance, and in quantum mechanics the notion of operational definitions is closely related to the idea of observables. This point was vigorously defended by Popper [32], who developed his own form of anti-essentialism and argued that modern science (and, in particular, physics) was able to make real progress only when it abandoned altogether the pretension of making essentialist assertions, and turned away from "what-is" questions of Aristotelian-scholastic flavour.

In biology, the publication of Darwin's Origin of Species in 1859 had a devastating effect on the then dominating paradigm based on the static, Aristotelian view of species, and shattered two thousand years of research which culminated in the monumental Linnaean system of taxonomic classification. According to Mayr, essentialism "dominated the thinking of the western world to a degree that is still not yet fully appreciated by the historians of ideas. [...] It took more than two thousand years for biology, under the influence of Darwin, to escape the paralyzing grip of essentialism" [24, p.87].

More recently, motivated by totally different considerations, cognitive scientists have come to a similar discontent towards essentialist explanations. Indeed, since Wittgenstein's well-known family resemblance argument, it has become increasingly clear that the classical essentialist, feature-based approach to categorization is too restrictive to be able to characterize the intricacies and the multifaceted nature of real-world categories. This culminated in the 1970's in Rosch's now classical "protoype theory" which is generally recognized as having revolutionized the study of categorization within experimental psychology; see [21] for an extensive account, and [41] for a recent evocation in the machine learning literature.

The above discussion seems to support Popper's claim that every scientific discipline "as long as it used the Aristotelian method of definition, has remained arrested in a state of empty verbiage and barren scholasticism, and that the degree to which the various sciences have been able to make any progress depended on the degree to which they have been able to get rid of this essentialist method" [31, p. 206].

5   Signs of a Transition?

It is now natural to ask: what is the current state of affairs in machine learning? As mentioned above, the field has been dominated since its inception by the notion of "essential" properties (i.e., features) and traces of essentialism can also be found, to varying degrees, in modern approaches which try to avoid the direct use of features (e.g., kernel methods). This essentialist attitude has had two major consequences which greatly contributed to shape the field in the past few decades. On the one hand, it has led the community to focus mainly on feature-vector representations. On the other hand, it has led researchers to maintain a reductionist position, whereby objects are seen in isolation and which therefore tends to overlook the role of relational, or contextual, information.

However, despite the power of vector-based representations, there are numerous application domains where either it is not possible to find satisfactory features or they are inefficient for learning purposes. This modeling difficulty typically occurs in cases when experts cannot define features in a straightforward way (e.g., protein descriptors vs. alignments), when data are high dimensional (e.g., images), when features consist of both numerical and categorical variables (e.g., person data, like weight, sex, eye color, etc.), and in the presence of missing or inhomogeneous data. But, probably, this situation arises most commonly when objects are described in terms of structural properties, such as parts and relations between parts, as is the case in shape recognition [2]. This led in 1960's to the development of the structural pattern recognition approach, which uses symbolic data structures, such as strings, trees, and graphs for the representation of individual patterns, thereby, reformulating the recognition problem as a pattern-matching problem.

Note that, from a technical standpoint, by departing from vector-space representations one is confronted with the challenging problem of dealing with (dis)similarities that do not necessarily possess the Euclidean behavior if there exists a configuration of points in some Euclidean space whose interpoint distances are given by D, or not even obey the requirements of a metric. The lack of the Euclidean and/or metric properties undermines the very foundations of traditional pattern recognition theories and algorithms, and poses totally new theoretical/computational questions and challenges. In fact, this situation arises frequently in practice. For example, non-Euclidean or non-metric (dis)similarity measures are naturally derived when images, shapes or sequences are aligned in a template matching process. In computer vision, non-metric measures are preferred in the presence of partially occluded objects [15]. As argued in [15], the violation of the triangle inequality is often not an artifact of poor choice of features or algorithms, and it is inherent in the problem of robust matching when different parts of objects (shapes) are matched to different images. The same argument may hold for any type of local alignments. Corrections or simplifications may therefore destroy essential information.

As for the reductionist position, in retrospect it is surprising that little attention has typically been devoted to contextual information. Indeed, it is a common-sense observation that in the real world objects do not live in a vacuum, and the importance of context in our everyday judgments and actions can hardly be exaggerated, some having gone so far as to maintain that all attributions of knowledge are indeed context-sensitive, a view commonly known as contextualism [33]. Admittedly, the use of contextual constraints in pattern recognition dates back to the early days of the field, especially in connection to optical character recognition problems and it reached its climax within the computer vision community in the 1980's with the development of relaxation labeling processes and Markov random fields [14]. However, all these efforts have soon fallen into oblivion, mainly due to the tremendous development of statistical learning theory, which proved to be so elegant and powerful. Recently, the computer vision community is paying again increasing attention to the role played by contextual information in visual perception, especially in high-level problems such as object recognition (see, e.g., [25]), and neuroscientists have started understanding how contextual processing takes actually place in the visual cortex.

It is clearly open to discussion to what extent the lesson learnt from the historical development of other disciplines applies to machine learning and pattern recognition, but it looks at least like that today's research in these fields is showing an increasing propensity towards anti-essentialist/relational approaches (see [28,29] for recent accounts). Indeed, in the last few years, interest around purely similarity-based techniques has grown considerably. For example, within the supervised learning paradigm (where expert-labeled training data is assumed to be available) the now famous "kernel trick" shifts the focus from the choice of an appropriate set of features to the choice of a suitable kernel, which is related to object similarities [39]. However, this shift of focus is only partial as the classical interpretation of the notion of a kernel is that it provides an implicit transformation of the feature space rather than a purely similarity-based representation. Similarly, in the unsupervised domain, there has been an increasing interest around pairwise algorithms, such as spectral and graph-theoretic clustering methods, which avoid the use of features altogether [40,26]. Other attempts include Balcan et al.'s theory of learning with similarity functions [1], and the so-called collective classification approaches, which are reminiscent of relaxation labeling and similar ideas developed in computer vision back in the 1980's (see, e.g., [37] and references therein).

Despite its potential, presently the similarity-based approach is far from seriously challenging the traditional paradigm. This is due mainly to the sporadicity and heterogeneity of the techniques proposed so far and the lack of a unifying perspective. On the other hand, classical approaches are inherently unable to deal satisfactorily with the complexity and richness arising in many real-world situations. This state of affairs hinders the application of machine learning techniques to a whole variety of real-world problems. Hence, progress in similarity-based approaches will surely be beneficial for machine learning as a whole and, consequently, for the long-term enterprise of building "intelligent" machines.

6   Conclusion

How are we to respond to the question which motivated the present study? Clearly the answer depends on the scope of the notion of a paradigm chosen (narrow *vs.* broad). If we stick to the narrow interpretation, we easily arrive at the conclusion, with Cristianini [7], that the fields of machine learning and pattern recognition are indeed mature ones, so much so that in their (short) history we have had a whole succession of paradigms, intended as specific achievements which attracted the attention of a large and enduring fraction of the community.

Our study, however, focused on the broad interpretation of the term and this led us to discuss the philosophical underpinnings of much of contemporary machine learning research. Our analysis has shown that the community has traditionally adhered, by and large, to an essentialist worldview, where objects are characterized and represented in terms of intrinsic, essential features. This view has long been abandoned by modern science and has been in fact considered an impediment to its development. *Mutatis mutandis*, nowadays we are witnessing an increasing discontent towards essentialist representations in the machine learning community [28,29]. Hence, although using Kuhn's view, we might say that the field has reached a satisfactory level of maturity even using his broad interpretation, there are signs which make us think that there is a need to bring to full maturation a paradigm shift that is just emerging, where researchers are becoming increasingly aware of the importance of similarity and relational information *per se*, as opposed to the classical feature-based (or vectorial) approach. Indeed, the notion of similarity (which appears under different names such as proximity, resemblance, and psychological distance) has long been recognized to lie at the very heart of human cognitive processes and can be considered as a connection between perception and higher-level knowledge, a crucial factor in the process of human recognition and categorization [13].

We conclude by noticing that according to Kuhn's picture, at any particular time a scientific field is supposed to have only *one* paradigm guiding it. Applied to machine learning, this interpretation seems too restrictive as none of the paradigms mentioned above (either broad or narrow) has really guided the research of the *whole* community. More recent developments of Kuhn's thought, which allow for multiple competing paradigms per time, can be found in Lakatos' and Laudan's work who talked about "research programmes" or "research traditions," respectively [20,23]. It is therefore tempting to explore whether, in order to provide a more faithful picture of the status of the machine learning field, we need to resort to these more sophisticated conceptual tools.

Acknowledgement. We would like to thank Terry Caeli, Nello Cristianini, Bob Duin and Viola Schiaffonati for their useful feedback.

References

1. Balcan, M., Blum, A., Srebro, N.: A theory of learning with similarity functions. Machine Learning 72, 89–112 (2008)
2. Biederman, I.: Recognition-by-components: A theory of human image understanding. Psychological Review 94, 115–147 (1987)
3. Bishop, C.: Pattern Recognition and Machine Learning. Springer, New York (2006)
4. Bridgman, P.W.: The Logic of Modern Physics. MacMillan, New York (1927)
5. Bunge, M.: Technology as applied science. Technology and Culture 7, 329–347 (1966)
6. Crick, F.: The recent excitement about neural networks. Nature 337, 129–132 (1989)
7. Cristianini, N.: On the current paradigm in artificial intelligence. AICom (in press)
8. Duda, R.O., Hart, P.E., Stork, D.G.: Pattern Classification. J. Wiley & Sons, New York (2000)
9. Duin, R.: Machine learning and pattern recognition, http://www.37steps.com
10. Duin, R., Pekalska, E.: The science of pattern recognition: Achievements and perspectives. In: Duch, W., Mandziuk, J. (eds.) Challenges for Computational Intelligence. SCI, vol. 63, pp. 221–259. Springer, Heidelberg (2007)
11. Feyerabend, P.: Against Method. New Left Books, London (1975)
12. Franssen, M., Lokhorst, G.J., van de Poel, I.: Philosophy of technology. In: Zalta, E.N. (ed.) The Stanford Encyclopedia of Philosophy (2010)
13. Goldstone, R.L., Son, J.Y.: Similarity. In: Holyoak, K., Morrison, R. (eds.) The Cambridge Handbook of Thinking and Reasoning, pp. 13–36. Cambridge University Press, Cambridge (2005)
14. Hummel, R.A., Zucker, S.W.: On the foundations of relaxation labeling processes. IEEE Trans. Pattern Anal. Machine Intell. 5, 267–287 (1983)
15. Jacobs, D.W., Weinshall, D., Gdalyahu, Y.: Classification with nonmetric distances: Image retrieval and class representation. IEEE Trans. Pattern Anal. Machine Intell. 22, 583–600 (2000)
16. Jain, R.C., Binford, T.O.: Ignorance, myopia, and naiveté in computer vision systems. Computer Vision, Graphics, and Image Processing: Image Understanding 53(1), 112–117 (1991)

17. James, W.: The Principles of Psychology. Harvard University Press, Cambridge (1983); Originally published in 1890
18. Kuhn, T.S.: The Structure of Scientific Revolutions, 2nd edn. The University of Chicago Press, Chicago (1970)
19. Kuhn, T.S.: What are scientific revolutions? In: The Road Since Structure, pp. 13–32. The University of Chicago Press, Chicago (2000)
20. Lakatos, I.: The Methodology of Scientific Research Programmes. Cambridge University Press, Cambridge (1978)
21. Lakoff, G.: Women, Fire, and Dangerous Things: What Categories Reveal about the Mind. The University of Chicago Press, Chicago (1987)
22. Langley, P.: Machine learning as an experimental science. Mach. Learn. 3, 5–8 (1988)
23. Laudan, L.: Progress and Its Problems: Towards a Theory of Scientific Growth. University of California Press, Berkeley (1978)
24. Mayr, E.: The Growth of Biological Thought. Harvard University Press, Cambridge (1982)
25. Oliva, A., Torralba, A.: The role of context in object recognition. Trend Cognit. Sci. 11, 520–527 (2007)
26. Pavan, M., Pelillo, M.: Dominant sets and pairwise clustering. IEEE Trans. Pattern Anal. Machine Intell. 29(1), 167–172 (2007)
27. Pavlidis, T.: 36 years on the pattern recognition front: Lecture given at ICPR'2000 in Barcelona, Spain on the occasion of receiving the K.S. Fu prize. Pattern Recognition Letters 24, 1–7 (2003)
28. Pekalska, E., Duin, R.: The Dissimilarity Representation for Pattern Recognition. Foundations and Applications. World Scientific, Singapore (2005)
29. Pelillo, M. (ed.): Similarity-Based Pattern Analysis and Recognition. Springer, London (in press)
30. Pelillo, M., Scantamburlo, T., Schiaffonati, V.: Computer science between science and technology: A red herring? In: 2nd Int. Conf. on History and Philosophy of Computing, Paris, France (2013)
31. Popper, K.R.: The Open Society and Its Enemies. Routledge, London (1945)
32. Popper, K.R.: Conjectures and Refutations: The Growth of Scientific Knowledge. Routledge, London (1963)
33. Price, A.W.: Contextuality in Practical Reason, Oxford (2008)
34. Resnik, M.D.: Mathematics as a Science of Patterns. Clarendon Press, Oxford (1997)
35. Rorty, R.: Philosophy and Social Hope. Penguin Books, London (1999)
36. Sanborn, A.N., Griffiths, T.L., Navarro, D.J.: Rational approximations to rational models: alternative algorithms for category learning. Psychological Review 117, 1144–1167 (2010)
37. Sen, P., Namata, G., Bilgic, M., Getoor, L., Gallagher, B., Eliassi-Rad, T.: Collective classification in network data. AI Magazine 29, 93–106 (2008)
38. Serra, J.: Is pattern recognition a physical science? In: Proc. 15th International Conference on Pattern Recognition (ICPR), Barcelona, Spain, pp. 33–40 (2000)
39. Shawe-Taylor, J., Cristianini, N.: Kernel Methods for Pattern Analysis. Cambridge University Press, Cambridge (2004)
40. Shi, J., Malik, J.: Normalized cuts and image segmentation. IEEE Trans. Pattern Anal. Machine Intell. 22(8), 888–905 (2000)
41. von Luxburg, U., Williamson, R.C., Guyon, I.: Clustering: Science or art? In: JMLR: Workshop and Conference Proceedings, vol. 27, pp. 65–79 (2012)
42. Watanabe, S.: Pattern Recognition: Human and Mechanical. John Wiley & Sons, New York (1985)

Enhance User-Level Sentiment Analysis on Microblogs with Approval Relations

Federico Alberto Pozzi, Daniele Maccagnola,
Elisabetta Fersini, and Enza Messina

University of Milano-Bicocca
Viale Sarca, 336 - 20126 Milan, Italy
{federico.pozzi,d.maccagnola,fersini,messina}@disco.unimib.it

Abstract. Sentiment Analysis for polarity classification on microblogs is generally based on the assumption that texts are independent and identically distributed (i.i.d). Although these methods are aimed at handling the complex characteristics of natural language, usually they do not consider microblogs as networked data. Early approaches for overcoming this limitation consist in exploiting friendship relationships, since connected users may be more likely to hold similar opinions (Homophily and Social Influence). However, the assumption about the friendship relations does not reflect the real world, where two connected users could have different opinions about the same topic. In order to overcome these shortcomings, we propose a semi-supervised framework that estimates user polarities about a given topic by combining post contents and weighted approval relations, which are intended to better represent the contagion on social networks. The experimental investigation reveals that incorporating approval relations can lead to statistically significant improvements over the performance of complex supervised classifiers based only on textual features.

1 Introduction

According to the definition reported in [1], sentiment *"suggests a settled opinion reflective of one's feelings"*. The aim of Sentiment Analysis (SA) is therefore to define automatic tools able to extract subjective information, such as opinions and sentiments from texts in natural language, in order to create structured and actionable knowledge to be used by either a decision support system or a decision maker [2]. Most of the works in SA [3,4] are merely based on textual information expressed in microblogs. Go et al. [5] presented the results of machine learning algorithms for classifying the sentiments of Twitter messages using distant supervision (emoticons are considered as ground truth for tweet labelling), while Barbosa and Feng [6] explored the linguistic characteristics of how tweets are written and the meta-information of words for sentiment classification. However, all of these approaches do not consider that microblogs are actually networked environments. Early studies for overcoming this limitation exploit the principle of homophily [7] for dealing with user connections.

M. Baldoni et al. (Eds.): AI*IA 2013, LNAI 8249, pp. 133–144, 2013.

This principle could suggest that users connected by a personal relationships may tend to hold similar opinions. According to this social principle, friendship relations have been considered in few recent studies. In [8], Speriosu et al. proposed to enrich the content representation by including user followers as additional features. In [9,10], authors estimate user-level sentiment by exploring tweet contents and friendship relationships among users. However, in [9] tweet sentiment is not estimated but always assigned matching the real user label, and the mention network (a user mentions another one using the Twitter @-convention) is also considered following the motivation that users will mention those who they agree with.

Although social relations play a fundamental role in SA on microblogs, we argue that considering friendship connections is a weak assumption for modelling homophily: two friends might not share the same opinion about a given topic. For this reason, we propose a framework to model user-label polarity classification by integrating post contents with approval relations (e.g. 'Like' button on Facebook and 'Retweet' on Twitter) that could better represent the principle of homophily. In this paper we focus on a semi-supervised learning paradigm that, given a small proportion of users already labeled in terms of polarity, predicts the remaining unlabelled user sentiments.

2   Approval Network

In order to represent the principle of homophily, we assume that a user who approves (e.g. by likes or retweets) a given message will likely share the same opinion of the original author[1].

Before introducing the model for inferring user-level polarities, we formally define the key components that allow us to explicitly use approval relations:

DEF. 1 *Given a topic of interest q, a **Directed Approval Graph** is a quadruple $DAG_q = \{V_q, E_q, \mathbf{X}_q^V, \mathbf{X}_q^E\}$, where $V_q = \{v_1, ..., v_n\}$ represents the set of active users about q; $E_q = \{(v_i, v_j)|v_i, v_j \in V_q\}$ is the set of approval edges, meaning that v_i approved v_j's posts; $\mathbf{X}_q^E = \{k_{i,j}|(v_i, v_j) \in E_q\}$ is the set of weights assigned to approval edges, indicating that v_i approved $k_{i,j}$ posts of v_j about q; $\mathbf{X}_q^V = \{w_i|v_i \in V_q\}$ is the set of coefficients related to nodes, where w_i represents the total number of posts of v_i about q.*

Given a DAG_q, we can define the Normalised Directed Approval Graph:

DEF. 2 *Given an Approval Graph $DAG_q = \{V_q, E_q, \mathbf{X}_q^V, \mathbf{X}_q^E\}$, a **Normalised Directed Approval Graph** is derived as a triple $N\text{-}DAG_q = \{V_q, E_q, \mathbf{C}_q^E\}$, where $\mathbf{C}_q^E = \{c_{i,j} = \frac{k_{i,j}}{w_j}|k_{i,j} \in \mathbf{X}_q^E, w_j \in \mathbf{X}_q^V\}$ is the set of normalised weights of approval edges.*

---

[1] Note that the Facebook's share tool is not properly an approval tool since it is often used to express a disagreement. This is possible because it allows to add a personal comment, while a retweet does not.

We extend the N-DAG$_q$ by defining an heterogeneous graph on topic q as a unique representation both for user-user and user-post relationships.

DEF. 3 *Given a N-DAG$_q$ = $\{V_q, E_q, \mathbf{C}_q^E\}$, let $P_q = \{p_1, \cdots, p_m\}$ be the set of nodes representing posts about q and $A_q^P = \{(v_i, p_t)|v_i \in V_q, p_t \in P_q\}$ be the set of arcs that connect the user v_i and the post p_t. A **Heterogeneous Normalised Directed Approval Graph** is a quintuple HN-DAG$_q$ = $\{V_q, E_q, C_q^E, P_q, A_q^P\}$.*

A graphical representation of HN-DAG is reported in Fig. 1.
In the following, topic q is intended to be fixed and therefore omitted.

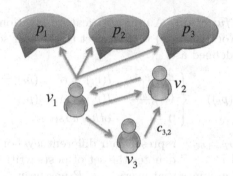

Fig. 1. Example of HN-DAG representing user-post and user-user (approval) dependencies

3   Approval Model

Given a HN-DAG, we introduce a vector of labels $\mathbf{L}^V = \{l(v_i) \in \{+, -\}|v_i \in V\}$ that defines each user as either "positive" (+) or "negative" (-) and an analogous vector of labels $\mathbf{L}^P = \{l_{v_i}(p_t) \in \{+, -\}|v_i \in V, p_t \in P\}$ that represents the polarity label of each post p_t written by user v_i. Our model is intended to obey to the Markov assumption: the sentiment $l(v_i)$ of the user v_i is influenced by the sentiment labels $l_{v_i}(p_t)$ of his posts and the sentiment labels of the directly connected neighbours $N(v_i)$. This assumption leads us to adapt the probabilistic model defined in [9] to combine user-post and user-user (approval) relations:

$$\log P(\mathbf{L}^V) = \left(\sum_{v_i \in V} \left[\sum_{p_t \in P, \alpha, \beta} \mu_{\alpha,\beta} f_{\alpha,\beta}(l(v_i), l_{v_i}(p_t)) \right. \right.$$

$$\left. \left. + \sum_{v_j \in N(v_i), \alpha, \beta} \lambda_{\alpha,\beta} g_{\alpha,\beta}(l(v_i), l(v_j)) \right] \right) \tag{1}$$

$$- \log Z$$

where $\alpha, \beta \in \{+, -\}$, $f_{\alpha,\beta}(\cdot, \cdot)$ and $g_{\alpha,\beta}(\cdot, \cdot)$ are feature functions that evaluate user-post and user-user relations respectively, and $\mu_{\alpha,\beta}$ and $\lambda_{\alpha,\beta}$ are parameters to be estimated. In particular, $\mu_{\alpha,\beta}$ represent the weights considering the setting where a user with label α posts a message with label β, while $\lambda_{\alpha,\beta}$ denote the weights considering the setting where a user with label α is connected to a user with label β. Z is the normalisation factor that enables a coherent probability distribution of $P(\mathbf{L}^V)$.

Before introducing how modelling user-post and user-user relations through the corresponding feature functions, we need to distinguish two categories of users: labeled (**black nodes** in the following), where the polarity labels are known, and unlabelled (**white nodes**), where polarity labels are unknown.

User-post feature function. A user-post feature function evaluates whether post polarity agrees (or disagrees) with respect to the user sentiment. Formally, $f_{\alpha,\beta}(l(v_i), l_{v_i}(p_t))$ is defined as:

$$f_{\alpha,\beta}(l(v_i), l_{v_i}(p_t)) = \begin{cases} \frac{\rho_{T-black}}{|P_{v_i}|} & l(v_i) = \alpha, l_{v_i}(p_t) = \beta, v_i \text{ black} \\ \frac{\rho_{T-white}}{|P_{v_i}|} & l(v_i) = \alpha, l_{v_i}(p_t) = \beta, v_i \text{ white} \\ 0 & otherwise \end{cases} \quad (2)$$

where $\rho_{T-black}$ and $\rho_{T-white}$[2] represent our different level of confidence in black and white users, and $P_{v_i} \subset P$ denotes the set of posts written by user v_i.
We point out that f assumes that every $p_t \in P$ has been classified. A methodology for automatically labelling posts is described in Sect. 5.

User-user feature function. A user-user feature function evaluates whether the polarity of a given user agrees (or disagrees) with his neighbour's sentiment. More specifically, we redefine the original feature function introduced in [9] in order to deal with approval networks. Given a HN-DAG we can formally define $g_{\alpha,\beta}(l(v_i), l(v_j))$ as follows:

$$g_{\alpha,\beta}(l(v_i), l(v_j)) = \begin{cases} \frac{\rho_{neigh} \cdot c_{i,j}}{\sum\limits_{v_k \in N(v_i)} c_{i,k}} & l(v_i) = \alpha, l(v_j) = \beta \\ 0 & otherwise \end{cases} \quad (3)$$

where ρ_{neigh}[2] represents the level of confidence in relationships among users.

4 Parameter Estimation and Prediction

We now address the problem of estimating μ and λ for inferring the assignment of user sentiment labels which maximises $\log P(\mathbf{L}^V)$. Starting from a small set of labeled data, we can initialise the values of μ and λ using the following approach:

$$\mu_{\alpha,\beta} = \frac{\sum\limits_{(v_i, p_t) \in E_{BU}} I(l(v_i) = \alpha, l_{v_i}(p_t) = \beta)}{\sum\limits_{(v_i, p_t) \in E_{BU}} I(l(v_i) = \alpha, l_{v_i}(p_t) = +) + I(l(v_i) = \alpha, l_{v_i}(p_t) = -)} \quad (4)$$

---

[2] Note that $\rho_{T-black}$, $\rho_{T-white}$ and ρ_{neigh} are empirically estimated (see Sect. 6.3).

$$\lambda_{\alpha,\beta} = \frac{\displaystyle\sum_{(v_i,v_j)\in E_{BP}} I(l(v_i) = \alpha, l(v_j) = \beta)}{\displaystyle\sum_{(v_i,v_j)\in E_{BP}} I(l(v_i) = \alpha, l(v_j) = +) + I(l(v_i) = \alpha, l(v_j) = -)} \tag{5}$$

where $I(\cdot)$ is the indicator function, while E_{BU} and E_{BP} are the subsets of user-user and user-post edges where users are labeled. The initial values, estimated according to Eq. (4)-(5), can be used to derive the optimal λ and μ that maximise $\log P(\mathbf{L}^V)$.

For seek of simplicity, we introduce a change of notation: we decompose $\log P(\mathbf{L}^V)$ in $\phi \cdot \Psi(\mathbf{L}^V)$, where

$$\phi = \{\mu_{\alpha,\beta}, \lambda_{\alpha,\beta}\} \tag{6}$$

and

$$\Psi(\mathbf{L}^V) = \{ \sum_{v_i \in V} \sum_{p_t \in P, \alpha, \beta} f_{\alpha,\beta}(l(v_i), l_{v_i}(p_t)),$$
$$\sum_{v_i \in V} \sum_{v_j \in N(v_i), \alpha, \beta} g_{\alpha,\beta}(l(v_i), l(v_j)) \} \tag{7}$$

In order to find the optimal values of ϕ we employed the SampleRank Algorithm [11]:

Algorithm 1. SampleRank

Input	: HN-DAG
Initialisation:	Learning rate η, initial parameters ϕ
Output	: Final parameter values ϕ and full label-vector $\mathbf{L}^V$

1 Randomly initialise $\mathbf{L}^V$;
2 **for** $step \leftarrow 1$ **to** $MaxNumberOfSteps$ **do**
3 $\quad$ $\mathbf{L}_{new}^V \leftarrow Sample(\mathbf{L}^V)$;
4 $\quad$ $\bigtriangledown \leftarrow \Psi(\mathbf{L}_{new}^V) - \Psi(\mathbf{L}^V)$;
5 $\quad$ **if** $\phi \cdot \bigtriangledown > 0 \wedge \mathbb{P}(\mathbf{L}_{new}^V, \mathbf{L}^V) < 0$ // *performance is worse but the objective function is higher*
6 $\quad$ **then** $\phi \leftarrow \phi - \eta\bigtriangledown$;
7 $\quad$ **if** $\phi \cdot \bigtriangledown < 0 \wedge \mathbb{P}(\mathbf{L}_{new}^V, \mathbf{L}^V) > 0$ // *performance is better but the objective function is lower*
8 $\quad$ **then** $\phi \leftarrow \phi + \eta\bigtriangledown$;
9 $\quad$ **if** $Convergence$ **then** break;
10 $\quad$ **if** $\mathbb{P}(\mathbf{L}_{new}^V, \mathbf{L}^V) > 0 \vee (\mathbb{P}(\mathbf{L}_{new}^V, \mathbf{L}^V) = 0 \wedge \phi \cdot \bigtriangledown > 0)$ **then**
11 $\quad\quad$ $\mathbf{L}^V \leftarrow \mathbf{L}_{new}^V$;

In Algorithm 1, *Sample* is a sampling function that randomly chooses an element of $\mathbf{L}^V$ and reverts its polarity. $\mathbb{P}(\mathbf{L}_{new}^V, \mathbf{L}^V)$ is the Accuracy[3] difference between $\mathbf{L}_{new}^V$ and $\mathbf{L}^V$ (only on the black nodes).

The algorithm converges when both the objective function $\phi \cdot \Psi(\cdot)$ and $\mathbb{P}(\mathbf{L}_{new}^V, \mathbf{L}^V)$ do not increase for a given number of steps.

5   Message Polarity Classification

Since the user-post feature function $f_{\alpha,\beta}$ assumes that every post $p_t \in P$ has to be classified, a sentiment classification methodology for posts is required.

The main polarity classification approaches are focused on identifying the most powerful model for classifying the polarity of a text source. However, an ensemble of different models could be less sensitive to noise and could provide a more accurate prediction [12].

For this reason we exploit the ensemble method proposed in [2], where the weighted contribution of each classifier is used to make a final label prediction. The work improves the original Bayesian Model Averaging (BMA) [13] by explicitly taking into account the marginal distribution of each classifier prediction and its overall reliability when determining the optimal label. Given a set of $\mathcal{R}$ classifiers, the approach assigns to a post p_t the label $l_{v_i}(p_t)$ that maximises:

$$P(l_{v_i}(p_t)|\mathcal{R}, \mathcal{D}) = \sum_{r \in \mathcal{R}} P(l_{v_i}(p_t)|r)P(r|\mathcal{D})$$

$$= \sum_{r \in \mathcal{R}} P(l_{v_i}(p_t)|r)P(r)P(\mathcal{D}|r) \qquad (8)$$

where $P(l_{v_i}(p_t)|r)$ is the marginal distribution of the label predicted by classifier r, while $P(\mathcal{D}|r)$ represents the likelihood of the training data $\mathcal{D}$ given r. The distribution $P(\mathcal{D}|r)$ can be approximated by using the F_1-measure obtained during a preliminary evaluation of classifier r:

$$P(\mathcal{D}|r) \propto \frac{2 \times P_r(\mathcal{D}) \times R_r(\mathcal{D})}{P_r(\mathcal{D}) + R_r(\mathcal{D})} \qquad (9)$$

where $P_r(\mathcal{D})$ and $R_r(\mathcal{D})$ denotes Precision and Recall[3] obtained by classifier r.

The set of baseline classifiers used by BMA is composed of dictionary-based classifier, Naïve Bayes (NB), Maximum Entropy, Support Vector Machines (SVM) and Conditional Random Fields (CRF). The TF weighting schema has been considered for NB and SVM. For further details, see [2].

6   Experiments

In this section, we present a case study to validate the proposed model based on approval networks. In particular, we focused our experimental investigation on connections derived from Twitter and we compared the ability of inferring user-level polarity classification with traditional approaches based only on textual features.

---

[3] Performance measures are defined in Sect. 6.4

6.1   Dataset

In order to evaluate the proposed model, we need a dataset composed of:

1. A set of users (V) and their manually tagged sentiment labels about a specific topic q;
2. Tweets written by users $\in V$ about a specific topic q with their manually tagged sentiment labels;
3. Retweet network about a specific topic q.

To the best of our knowledge, no datasets containing all the above information are available. In order to easily create a reliable toy dataset[4], the following steps have been performed:

- Crawling of a set of 2500 users from Twitter who tweeted about the topic 'Obama' during the period 8-10 May 2013;
- Download of the last 3200[5] tweets for each user;
- Filtering of out-of-topic posts by removing messages that do not match the regular expression "obama|barack";
- Selection of the most active users by considering those authors who emitted at least 50 tweets about Obama[6];
- Manual annotation of each tweet and user by 3 annotators for labelling the corresponding polarity (62 users and 159 tweets). Only positive and negative tweets have been considered;
- Simulation of the underlying approval network according to a Poisson distribution: two users with the same polarity have an higher probability to be connected by one or more retweets than two users with a different polarity; moreover, a user that posts a high number of tweets has a higher probability to have incoming connections.

Considering that our model has been defined for dealing with a semi-supervised environment, we need to distinguish labeled (black) from unlabelled (white) users. As black nodes we considered those users whose bio (description on Twitter) or name clearly state a positive or negative opinion about the topic 'Obama'. For instance, a positive user's bio could report *"I like football, TV series and Obama!"* and/or the name could be *"ObamaSupporter"*.

6.2   BMA Settings

The BMA model has been trained by using positive and negative tweets of the *Obama-McCain Debate (OMD)*[7] dataset. As a test set, we employed the 159 tweets written by the users of our network.

---

[4] The dataset can be downloaded at http://www.mind.disco.unimib.it
[5] Due to a limit imposed by Twitter
[6] 50 is the average number of tweets posted by the users.
[7] https://bitbucket.org/speriosu/updown/src/5de483437466/data/

Since much of the tweet are similar to SMS messages, the writing style and the lexicon of tweets is widely varied. Moreover, tweets are often highly ungrammatical, and filled with spelling errors. In order to clean the dataset, we captured a set of patterns, which are detected using dictionaries a priori defined and regular expressions. The normalised tweet is then used to train the text-based BMA. The applied filters are:

- **HTML Links**: All tokens matching the following REGEXP are deleted: `(https?|ftp|file)://[-a-zA-Z0-9+&@#/%?=~_|!:,.;]*[-a-zA-Z0-9+&@#/%=~_|]`;
- **Hashtags**: The symbol # is removed from all the tokens;
- **Mention Tags**: The tokens corresponding to a mention tag, identified through the REGEXP @(.+?), are removed;
- **Retweet Symbols**: All the tokens matching the expression RT are removed;
- **Laughs**: If a token has a sub-pattern matching `((a|e|i|o|u)h|h(a|e|i|o|u))\\1+|(ahha|ehhe|ihhi|ohho|uhhu)+`, then the whole token is replaced with `TAGLAUGHT`;
- **Emoticons**: In order to detect positive and negative emoticons, two dictionaries have been defined. If a token appears in the dictionary of positive emoticons then it is replaced with `POSEMOTICON`, otherwise with `NEGEMOTICON`;
- **Sentiment Expressions**: Several sentiment expressions are used in English. Expressions such as 'ROFL','LMAOL','LMAO','LMAONF' represent positive expressions. In order to facilitate Term-Frequency (TF) counting, they are replaced with `POSEMOTICON` and `NEGEMOTICON`, as well;
- **Slang correction**: A dictionary of a priori defined slang expressions with their meaning, such as 'btw' (by the way), 'thx' (thanks), 'any1' (anyone) and u (you) has been built. In order to facilitate TF counting, each found slang expression is replaced with its meaning;
- **Onomatopoeic expressions**: as the previous point, a mapping dictionary has been defined for onomatopoeic expressions, such as 'bleh' (`NEGEMOTICON`) and 'wow' (`POSEMOTICON`).

In addition to filters, misspelled tokens have been corrected using the Google's Spell Checker API[8]. Since the Google's algorithm takes the neighbourhood (context) of a misspelled token into account in suggesting the correction, the whole previously filtered tweet is considered as a query rather than the single token. Note that the adaptation and modifications of the REGEXPs adopted in these filters to other microblogs is straightforward.

6.3   SampleRank and Approval Model Settings

Regarding SampleRank Algorithm, we set the maximum number of steps to 10000 and we assume that convergence is reached when results are persistent for 500 steps.

---

[8] https://code.google.com/p/google-api-spelling-java/

Considering that SampleRank results depend on a sampling function, we performed $k = 1, 5, 11, 15, 21, 101$ runs to get k predictions (votes) and take a majority vote among the k possible labels for each user. For each k, we performed 500 experiments to compute the average performance.

SampleRank and the approval model need parameters to be fixed a priori: the learning rate η, $\rho_{T-black}$, $\rho_{T-white}$ and ρ_{neigh}. In order to select the optimal combination, we empirically varied their values[9] and hold those which have exhibited the highest performance: $\eta = 0.01$, $\rho_{T-black} = 1$, $\rho_{T-white} = 0.2$ and $\rho_{neigh} = 0.9$.

Since SampleRank is a time expensive algorithm, it has been implemented using Fracture[10], a Java Multi Core library that allows to parallelise the computation on all the machine cores.

6.4   Results

The performance of the proposed model has been evaluated by measuring the well known Precision, Recall and F-Measure:

$$P = \frac{TP}{TP + FP} \quad R = \frac{TP}{TP + FN} \quad F1 = \frac{2 \cdot P \cdot R}{P + R} \tag{10}$$

both for the positive and negative labels (in the sequel denoted by P_+, R_+, $F1_+$ and P_-, R_-, $F1_-$ respectively). We also measured Accuracy as:

$$Acc = \frac{TP + TN}{TP + FP + FN + TN} \tag{11}$$

Concerning BMA, the performance computed on the 159 tweets achieves 60.37% of Accuracy, compared with the 58.49% obtained by the baseline classifier which reaches the highest performance (CRF). These results confirm the performance improvement stated in [2] when using BMA. In order to estimate the user polarity by using their tweets, we considered a majority voting mechanism to infer the final user label. For instance, if a user posts 3 positive and 2 negative tweets, the final user label will be 'positive'. When inferring labels of white nodes on the collected dataset, this heuristic reaches a 66.66% of Accuracy.

Regarding the proposed model based on approval relationships, we firstly investigate its robustness by monitoring results according to the considered k votes. In order to select the best value of k, we tried to minimise the model bias by considering tweets manually labeled.

Fig. 2 shows the negative correlation between the number of votes and the Accuracy variability, suggesting that a high number of votes ensures the stability of the model. The average Accuracy values and the confidence intervals (with a confidence level of 0.01) are reported in Table 1.

---

[9] Parameters ρ have been varied from 0 to 1 with an increment of 0.1, while η has been empirically fixed to 0.01

[10] http://kccoder.com/fracture/

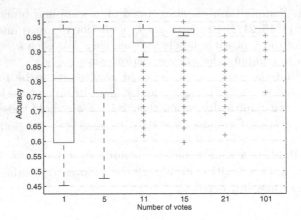

Fig. 2. Negative correlation between the number of votes and the Accuracy variability

Table 1. Confidence intervals achieved on different experiments

k	1	5	11	15	21	101
Interval	0.783 ± 0.022	0.881 ± 0.015	0.925 ± 0.011	0.936 ± 0.010	0.95 ± 0.007	0.976 ± 0.001

Once determined the optimal number of k, we accordingly set up our model by using the tweet labels classified by the BMA approach.

Table 2 reports performance achieved by our approach in respect of the BMA one. Note that our model is able to reach 93.8% of Accuracy, significantly outperforming the BMA approach. Approval relations enclosed in our model ensures an improvement of 27% with respect to the text-only method.

Table 2. Comparison between BMA (only tweets) and Relations (SampleRank) + Tweets (BMA labeled)

	P_-	R_-	$F1_-$	P_+	R_+	$F1_+$	Acc
BMA	0.655	0.826	0.731	0.692	0.474	0.563	0.666
Relations + Tweets	0.933	0.963	0.945	0.958	0.907	0.925	0.938

We report in Fig. 3 a snapshot that compares the ground truth network, the prediction provided by BMA and the one obtained by our model. Since BMA does not take into account any kind of relationship, the correct prediction of a user does not have any effect on adjoining users. In our case, the prediction of each user has impact on all the other nodes by a "propagation" effect, smoothing each predicted label according to adjoining nodes. This investigation finally confirms that the inclusion of relationships in predictive models, as suggested in other studies [14,15,16], leads to improve recognition performance when dealing with non-propositional environments.

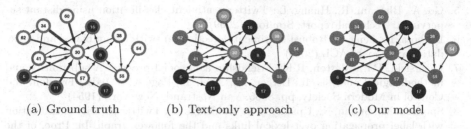

(a) Ground truth   (b) Text-only approach   (c) Our model

Fig. 3. Example of the studied approval network. (a) is the Ground truth derived from human annotation, (b) the BMA results and (c) results of our model. The real user labels are indicated by the border colour: green for positive and red for negative. Background colour denotes if a node is black or white and the edge thickness reflects the normalised weight. Grey nodes in (b) and (c) represent the misclassified users.

7   Conclusion and Future Work

In this work, we proposed a user-level polarity classification approach which exploits approval relationships to represent the principle of homophily. The computational results show that our model ensures stable performance, significantly outperforming the BMA (text-only based) approach. In this work we address the problem in the context of Twitter because it is easier to retrieve data, although adaptation of this framework to other microblogs is straightforward (using other microblog-dependent approval tools, e.g. 'likes' count on Facebook). Exploring relationships available in microblogs, in addition to contents analysis, represents an interesting research direction in Social Network Analysis.

Concerning ongoing research, we plan to compare our BMA approach with other text-based classifiers. Moreover, we are now focusing on a larger dataset in order to extend the considered heterogeneous graph on multiple topics.

Acknowledgment. This work has been partially supported by the grant "Dote Ricercatori": FSE, Regione Lombardia.

References

1. Pang, B., Lee, L.: Opinion mining and sentiment analysis. Foundations and Trends in Information Retrieval 2, 1–135 (2008)
2. Pozzi, F.A., Fersini, E., Messina, E.: Bayesian model averaging and model selection for polarity classification. In: Métais, E., Meziane, F., Saraee, M., Sugumaran, V., Vadera, S. (eds.) NLDB 2013. LNCS, vol. 7934, pp. 189–200. Springer, Heidelberg (2013)
3. Wang, S., Manning, C.D.: Baselines and bigrams: simple, good sentiment and topic classification. In: Proceedings of the 50th Annual Meeting of the Association for Computational Linguistics: Short Papers, ACL 2012, vol. 2, pp. 90–94 (2012)
4. Maas, A.L., Daly, R.E., Pham, P.T., Huang, D., Ng, A.Y., Potts, C.: Learning word vectors for sentiment analysis. In: Proceedings of the 49th Annual Meeting of the Association for Computational Linguistics: Human Language Technologies, HLT 2011, vol. 1, pp. 142–150 (2011)

5. Go, A., Bhayani, R., Huang, L.: Twitter sentiment classification using distant supervision. Technical report, Stanford (2009)
6. Barbosa, L., Feng, J.: Robust sentiment detection on twitter from biased and noisy data. In: Proc. of ACL (2010)
7. Lazarsfeld, P.F., Merton, R.K.: Friendship as a social process: A substantive and methodological analysis. In: Berger, M., Abel, T., Page, C.H. (eds.) Freedom and Control in Modern Society, pp. 8–66. Van Nostrand, New York (1954)
8. Speriosu, M., Sudan, N., Upadhyay, S., Baldridge, J.: Twitter polarity classification with label propagation over lexical links and the follower graph. In: Proc. of the First Workshop on Unsupervised Learning in NLP (2011)
9. Tan, C., Lee, L., Tang, J., Jiang, L., Zhou, M., Li, P.: User-level sentiment analysis incorporating social networks. In: Proc. of the 17th ACM SIGKDD International Conference on Knowledge Discovery and Data Mining, KDD 2011, pp. 1397–1405 (2011)
10. Hu, X., Tang, L., Tang, J., Liu, H.: Exploiting social relations for sentiment analysis in microblogging. In: Proceedings of the Sixth ACM International Conference on Web Search and Data Mining, WSDM 2013, pp. 537–546. ACM (2013)
11. Wick, M., Rohanimanesh, K., Culotta, A., McCallum, A.: Samplerank: Learning preferences from atomic gradients. In: NIPS Workshop on Advances in Ranking (2009)
12. Dietterich, T.G.: Ensemble learning. In: The Handbook of Brain Theory and Neural Networks, pp. 405–508. Mit Pr. (2002)
13. Hoeting, J.A., Madigan, D., Raftery, A.E., Volinsky, C.T.: Bayesian model averaging: A tutorial. Statistical Science 14(4), 382–417 (1999)
14. Fersini, E., Messina, E., Archetti, F.: A probabilistic relational approach for web document clustering. Information Processing and Management 46(2), 117–130 (2010)
15. Sharara, H., Getoor, L., Norton, M.: Active surveying: A probabilistic approach for identifying key opinion leaders. In: IJCAI, pp. 1485–1490 (2011)
16. Fersini, E., Messina, E.: Web page classification through probabilistic relational models. International Journal of Pattern Recognition and Artificial Intelligence (2013)

Abstraction in Markov Networks

Lorenza Saitta

Università del Piemonte Orientale
Dipartimento di Informatica
Viale Teresa Michel 11, 15121 Alessandria, Italy
saitta@mfn.unipmn.it

Abstract. In this paper a new approach is presented for taming the complexity of performing inferences on Markov networks. The approach consists in transforming the network into an abstract one, with a lower number of vertices. The abstract network is obtained through a partitioning of its set of cliques. The paper shows under what conditions exact inference may be obtained with reduced cost, and ways of partitioning the graph are discussed. An example, illustrating the method, is also described.

Keywords: Abstraction, Graphical models, Markov networks, Approximate inference.

1 Introduction

Markov networks are graphical models of joint probability distributions over a set of N variables [2]. In the graph nodes are associated to random variables, and edges connect pairs of statistically dependent variables. The joint probability distribution over the variables is expressed as a normalized product of *potential* functions, defined over the cliques of the graph.

One of the major issue in general Markov networks is the computational complexity of inference; in this paper we consider the problem of computing the marginal probability distribution over a set of variables. Several techniques have been developed to reduce the complexity of inference, either relying on particular structures of the graph to compute exact inference, or using sampling techniques (Markov Chain Monte Carlo sampling, Gibbs sampling) for computing approximate inference [8]. For instance, Bishop presents the sum-product algorithm [2] for undirected tree: it relies on the notion of *factor graph*.

Another approach, which is gaining interest, is the use of *variational approximations*. In this case, the true joint probability distribution $P(\mathbf{X})$ over the variables $\mathbf{X}$ of the network is approximated by a totally or partially factorized one, $Q(\mathbf{X})$. The distribution $Q(\mathbf{X})$ is parametrized, and the parameters $\boldsymbol{\theta}$ are chosen in such a way to minimize the Kulback-Leibler divergence between $P(\mathbf{X})$ and $Q(\mathbf{X})$. Using a totally factorized approximate distribution corresponds to the so-called *Mean Field Approximation* [12].

In this paper, we describe a different approach, based on the use of *abstraction* for reducing the complexity of inference in Markov networks. Abstraction, in this

M. Baldoni et al. (Eds.): AI*IA 2013, LNAI 8249, pp. 145–156, 2013.

context, aims at transforming the original network into a smaller/simpler one, which can be handled with a reduced complexity, still preserving some required properties. Graph abstraction for specific tasks have been proposed early in the literature. For instance, Holte et al. [6] have described a STAR abstraction operator that allows significant speed up in problem solving. Clustering of nodes in a graph, detection of "communities", and building up hierarchies and multi-scale views can all be considered as forms of abstraction on graphs [5,4,1,10,3].

The theoretical basis of this work was presented earlier, in a paper oriented to abstraction in Logical Markov Networks [13]. In this paper we target general Markov networks, and report some experimental results. In order to make the paper self-contained, we report here some details of the method.

2 Markov Networks with Boolean Variables

Given a vector $\mathbf{X} = (X_1, X_2, ...X_N)$ of stochastic Boolean variables, which assumes values $\mathbf{x} = (x_1, x_2, ..., x_N) \in \mathcal{X} = \{0, 1\}^N$, a Markov network $\mathcal{G}$ represents the joint probability distribution over $\mathbf{X}$. The network has N nodes and M cliques. An assignment of values to $\mathbf{x}$ is called a *world*. The set of all worlds, $\mathcal{X}$, contains 2^N elements. The restriction to Boolean variables is only for computational simplicity reasons, as the method works unchanged for variables with different domains.

The probability over $\mathbf{X}$ can be expressed as a product of M *potential functions*, each one associated to a (maximal) clique of $\mathcal{G}$:

$$P(\mathbf{X} = \mathbf{x}) =_{def} P(\mathbf{x}) = \frac{1}{Z} \prod_{k=1}^{M} \varphi_k(\mathbf{x}) \tag{1}$$

In (1) Z is the *partition function*, which is a normalization factor defined by:

$$Z = \sum_{\mathbf{x} \in \mathcal{X}} \prod_{k=1}^{M} \varphi_k(\mathbf{x}) \tag{2}$$

The potential functions are arbitrary, except for the constraint to be everywhere different from zero. It can be proved that (1) can represent any probability distribution. In order to estimate the computational complexity of evaluating Z, let us assume, as a unit, the complexity required to compute one potential function. Then:

$$C(N, M) = M \cdot 2^N \tag{3}$$

3 Partitions of a Markov Network

Given a Markov network $\mathcal{G}$, let us partition the *cliques* of $\mathcal{G}$ into two sets, corresponding to (overlapping) subgraphs $\mathcal{G}_1$ and $\mathcal{G}_2$, where $\mathcal{G}_1$ contains h cliques, and $\mathcal{G}_2$ contains $(M - h)$ cliques. The nodes in $\mathcal{G}_1$ and $\mathcal{G}_2$ can be partitioned into three sets, namely, $\mathbf{U}$, containing the nodes occurring only in $\mathcal{G}_1$, $\mathbf{V}$, containing the nodes occurring only in $\mathcal{G}_2$, and $\mathbf{T}$, containing the nodes shared by $\mathcal{G}_1$ and $\mathcal{G}_2$. The values taken on by $\mathbf{U}$, grouping r variables, by $\mathbf{T}$, grouping s variables, and by $\mathbf{V}$, grouping $(N - r - s)$ variables, are contained in three subvectors of $\mathbf{x}$, *i.e.*, $\mathbf{u} \in \mathcal{X}_u$, $\mathbf{t} \in \mathcal{X}_t$, and $\mathbf{v} \in \mathcal{X}_v$, respectively. We have:

$$|\mathcal{X}_u| = 2^r \qquad |\mathcal{X}_t| = 2^s \qquad |\mathcal{X}_v| = 2^{N-r-s}$$

Without loss of generality we may assume that $\mathbf{u}$ contains the first r variables $(X_1, ..., X_r)$, $\mathbf{t}$ contains the subsequent s variables $(X_{r+1}, ..., X_{r+s})$, and $\mathbf{v}$ contains the last $(N - r - s)$ variables $(X_{r+s+1}, ..., X_N)$:

$$\mathbf{u} = (x_1, x_2, ..., x_r)$$
$$\mathbf{t} = (x_{r+1}, ..., x_{r+s})$$
$$\mathbf{v} = (x_{r+s+1}, ..., x_N)$$

Let us now compute the probability of a generic world $\mathbf{x}$, by taking into account the partition of $\mathcal{G}$ into $\mathcal{G}_1$ and $\mathcal{G}_2$:

$$P(\mathbf{x}) = \frac{1}{Z} \prod_{k=1}^{h} \varphi_k(\mathbf{u}, \mathbf{t}) \prod_{k=h+1}^{M} \varphi_k(\mathbf{v}, \mathbf{t}) \tag{4}$$

From (4) we want to compute the partition function Z:

$$Z = \sum_{\mathbf{x} \in \mathcal{X}} \left(\prod_{k=1}^{h} \varphi_k(\mathbf{u}, \mathbf{t}) \prod_{k=h+1}^{M} \varphi_k(\mathbf{v}, \mathbf{t}) \right) \tag{5}$$

We can split the sum in (5) as follows:

$$Z = \sum_{\mathbf{t} \in \mathcal{X}_t} \left[\sum_{\mathbf{u} \in \mathcal{X}_u} \prod_{k=1}^{h} \varphi_k(\mathbf{u}, \mathbf{t}) \sum_{\mathbf{v} \in \mathcal{X}_v} \prod_{k=h+1}^{M} \varphi_k(\mathbf{v}, \mathbf{t}) \right] \tag{6}$$

In expression (6) the two internal sums are independent, for each value of $\mathbf{t}$. Then, $\mathbf{u}$ and $\mathbf{v}$ are stochastically independent, given $\mathbf{t}$.

In order to compute Z from (6) we obtain a complexity (in the number of evaluations of potential functions):

$$\mathcal{C}(N, M, r, s, h) = \mathcal{O}\left(2^{r+s}h + 2^{N-r}(M - h)\right)$$

In equation (6), let us define:

$$\Phi_1(\mathbf{u}, \mathbf{t}) = \prod_{k=1}^{h} \varphi_k(\mathbf{u}, \mathbf{t}) \text{ and } \Phi_2(\mathbf{v}, \mathbf{t}) = \prod_{k=h+1}^{M} \varphi_k(\mathbf{v}, \mathbf{t}) \tag{7}$$

Moreover, let:

$$\Psi_1(\mathbf{t}) = \sum_{\mathbf{u} \in \mathcal{X}_u} \Phi_1(\mathbf{u}, \mathbf{t}) \text{ and } \Psi_2(\mathbf{t}) = \sum_{\mathbf{v} \in \mathcal{X}_v} \Phi_2(\mathbf{v}, \mathbf{t}) \tag{8}$$

Using equations (8), the marginal probability distribution over the set of variables $\mathbf{t}$ can be expressed as:

$$P(\mathbf{t}) = \Psi_1(\mathbf{t})\dot{\Psi}_2(\mathbf{t}) \tag{9}$$

If we think in terms of message passing, the two subgraphs $\mathcal{G}_1$ and $\mathcal{G}_2$ send to $\mathbf{t}$ a massage each, and $\mathbf{t}$ receives the messages and multiplies them to find its own probability distribution. Finally, equation (5) can be rewritten as:

$$Z = \sum_{\mathbf{t}\in\mathcal{X}_t} \Psi_1(\mathbf{t})\cdot\Psi_2(\mathbf{t}) \tag{10}$$

The above results can be generalized to the case in which the product of potential functions is partitioned into R clique sets, each one corresponding to a subgraph $\mathcal{G}_k$ ($1 \leqslant k \leqslant R$); $\mathcal{G}_k$ contains a different set of h_k cliques. We have then:

$$\mathcal{G} = \bigcup_{k=1}^{R} \mathcal{G}_k \quad \text{and} \quad \sum_{k=1}^{R} h_k = M$$

Generalizing formula (4), the probability of a world $\mathbf{x}$ can be written as follows:

$$P(\mathbf{x}) = \frac{1}{Z}\prod_{k=1}^{R} \Phi_k(\mathbf{x}^{(k)}) \tag{11}$$

where $\mathbf{x}^{(k)}$ contains the variables that appear in $\mathcal{G}_k$. Each subgraph contributes to the probability distribution with the product:

$$\Phi_k(\mathbf{x}^{(k)}) = \prod_{j=h_{k-1}+1}^{h_k} \varphi_j(\mathbf{x}^{(k)}) \quad (1 \leqslant k \leqslant R) \tag{12}$$

In order to compute Z, we have to use equation (5). In each $\Phi(\mathbf{x}^{(k)})$ let us partition the set of variables into two groups: $\mathbf{u}^{(k)} \in \mathcal{X}_u^{(k)}$ and $\mathbf{t}^{(k)} \in \mathcal{X}_t^{(k)}$, such that the variables in $\mathbf{u}^{(k)}$ only occur inside $\mathcal{G}_k$, whereas the variables in $\mathbf{t}^{(k)}$ may also occur in some other $\mathcal{G}_j$. We may notice that the sets of variables $\mathbf{u}^{(k)}$ are all disjoint, by definition, whereas the sets $\mathbf{t}^{(k)}$ may share some variables among them. Then, let us define:

$$\mathbf{t} = \bigcup_{k=1}^{R} \mathbf{t}^{(k)} \in \mathcal{X}_t$$

As a consequence, Z can be rewritten as follows:

$$Z = \sum_{\mathbf{t}\in\mathcal{X}_t} \left(\sum_{\mathbf{u}^{(1)}\in\mathcal{X}_u^{(1)}} \Phi_1(\mathbf{u}^{(1)},\mathbf{t}) \; \dots \; \sum_{\mathbf{u}^{(R)}\in\mathcal{X}_u^{(R)}} \Phi_R(\mathbf{u}^{(R)},\mathbf{t}) \right)$$

For each assignment to the common variables $\mathbf{t}$, the internal sums in Z are independent. We can define:

$$\Psi_k(\mathbf{t}) = \sum_{\mathbf{u}^{(k)} \in \mathcal{X}_\mathbf{u}^{(k)}} \Phi_k(\mathbf{u}^{(k)}, \mathbf{t}) \tag{13}$$

Finally, the partition function becomes:

$$Z = \sum_{\mathbf{t} \in \mathcal{X}_t} \prod_{k=1}^{R} \Psi_k(\mathbf{t}) \tag{14}$$

The complexity for computing Z through equation (14) can be evaluated by defining:

$$h = \max_{1 \leqslant k \leqslant R} h_k \quad \text{and} \quad s = |\mathbf{t}| \quad \text{and} \quad r = \max_{1 \leqslant k \leqslant R} |\mathbf{u}^{(k)}|$$

Then:

$$\mathcal{C}(N, M, R, r, s, h) = \mathcal{O}\left(2^s \sum_{k=1}^{R} h_k e^{|\mathbf{u}^{(k)}|}\right) = \mathcal{O}(R \, h \, 2^{r+s})$$

Notice that $(r + s) < n$. The above formula also applies to the case of $R = 2$.

4   Abstraction in Markov Networks

As inference is exponential in the size of a Markov network, one way to make it less costly is to apply to it some abstraction operator, whose effect is to reduce its size. In Markov networks abstraction has two components: a *syntactic* one, which acts on the networks structure, and a *semantic* one, which involves probability distributions.

Let us consider a network $\mathcal{G}$ partitioned into R subnetworks, as in Section 3, and let $\mathcal{G}_k$ be one of such subnetwork. Syntactically, we let the nodes that occur only in $\mathcal{G}_k$ collapse into a single Boolean node $\xi_k \in \{0, 1\}$. Semantically, we have to link some function of the collapsed nodes with the ξ_ks. More precisely:

$$\xi_k = \alpha_k(\mathbf{u}^{(k)}) \qquad (1 \leqslant k \leqslant R) \tag{15}$$

Let us define:

$$D_1^{(k)} = \{\mathbf{u}^{(k)} \mid \xi_k = 1\} \tag{16}$$

$$D_0^{(k)} = \{\mathbf{u}^{(k)} \mid \xi_k = 0\} \tag{17}$$

By applying the abstraction functions α_k, the network $\mathcal{G}$ with N nodes becomes a network $\mathcal{G}'$ with $N' = (R+s)$ nodes; in fact, each set $\mathbf{u}^{(k)}$ is mapped onto a single node ξ_k, whereas the nodes corresponding to $\mathbf{t}$ remain the same. Let us notice that the subdivision of $\mathcal{G}$ into the $\mathcal{G}_k$'s was done by separating sets of *cliques*, not sets of nodes. Then, each new node ξ_k still forms (possibly collapsed) cliques with the nodes in $\mathbf{t}$. We may think that ξ_k and the nodes $\mathbf{t}$ form a unique "clique", which contributes to the abstract probability distribution $Q(\xi_1, ..., \xi_R, \mathbf{t})$ a single potential function $\Phi'_k(\xi_k, \mathbf{t})$; then:

$$Q(\xi_1, ..., \xi_R, \mathbf{t}) = \frac{1}{Z'} \prod_{k=1}^{R} \Phi'_k(\xi_k, \mathbf{t}) \tag{18}$$

Let us now compute the marginal probability distribution of a generic ξ_k; from (18) we obtain:

$$Q(\xi_k) = \frac{1}{Z'} \sum_{\mathbf{t} \in \mathcal{X}_t} \sum_{\xi_j \neq \xi_k} \prod_{j=1}^{R} \Phi'_k(\xi_j, \mathbf{t})$$

$$= \frac{1}{Z'} \sum_{\mathbf{t} \in \mathcal{X}_t} \left(\Phi'_k(\xi_k, \mathbf{t}) \prod_{j=1, j \neq k}^{R} \Psi'_j(\mathbf{t}) \right)$$

We may wonder whether there exists a definition of the $\Phi'_k(\xi_k, \mathbf{t})$ $(1 \leqslant k \leqslant R)$ that makes $Q(\xi_k)$ (distribution of ξ_k computed in the abstract network) equal to $P(\xi_k)$ (distribution of ξ_k computed in the ground network). Using expression (11), we obtain:

$$P(\xi_k) = \frac{1}{Z} \sum_{\mathbf{t} \in \mathcal{X}_t} \prod_{j=1, j \neq k}^{R} \Psi_j(\mathbf{t}) \sum_{\mathbf{u}^{(k)} \in D_{\xi_k}^{(k)}} \Phi_k(\mathbf{u}^{(k)}, \mathbf{t})$$

where $D_{\xi_k}^{(k)} = D_1^{(k)}$ if $\xi_k = 1$ and $D_{\xi_k}^{(k)} = D_0^{(k)}$ if $\xi_k = 0$.

By comparing expressions $P(\xi_k)$ and $Q(\xi_k)$ we see that $P(\xi_k) = Q(\xi_k)$ if we define:

$$\Phi'_k(\xi_k, \mathbf{t}) = \sum_{\mathbf{u}^{(k)} \in D_{\xi_k}^{(k)}} \Phi_k(\mathbf{u}^{(k)}, \mathbf{t}) \tag{19}$$

By summing equation (19) for $\xi_k = 0$ and $\xi_k = 1$, we obtain, by definition, that:

$$\Psi'_k(\mathbf{t}) = \Psi_k(\mathbf{t}) \quad \rightarrow \quad Z = Z' \tag{20}$$

We may notice that equality (20) allows Z' to be computed exactly on the abstract network, with a reduced complexity. The partitioning process may be repeated, obtaining thus multiple layers of abstraction. Clearly, the abstract network contains less information than the ground one. This can be seen when the complete probability distribution $P(\mathbf{x})$ must be recovered. In fact, from the abstract network we only know that:

$$\sum_{\mathbf{u}^{(k)} \in D_1^{(k)}} P(\mathbf{u}^{(k)}) = Q(\xi_k = 1) = Q_k(1) \tag{21}$$

$$\sum_{\mathbf{u}^{(k)} \in D_0^{(k)}} P(\mathbf{u}^{(k)}) = Q(\xi_k = 0) = Q_k(0) \tag{22}$$

Moreover, equations (21) and (22) are not independent, because, by summing up them, both sides must be equal to 1.

If the α_k functions are not invertible (as it is the usual case in abstraction), we may think that they are dependent on a set of variational parameters θ_k, which can be used to minimize the Kulback-Leibler divergence between the true $P(\mathbf{u}^{(k)})$ and the one obtained inverting the $\alpha_k(\mathbf{u}^{(k)})$'s. Obviously, it is not necessary to abstract all the subgraphs $\mathcal{G}_k$; we may abstract R_1 of them, leaving the remaining $R_2 = R - R_1$ untouched. In this case, the probability distribution on the abstract graph is given by:

$$Q(\xi_1, ..., \xi_{R_1}, \mathbf{t}, \mathbf{v}^{(R_1+1)}, ..., \mathbf{v}^{(R)}) = \prod_{k=1}^{R_1} \Phi'_k(\xi_k, \mathbf{t}) \prod_{k=R_1+1}^{R} \Phi_k(\mathbf{v}^{(k)}, \mathbf{t}) \qquad (23)$$

By exploiting equations (19) and (20), it is easy to show that:

$$P(\mathbf{t}) = Q(\mathbf{t}) \qquad (24)$$
$$P(\mathbf{v}^{(k)}) = Q(\mathbf{v}^{(k)}) \qquad (R_1 + 1 \leqslant k \leqslant R) \qquad (25)$$
$$P(\mathbf{v}^{(k)}|\mathbf{t}) = Q(\mathbf{v}^{(k)}|\mathbf{t}) \qquad (R_1 + 1 \leqslant k \leqslant R) \qquad (26)$$

Then, the marginal probability distributions of the non abstracted variables can be computed from the abstract graph with reduced complexity. For instance, let us suppose that we want to compute the probability distribution of a query q in presence of an evidence e. For the sake of simplicity, let us assume that $q \equiv x_j$ (q is a single variable), and that the evidence $e \equiv (x_i = 1)$ is the assignment of the value 1 to another single variable x_i. We want to compute:

$$P(q|e) = \frac{P(q, e)}{P(e)} = \frac{P(x_j, x_i = 1)}{P(x_i = 1)} \qquad (27)$$

Usually, it is too expensive to compute (27) directly, because it involves sums over possibly large set of worlds; then an approximate value is estimated through Gibbs sampling, which consists of sampling each node's value given its Markov blanket [9]. Using the abstract network $\mathcal{G}'$, either an exact value can be computed with lower complexity, or an approximate estimation can be made, depending on the positions of the nodes x_j and x_i in $\mathcal{G}'$. There are three cases:

- The query q and the evidence e belong both to the same the vector $\mathbf{u}^{(k)}$ or to $\mathbf{t}$
- The query q and the evidence e belong one to some $\mathcal{G}_k$ and one to $\mathbf{t}$
- The query q and the evidence e belong to two different subgraphs $\mathcal{G}_k$ and $\mathcal{G}_\ell$.

The case in which both q and e belong to $\mathbf{t} = (x_j, x_i, x_{i_1}, ..., x_{i_s})$ is the most favorable, because all the subgraphs $\mathcal{G}_k$ ($1 \leqslant k \leqslant R$) can be abstracted to single nodes ξ_k. In this case we only need to compute the distribution $Q(\mathbf{t})$ by marginalizing over the R variables ξ_k. The case in which both q and e belong to $\mathbf{u}^{(k)} = (x_j, x_i, x_{i_1}, ..., x_{i_s})$ is analogous, but in this case we may only abstract

$(R - 1)$ subgraphs. In fact, subgraph $\mathcal{G}_k$ cannot be abstracted, because, in this case, we would only obtain from the abstract graph the probability $Q(\xi_k) = P(\xi_k)$, which is an aggregated probability over the set of variables in $\mathbf{u}^{(k)}$.

If the query q and the evidence e belong one to some $\mathcal{G}_k$ and one to $\mathbf{t}$, we cannot abstract $\mathcal{G}_k$, and we need the joint probability distribution over $\mathbf{u}^{(k)}$ and $\mathbf{t}$. Finally, when the query q and the evidence e belong to two different subgraphs $\mathcal{G}_k$ and $\mathcal{G}_\ell$, we need to leave both $\mathcal{G}_k$ and $\mathcal{G}_\ell$ ground, and we have to compute the conditional probabilities of the variables $\mathbf{u}^{(k)}$ and of the variables $\mathbf{u}^{(\ell)}$ with respect to $\mathbf{t}$. In fact, knowing $\mathbf{t}$ makes the $\mathbf{u}^{(k)}$'s and the $\mathbf{u}^{(\ell)}$'s conditionally independent.

If the evidence and/or the query consists of the conjunction of more than one nodes, more subgraphs cannot be abstracted, by reducing the effectiveness of the approach. In this case it would be maybe better to accept an approximate answer to the query, which can be obtained in two ways: one is by sampling the relevant $\mathbf{u}^{(k)}$, assuming known the $\mathbf{t}$ (an extension of the method of the Markov blanket), or by defining an approximate inversion of the abstraction function.

If we have to perform abstraction a priori, before knowing which the query and evidence are, in order to find a suitable subdivision of the ground graph, we use the dual graph $\mathcal{D}$, whose nodes correspond to the cliques of $\mathcal{G}$. In $\mathcal{D}$ two nodes are connected iff the corresponding cliques share at least one variable. On this graph we may apply any algorithm for partitioning the nodes (Mincut, Community detection, and so on) in such a way to minimize the number of variable in $\mathbf{t}$, increasing thus conditional independence among sets of variables.

On the contrary, is we may abstract $\mathcal{G}$ after the query q and evidence e are provided, we have to try to include both q and e in the same $\mathcal{G}_k$. This is possible by considering the minimum set of cliques that includes both.

5 Experiments

In order to test the effectiveness of the approach, we have built up a set of artificial Markov networks using four generative models: Erdös-Renyi's to build random graphs [14], Watts and Strogatz's to build small-world graphs [15], Barabàsi and Albert to build scale-free graphs [16], and Logical Markov Networks [9].

For each graph we have extracted the maximal cliques, and we have associated to each clique an exponential potential function. More precisely, if $x_1, ..., x_r$ are the nodes in a clique, the associated potential function has the following structure:

$$\varphi(x_1, ..., x_r) = \exp\left(\sum_{i=1}^{r} a_i x_i\right) \quad (a_i \in \{-5, ..., -1\} \cup \{1, ..., 5\}) \qquad (28)$$

Potential function (28) is never equal to 0, and all the clique's variables occur in it. The coefficients a_i are extracted randomly from the set of integers $\{-5, -4, -3, -2, -1, 1, 2, 3, 4, 5\}$. After removing the possible isolated vertices, we have run an algorithm for detecting overlapping communities [17], and then we have selected a query q of the form:

$$q = P(x_{i_1} = b_1, x_{i_2} = b_2, \cdots, x_{i_h} = b_h) \quad \text{with } b_j \in \{0,1\} \text{ for } (1 \leqslant j \leqslant h) \quad (29)$$

We do not consider here any evidence. This is not restrictive, because the presence of evidence only helps reducing the computational complexity of the inference process. Fig. 1 reports some of the obtained experimental results.

Model	N	M	R	$|q|$	G	ρ
ER	150	72	4	10	1	0.372
ER	200	112	6	10	1	0.296
ER	200	98	5	10	2	0.653
ER	250	83	12	20	4	0.647
ER	300	120	15	20	4	0.843
ER	300	230	18	20	6	1.364
ER	1000	320	34	10	1	0.043
ER	2000	532	28	20	2	0.005
ER	3000	780	31	30	4	0.002
SW	150	54	4	10	1	0.269
SW	200	74	5	10	1	0.318
SW	200	88	7	10	2	0.498
SW	250	96	6	20	4	0.925
SW	300	107	12	20	4	0.655
SW	300	102	10	20	6	1.229
SW	1000	35	15	10	1	0.082
SW	2000	77	29	20	2	0.014
SW	3000	82	22	30	4	0.008
BA	150	98	12	10	1	0.361
BA	200	162	11	10	1	0.325
BA	200	149	15	10	2	0.459
BA	250	222	13	20	4	0.652
BA	300	265	22	20	12	1.572
BA	300	240	20	20	20	1.000
BA	2000	320	16	20	4	0.007
BA	3000	562	32	30	12	0.002
LMN	15	12	4	3	1	0.0017
LMN	200	112	10	4	1	0.001
LMN	200	112	10	4	3	0.006

Fig. 1. Excerpt from the experimental results. In the table, $|q|$ denotes the number of variables involved in the query, and G is the number of subgraphs over which the query is spread. ρ is the ratio between the time required to obtain an answer to q in the abstract network and the time required to answer it in the ground one. The former includes the time for partitioning the original graph and the time to compute the abstract potential functions. For this reason some of the entries in column ρ may be greater than 1. The computational complexity required to find the exact answer in the ground networks limits the possibility of experimentation. For this reason, the values reported for the networks with at least 1000 nodes are approximate values found with Gibbs sampling.

From Fig. 1 it appears that abstracting the network is not always useful, as it is, in fact, well known. This happens when the query involves variables that are spread over several subgraphs, and/or when the cost for obtaining the clustering is high.

A particularly favorable case involves abstracting Logical Markov Networks [9,13], because the nature of the problem itself suggests very useful partitions.

Whatever the graph's structure, increasing the number of nodes increases the usefulness of abstraction. This is due to the fact that, usually, the size of the query does not increase with the size of the network, and then it is more likely that the nodes of the query are (or can be) grouped into a small number of the subgraphs in the partition.

6   Conclusions

In this paper we have described an approach, based on the use of abstraction operators, to reduce the complexity of inference in Markov network. In our approach abstraction has a syntactic part and a semantic one, the former consisting in collapsing nodes together, and the latter in linking probabilities across abstraction levels. Other types of abstraction could be considered as well. A related approach has been proposed for abstracting generic CSPs [7]. Relations also exist with the paper by Shavlik and Natarajan [11]. In their paper the authors describe a method for speeding up inference in ground Logical Markov Networks by avoiding computing groundings that are certainly true/false.

In order to find conditions under which abstraction is most likely useful, much more experiments and on much larger networks are needed. The problem with large networks, however, is that the amount of computation needed to handle exactly the ground graph may be prohibitive. On the there hand, we may notice that it is not necessary to go to size such that of social networks, because in an inference problem a few hundreds of variables are usually more than sufficient.

At the end, we would like to mention a possible evolution of this work, which we find challenging. Abstraction is a process that, starting from a detailed description of a system, builds up a simplified one, which is then used to solve a problem. One might wonder about the possibility of directly acquiring from the world the "abstracted" representation, trying to solve the same given problem. In this case, the cost of abstracting could be avoided, but there is the difficulty of "reconstructing" the ground representation, if needed.

The same problem has been tackled recently in signal processing, under the name of *compressed sensing*, which, in some sense, can be considered as the reverse of abstraction. Compressed sensing [18,19] tries exactly to do this, in that it acquires limited information about a signal and tries to (exactly or approximately) reconstruct it. Clearly, as information cannot be created, some hypotheses are necessary:

- The ground "system" consists of a vector $\mathbf{x}$ with N components, which is sparse, i.e., it has only $k < N$ components different from 0[1].

---

[1] It may be that not the signal itself, but some transformation of it (for instance Fourier transform) is sparse.

– The measurements are collected into a vector $\mathbf{y}$ with $M < N$ components.
– The measurement matrix $\mathbf{A}_{M \times N}$ such that $\mathbf{y} = \mathbf{A}\,\mathbf{x}$ is known.

As the system $\mathbf{y} = \mathbf{A}\,\mathbf{x}$ is linear, with M equations and N unknowns, it has (in general) infinite solutions. Then, finding $\mathbf{x}$ has to be cast as an optimization problem, namely:

$$\mathbf{x} = ArgMin\|\mathbf{x}'\|_{\ell_1} \quad \text{under the constraint} \quad \mathbf{y} = \mathbf{A}\,\mathbf{x} \qquad (30)$$

Problem (30) can only be solved if $k < M < N$. Actually, it has a simple solution: try all k positions out of the N in $\mathbf{x}$ for locating the elements different from 0. However, this solution is exponential in N, and compressed sensing aims at a feasible solution instead. In practice, the situation is even more complex. In fact, let us denote $\rho = k$ and $\alpha = M$, and, for each pair (ρ, α) let us generate an ensemble of random matrices $\mathcal{A}$, built up by extracting each entry of $\mathbf{A} \in \mathcal{A}$ according to a Gaussian distribution. Then, the plane (ρ, α) is divided into three regions. If $\rho > \alpha$, reconstruction is impossible; on the contrary, the triangle where $\rho < \alpha$ consists of two subregions: in one reconstruction is polynomial, whereas, in the other, reconstruction is exponential. The boundary between reconstructability and non reconstructability is a *phase transition* [19].

Linking compressed sensing to abstraction would open new ways to approach the difficult problem to selecting a "good" abstraction for a given problem. The challenge is to transfer concepts born inside a numerical setting to a symbolic one.

References

1. Arenas, A., Fernandez, A., Gomez, S.: Analysis of the structure of complex networks at different resolution levels. New Journal of Physics 10, 053039 (2008)
2. Bishop, C.M.: Pattern Recognition and Machine Learning. Springer (2006)
3. Bulitko, V., Sturtevant, N., Lu, J., Yau, T.: Graph abstraction in real-time heuristic search. Journal of Artificial Intelligence Research 30, 51–100 (2007)
4. Clauset, A., Moore, C., Newman, M.E.J.: Hierarchical structure and the prediction of missing links in networks. Nature 453, 98–101 (2008)
5. Epstein, S.L., Li, X.: Cluster graphs as abstractions for constraint satisfaction problems. In: Proc. Symposium on Abstraction, Reformulation and Approximation, Lake Arrowhead, CA, pp. 58–65 (2009)
6. Holte, R.C., Mkadmi, T., Zimmer, R.M., MacDonald, A.J.: Speeding up problem solving by abstraction: A graph oriented approach. Artificial Intelligence 85, 321–361 (1996)
7. Lecoutre, C., Merchez, S., Boussemart, F., Grégoire, É.: A CSP Abstraction Framework. In: Choueiry, B.Y., Walsh, T. (eds.) SARA 2000. LNCS (LNAI), vol. 1864, pp. 326–327. Springer, Heidelberg (2000)
8. Poon, H., Domingos, P.: Sound and efficient inference with probabilistic and deterministic dependencies. In: Proc. of the National Conference on Artificial Intelligence, Boston, MA, pp. 458–463 (2006)
9. Richardson, M., Domingos, P.: Markov logic networks. Machine Learning 62, 107–136 (2006)

10. Saitta, L., Henegar, C., Zucker, J.D.: Abstracting complex interaction networks. In: Proc. Symposium on Abstraction, Reformulation and Approximation, Lake Arrowhead, CA, pp. 190–193 (2009)
11. Shavlik, J., Natarajan, S.: Speeding up inference in Markov logic networks by preprocessing to reduce the size of the resulting grounded network. In: Proc. Intern. Joint Conf. on Artificial Intelligence, Pasadena, CA, pp. 1951–1956 (2009)
12. Wiegerinck, W.: Variational approximations between mean field theory and the junction tree algorithm. In: Proc. of the 16th Conf. on Uncertainty in Artifical Intelligence, Stanford, CA, USA, pp. 626–633 (2000)
13. Saitta, L., Vrain, C.: Abstracting Markov networks. Presentation to the Symposium on Abstraction, Reformulation and Approximation, Cardona, Spain (2010)
14. Erdös, P., Rényi, P.: On Random Graphs. Publ. Math. Debrecen 6, 290–297 (1959)
15. Watts, D.J., Strogatz, S.H.: Collective dynamics of *small-world* networks. Nature 393, 440–442 (1998)
16. Barabási, A.L., Albert, R.: Emergence of scaling in random networks. Science 159, 509–512 (1999)
17. Xie, J., Kelley, S., Szymanski, B.K.: Overlapping Community Detection in Networks: the State of the Art and Comparative Study. ACM Computing Surveys 45, Article 43 (2013)
18. Krzakala, F., Mézard, M., Sausset, L., Sun, Y., Zdeborova, L.: Probabilistic Reconstruction in Compressed Sensing: Algorithms, Phase Diagrams, and Threshold Achieving Matrices. J. Stat. Mech., P08009 (2012)
19. Barbier, J., Mézard, Zdeborova, L.: Compressed Sensing of Approximately-Sparse Signals: Phase Transitions and Optimal Reconstruction. In: Proc. of the 50th Annual Conf. on Communication, Control, and Computing, Allerton, USA, pp. 800–807 (2012)

Improving the Structuring Capabilities of Statistics–Based Local Learners

Slobodan Vukanović, Robert Haschke, and Helge Ritter

Cognitive Interaction Technology – Center of Excellence (CITEC),
Bielefeld University, Bielefeld, Germany

Abstract. Function approximation, a mainstay of machine learning, is a useful tool in science and engineering. Local learning approches subdivide the learning space into regions to be approximated locally by linear models. An arrangement of regions that conforms to the structure of the target function leads to learning with fewer resources and gives an insight into the function being approximated. This paper introduces a covariance–based update for the size and shape of each local region. An evaluation shows that the method improves the structuring capabilities of state–of–the–art statistics–based local learners.

Keywords: Function Approximation, Locally Weighted Regression, Input Space Structuring.

1 Introduction

Locally weighted regression (LWR) approximates the target function with a population of linear models, each localized in the input space. The output to a given query is computed as a weighted (usually Gaussian) combination of the outputs of the local models [1]. The principal advantage that LWR offers over classical function approximation is its independence from the choice of the class of the approximating function. This leads to learning systems that are robust to destructive interference, and whose complexity can be increased incrementally by adding new models. LWR is thus popular for learning in online scenarios, such as the ones often encountered in robotics [7].

The locality of each model is defined by its position in the input space and a distance metric, which is usually adapted as the learning progresses. The adaption mechanism often aims to structure the input space by exploiting the linear subspaces that are local to each model. A reasonable structuring of the input space offers insights into the structure of the function being approximated, and it enables simpler functions to be learned with fewer models.

It has been demonstrated in [8] that the statistics–based Locally Weighted Projection Regression (LWPR, [9]), a popular local learning system [7], is not as capable of structuring the input space as the competing, genetics–based XCSF Learning Classifier System [10]. LWPR produces tight–fitting, highly non–overlapping spherical populations from which the structure is difficult to grasp.

M. Baldoni et al. (Eds.): AI*IA 2013, LNAI 8249, pp. 157–168, 2013.

In contrast, XCSF shapes its models to conform to the structure of the target function.

This paper introduces a weighted covariance distance metric update for statistics–based LWR approaches with ellipsoidal neighborhoods, such as LWPR. Training samples are weighted according to their current prediction error and the proximity to each model, and are either included in the distance metric update or ignored. We show that such an update improves the structuring capabilities over the default LWPR distance metric update, producing populations structured similarly to those in XCSF. Additionally, we show that the improved structuring obtained by our local method can lead to reducing population sizes, without the need for a global model deletion mechanism used by XCSF.

The following section gives an overview of well–known LWR approaches (including LWPR and XCSF), focussing on distance metric updates. Section 3 details our weighted covariance distance metric update method. Section 4 provides an experimental comparison between our approach and the defaults in LWPR and XCSF. Section 5 concludes.

2   Related Work

The default implementation of Local Linear Maps (LLM, [5]) uses a fixed number of models whose positions in the input space can be adapted during training, e.g. by vector quantization. The local models learn to approximate the underlying function using ordinary least squares regression. No information about the output space is given to the distance metric update, thus the populations of the models after training do not resemble the structure of the target function. Another approach that considers only the input space, but in which the population size is allowed to incrementally change, is the Locally Weighted Interpolating Growing Neural Gas (LWIGNG, [3]).

LWPR, a descendant of the Receptive Field Weighted Regression (RFWR, [6]), performs input dimensionality reduction through partial least squares (PLS) regression. The population size is incrementally increased by allocating models in parts of the input space not covered by existing models. The *receptive field* (RF) of each local model is described by an ellipsoid fixed at a point $\mathbf{c}$ in the input space and a distance metric $\mathbf{D}$, a positive semi–definite matrix:

$$(\mathbf{x} - \mathbf{c})^T \mathbf{D} (\mathbf{x} - \mathbf{c}) = 1 \tag{1}$$

A function of $\mathbf{c}$, $\mathbf{D}$, and an incoming training sample returns an *activation weight* $w \in [0, 1]$ for that sample. The higher the activation w, the more influence the sample has on the model during training. To try to adapt a receptive field's shape and size to the local linear region, $\mathbf{D}$ is Cholesky–decomposed into $\mathbf{M}$, which is then trained by minimizing the cost function J of the leave–one–out cross–validation error for each iteration step t and a given learning rate α:

$$\mathbf{M}_t = \mathbf{M}_{t-1} + \alpha \frac{\partial J}{\partial \mathbf{M}_{t-1}}, \mathbf{D}_t = \mathbf{M}_t^T \mathbf{M}_t \tag{2}$$

Local Gaussian Process Regression (LGP, [4]) uses the same model allocation mechanism as LWPR but, in contrast, adapts the centers and not the distance metrics of its receptive fields.

The XCSF Learning Classifier System combines reinforcement learning mechanisms and genetic algorithms. It is not a specific LWR algorithm but a framework that can employ different kinds of regression models (linear, quadratic, etc) and distance metrics (rectangular and ellipsoidal). It can, however, be tuned for comparison with LWPR [8]. The crossover and mutation operators are applied to the center of each receptive field and its distance metric, and the results are inserted into the model population. Once a maximum population size is reached, a global deletion mechanism removes receptive fields from overcrowded regions. An advantage that the evolutionary mechanism has over the statistics–based LWPR is that the optimal structuring of the input space can be found and that the prediction is often more accurate [8]. A disadvantage is the greater number of receptive fields required, which incurs a higher computational cost.

3   The Distance Metric Update

We consider the representation of the receptive fields identical to LWPR (Equation 1). Given a training sample $(\mathbf{x}_t, \mathbf{y}_t)$ at step t, we model the distance metric $\mathbf{D}_t$ of each receptive field as the inverse of the incremental covariance $\boldsymbol{\Sigma}_t$:

$$\mathbf{D}_t = (\boldsymbol{\Sigma}_t)^{-1}, \boldsymbol{\Sigma}_t = \frac{\mathbf{S}_t}{\varPhi_t} \tag{3}$$

$\mathbf{S}_t$ is the matrix of weighted incremental sums (employing the stable covariance update from [2]):

$$\mathbf{S}_t = (1 - \gamma) \cdot \mathbf{S}_{t-1} + \gamma \cdot \phi(\mathbf{p}) \cdot (\mathbf{x}_t - \mathbf{c})(\mathbf{x}_t - \mathbf{c})^T \tag{4}$$

where $\gamma \in [0, 1]$ is the forgetting factor, $\phi(\mathbf{p})$ is the weighting function with parameters $\mathbf{p}$, and $\mathbf{c}$ is the center of the receptive field. $\varPhi_t$ is the sum of weights:

$$\varPhi_t = (1 - \gamma) \cdot \varPhi_{t-1} + \gamma \cdot \phi(\mathbf{p}) \tag{5}$$

Figure 1(a) shows an example of a simple $\mathbb{R}^2 \to \mathbb{R}^1$ sigmoid function to be modeled, containing two flat areas separated by a steep slope. A single receptive field covering the lower flat area (viewed from above) is shown in Figure 1(b). The distance metric of the receptive field has been set by hand. Because of its coverage and a sufficiently low error, we use this field as a model when discussing updates in the remainder of the section.

The weighting function ϕ should be such that the local training samples that produce a small error are included in the covariance update, and the local samples that produce a large error are excluded. The derivation of a weighting function with these characteristics is given below, followed by a discussion on the values of the parameter vector $\mathbf{p}$.

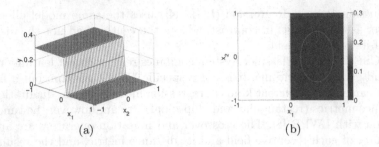

Fig. 1. (a) An example function to be modeled by the network. (b) A single receptive field centered at $(0.5, 0.0)$ in the input space.

3.1   The Weighting Function and Its Parameters

The covariance update will prompt a receptive field to shrink to include samples that have large weights assigned by ϕ and a high activation w. Similarly, a receptive field will grow to include samples that have large weights and a low w. The samples with very small weights will be excluded from the covariance update regardless of their location, prompting no change in the distance metric. These dynamics must be taken into account for weight assignment: ϕ needs to be a function of both the current prediction error and the activation.

Given the magnitude of the error err (low or high) and the proximity $g = 1 - w$ of a training sample to the center of the field (high, low, or not local), we identify five cases $A_1, \ldots, A_5$ to which that sample can belong. These are shown in Figure 2.

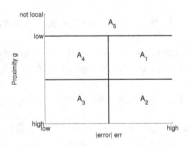

Fig. 2. Regions in which a training sample may fall: high error and low proximity (A_1), high error and high proximity (A_2), low error and high proximity (A_3), low error and low proximity (A_4), or not local (A_5)

Let σ^+ and σ^- be one–dimensional logistic functions:

$$\sigma^+(x, s, k) = \frac{1}{1 + e^{-s(|x| - k)}}$$
$$\sigma^-(x, s, k) = \frac{-1}{1 + e^{-s(|x| - k)}} + 1$$

(6)

where s is the slope and k is the position of each sigmoid. The functions $A_i(\mathbf{p}_i)$ with parameters $\mathbf{p}_i$ are products of the sigmoids:

$$
\begin{aligned}
A_1(\text{err}, s_1^{\text{err}}, k_1^{\text{err}}, g, s_1^g, k_1^g) &= \sigma^+(\text{err}, s_1^{\text{err}}, k_1^{\text{err}})\sigma^+(g, s_1^g, k_1^g) \\
A_2(\text{err}, s_2^{\text{err}}, k_2^{\text{err}}, g, s_2^g, k_2^g) &= \sigma^+(\text{err}, s_2^{\text{err}}, k_2^{\text{err}})\sigma^-(g, s_2^g, k_2^g) \\
A_3(\text{err}, s_3^{\text{err}}, k_3^{\text{err}}, g, s_3^g, k_3^g) &= \sigma^-(\text{err}, s_3^{\text{err}}, k_3^{\text{err}})\sigma^-(g, s_3^g, k_3^g) \\
A_4(\text{err}, s_4^{\text{err}}, k_4^{\text{err}}, g, s_4^g, k_4^g) &= \sigma^-(\text{err}, s_4^{\text{err}}, k_4^{\text{err}})\sigma^+(g, s_4^g, k_4^g) \\
A_5(g, s_5^g, k_5^g) &= \sigma^-(g, s_5^g, k_5^g)
\end{aligned}
\tag{7}
$$

where $s_i^{\text{err}}, k_i^{\text{err}}$ are the slope and the position in the error direction and s_i^g, k_i^g are the slope and the position in the proximity direction of A_i. Each function $A_1(\mathbf{p}_1), \ldots, A_4(\mathbf{p}_4)$ has the value 1 in the corresponding region $A_1, \ldots, A_4$ in Figure 2, while the function $A_5(\mathbf{p}_5)$ has the value 0 in the region A_5.

The weighting function ϕ is written as a combination of $A_1(\mathbf{p}_1), \ldots, A_4(\mathbf{p}_4)$ multiplied by $A_5(\mathbf{p}_5)$:

$$
\phi(\mathbf{p}) = A_5(\mathbf{p}_5) \sum_{i=1}^{4} a_i A_i(\mathbf{p}_i)
\tag{8}
$$

where $a_1, \ldots, a_4$ determine the magnitude of the weights in the corresponding regions.

The left panels of Figure 3(a) and 3(b) show two examples of weighting functions from above (cf. Figure 2). The right panels show the receptive field after 20,000 noiseless updates taken from a $(x_1, x_2) \in [-1, 1]^2$ uniform random distribution. The weighting function in Figure 3(a) will assign the value 0 to many weights, i.e. it will exclude a lot of training samples from the distance metric update. Samples with low proximity that have a low error (i.e. along the vertical x_2 direction in the right panel in Figure 3(a)) expand the field too much (cf. Figure 1(b)), and there are not enough high proximity samples weighted to counter this expansion. The weighting function in Figure 3(b) will assign the value 1 to many weights, i.e. it will include a lot of training samples in the distance metric update. Regardless of their error, samples with high proximity will dominate the update, preventing the field from growing to cover the desired region (cf. Figure 1(b)).

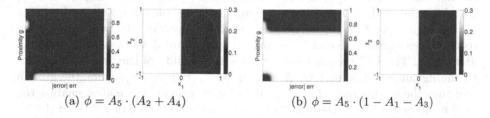

(a) $\phi = A_5 \cdot (A_2 + A_4)$ (b) $\phi = A_5 \cdot (1 - A_1 - A_3)$

Fig. 3. Two weighting functions and the corresponding receptive fields after 20,000 updates

The combinations of $A_i(\mathbf{p}_i)$ and the parameter values that produce extreme inclusion and exlusion shown in Figure 3 should be avoided. Ideally, ϕ should assign large weights to high proximity samples producing high errors to trigger shrinking (region A_2, $a_2 \approx 1$), it should assign large weights to low proximity samples producing low errors to trigger growth (region A_4, $a_4 \approx 1$), and it should assign zero weights in all other cases (regions A_1 and A_3, $a_1 = a_3 = 0$):

$$\phi(\mathbf{p}) = A_5(\mathbf{p}_5) \cdot (a_2 A_2(\mathbf{p}_2) + a_4 A_4(\mathbf{p}_4)) \tag{9}$$

Parameter values that merge or produce gaps between A_2 and A_4 are avoided by grouping like parameters: s_2^{err} and s_4^{err} into s^{err}, k_2^{err} and k_4^{err} into k^{err}, s_2^g, s_4^g, and s_5^g into s^g, k_2^g and k_4^g into k^{in}, and k_5^g into k^{out}. The parameters are thus:

$$\begin{aligned}
\mathbf{p} &= (\text{err}, s^{\text{err}}, k^{\text{err}}, g, s^g, k^{\text{in}}, k^{\text{out}}) \\
\mathbf{p}_2 &= (\text{err}, s^{\text{err}}, k^{\text{err}}, g, s^g, k^{\text{in}}) \\
\mathbf{p}_4 &= (\text{err}, s^{\text{err}}, k^{\text{err}}, g, s^g, k^{\text{in}}) \\
\mathbf{p}_5 &= (g, s^g, k^{\text{out}})
\end{aligned} \tag{10}$$

3.2   Field Dynamics

Given the weighting function $\phi(\mathbf{p})$ (Equation 9) and a reasonable choice of the values for parameters $\mathbf{p}$, a receptive field can shape itself to cover a region up to an error of k^{err}. Because of the mutual dependency between the shape of the field and the regression model, initially large fields will shrink until the model is trained. Expansion in the linear regions may follow. We now discuss what a reasonable choice for the values $\mathbf{p}$ is.

Following the layout and the legend of Figure 3, Figure 4 shows the resulting receptive field after 20,000 noiseless updates with different values of the parameters $\mathbf{p}$. The k^{err} parameter determines which samples to include in the update based on their error. A smaller value (Figure 4(a)) results in a smaller field that approximates the underlying region well, while a larger value (Figure 4(b)) includes more points, making the field cover more space. k^{err} is a parameter that requires tuning, as it affects the size of the fields (and thus their number, given the model allocation strategy in LWPR). The k^{in} parameter determines the proximity at which the weights change between exclusion and inclusion. A smaller value (Figure 4(c)) causes inclusion of the points with a high proximity, shrinking the field. A larger value (Figure 4(d)) assigns large weights to points with a high error that are at a lower proximity, making the field grow incorrectly. A reasonable value of k^{in} is around 1 standard deviation, in the interval $[0.30, 0.40]$. The k^{out} parameter determines the locality of the field with respect to the distance metric update. A smaller value (Figure 4(e)) voids the growth and thus the field is only shrunk, while a larger value (Figure 4(f)) enables the field to observe more of the input space. A reasonable value of k^{out} is in the interval $[0.80, 0.95]$. Finally, the slope parameters s^{err} and s^g determine the rate of the transition of other parameters. A smaller value (Figure 4(g)) cause smoother, slower updates to the distance metric, while a larger value (Figure 4(h)) leads

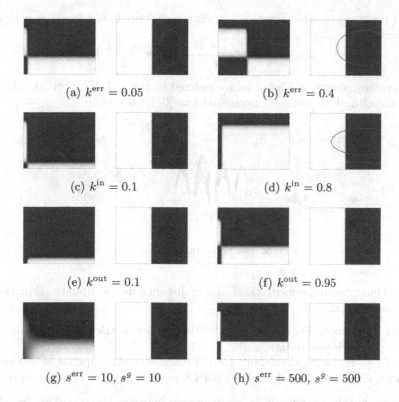

(a) $k^{\text{err}} = 0.05$ (b) $k^{\text{err}} = 0.4$

(c) $k^{\text{in}} = 0.1$ (d) $k^{\text{in}} = 0.8$

(e) $k^{\text{out}} = 0.1$ (f) $k^{\text{out}} = 0.95$

(g) $s^{\text{err}} = 10$, $s^g = 10$ (h) $s^{\text{err}} = 500$, $s^g = 500$

Fig. 4. The effect of parameter values of the weighting function on field dynamics. Unless stated differently, the parameters have the following values: $s^{\text{err}} = 100$, $k^{\text{err}} = 0.05$, $s^g = 30$, $k^{\text{in}} = 0.32$, $k^{\text{out}} = 0.85$.

to quick, abrupt updates. A reasonable setting for s^{err} is ≥ 100, since a sharp threshold between the errors is required. s^g should be set in the range $[30, 50]$, as abrupt updates are not desirable.

4   Evaluation

The evaluation of the distance metric update consists of two experiments. Experiment One is a replication of the study from [8], a comparison between LWPR and XCSF on learning three two–dimensional functions:

1. the *Crossed Ridge* function, containing a mixture of linear and non–linear regions

$$y = \max[\exp(-10x_1^2), \exp(-50x_2^2), 1.25\exp(-5(x_1^2 + x_2^2))]$$

2. the *Sine* function, which is linear in the $(1, -1)$ direction but highly non–linear in the perpendicular $(1, 1)$ direction

$$y = \sin(2\pi(x_1 + x_2))$$

3. and the *Sine–in–Sine* function, an extremely non–linear function with high curvature

$$y = \sin\left(4\pi\left(\frac{x_1+1}{2} + \sin\left(\pi\frac{x_2+1}{2}\right)\right)\right)$$

The functions, shown in Figure 5, are defined in the $[-1,1]^2$ interval, which is, due to algorithmic reasons [8], normalized to $[0,1]^2$ for XCSF.

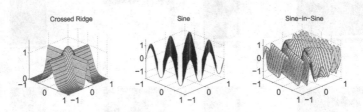

Fig. 5. The three target functions

We include two implementations of our distance metric update (Equation 3) in the comparison:

1. *CovLWPR*, replacing the default LWPR distance metric update (Equation 2) by our covariance update, and
2. *CovLLM*, replacing the default LWPR distance metric update by our covariance update and the default LWPR PLS regression with the LLM regression.

Of interest is the comparison in performance between the statistics–based approaches (CovLWPR/CovLLM and LWPR), and the comparison of the input space structuring between all four approaches. As in [8], the metrics include the mean absolute error (MAE), the population size, the visually–judged structuring of the input space, and the averege field volume, termed *generality* in [8]. Due to the difference in field activation mechanisms in LWPR and XCSF, the average field volume of the two is not directly comparable [8]. We are able, however, to compare the generality of LWPR to that of CovLWPR and CovLLM after training.

Experiment Two studies how fields shaped to exploit linear subspaces in the target function can improve the learning quality over those that do not in statistics–based methods. In contrast to Experiment One, which replicates exactly the conditions for comparison between LWPR and XCSF used in [8], we compare the populations of LWPR with the similarly–sized populations of CovLLM, the better–performing of the two covariance distance metric update implementations. The learning problem is the Sine, the function that has the most structure that can be exploited.

4.1   Experiment One

Ten independent runs of all four systems on each function were performed. Each run consisted of 100,000 learning iterations of noiseless samples from a uniform–random distribution in the input space. The values for the non–default parameters used are shown in Figure 6, and mirror those from [8]. In addition, we lower the w_prune parameter in LWPR from the default value of 1 (no pruning) to 0.75, in order to remove highly overlapping fields. The values of k^{err} used in CovLLM and CovLWPR were 0.0050 for the Crossed Ridge, 0.0200 for the Sine, and 0.0500 for the Sine–in–Sine.

	Value	Meaning
N	2500	Maximum population size
		for the Crossed Ridge and
		the Sine
	3950	Maximum population size
		for the Sine–in–Sine

(a) **XCSF**

	Value	Meaning
init_D	500	Initial size of RFs
alpha	1000	Distance metric learning rate
penalty	10^{-9}	Penalty term to avoid
		indefinite shrinking
w_gen	0.2	RF insertion treshold
w_prune	0.75	RF removal treshold

(b) **LWPR**

	Value
k^{in}	0.4
k^{out}	0.8
s^{err}	500
s^{g}	30
γ	5×10^{-4}

(c) **The weighting function ϕ**

	Value	Meaning
ϵ_A	0.45	Linear model learning rate
ϵ_{out}	0.45	Output weights learning rate

(d) **LLM**

Fig. 6. Values for the non–default parameters used in Experiment One

The resulting population plots for the three functions are shown in Figure 7 (for the ease of comparison, fields produced by XCSF have been scaled to 20% of their true sizes), and the performance is summarized in Figure 8.

We note that the results obtained for LWPR and XCSF are identical to [8]. XCSF produces a lower MAE on the Sine and the Sine–in–Sine, while the two perform equally well on the Crossed Ridge. The population sizes after learning are almost equal, while the structure of the functions are clearly visible from XCSF populations but not from LWPR populations.

For all functions, CovLLM and CovLWPR produce smaller final population sizes than LWPR and XCSF (Figure 8). Similarly, the average field volume in the final CovLLM and CovLWPR populations is greater than that in the LWPR population. This is due to the growth resulting from our distance metric update and the consequent pruning of overlapping fields. The final structures of the CovLLM and CovLWPR populations resemble the learned functions much more than the structures of LWPR populations, and are visually similar to the final structures of the XCSF populations.

Although the difference is not drastic, CovLLM produces a slightly higher error than LWPR on all functions in Experiment One. Elongated ellipsoidal fields

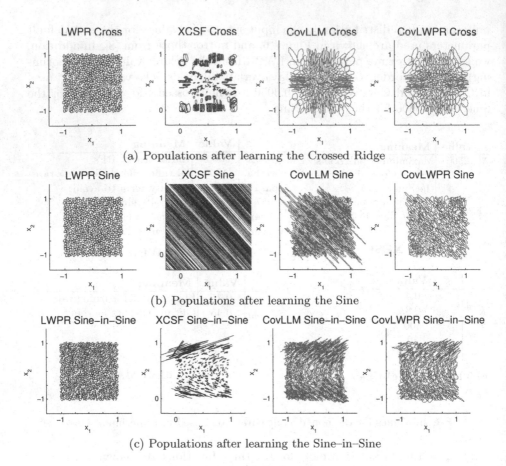

(a) Populations after learning the Crossed Ridge

(b) Populations after learning the Sine

(c) Populations after learning the Sine–in–Sine

Fig. 7. Population plots for Experiment One

in CovLLM's populations (most evident in the Sine population plot in Figure 7(b)) are more prone to errors when subjected to rotation than the spherical fields in LWPR. This is especially true for extremely long fields, where even slight rotations produce large errors in the extremities along the major axis.

CovLLM consistently outperforms CovLWPR throughout Experiment One. Unlike the regression in LLM, the PLS regression retains the history of the training samples seen previously, making it difficult to recover from incorrect shape updates while the regression is still being trained. This restrains the system from learning responsively, and is illustrated both by the higher error and the population plots. The resulting fields are not as slender in the non–linear directions, they are not as long in the linear directions, and some fields are misaligned. The amount of history kept can be reduced but, unfortunately, the PLS regression in such cases becomes unreliable, causing the mutual dependency between the shape and the regression to interfere with the ability to learn.

	MAE	Number of RFs	Generality
LWPR	0.0056 ± 0.0002	439.2 ± 6.58	0.0062 ± 0.000001
XCSF	0.0051 ± 0.0006	438.5 ± 8.96	0.0599 ± 0.0071
CovLLM	0.0075 ± 0.0004	327.7 ± 5.36	0.0107 ± 0.0003
CovLWPR	0.0084 ± 0.0005	344.8 ± 6.05	0.0103 ± 0.0003

(a) The Cross

	MAE	Number of RFs	Generality
LWPR	0.0127 ± 0.0003	460.2 ± 5.31	0.0057 ± 0.000006
XCSF	0.0040 ± 0.0018	452.2 ± 11.97	0.9210 ± 0.4875
CovLLM	0.0148 ± 0.0016	278.8 ± 7.45	0.0131 ± 0.0004
CovLWPR	0.0273 ± 0.0028	373.5 ± 10.00	0.0088 ± 0.0005

(b) The Sine

	MAE	Number of RFs	Generality
LWPR	0.0500 ± 0.0005	606.2 ± 7.07	0.0041 ± 0.00002
XCSF	0.0279 ± 0.0028	545.5 ± 20.00	0.0108 ± 0.0071
CovLLM	0.0762 ± 0.0050	585.7 ± 10.64	0.0054 ± 0.0001
CovLWPR	0.1408 ± 0.0104	474.0 ± 9.39	0.0065 ± 0.0002

(c) The Sine–in–Sine

Fig. 8. The prediction errors, population sizes, and generality measurements ($E \pm \sigma$) for Experiment One after learning

4.2 Experiment Two

We performed ten independent runs of CovLLM and LWPR on the Sine. As in Experiment One, each run consisted of 100,000 learning iterations of noiseless samples from a uniform–random distribution in the input space. In order to obtain comparable population sizes, we set init_D to 10, w_gen to 0.01, and init_alpha to 550 in both systems, and set k^{err} to 0.04 in CovLLM. The other parameters were set as shown in Figure 6(b), (c), and (d). The results of learning are summarized in Figure 9.

	MAE	Number of RFs	Generality
LWPR	0.0636 ± 0.0030	70.5 ± 3.75	0.0160 ± 0.0014
CovLLM	0.0387 ± 0.0021	72.2 ± 4.78	0.0319 ± 0.0020

Fig. 9. The results of Experiment Two: the prediction errors, population sizes, and generality measurements ($E \pm \sigma$) after learning

The receptive fields trained by CovLLM are larger, having an elongation along the linear (1, -1) direction. Thus, CovLLM exploits the linear subspaces in the Sine to produce a 40% lower error than LWPR on an almost identical population size.

5   Conclusion

A weighted covariance distance metric update for statistics–based LWR algorithms was introduced. It operates locally on each model, waiving the requirement for a global receptive field pruning mechanism. The method allows the fields to exploit the linear substructures of the target functions, which can lead to accurate learning with smaller populations. Unlike the update present in LWPR, it enables the structure of the function to be inferred from model populations.

References

[1] Atkeson, C.G., Moore, A.W., Schaal, S.: Locally weighted learning for control. Artificial Intelligence Review 11, 11–73 (1997)
[2] Finch, T.: Incremental calculation of weighted mean and variance. University of Cambridge (2009)
[3] Flentge, F.: Locally weighted interpolating growing neural gas. IEEE Transactions on Neural Networks 17(6), 1382–1393 (2006)
[4] Nguyen-tuong, D., Peters, J.: Local gaussian process regression for real-time model-based robot control. In: Proc. IEEE/RSJ International Conference on Intelligent Robots and Systems, IROS 2008, pp. 380–385 (2008)
[5] Ritter, H., Martinetz, T., Schulten, K.: Neural computation and self-organizing maps - an introduction. Computation and neural systems series. Addison-Wesley (1992)
[6] Schaal, S., Atkeson, C.G.: Constructive incremental learning from only local information. Neural Computation 10, 2047–2084 (1997)
[7] Sigaud, O., Salaun, C., Padois, V.: On-line regression algorithms for learning mechanical models of robots: a survey. Robotics and Autonomous Systems 59(12), 1115–1129 (2011)
[8] Stalph, P.O., Rubinsztajn, J., Sigaud, O., Butz, M.V.: A comparative study: function approximation with LWPR and XCSF. In: GECCO (Companion), pp. 1863–1870 (2010)
[9] Vijayakumar, S., D'Souza, A., Schaal, S.: Incremental online learning in high dimensions. Neural Computation 17, 2602–2634 (2005)
[10] Wilson, S.W.: Classifiers that approximate functions. Natural Computing 1(2-3), 211–234 (2002)

Kernel-Based Discriminative Re-ranking for Spoken Command Understanding in HRI

Roberto Basili[1], Emanuele Bastianelli[2], Giuseppe Castellucci[3],
Daniele Nardi[4], and Vittorio Perera[4]

[1] Dept. of Enterprise Engineering,
[2] Dept. of Civil Engineering and Computer Science Engineering,
[3] Dept. of Electronic Engineering,
University of Roma Tor Vergata,
Rome, Italy
basili@info.uniroma2.it,
{bastianelli,castellucci}@ing.uniroma2.it
[4] Dept. of Computer, Control, and Management Engineering,
University of Roma La Sapienza,
Rome, Italy
{nardi,perera}@dis.uniroma1.it

Abstract. Speech recognition is being addressed as one of the key technologies for a natural interaction with robots, that are targeting in the consumer market. However, speech recognition in human-robot interaction is typically affected by noisy conditions of the operational environment, that impact on the performance of the recognition of spoken commands. Consequently, finite-state grammars or statistical language models even though they can be tailored to the target domain exhibit high rate of false positives or low accuracy. In this paper, a discriminative re-ranking method is applied to a simple speech and language processing cascade, based on off-the-shelf components in realistic conditions. Tree kernels are here applied to improve the accuracy of the recognition process by re-ranking the n-best list returned by the speech recognition component. The rationale behind our approach is to reduce the effort for devising domain dependent solutions in the design of speech interfaces for language processing in human-robot interactions.

1   Introduction

As robots are being marketed for consumer applications (viz. telepresence, cleaning, entertainment, ...), natural language interaction is expected to make them more appealing and accessible to the general user. The latest technologies in speech understanding are now available on cheap computing devices or through the cloud and have been successfully deployed on a variety of consumer products, such as mobile phones or car navigation systems. The first consumer robots with speech interfaces have already appeared in the market.

The design of vocal interfaces for robots brings about a number of technological challenges in engineering state of the art technologies into solutions, that are suited for controlling the actions of complex devices, such as robots. Specifically, sound acquisition must take into account the position of the microphone, environmental noise, as well as distance and orientation towards the speaker. Speech recognition must provide

M. Baldoni et al. (Eds.): AI*IA 2013, LNAI 8249, pp. 169–180, 2013.
© Springer International Publishing Switzerland 2013

a robust and reliable performance, while natural language understanding should allow for a large and unconstrained language. Moreover, the interaction with robots becomes really challenging, because of the relationships that the language and robot behaviors have with the operational environment (i.e. the grounding problem).

In this paper, we focus on the design of the component of the interface that takes care of extracting commands for the robot from speech. While a full-fledged natural language interface would require additional components such as speech generation, dialog management and, possibly, the integration with other communication modalities, the understanding of spoken commands definitely plays a key role in the interaction. Our work is based on data acquired in a real set-up developed within the project S4R[1], where a wheeled robot executes spoken commands in a home environment.

While state of the art ASRs (Automatic Speech Recognition) no longer requires the construction of ad hoc acoustic models in order to achieve good performances, a key step is the characterization of the language model, which defines the language recognized by the ASR. The construction of the language model still represents a tough challenge for the interface designer. Consequently, the general purpose recognizers that recently became popular have attracted the interest of several researchers working on talking robots [10]. Unfortunately, also because of the low quality of the signal in robotic applications, these general purpose recognizers often fail to provide a correct representation of the spoken sentence and this makes the interpretation of the user commands error-prone.

In the literature, there are a few examples of speech interfaces for robots that rely on this kind of ASRs for speech understanding. Moreover, they typically tend to use the best output returned by the recognizer and focus on the grounding process. The goal of our work is to select the best candidate to be used in a spoken command understanding process through a learning to rank process. To this end, we have addressed several variants of the problem formulation and several approaches to the machine learning process. Specifically, we discuss two problem settings: the first one considering simple robot commands, while the second one considering more articulated commands. As for the machine learning techniques, we applied Tree Kernel methods, Lexical similarity (through Latent Semantic Analysis) and combinations of them.

Our experiments show that learning how to re-rank the multiple interpretations of the ASR can be beneficial, in particular in a context, such as the robotic one, where the sound signal can be affected by noise.

The paper is organized as follows. In the next section, we provide the background related work; then, we present the problem formulation, the proposed approach and our experimental results; we conclude the paper with a discussion of the contributions of the proposed approach and of directions for future research.

2 Motivations and Related Works

In this section we survey the related work as follow: first we look at different systems implementing vocal interfaces for HRI, then we look at works, from different fields, dealing with the re-ranking in Natural Language Processing.

---

[1] Speaky for Robots experiment of ECHORD European Cleaning House for Open Robotics Development http://www.echord.info/wikis/website/speaky

As above mentioned, the interest in **vocal interfaces for robots** is rapidly increasing. The tasks for which these interfaces have been developed range from supports to museums guided tours to demos of commercial products, from playing with children to acting as bartender. In the present work, a wheeled robot specialized to perform tasks in a home environment is targeted.

The *Language Model* for the ASR was early defined using grammars; more recently, *dictation engines* are becoming more popular in Vocal Robot Interfaces (VRIs). In [6] the two approaches are combined in order to have more consistent results; here we focus on the works using dictation engines.

In [12], [9] and [20] works on vocal interaction with robots are presented. Robots here can be instructed to follow route instructions or to perform commands as *"Put the tire pallet on the truck"*. These are grounded using the *Spatial Description Clauses* and the *Generalized Grounding Graph* approach, respectively. In all these cases, the quality of the ASR output has a crucial impact on the accuracy of the command understanding task.

In [7] and [2] the speech recognizer is provided by SPHINX[2] and the audio signal is registered by a microphone connected to the on-board laptop of the robot. Using the DIARC architecture, a fully incremental system is obtained that determines the meaning of the command received, generate the corresponding goal and actions, when no conflict with previous goals is raised.

The speech interface presented in [21] is tested on two different robots: a PR2 and a Turtlebot; in both cases the audio is acquired through the robot's microphone and then analyzed using the Android API. The output of the speech recognizer is parsed by the Stanford Parser, and then *Semantic Frames* are extracted. These represent a scene understanding paradigm, that in turn corresponds to the robot's executable plan, thus providing an integrated model for language processing and robot actions. Our work foresees a similar chain between the ASR and the parser.

To the best of our knowledge, all the above HRI approaches restrict their analysis to the first (i.e. best) output of the speech recognizer, which is not always correct since audio signals for robotic platforms can be largely affected by noise. A re-ranking approach taking into account several alternatives supports a more informed, global, optimization, jointly looking to wider evidence. This follows a general research line suggested in information retrieval ([19]), spoken language translation ([17]) and parsing ([3]).

Ranking multiple solutions is one of the main tasks commonly performed by information retrieval systems ([19]). Given a query and a list of relevant documents, the ranking stage produces a *new* ordered list: semantic properties of the query or user-specific modeling are here typically used as criteria to assess document relevance. **Learning to rank**, also called *re-ranking*, is the application of learning algorithms to the ranking. Two candidates in a pair are compared and the selection of the "best" one is induced from annotated data. When all pairs from an n-best list are classified, the final ranking is proportional to the number of matches won by a candidate.

In Natural Language Processing, re-ranking has been largely used such as in parse tree re-ranking [3] to improve the parsing accuracy, or Machine Translation (as in [17,22]). In these last approaches, a joint optimization of the Language (LM) and

---

[2] http://cmusphinx.sourceforge.net/

Acoustic Modeling (AM) with the Statistical alignment (SMT) model is suggested, with significant impact on the performance due to the correlations between the speech level (e.g. user adaptation) with the semantic level optimization. All these methods apply to the full set of LM, ASR and SMT parameters: this is not always guaranteed in HRI architectures where ASR and SLU are usually independently carried out.

Discriminative methods applied to re-ranking include [14], where joint modeling of speech recognition and language understanding through a perceptron classifier is defined in a conversational dialogue system. Multiple speech recognition and utterance classification hypotheses are merged into one list and the command understanding results are learned from the training set. Several discriminative approaches use of complex kernels ([16]). Tree kernels are applied to grammatical parsing in [3]. In spoken language understanding, in [5], a discriminative approach for the re-ranking of semantic hypotheses is applied. It is based on SVM and on the tree kernel formulation [3] while a "Preference Kernel" [18] is used as the final re-ranking metrics. All these works are close to our investigation. However, our method does not refer to any semantically annotated resource, as no information about commands or speech acts is used. This makes it applicable to a wider set of operational contexts in HRIs. In the next section, a flexible learning to rank model extending a general language and speech processing cascade for VRIs is presented. Command Recognition, usually acting on the solution output by the ASR, is here showed to be improved through accurate re-ranking. The novel aspect of our approach is that the adopted kernel integrates grammatical and lexical information without relying on domain dependent and manually engineered features, with significant performances and a very large applicability.

3   Re-ranking Interpretations

The adoption of general purpose language processing tools facilitates the design of general speech processing cascades in VRIs. In our proposed setting, a general purpose ASR system is first applied and the n-best list of candidate transcriptions is derived. Then, parsing of individual candidates is carried out through a standard NLP parser, able to derive parse trees from individual utterances. Syntactic trees of individual hypothesis are thus used as input to a kernel-based re-ranking step that rejects some and computes confidence scores for the final set of remaining candidates.

The selection of the correct command is a function over the set of hypotheses generated by local stochastic models, such as Stochastic Grammars (SGs) or Conditional Random Fields (CRF). It makes use of a global model over the adopted local models able to assign a more informed preference to hypothesis and thus discard the wrong ones. Global re-ranking models in fact can encode a larger number of properties, ranging from lexical features to grammatical dependencies obtained through the representation of individual examples in the form of trees or sequences. For example, given a spoken sentence such as "*Please bring me the cup of coffee that is on the table*", the following list of candidate hypotheses can be obtained from the ASR:

$i)$ *Please bring me the cup of coffee date on the table*
$ii)$ *Please bring me the cup of coffee that is on the table*
$iii)$ *Please bring me to cup of coffee that is on the table*

While ii) is the correct interpretation, it does not receive the best confidence and re-ranking aims at measuring that it violates the lowest number of grammatical constraints with respect to i) and iii). Notice that this preference depends on the grammar, as "... *the cup ... that is on the table ...*" corresponds to a properly formed fragment as a referring expression, while "*bring ... to cup ...*" is ungrammatical. However, lexical information is also important as "... *cup of coffee date ...*" in i) is implausible, as it lacks of a specific semantic interpretation given the lexical senses of the involved words. Learning the preference function thus means learning these patterns at both lexical and grammatical level.

Kernel methods have been successfully applied in machine learning to decouple the representation problem from the inductive algorithmics [16]. Kernels are used to map the current instances in a complex representation space where convergence of the training algorithm is guaranteed while more expressive features are made available. Syntactic kernels such as tree kernels compute similarity as the count of all meaningful struc-tures and substructures shared between two instances, i.e. between their corresponding parse trees (e.g. [3]). Starting from what said above, syntax can play a crucial role in the reranking process. The syntactic regularities seen in the training set can be captured and modeled by the tree kernel and, together with lexical semantics, they can drive the reranking process. In fact, the syntactic similarities between seen and unseen trees will assigning a lower score to those sentence among the candidate ones that present an irreg-ular syntactic structure. We will see in Section 3.2 how they are mathematically defined, while concentrating hereafter on the adopted re-ranking scheme.

3.1   Features and Grammatical Models for SVM-Based Ranking

In our VRI application, re-ranking aims essentially at classifying individual hypothesis pairs (H_i, H_j), where both H_i and H_j are included in the n-best list obtained from the ASR output. Each pair is an instance of the targeted total order among pairs: we want to learn a ranking function to acquire the hyperplane separating positive pairs (that reflects the correct ranking) from negative ones (i.e. pairs determining the inverted order). The classifier is expected to learn **if H_i is more accurate than** H_j as an interpretation of the spoken utterance. It can exploit the whole utterance transcription and a larger features set than the one used by local models (such as the HMMs largely adopted for ASR).

During training, given the n-best list of transcriptions H, positive pairs (H_i, H_k) can be derived, for $k \in [1...n]$ with $i \neq k$ and H_i the correct transcription. Instead, negative instances are obtained as the inverse of the positives (H_k, H_i). At classification time, hypotheses cannot refer to the gold standard, and all possible pairs (H_i, H_k), with $i, k \in [1...n]$ and $i \neq k$, are simply compared[3]. Given a pair of hypotheses $e_1 = (H_{11}, H_{12})$, and $e_2 = (H_{21}, H_{22})$, the re-ranking kernel K_R [18] over e_1 and e_2 is thus defined by:

$$K_R(e_1, e_2) = K(H_{11}, H_{21}) + K(H_{12}, H_{22}) - \\ K(H_{11}, H_{22}) - K(H_{12}, H_{21}) \tag{1}$$

where K can be any kernel function. It is worth noticing here that the definition of kernel K can combine other, more specific, kernels. For example, lexical features (such

---

[3] In this setting, just $n(n-1)/2$, and not n^2, pairs are compared as data are symmetric, and for the final model $K(H_i, H_j) = K(H_j, H_i)$.

as the conventional bag-of-words, BoW, representations) can be captured by a lexical kernel K_{lex}, while a tree kernel (as a generic TK K_{TK}) could be integrated through kernel composition (e.g. linear combinations), as:

$$K(u_1, u_2) = K_{lex}(u_1, u_2) + K_{TK}(u_1, u_2), \tag{2}$$

where utterances are individual sentences (i.e. hypotheses such as H_{ij}). K_{lex} insists on traditional feature vectors for u_i, while K_{TK} exploits the grammatical dependencies as expressed into the parse trees of its u_i's arguments.

3.2 Lexical Semantics and Tree Kernels for Re-ranking

Complex kernels are able to compose heterogenous sources of evidence, i.e. lexical and grammatical information about sentences. In our re-ranking model, we exploit a syntactic similarity model that is extended through lexical similarity, based on Latent Semantic Analysis (LSA, [13]). This is due to the adoption of Smoothed Partial Tree Kernels (SPTKs) [4]. SPTK measure similarity between two trees in terms of the number of shared subtrees, but the contribution to the kernel is also given by lexical nodes, proportionally to the lexical similarity (LSA) between the pairs of corresponding words.

LSA as a Lexical Similarity Model. LSA is inspired by dimensionality reduction methods. In LSA, a word-by-context matrix M obtained through large scale corpus analysis is decomposed through Singular Value Decomposition (SVD) [8,13] into the product of three new matrices: U, S, and V so that S is diagonal and $M = USV^T$. M is then approximated by $M_k = U_k S_k V_k^T$, where only the first k columns of U and V are used, corresponding to the first k greatest singular values. This approximation projects a generic word w_i into the k-dimensional space using $W = U_k S_k^{1/2}$, where each row corresponds to the representation vectors $\boldsymbol{w_i}$. The original statistical information about M is captured by the new k-dimensional space, which preserves the global structure while removing low-variant dimensions, i.e., distribution noise. Given two words w_1 and w_2, their similarity function σ is estimated as the cosine similarity between the corresponding projections $\boldsymbol{w_1}, \boldsymbol{w_2}$ in the LSA space, i.e $\sigma(w_1, w_2) = \frac{\boldsymbol{w_1} \cdot \boldsymbol{w_2}}{\|\boldsymbol{w_1}\| \|\boldsymbol{w_2}\|}$. This measure captures second order relations between words and is an effective model for paradigmatic relations. It captures synonymy and co-hyponymy between words, such as in "*hall, room* vs. *kitchen*" or "*line, direction* vs. *path*". Notice that σ is a valid kernel and can be used as a more effective form of lexical kernel, e.g. K_{lex}, in Eq. 2.

Tree Kernels Driven by Semantic Similarity. Tree kernels exploit syntactic similarity through the idea of convolutions among substructures. Any tree kernel computes the number of common substructures between two trees T_1 and T_2 without explicitly considering the whole fragment space. The general equation is:

$$TK(T_1, T_2) = \sum_{n_1 \in N_{T_1}} \sum_{n_2 \in N_{T_2}} \Delta(n_1, n_2), \tag{3}$$

where N_{T_1} and N_{T_2} are the sets of the T_1's and T_2's nodes, respectively and $\Delta(n_1, n_2)$ is equal to the number of common fragments rooted in the n_1 and n_2 nodes[4]. It is

---

[4] To have a similarity score between 0 and 1, a normalization in the kernel space, i.e. $\frac{TK(T_1,T_2)}{\sqrt{TK(T_1,T_1) \times TK(T_2,T_2)}}$ is applied.

worth noticing that the function Δ determines the nature of the kernel space and is used to model different kernels. Syntactic tree kernel (STK) are used to model complete context free rules as in [3]. Partial tree kernel (PTK, [15]) allows partial matches rules to be matched against the non terminal nodes of the parse trees. The algorithm for SPTK puts for more emphasis on lexical nodes. The Δ function is the following: if n_1 and n_2 are leaves then $\Delta_\sigma(n_1, n_2) = \mu\lambda\sigma(n_1, n_2)$; else

$$\Delta_\sigma(n_1, n_2) = \mu\sigma(n_1, n_2) \times \left(\lambda^2 + \sum_{I_1, I_2, l(I_1) = l(I_2)} \lambda^{d(I_1) + d(I_2)} \prod_{j=1}^{l(I_1)} \Delta_\sigma(c_{n_1}(I_{1j}), c_{n_2}(I_{2j}))\right),$$

(4)

where (1) σ is any similarity function between nodes, e.g., between their lexical labels; (2) $\lambda, \mu \in [0, 1]$ are decay factors; (3) $c_{n_1}(h)$ is the hth child of the node n_1; (4) I_1 and I_2 are two sequences of indexes, i.e., $I = (i_1, i_2, .., l(I))$, with $1 \leq i_1 < i_2 < .. < i_{l(I)}$; and (5) for the spans of matching sequences $d(I_1)$ and $d(I_2)$ the following holds: $d(I_1) = I_{1l(I_1)} - I_{11} + 1$ and $d(I_2) = I_{2l(I_2)} - I_{21} + 1$. The above SPTK kernel recursively matches tree structures and lexical nodes, thus generalizing grammatical and lexical information in training data.

4 Experimental Results

The validation of the proposed approach has been performed against the experimental setting developed for the S4R project[5], where a prototype speech interface has been implemented on a wheeled robot operating in a home environment. The adopted set-up, i.e. resources and specific feature engineering, as well as the outcomes of the experiments are hereafter presented.

Experimental data are audio files of spoken sentences that result into one or more robot actions and that are associated to a manually validated transcription. We considered two data sets with different characteristics. The first one (hereafter Simple Sentences, SS) includes commands that correspond to a unique robot action (e.g. "Go to the bathroom"). Sentences in SS have been automatically generated using the ASR grammars defined in the S4R prototype. They consist in 304 single commands, recorded through the built-in microphone of a laptop by 3 speakers. The second data set, named Realistic Set (RS), include more complex commands (e.g. "Go to the kitchen and take the cup of coffee"). These 143 sentences are free, more colloquial (e.g."Could you please go ... ") and correspond to commands recorded both through the built-in microphone of a laptop and through the S4R microphone. Finally, while SS sentences are recorded in a controlled environment, RS sentences are recorded by 12 speakers in more realistic environmental conditions. The RS set is thus representative of a more realistic setting than SS in terms of linguistic variability and audio quality.

All audio files are analyzed through the Google ASR web tool[6]. The service returns at most 10 different hypotheses for each recorded sentence. Google ASR also reports a confidence score ν_0 relative to the first interpretation only. In the experiments an ad-hoc

---

[5] http://labrococo.dis.uniroma1.it/?q=s4r
[6] Via an HTTP request to:
  https://www.google.com/speech-api/v1/recognize?

scaling function ($\nu_k = \frac{0.75}{k-1} \cdot \nu_0$) is applied to estimate scores characterizing the other interpretations from the Google ASR output.

As some output sentences were negligibly different from the others, lemmatization and duplicate removal have been applied to the transcribed sentences. In order to reduce the evaluation bias to errors, only those commands with an available solution within the input 10 candidates were retained for the experiments. The Final Simple Set (FSS) and the Final Realistic Set (FRS) obtained after this pre-processing stage, include thus 92 and 58 sentences, respectively. On average, we found about 6 interpretations per utterance.

Different experimental settings are used to verify the validity of the re-ranking approach. In a first run, the validity of the Preference Kernel formulation is experimented on the two datasets separately. Then, different combinations of the two datasets are exploited to better verify the generalization capability of the presented model. In all settings, *Coverage at k*, i.e. the percentage of sentences for which the correct transcription is produced within the first k hypotheses, is used to evaluate the re-ranker quality against two baselines: the original output of the Google ASR and a 'dummy' classifier based on a simple Bag-Of-Words (BOW) representation.

Feature Engineering. The Preference Kernel formulation (Eq. 1) is used as the global model for re-ranking, as discussed in Section 3. Tree kernels combined with other features have been here tested. Constituent based parse trees, obtained with the Stanford Parser ([11]), are used as the representation formalism of each example.

The adopted tree kernels are the standard PTK, as in [15], as well as the SPTK as in Eq. 4, where the similarity function is modeled through LSA. LSA vectors are derived from a Word Space built from the analysis of the ukWaC data set [1]. Notice that this Web corpus is not specific to the S4R domain. In the Word Space construction, POS tagging is first applied and co-occurrence vectors are computed over small windows of size $n = \pm 3$, that better captures paradigmatic lexical properties of words. LSA vectors help lexical generalization as they capture word similarity (or dissimilarity): examples with the word *room* are more similar to examples with *kitchen* than to *table*, while *table* is similar to *chair*. In the generation of the Word Space, the 20,000 most frequent words are used to build the source co-occurrence M matrix. Vector components are weighted through the point-wise mutual information scores. The SVD reduction was applied to M, with a dimensionality cut of $k = 250$. The LSA vectors are used to compute the σ function in Eq. 4.

The tree kernel is combined in several experiments with another feature vector each used to capture lexical and other properties of an instance. This gives rise to two kernels as suggested in Eq. 2. The feature vector considered in the linear kernel (K_{lex}), is obtained by different configurations of the following:

- LSA refers to the vector representing the candidate sentence in the LSA space, as the linear combination of single word vectors composing it.
- CONF refers to the vector representing the numerical score corresponding to the confidence output by the ASR (or computed through the scaling function previously defined).

Due to the lack of a gold standard ranking, in training all the *gold vs. other* interpretation pairs are used and their inverse are used for negative examples. The final rank over the n-best list of candidates for a test sentence associates a score $\pi(e_i) = \frac{1}{n}\sum_{j=1}^{n} K_R(e_i, e_j)$ to each interpretation, proportional to its comparisons with all other entries. K_R is the score of the kernel function (Eq. 1) evaluated over examples (e_i, e_j) and computed by the SVM-Light-TK implementation[7].

In the first experiment, the re-ranking approach is tested on the two datasets separately. These settings are useful to verify how the re-ranking approach can be beneficial in a in-domain scenario.

Simple Set. In the first experiment only the simple dataset FSS is used. A 70%-30% training-test split is adopted and 5-fold parameter tuning over the training set has been applied. Figure 1 reports results in terms of the coverage achieved when k best interpretations are admitted. On the abscissas the positions are reported, while in the ordinates the coverage is shown.

The outcome of this first experiment shows substantial improvement. Notice that the best re-ranking configuration is the $SPTK+CONF$, where K_{TK} is the SPTK and K_{lex} is applied to the $CONF$ vector on the first two positions. Moreover, the general purpose architecture controlled via SVM-based re-ranking is outperforming other methods. Finally, the specific nature of the SPTK kernel, that combines the syntactic constraints with lexical semantic information (through the LSA vectors), seems to adequately model the relevant aspects of the input commands. Notice that, the proposed approach better captures the concept of order between interpretations with respect to a simple Bag-Of-Word (BOW) model. In fact, the BOW model has a poorer performance than the Google one. Moreover, other tests showed that the $LSA+CONF$ setting does not achieve the same performance of the tree kernel.

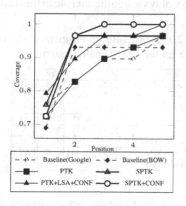

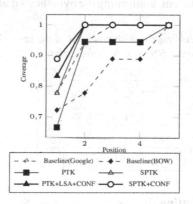

Fig. 1. Coverage at k-th positions for the SS vs. SS configuration

Fig. 2. Coverage at k-th positions for the RS vs. RS configuration

Realistic Set. In this experiment the realistic dataset FRS only is used. Again, a 70%-30% training-test split is adopted and 5-fold parameter tuning over the training set has been applied. Figure 2 reports a plot of the results in term of the coverage achieved when

---

[7] Available at http://disi.unitn.it/moschitti/Tree-Kernel.htm

k best interpretations are admitted. Again, on the abscissas the positions are reported, while in the ordinates the coverage is shown.

In this case, Google achieves a very high performance. This is probably due to the fact that acoustic models used by Google are very sensitive to the pronunciation. Notice that the Realistic Set is composed by audio files recorded by native speakers (i.e. better pronunciation), while Simple Set audio files are recorded mainly by non-native speakers: it influences the ASR performance and consequently the NL parser used to produce the parse trees. Even if the pure Google performance is remarkable, our model is able to improve the overall performance positioning the correct utterance at most in the 2nd position. This is practically relevant, since allows subsequent processes, e.g. a Semantic Parser, to consider fewer alternatives. As in the previous setting, the configuration without trees is unable to reach the same performance as the ones that use them.

In this setting the generalization capability of the re-ranking algorithm is experimented. Two main configuration are here explored. First, a SS vs RS is discussed. Finally, the second configuration is obtained by mixing the two dataset.

Simple Set vs. Realistic Set. In this experiment both the simple and the realistic dataset are used. In this case, the FSS dataset is used for training, while the RSS dataset is used for testing. Again, a 5-fold parameter tuning over the training set has been applied. This seems to better represent realistic conditions, where most simple commands are available to train the system, and only a smaller percentage of more complex structures contributes to learning. Table 1 reports the results in terms of the coverage achieved when k best interpretations are admitted. Also in this case, the re-ranking model is able to improve the Google baseline. The $SPTK+LSA+CONF$ configuration is able to move to the very first positions (at most 2) the correct interpretation. Again, the $LSA+CONF$ configuration is not sufficient, confirming the hypothesis that syntax plays a significant role in this task.

Table 1. Coverage at k-th positions of the re-ranking for the SS vs. RS configuration

System	Cov_1	Cov_2	Cov_3	Cov_4
Baseline (Google)	0.82	0.89	0.96	0.98
Baseline (BOW)	0.58	0.81	0.89	0.93
LSA+CONF	0.87	0.93	0.98	1.0
PTK	0.74	0.86	0.89	0.94
SPTK	0.86	0.93	0.96	0.98
PTK+LSA+CONF	0.86	0.94	1.0	1.0
SPTK+CONF	0.86	0.96	1.0	1.0
SPTK+LSA+CONF	0.87	0.96	1.0	1.0

Table 2. Coverage at k-th positions of the re-ranking for the Mixed configuration

System	Cov_1	Cov_2	Cov_3	Cov_4
Baseline (Google)	0.76	0.87	0.93	0.93
Baseline (BOW)	0.78	0.93	0.95	0.97
LSA+CONF	0.82	0.91	0.93	0.93
PTK	0.74	0.85	0.89	0.93
SPTK	0.72	0.95	0.95	0.97
PTK+LSA+CONF	0.80	0.89	0.95	0.95
SPTK+CONF	0.82	0.93	0.97	0.97
SPTK+LSA+CONF	0.80	0.89	0.93	0.97

Mixed Setting. In this experiment the Simple Set and the Realistic Set are mixed. In particular, training data are obtained by merging the 70% of both datasets, and testing data are obtained by considering the remaining 30% of both datasets. A 5-fold parameter tuning over the training set has been applied. Table 2 reports the performances for the mixed configuration. In this case, the re-ranking model is able to improve the Google baseline as well as the BOW baseline. The $SPTK+LSA+CONF$ and the $SPTK+CONF$ configurations are again the best performing with this setting. Notice that this experimental set-up is more difficult with respect the others, as for the more heterogeneous operational conditions and the reduced size of the two training datasets.

The aim of our experiments was to demonstrate that a re-ranking approach could be successfully applied to the different interpretations that an ASR engine produces in voice interaction with a service robot. In particular, our data setting is constructed to reflect the variability of train/test data that better fits the robot operational requirements, with colloquial forms and multiple robot commands for one single utterance.

In the experiments, we considered several additional variants of features and configurations. Overall, our experiments suggest the following conclusions. Syntactic features alone (as provided by the PTK) seem not sufficient for accurate re-ranking. Different ASR transcriptions are very similar to each other from a syntactic point of view and do not allow to discriminate examples. In fact, if a PTK similarity is applied to all trees in the training and test set, respectively, an average of 0.72-0.73 similarity score is obtained. Speech confidence features and, mostly, LSA are significantly beneficial to the re-ranker accuracy when PTK is involved. Moreover, when using SPTK, the LSA features are embedded and, in most cases, do not need to be added to the vector. Finally, none of the proposed approaches requires manual feature engineering, thus making them applicable with low effort.

5 Conclusions

In this paper, standard processing chains and state-of-the-art learning algorithms for speech and language interpretation in VRIs are proposed. Specifically, we address a general purpose ASR for a wheeled robot speech interface operating in home environment applications. A discriminative re-ranking methodology is thus proposed to select the best interpretation hypothesis within the n-best list of candidates returned by the ASR component. Syntactic properties of the utterances are exploited through the use of advanced tree kernels. Several combinations of lexical, syntactic and even basic audio features (e.g. the confidence of the ASR system) have been experimented over domain specific data sets. The results confirm that SVM re-ranking is applicable and very effective (up to 20% error reduction). Moreover, the proposed kernels make no use of domain specific features, thus confirming their wide applicability in quite different domains and operational conditions. More experimentation is needed to assess some of these outcomes, such as the test of other ASR sytems or datasets, with a larger variety of speakers and conditions.

The increased performance of general purpose ASR, also supported by the improvement through re-ranking suggested in this paper, enables for a wider use of off-the-shelf language and speech processing tools are in modern robotic systems. They would allow to use larger lexicons and grammars in support of more complex interactions (e.g. dialogue) with realistic costs and effort.

References

1. Baroni, M., Bernardini, S., Ferraresi, A., Zanchetta, E.: The wacky wide web: a collection of very large linguistically processed web-crawled corpora. LRE 43(3), 209–226 (2009)
2. Cantrell, R., Scheutz, M., Schermerhorn, P., Wu, X.: Robust spoken instruction understanding for HRI. In: 2010 5th ACM/IEEE International Conference on HRI, pp. 275–282 (March 2010)

3. Collins, M., Duffy, N.: New ranking algorithms for parsing and tagging: kernels over discrete structures, and the voted perceptron. In: Proc. of the 40th Annual Meeting on ACL, ACL 2002, pp. 263–270. Association for Computational Linguistics, Stroudsburg (2002), http://dx.doi.org/10.3115/1073083.1073128
4. Croce, D., Moschitti, A., Basili, R.: Structured lexical similarity via convolution kernels on dependency trees. In: EMNLP, pp. 1034–1046 (2011)
5. Dinarelli, M., Moschitti, A., Riccardi, G.: Discriminative reranking for spoken language understanding. IEEE Transactions on Audio, Speech & Language Processing 20(2), 526–539 (2012)
6. Doostdar, M., Schiffer, S., Lakemeyer, G.: A robust speech recognition system for service-robotics applications. In: Iocchi, L., Matsubara, H., Weitzenfeld, A., Zhou, C. (eds.) RoboCup 2008. LNCS, vol. 5399, pp. 1–12. Springer, Heidelberg (2009)
7. Dzifcak, J., Scheutz, M., Baral, C., Schermerhorn, P.: What to do and how to do it: Translating natural language directives into temporal and dynamic logic representation for goal management and action execution. In: Proc. ICRA 2009, Kobe, Japan (May 2009)
8. Golub, G., Kahan, W.: Calculating the singular values and pseudo-inverse of a matrix. J. Soc. Ind. Appl. Math.: Series B, Numerical Analysis (1965)
9. Huang, A.S., Tellex, S., Bachrach, A., Kollar, T., Roy, D., Roy, N.: Natural language command of an autonomous micro-air vehicle. In: IROS, Taipei, Taiwan, pp. 2663–2669 (October 2010)
10. Special Issue: Dialogue with Robots, vol. 34(2) (2011)
11. Klein, D., Manning, C.D.: Accurate unlexicalized parsing. In: Proc. of ACL 2003, pp. 423–430 (2003)
12. Kollar, T., Tellex, S., Roy, D., Roy, N.: Toward understanding natural language directions. In: 2010 5th ACM/IEEE International Conference on HRI (2010)
13. Landauer, T., Dumais, S.: A solution to plato's problem: The latent semantic analysis theory of acquisition, induction and representation of knowledge. Psychological Review 104 (1997)
14. Morbini, F., Audhkhasi, K., Artstein, R., Segbroeck, M.V., Sagae, K., Georgiou, P.G., Traum, D.R., Narayanan, S.S.: A reranking approach for recognition and classification of speech input in conversational dialogue systems. In: SLT, pp. 49–54 (2012)
15. Moschitti, A.: Efficient convolution kernels for dependency and constituent syntactic trees. In: Fürnkranz, J., Scheffer, T., Spiliopoulou, M. (eds.) ECML 2006. LNCS (LNAI), vol. 4212, pp. 318–329. Springer, Heidelberg (2006)
16. Shawe-Taylor, J., Cristianini, N.: Kernel Methods for Pattern Analysis. Cambridge University Press (2004)
17. Shen, L., Sarkar, A., Och, F.: Discriminative reranking for machine translation. In: Proc. of HLT-NAACL 2004, pp. 177–184 (2004)
18. Shen, L., Joshi, A.K.: An svm based voting algorithm with application to parse reranking. In: Proc. of HLT-NAACL 2003, CONLL 2003, vol. 4, pp. 9–16. ACL, Stroudsburg (2003)
19. Taylor, M., Zaragoza, H., Craswell, N., Robertson, S., Burges, C.: Optimisation methods for ranking functions with multiple parameters. In: Proc. of the 15th ACM International CIKM, CIKM 2006, pp. 585–593. ACM, New York (2006)
20. Tellex, S., Kollar, T., Dickerson, S., Walter, M.R., Banerjee, A.G., Teller, S., Roy, N.: Understanding natural language commands for robotic navigation and mobile manipulation. In: Proc. of AAAI, San Francisco, CA, pp. 1507–1514 (2011)
21. Thomas, B., Jenkins, O.C.: Verb semantics for robot dialog. In: Robotics: Science and Systems Workshop on Grounding Human-Robot Dialog for Spatial Tasks, Los Angeles, CA, USA (June 2011)
22. Wang, W., Stolcke, A., Zheng, J.: Reranking machine translation hypotheses with structured and web-based language models. In: IEEE Workshop on ASRU 2007, pp. 159–164. IEEE (2007)

A Natural Language Account
for Argumentation Schemes

Elena Cabrio[1], Sara Tonelli[2], and Serena Villata[1]

[1] INRIA Sophia Antipolis, France
firstname.lastname@inria.fr
[2] Fondazione Bruno Kessler, Trento, Italy
satonelli@fbk.eu

Abstract. One of the essential activities carried out by humans in their everyday linguistic interactions is the act of drawing a conclusion from given facts through some forms of reasoning. Given a sequence of statements (i.e. the premises), humans are able to infer or derive a conclusion that follows from the facts described in the premises. In the computational linguistics field, discourse analyses have been conducted to identify the discourse structure of connected text, i.e. the nature of the discourse relationships between sentences. In parallel, research in argumentation theory has proposed argumentation schemes as structures for defining various kinds of arguments. Although the two fields of study are strongly intertwined, only a few works have put them into relation. However, a clear natural language account for argumentation schemes is still missing. To address this open issue, our work analyses how argumentation schemes fit into the discourse relations in the Penn Discourse Treebank.

1 Introduction

Argumentation theory [19] has been proposed to tackle a variety of problems in Artificial Intelligence. In particular, reasoning systems have to interact not only with intelligent agents but also with humans. This means that they should be able to reason not only in a purely deductive monotonic way, but they need to carry out presumptive, defeasible reasoning. Moreover, the arguments behind this reasoning must be expressed in a dialogical form such that they can be consumed by humans too. Argumentation schemes [26] have been introduced to capture reasoning patterns which are both non-deductive and non-monotonic as used in everyday interactions. The issue of representing the structure of the arguments used by humans in everyday interactions has been addressed also in the computational linguistics field where discourse analysis aims at identifying the discourse structure of connected text, i.e. the nature of the discourse relationships between sentences [13]. Even if the two research lines are connected, a clear account of the relation linking the two is still missing. The research question we answer in this paper is: *How to bridge the argument patterns proposed in argumentation schemes and in discourse analysis towards a better account of natural language arguments?* In other words, we aim at studying what is the connection between the argumentation schemes and the discourse relations detected in discourse analysis.

M. Baldoni et al. (Eds.): AI*IA 2013, LNAI 8249, pp. 181–192, 2013.

We select five argumentation schemes, and we map these patterns to the categories of the discourse relations in the Penn Discourse Treebank (PDTB) [18]. In particular, we highlight which relations can be annotated with the corresponding scheme, and we extract the connectives characterizing each scheme in natural language (NL) data.

The advantage of this analysis is twofold. First, the dataset resulting from this investigation, where the categories of the PDTB are annotated with the schemes they are associated with, represents a rich training corpus fundamental for the improvement of the state of research in argumentation in computational linguistics, as highlighted by Feng and Hirst [9]. Second, this mapping between argumentation schemes and PDTB relations can be fruitfully used to support automated classification [9] or argument processing [3]. In this paper, we do not use NL semantics for a better understanding of critical questions in argumentation schemes [28].

The layout of the paper is as follows. In Section 2 we provide the basic idea behind the definition of argumentation schemes, as well as the description of the schemes we consider. After introducing the PDTB, Section 3 presents our analysis on how argumentation schemes are represented in the PDTB. In Section 4, we summarize the related research comparing it with the proposed approach.

2   Argumentation Schemes

Argumentation schemes [26] are argument forms that represent inferential structures of arguments used in everyday discourse. In particular, argumentation schemes are exploited in contexts like legal argumentation [10], inter-agent communication [22,15], and pedagogy [24]. They are motivated by the observation that most of the schemes that are of central interest in argumentation theory are forms of plausible reasoning that do not fit into the traditional deductive and inductive argument forms [19]. Each scheme is associated with a set of so called *critical questions* (CQ), which represent standard ways of critically probing into an argument to find aspects of it that are open for criticism. In particular, the combination of argumentation scheme and critical questions is used to *evaluate* the argument in a particular case: the argument is evaluated by judging if all the premises are supported by some weight of evidence. In this case, the weight of acceptability is shifted towards the conclusion of the argument which is further subject to a rebuttal by means of the appropriate critical question.

Let us consider the following argumentation scheme:

Argument from Example

Premise: In this particular case, the individual a has property F and also property G.
Conclusion: Therefore, generally, if x has property F, then it also has property G.

This scheme corresponds to one of the most common types of reasoning in debates [12] since it is used to support some kind of generalization. The argument from example is a weak form of argumentation that do not confirm a claim in a conclusive way, or associate it with a certain probability, but it gives only a small weight of presumption to support the conclusion. Three examples of critical questions for the *Argument from Example* scheme are the following:

CQ1: Is the proposition presented by the example in fact true?
CQ2: Does the example support the general claim it is supposed to be an instance of?
CQ3: Is the example typical of the kinds of cases that the generalization ranges over?

For the purpose of this paper, we do not consider all the 65 argumentation schemes presented by Walton and colleagues [26] since some of them, like the *Argument from Position to Know*, deal with argument patterns which involve the information sources. Reasoning about the information source using argumentation schemes [16] is out of the scope of this paper. Beside the above presented *Argument from Example*, the following argumentation schemes will be the focus of the analysis we carry out in this paper.

Argument from Cause to Effect

Major Premise: Generally, if A occurs, then B will (might) occur.
Minor Premise: In this case, A occurs (might occur).
Conclusion: Therefore, in this case, B will (might) occur.

Argument from Effect to Cause

Major Premise: Generally, if A occurs, then B will (might) occur.
Minor Premise: In this case, B did in fact occur.
Conclusion: Therefore, in this case, A also presumably occured.

Practical Reasoning

Major Premise: I have a goal G.
Minor Premise: Carrying out action A is a means to realize G.
Conclusion: Therefore, I ought (practically speaking) to carry out this action A.

Argument from Inconsistency

Premise: If a is committed to proposition A (generally, or in virtue of what she said in the past)
Premise: a is committed to proposition $\neg A$, which is the conclusion of the argument α that a presently advocates.
Conclusion: Therefore, a's argument α should not be accepted.

Argumentation schemes have been used in the Araucaria system [23] to mark instantiations of such schemes explicitly, providing in this way an online repository of arguments.[1] This annotated corpus contains approximately 600 arguments, manually annotated, extracted from various sources such as US Congress Congressional Record, and New York Times. Although, up to our knowledge, Araucaria is the best argumentation corpus available to date, it still has some drawbacks. First, Araucaria is rather small if compared for instance with the PDTB. Moreover, given that the final aim of this paper is to bridge discourse in Natural Language Processing (NLP) and argumentation schemes, we need a corpus like the PDTB, which is a well-established, standard reference in NLP, and where the discourse relations are already annotated. We limit our

---

[1] http://araucaria.computing.dundee.ac.uk/

analysis to the above mentioned five schemes because they naturally fit the discourse relations used in the PDTB given their high degree of generality, and the fact that they do not involve contextual elements, i.e., some kind of knowledge about the situation where the reasoning proposed by the scheme is addressed. It is precisely for these reasons that in the present pilot study, we do not consider the remaining argumentation schemes proposed by Walton [26].

3   Bridging Argumentation Schemes and Discourse Relations

In this section we position and analyze the work carried out in the computational linguistics field on discourse analysis, under the perspective of argumentation schemes. We rely on the PDTB (Section 3.1) as the reference resource of NL text annotated with discourse relations. In Section 3.2 we start from the argumentation schemes, and we analyze how they fit into the categories of the discourse relations in PDTB. Examples in natural language support us in bringing to light the similarities and the discrepancies between the classifications sketched by the two research fields. Section 3.3 describes the feasibility study we carried out on the PDTB, and the annotation results.

3.1   The Penn Discourse Treebank

The PDTB is a resource built on top of the Wall Street Journal corpus [14] consisting of a million words annotated with discourse relations by human annotators. Discourse connectives are seen as discourse predicates taking two text spans as arguments, that correspond to propositions, events and states. In the PDTB, relations can be explicitly signaled by a set of lexically defined connectives (e.g. "because", "however"). In these cases, the relation is overtly marked, which makes it relatively easy to detect using NLP techniques [17]. A relation between two discourse arguments, however, does not necessarily require an explicit connective, because it can be inferred also if a connective expression is missing. These cases, referred to as *implicit relations*, are annotated in the PDTB only between adjacent sentences within paragraphs. If the connective is not overt, PDTB annotators were asked to insert a connective to express the inferred relation.

The abstract objects involved in a discourse relation are called Arg1 and Arg2 according to syntactic criteria, and each relation can take two and only two arguments. Examples 1 (a) and (b) represent sentences connected, respectively, by an explicit and an implicit relation. Arg1 and Arg2 are reported in italics and in bold respectively.

Example 1
(a) Explicit: *The federal government suspended sales of U.S. savings bonds* <u>because</u> **Congress hasn't lifted the ceiling on government debt.**
(b) Implicit: *The projects already under construction will increase Las Vegas's supply of hotel rooms by 11,795, or nearly 20%, to 75,500.* **By a rule of thumb of 1.5 new jobs for each new hotel room, Clark County will have nearly 18,000 new jobs.**

While in Example 1(a) the connective "because" explicitly signals a causal relation holding between Arg1 and Arg2, in (b) no connective was originally expressed.

A consequence relation is inferred between '*the increase in the number of rooms*' and '*the increase in the number of jobs*', though no *explicit* connective expresses this relation.

Each discourse relation is assigned a sense label based on a three-layered hierarchy of senses. The top-level, or *class level*, includes four major semantic classes, namely TEMPORAL, CONTINGENCY, COMPARISON and EXPANSION. For each class, a more fine-grained classification has been specified at *type* level (see Figure 1). For instance, the relation in Example 1(a) belongs to the CONTINGENCY class and the *Cause* type. A further level of *subtype* has been introduced to specify the semantic contribution of each argument. *Cause*, for instance, comprises the *reason* and the *result* subtypes. The former applies when the situation described in Arg2 is the cause of the situation in Arg1 (Example 1 (a)), while the latter indicates that the situation in Arg2 results from the situation in Arg1. The annotation scheme was developed and refined by the PDTB group in a bottom-up fashion, following a lexically grounded approach to annotation.

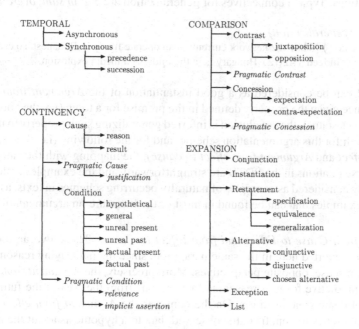

Fig. 1. Sense tags [The PDTB Research Group, 2008]

While in the PDTB framework, arguments are not considered "logical arguments", for convention in our work we represent them in the standard format of a logical argument, where Arg1 is a (set of) premise(s), and Arg2 is the conclusion.

3.2 Mapping Argumentation Schemes to PDTB

In the following, we investigate how the argumentation schemes described in [26] (see Section 2) fit into the discourse relations in the PDTB. The examples extracted from the PDTB for some categories of discourse relations perfectly represent instantiations of the argumentation schemes (e.g., the discourse relation EXPANSION:*Restatement:*"generalization" fits into the argumentation scheme *Argument from example*). On the contrary, for some other schemes the mapping is less straightforward, even if the relation definitions in the PDTB and the provided schemes are similar (see the PDTB Manual [25]).

Argument from Example. As introduced before, such argumentation scheme is used to support some kind of generalization. Its definition shows high similarity with the discourse relation EXPANSION:*Restatement:*"generalization". More specifically, "generalization" applies when the connective indicates that Arg2 (i.e. the conclusion) summarizes Arg1 (the premises), or in some cases expresses a conclusion based on Arg1 (as in Example 2). Differently from the argumentation schemes, where the standard format allows *therefore* as the only connective to introduce the conclusion, in NL different connectives can be used to this purpose, and can vary according to the expressed discourse relations. Typical connectives for generalization are e.g. *in sum, overall, finally.*

Example 2 (generalization)
PREMISE: (Arg1) While the network currently can operate freely in Budapest, so can others
CONCLUSION: **indeed** (Arg2) Hungary is in the midst of a media explosion.

Example 2 can be considered as a good instantiation of the *Argument from example* scheme, since given the property defined in the premise for a town (i.e. the good quality of the network status), the conclusion is inferred generalizing such property to the whole country. Both for this argumentation scheme, and for the following (i.e. *Argument from cause to effect* and *Argument from effect to cause*), the mapping with the categories of the discourse relations in the PDTB are straightforward, and the examples collected can be fruitfully considered as examples of naturally occurring schemes in texts, as opposed to ad-hoc examples that can be found in most of the literature on argumentation theory.

Argument from Cause to Effect and from Effect to Cause. These two argumentation schemes are reported here in the same paragraph, since the underlying reasoning steps address, in a sense, opposite perspectives. More precisely, the *Argument from cause to effect* is a predictive form of reasoning that reasons from the past to the future, based on a probabilistic generalization. On the contrary, the *Argument from effect to cause* is based on a retroduction, from the observed data to a hypothesis about the presumed cause of the data (abductive reasoning) [26]. Comparing these definitions with the definitions provided for the discourse relations in the PDTB, we can note that they are highly similar with the discourse relation: CONTINGENCY:*cause*, identified when the situations described in Arg1 and Arg2 are causally influenced, and the two are not in a conditional relation. Directionality is specified at the level of subtype: "reason" ($\|Arg2\| < \|Arg1\|^2$, see Example 3) and "result" ($\|Arg1\| < \|Arg2\|$, see Example 4) specifying which situation is the cause and which the effect. Both subtypes can be respectively mapped to the argumentation schemes *Argument from effect to cause*, and *Argument from cause to effect*. In the former (i.e."reason") the connective indicates that the situation described in Arg2 is the cause, and the situation in Arg1 is the effect.

---

[2] The symbol $<$ used in the PDTB categories means "causes".

The typical connective for such relation is indeed *because*. On the contrary, for the latter (i.e. "result") , the connective indicates that the situation in `Arg1` is the reason, and the situation in `Arg2` is the result. Typical connectives are *so that, thefore, as a result*.

Example 3 (reason)
CONCLUSION: (`Arg1`) she pleaded guilty.
PREMISE: **because** (`Arg2`) she was afraid of further charges

Example 4 (result)
PREMISE: (`Arg1`) Producers were granted the right earlier this year to ship sugar and the export licenses were expected to have begun to be issued yesterday
CONCLUSION: **as a result** (`Arg2`) it is believed that little or no sugar from the 1989-90 crop has been shipped yet

Note that, due to language variability, the sequence of premises and conclusion in NL arguments does not always follow the standard structure (where premises always come first), as e.g. in Example 3, where the conclusion is expressed at the beginning of the sentence. In the same example, the reasoning is carried out from effect to cause (i.e. the fact that she was afraid of further charges, generates the woman's reaction of declaring herself guilty). On the contrary, in Example 4, the reasoning is carried out from cause to effect (i.e. the fact that licenses were expected to have been issued the day before - but it did not happen - let to conclude that the sugar has not been shipped yet).

So far so good. In the following, we enter into a grey area, where the mapping between the argumentation schemes and the categories of the discourse relations is more blurry, and the examples collected in the PDTB do not always represent correct instantiations of such schemes. But since the goal of our work is to investigate all the possible connections between the two research fields, we allow us some simplifications with respect to the actual complexity that the reasoning step addressed by these argumentation schemes involve.

Practical reasoning. This argumentation scheme involves the general human capacity for resolving, through reflection, the question of what one is to do, given the goal that one has in mind. To fit such scheme into one discourse category, we need therefore to consider a relation that relies on some kind of pragmatic reasoning, and on common background knowledge. For this reason, we think that the most appropriate relation annotated in the PDTB is the CONTINGENCY:*Pragmatic condition*, used for instances of conditional constructions whose interpretation deviates from that of the semantics of *Condition*. In all cases, `Arg1` holds true independently of `Arg2`. The conditional clause in the "relevance" conditional (`Arg2`, i.e. the premise) provides the context in which the description of the situation in `Arg1`, i.e. the conclusion, is relevant (see Example 5). There is no causal relation between the two arguments.

Example 5 (relevance)
PREMISE (`Arg1`): here's the monthly sum you will need to invest to pay for four years at Yale, Notre Dame and University of Minnesota PREMISE : **if** (`Arg2`) you start saving for your child's education on Jan. 1, 1990

In Example 5 the major premise, i.e. the goal, is implicit (i.e., enthymeme [26]), and concerns the child education (in other words, the goal is to send the child to one of the best U.S. universities). The other two premises (i.e. Arg2 and Arg1) describe the action to be carried out to obtain the goal (i.e. given the amount of money you need, you can have it if you start saving from the beginning of 1990). Following the scheme's structure, also the conclusion is left implicit (i.e. *therefore, if you want to reach your goal, you should start saving*). Another interesting observation emerging from naturally occurring data is the fact that in human linguistic interactions a lot is left implicit, following [11]'s conversational *Maxim of Quantity* (i.e. do not make your contribution more informative than is required).

The tag "implicit assertion" applies in special rhetorical uses of if-constructions when the interpretation of the conditional construction is an implicit assertion.

Example 6 (implicit assertion)
PREMISE: **if** (Arg2) you want to keep the crime rates high
CONCLUSION (Arg1): O'Connor is your man

In Example 6 the conclusion, i.e. *O' Connor is your man*, is not a consequent state that will result if the condition expressed in the premise holds true. Instead, the conditional construction in this case implicitly asserts that O'Connor will keep the crime rates high (enthymeme), and requires a pragmatic reasoning step. For both subtypes, the typical connective expressing the discourse relation is *if*.

Argument from Inconsistency. The last argumentation scheme we consider in our inspection is the *Argument from inconsistencies*, where the inconsistency can be detected in an arguer's commitment set. Even if the mapping of such scheme with one of the discourse categories is far from being straightforward, after a careful analysis of both the definitions and the examples in the PDTB, the relation COMPARISON:*concession* seems to fall within such scheme. In fact, this relation applies when one of the arguments describes a situation A which causes C, while the other asserts (or implies) $\neg C$. Alternatively, the same relation can apply when one premise denotes a fact that triggers a set of potential consequences, while the other denies one or more of them, and still in this case it fits the definition of the above mentioned argumentation scheme. Formally, we have $A < C \wedge B \rightarrow \neg C$, where A and B are drawn from $\|Arg1\|$ and $\|Arg2\|$ ($\neg C$ may be the same as B, where $B \rightarrow B$ is always true). Two *concession* subtypes are defined in terms of the argument creating an expectation and the one denying it. Specifically, when Arg2 creates an expectation that Arg1 denies (A=$\|Arg2\|$ and B=$\|Arg1\|$), it is tagged as *expectation* (see Example 7). When Arg1 creates an expectation that Arg2 denies (A=$\|Arg1$ and B=$\|Arg2\|$), it is tagged as *contra-expectation* (see Example 8).

Example 7 (expectation)
PREMISE (Arg1): attorneys for the two sides apparently began talking again yesterday in attempt to settle the matter before Thursday
PREMISE: **although** (Arg2) settlement talks had been dropped

Example 8 (contra-expectation)
CONCLUSION: (Arg1) The demonstrators have been non-violent
PREMISE: **but** (Arg2) the result of their trespasses has been to seriously impair the rights of others unconnected with their dispute

In Example 7 we start from the evidence provided by the premise according to which the settlement talks between the attorneys have started, and we are pushed to conclude that they are still going on, while the conclusion provided by the arguer is inconsistent (i.e. settlement talks had been dropped). With the same reasoning step, in Example 8 we expect that no bad consequences are caused by the demonstrators thanks to their pacific attitude, but our expectation is wrong.

3.3 Argumentation Schemes in the PDTB

Table 1 reports some statistics on the PDTB relations considered in our study. We extract them from the PDTB and report the number of examples both of *implicit* and *explicit* relations. Since PDTB annotators were allowed to assign more than one relation label, we report only the relations whose *first* label is the one reported in the first column. Also, we consider only the examples in which Arg2 is not embedded in Arg1 (more than 90% of the overall examples), because we want to avoid that premises and conclusions according to argumentation schemes are expressed by discontinuous arguments. Next to each discourse subtype, we also list the three most-frequent connectives occurring in the *explicit* relations (for *Relevance*, only two connectives are found in the extracted examples). This confirms that, although *therefore* is the only connective usually employed in argumentation schemes to introduce the conclusion, corpus-based analysis shows more variability and a much richer repository of admissible connectives.

Table 1. Statistics about the extracted examples

Relation class.Type *Subtype* ('most-frequent connectives')	Number Explicit	Number Implicit
Expansion.Restatement		
Generaliz. ('in short', 'in other words')	16	190
Contingency.Cause		
Reason ('because', 'as', 'since')	1,201	2,434
Result ('so', 'thus', 'as a result')	617	1,678
Contingency.Pragm.Condition		
Relevance ('if', 'when')	21	1
ImplicitAssertion ('if', 'when', 'or')	46	0
Comparison.Concession		
Expectation ('although', 'though', 'while')	386	31
ContraExpectation ('but', 'still', 'however')	798	182

To verify the correctness of our initial hypothesis, i.e. that the sentences annotated with the above mentioned PDTB relations correspond to the argumentation schemes described in Section 3.2, we manually annotated a subset of examples from the PDTB explicit relations. More precisely, we annotated 50 randomly selected examples per argumentation scheme (16 from the Expansion.Restatement.*Generalization*), with respect to the fact that they correspond to the expected argumentation scheme or not. Table 2 reports the obtained results. The tag *INCORRECT* is assigned when the extracted example is not clear once out of context (i.e., the annotators were not able to judge it due to the lack of context). As can be seen, some PDTB relations perfectly match the expected argumentation schemes (e.g. *Argument from Effect to Cause*), while for some others (i.e. *Practical Reasoning*) the mapping is much less straightforward (reflecting the observations in Section 3.2), even if it was expected by their definitions.

Table 2. Outcome of the annotation phase on the PDTB

Argumentation scheme	# examples	YES	NO	INCORRECT
Argument from Example	16	8	8	0
Argument from Cause to Effect	50	31	8	11
Argument from Effect to Cause	50	42	6	2
Practical Reasoning	50	29	18	3
Argument from Inconsistency	50	34	8	8
TOTAL	216	144	48	24

To assess the validity of the annotation task, the same annotation task has been independently carried out also by a second annotator, so as to compute inter-annotator agreement. It has been calculated on a sample of 50 argument pairs (10 per each expected argumentation scheme, randomly extracted). While the percentage of agreement between the two annotators is 88%, Cohen's kappa [6] is 0.71. As a rule of thumb, this is a satisfactory agreement, confirming the reliability of the obtained resource.

4　Related Work

The need for coupling argumentation theory and NLP is becoming more and more important in the latest years as shown by the increasing number of online debate systems like Debategraph[3] and Debatepedia[4]. Some approaches have been proposed to address this issue in the two research communities. Chasnevar and Maguitman [7] propose a defeasible argumentation system to provide recommendations on language patterns to assist the language usage assessment. Wyner and van Engers [27] propose to couple NLP and argumentation to support policy makers. The user is asked to write input text using Attempt to Controlled English allowing for a restricted grammar and vocabulary, and then sentences are translated to First Order Logic. In this paper, we do not look for a translation in formal logic of NL arguments: we are interested in the structure of the arguments such as in argumentation schemes, where the relation among the premises and the conclusion is represented through the PDTB discourse relations. Cabrio and Villata [4] propose to use the textual entailment approach to extract from Debatepedia the arguments in NL and the relations among them. Then, the arguments are composed in a Dung-like [8] abstract argumentation framework to select the acceptable arguments. The authors look only at the relations among the arguments, while here we are more interested in the relation among premises and conclusion in NL arguments. Carenini

---

[3] http://debategraph.org
[4] http://dbp.idebate.org/

and Moore [5] present a computational framework for generating evaluative arguments. We use a different model of arguments, i.e. argumentation schemes, and we do not provide an automatic system for argument generation. Feng and Hirst [9] present an automatic system for classifying the argumentation schemes of NL arguments to infer enthymemes, using Araucaria as data set. Our analysis can be used to support this kind of automated classification task thanks to the mapping with the discourse relations we provide, and the resulting annotated arguments corpus can be used for training. Amgoud and Prade [1] start from an argumentation model presented in linguistics [2] and formalize it using formal argumentation. They envisage a comparison with argumentation schemes as future work. In this paper, we consider only such schemes to provide the parallel with NLP.

5   Concluding Remarks

We presented an analysis of the connections between two distinct research areas, namely discourse analysis in NLP, and argumentation schemes in argumentation. Following the idea of focusing first on models of NL schemes and then building formal systems [21], the rationale of this kind of analysis is to provide a first, but compulsory step towards the development of automatic techniques able to deal with the complexities present in natural language arguments. Even if recent approaches like [20,4,28,1] provide a first attempt to tackle the open problem of natural language argumentation, they show that a satisfiable result has still to be reached. As demonstrated in this paper, the development of automated systems going beyond applications like e.g. the one proposed by Cabrio and Villata [4], where only two relations among the arguments are considered and arguments are abstract, is much more complex. In our pilot study, we produced a small corpus of annotated data, to prove the feasibility of the proposed approach. We plan to extend it through a large-scale annotation task. Further future work includes the study of the emergence of other argumentation schemes starting from the discourse relations of PDTB, and the design and implementation of an automated framework able to detect not only the abstract arguments from natural language text, but also their internal structure [21] with the aim of verifying the coherence of such arguments before considering the (possible) relations with the other arguments. The bridge with discourse analysis enables us to carry out an in-depth study of the argument structures, relying on the data previously annotated with discourse relations, and now annotated also with the corresponding argumentation schemes. As an additional outcome of our work, we propose to release the annotation of the PDTB examples with the considered argumentation schemes, that can be fruitfully exploited as a training corpus in NLP applications.

References

1. Amgoud, L., Prade, H.: Can AI models capture natural language argumentation? Int. J. of Cognitive Informatics and Natural Intelligence (2013)
2. Apotheloz, D.: The function of negation in argumentation. J. of Pragmatics, 23–38 (1993)
3. Bex, F., Reed, C.: Dialogue templates for automatic argument processing. In: Procs of COMMA 2012, pp. 366–377 (2012)

4. Cabrio, E., Villata, S.: Natural language arguments: A combined approach. In: Procs of ECAI. Frontiers in Artificial Intelligence and Applications, vol. 242, pp. 205–210 (2012)
5. Carenini, G., Moore, J.D.: Generating and evaluating evaluative arguments. Artif. Intell. 170(11), 925–952 (2006)
6. Carletta, J.: Assessing agreement on classification tasks: the kappa statistic. Comput. Linguist. 22(2), 249–254 (1996)
7. Chesñevar, C.I., Maguitman, A.: An argumentative approach to assessing natural language usage based on the web corpus. In: Procs of ECAI, pp. 581–585 (2004)
8. Dung, P.: On the acceptability of arguments and its fundamental role in nonmonotonic reasoning, logic programming and n-person games. Artif. Intell. 77(2), 321–358 (1995)
9. Feng, V.W., Hirst, G.: Classifying arguments by scheme. In: Procs of ACL 2011, pp. 987–996 (2011)
10. Gordon, T.F., Walton, D.: Legal reasoning with argumentation schemes. In: ICAIL, pp. 137–146. ACM (2009)
11. Grice, H.P.: Logic and conversation. In: Cole, P., Morgan, J.L. (eds.) Syntax and Semantics. Speech Acts, vol. 3, pp. 41–58. Academic Press (1975)
12. Hastings, A.C.: A reformulation of the models of reasoning in argumentation. Ph.D. thesis (1963)
13. Mann, W., Thompson, S.: Rhetorical structure theory: Toward a functional theory of text organization. Text 8(3), 243–281 (1988)
14. Marcus, M.P., Santorini, B., Marcinkiewicz, M.A.: Building a Large Annotated Corpus of English: the Penn Treebank. Computational Linguistics 19(2), 313–330 (1993)
15. McBurney, P., Parsons, S.: Risk agoras: Dialectical argumentation for scientific reasoning. In: Procs of UAI, pp. 371–379 (2000)
16. Parsons, S., Atkinson, K., Haigh, K.Z., Levitt, K.N., McBurney, P., Rowe, J., Singh, M.P., Sklar, E.: Argument schemes for reasoning about trust. In: Procs of COMMA 2012, pp. 430–441 (2012)
17. Pitler, E., Nenkova, A.: Using syntax to disambiguate explicit discourse connectives in text. In: Procs of ACL 2009 (2009)
18. Prasad, R., Dinesh, N., Lee, A., Miltsakaki, E., Robaldo, L., Joshi, A., Webber, B.: The Penn Discourse TreeBank 2.0. In: Procs of LREC 2008 (2008)
19. Rahwan, I., Simari, G. (eds.): Argumentation in Artificial Intelligence. Springer (2009)
20. Reed, C., Grasso, F.: Recent advances in computational models of natural argument. Int. J. Intell. Syst. 22(1), 1–15 (2007)
21. Reed, C., Walton, D.: Towards a formal and implemented model of argumentation schemes in agent communication. Autonomous Agents and Multi-Agent Systems 11(2), 173–188 (2005)
22. Reed, C.: Dialogue frames in agent communication. In: Procs of ICMAS 1998, pp. 246–253. IEEE Computer Society (1998)
23. Reed, C., Rowe, G.: Araucaria: Software for argument analysis, diagramming and representation. International Journal on Artificial Intelligence Tools 13(4), 983 (2004)
24. Reed, C., Walton, D.: Applications of argumentation schemes. In: Procs of OSSA (2001)
25. The PDTB Research Group: The PDTB 2.0. Annotation Manual. Tech. Rep. IRCS-08-01, Institute for Research in Cognitive Science, University of Pennsylvania (2008)
26. Walton, D., Reed, C., Macagno, F.: Argumentation Schemes. Cambridge Univ. Press (2008)
27. Wyner, A., van Engers, T.: A framework for enriched, controlled on-line discussion forums for e-government policy-making. In: Procs of eGov 2010 (2010)
28. Wyner, A.: Questions, arguments, and natural language semantics. In: Procs of CMNA 2012 (2012)

Deep Natural Language Processing for Italian Sign Language Translation

Alessandro Mazzei[1], Leonardo Lesmo[1], Cristina Battaglino[1],
Mara Vendrame[2], and Monica Bucciarelli[2]

[1] Dipartimento di Informatica,
[2] Dipartimento di Psicologia,
Università degli Studi di Torino

Abstract. This paper presents the architecture of a translator from written Italian into Italian Sign Language. We describe the main features of the four modules of this architecture, i.e. a dependency parser for Italian, an ontology based semantic interpreter, a generator based on expert-systems and combinatory categorial grammars, a planner to position signs in space. The result of this translation chain is signed by a virtual character. Finally, we report the results of a first "intrinsic" experiment for the evaluation of translation quality.

1 Introduction

In the last years the computational linguistic community showed a growing interest toward sign languages, and a number of projects concerning the translation into a signed language (SL) have recently started. Some of these projects adopt symbolic techniques: English to British SL [1], English to American SL [2,3]. Some other projects adopt statistical techniques based on parallel corpora: English to Irish SL [4], Chinese to Chinese SL [5]. Indeed, developing large parallel vocal/signed corpora is a very hard process. First, building large collections of video is long and expensive. Second, there are theoretical difficulties concerning the extra-video annotation of the sign language. For instance, while there are standards for representation of phonological information of a sign there are no common practices to represent its morpho-syntactic inflections. So, classical machine translation architecture, that is based on symbolic (no statistical) processing of the language [6] is still appealing for translating into a SL.

In this paper, we present the main features of a symbolic translation architecture from Italian to Italian Sign Language (Lingua Italiana dei Segni, henceforth LIS), that is the SL used by the Italian deaf (signing) community [7]. This translation system has been used in the ATLAS project, that concerns the specific application domain of the weather forecasts[1]. LIS is a poorly studied language and linguists often do not agree on basic linguistic properties, e.g. sentence word order. As starting point of the ATLAS project, a group of linguists expert on LIS produced a (small) parallel corpus of Italian-LIS sentences extracted from TV news and concerning weather forecasts. This corpus consists of about 300 Italian sentences aligned to their LIS translation and it has been

---

[1] http://www.atlas.polito.it/

M. Baldoni et al. (Eds.): AI*IA 2013, LNAI 8249, pp. 193–204, 2013.

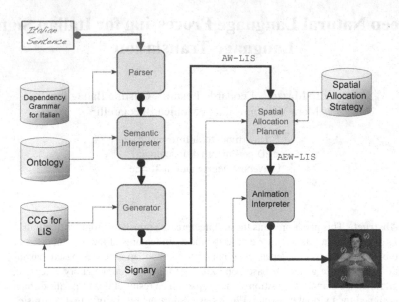

Fig. 1. The rule-based translation architecture

primarily used to produce an electronic lexicon consisting of about 1500 signs that can be accessed by a virtual interpreter [8].

Our translation system can be defined as knowledge-based and restricted-interlingua, since it uses extra-linguistic information and deals with only two languages [6]. The translation system is a chain composed by five distinct modules, that are: (1) a dependency parser for Italian; (2) an ontology based semantic interpreter; (3) a generator; (4) a spatial planner; (5) an animation interpreter that performs the synthesis of the sequence of signs, i.e. the final LIS sentence[2] (see Fig. 1). The parser and the semantic interpreter have been previously used in other projects [10,11]), while the other modules have been developed for this project. A distinctive feature of our system concerns the treatment of the hands position. We have split the generation process into two distinct steps: in the first step the generator produces a *spatially not assigned* sequence of signs; in the second step the planner produces spatial coordinates to each sign.

The paper is organized as follows: in Section 2 we describe the parser and the ontology based interpreter; in Section 3 we describe the LIS generator, composed by two sub-modules: a sentence planner based on expert systems and a realizer based on formal grammars; in Section 4 we describe the planner used for spatial allocation of the signs. Finally, in Section 5 we give some details on system evaluation and conclude the paper in Section 6.

2 Parsing and Interpretation

In this Section we give some details about the parser and the semantic interpreter. In the first processing step, the syntactic structure of the source language is obtained by

---

[2] The animation interpreter is described in [9]

Fig. 2. The (simplified) syntactic structure of the sentence "Oggi ultimo giorno del mese di giugno, con valori di temperatura superiori alla media" (*Today is the last day of the month of June, with temperature values exceeding the average*) produced by the TUP parser

the TUP parser, a rule-based parser for Italian that has supported the construction of the TUT treebank [12,13]. The TUP uses a morphological dictionary of Italian (about 25,000 lemmata) and a rule-based grammar. The final result is a *dependency tree*, that makes clear the structural syntactic relationships occurring between the words of the sentence [14]. Each word in the source sentence is associated with a node of the tree, and the nodes are linked via labeled arcs that specify the syntactic role of the dependents with respect to their head (the parent node). In Figure 2, we report the syntactic analysis for the sentence "Oggi ultimo giorno del mese di giugno, con valori di temperatura superiori alla media" (rough translation: *Today [is the] last day of the month of June, with temperature values exceeding the average*). We have two sub-sentences: the first sub-sentence "Oggi ultimo giorno del mese di giugno" is nominal since it is rooted by the noun "giorno" (*day*); the second sub-sentence "valori di temperatura superiori alla media" is nominal too, since is rooted by the noun "valore" (*value.*), and it is syntactically related to the first sub-sentence by the preposition "con" (*with*). The edge label "ARG" indicates a *ARG*ument relation, i.e. an obligatory relation from one head and its argument. The edge label "RMOD" indicates a *R*estricting *MOD*ifier relation, i.e. a non obligatory relation from the head to its dependent [13].

The interpreter is based on ontology [15] concerning the application domain, i.e. weather forecasts, as well as more general common knowledge about the world. A peculiarity of our approach is the ontological status of the language, that is part of the foundational ontology [16,17,18]. Starting from the lexical semantics of the words in the sentence and on the basis of the dependency structure, the interpreters looks for "connection paths" that are joined into a single structure that represents the meaning of the sentence (cf. [19]): semantic roles and other kind of semantic relations are encoded in this structure and could be straightforwardly translation into some form of logic, as First Order Logic (FoL) predicates. In particular similar to other approaches our ontological meaning representation is unscoped. In short, variables represent nodes, propositions represent node properties, modalities represents relation between nodes. In the following, we interchangeably use the terms semantic network and logic representation as well as node of the network and logic predicate. In Figure 3, we report the interpretation of the sentence "Oggi ultimo giorno del mese di giugno, con valori di temperatura superiori alla media". Each node in the network contains a variable that uniquely identifies the node. The nodes of the network contain instances (with prefix name £ in the Figure) and concepts (with prefix name ££ in the Figure) from the ontology. The prefix § is adopted to represent some special entities: for instance §SPEAKER+OTHERS represents the speaker and all the hearers of the sentence

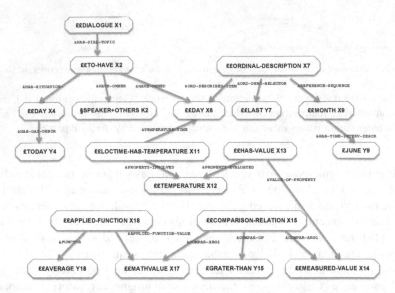

Fig. 3. The semantic network resulting from the interpretation of the sentence "Oggi ultimo giorno del mese di giugno, con valori di temperatura superiori alla media"

(i.e. the pronoun *we*). The relations (with prefix name **&** in the Figure) represent meaning relations between concepts/instances. Some relation are *reified*, i.e. translated into concepts: for example the concept HAS-VALUE and the binary relation VALUE-OF-PROPERTY are the result of the reification of an original relation between a more specific relation between the concepts TEMPERATURE and a MEASURED-VALUE.

Informally speaking, we can say that the semantic interpreter organizes the information into the semantic networks as a number of messages, a sort of "information clusters" that are weakly connected with the other parts of the network. In the network of Figure 3 we can distinguish six clusters of nodes, corresponding to six messages. A paraphrases of the meaning represented by the semantic network is: (message 1) the topic of the dialogue is that we (i.e. the speaker and the hearers) have that today is a day (message 2) involved into an order relation with the ordinal "last" and with the Month of June; this day has a temperature (message 3) with a value (message 4); this value is involved in a comparison (message 5) with a mathematical value that is the average (message 6). In the next section we describe how the microplanner manages this organization of the information.

3   The LIS Generator

The standard architecture for Natural Language Generation uses three distinct components: document planning, sentence planning and realization [20]. Our generator is composed by only two components: the sentence planner is called *SentenceDesigner*, that builds a tree representing the generic LIS lexical items and some generic syntactic relations among them; the realizer is implemented a formal grammar for LIS by

```
(defrule rule-COMPARISON-RELATION ()
  (semantic-state    (name ££COMPARISON-RELATION) (arg-1 ?X1))
  (semantic-relation (name &COMPAR-ARG1) (arg-1 ?X1) (arg-2 ?X2))
  (semantic-relation (name &COMPAR-ARG2) (arg-1 ?X1) (arg-2 ?X3))
  (semantic-relation (name &COMPAR-OP)   (arg-1 ?X1) (arg-2 ?X4))
  =>
  (assert (syntactic-relation (name SYN-SUBJ) (arg-1 ?X4) (arg-2 ?X2)))
  (assert (syntactic-relation (name SYN-OBJ)  (arg-1 ?X4) (arg-2 ?X3))))
```

```
(defrule rule-APPLIED-FUNCTION ()
  (semantic-state    (name ££APPLIED-FUNCTION) (arg-1 ?X1))
  (semantic-relation (name &FUNCTOR) (arg-1 ?X1) (arg-2 ?X2))
  (semantic-relation (name &APPLIED-FUNCTION-VALUE)
                     (arg-1 ?X1)  (arg-2 ?X3))
  =>
  (assert (syntactic-relation (name SYN-RMOD)  (arg-1 ?X3) (arg-2 ?X2))))
```

Fig. 4. Two rules of the knowledge-base used by the expert system for lexicalization

using OpenCCG [21]. SentenceDesigner basically performs a three-steps algorithm: in the first step SentenceDesigner identifies the messsages present in the sentence (see Figure 3), looking for nodes with no heading over: in Figure 3 these nodes are DIALOGUE, ORDINAL-DESCRIPTION, LOCTIME-HAS-TEMPERATURE, HAS-VALUE, COMPARISON-RELATION, APPLIED-FUNCTION. In the second step, that correspond to "lexicalization" [20], SentenceDesigner performs two distinct substeps for each message. The first substep introduces new prelexical nodes in the message by using a table-based heuristic. For example, the prelexical node value belonging to the class evaluable-entity is introduced corresponding to concept the MATH-VALUE. The second substep introduces of syntactic relations between prelexical nodes. Since this mapping from semantic to syntactic relations is based on linguistic knowledge provided by linguists, we used an expert system[3], that allows for a sharp modulation of the knowledge and speed-up revisions. All the rules have a common schema: the Left Hand Side (LHS) specifies the "semantic content" that the rule captures, the Right Hand Side (RHS) specifies the syntactic relations. In Figure 4 we report two rules that are "fired" by SentenceDesigner during the analysis of the messages in Figure 3. The first rule encodes the *comparison* semantic relation into one *subject* and one *object* syntactic relations; the second rule encodes the semantic relation concerning a mathematical value as a *modifier* relation.[4] The syntactic labels used in these rule are essentially the ones used in the annotation of TUT [12,13]. The third step of the algorithm concerns the simplification of the messages built in the previous steps: after the merging of the various syntactic relations among the various messages, we simplify the messages (often LIS is more compact than Italian), we remove duplicate among messages, we give a realization order to the messages. The final result of SentenceDesigner consists of a number of *abstract* syntax trees [20]. In Figure 5 there are the two abstract syntax trees obtained by the input given by the semantic network of Figure 3.

The abstract trees are used as input for the OpenCCG realizer, that is based on a combinatorial categorial grammars (CCG) [23] for LIS. CCG is a mildly context-sensitive formalism that is theoretically adequate to describe the complexity of natural language syntax (e.g. cross-serial dependencies, non-constituency coordination) and it has a very straight syntax-semantic interface. Realization accounts for a number of morpho-syntactic phenomena, that are *inflection, agreement, word order, function words*. LIS has no function words but, similar to all SLs, it has a peculiar and rich system of inflection and agreement. OpenCCG allows to encode an inflectional system

---

[3] Since SentenceDesigner is written in lisp, we used the LISA expert system. This is an implementation of the RETE algorithm compliant with Common lisp Specifications [22].

[4] The actual implementation of SentenceDesigner consists of 50 rules.

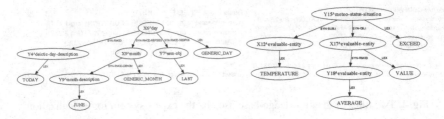

Fig. 5. The output of SentenceDesigner on the by the semantic network of Figure 3

Table 1. A fragment of the CCG for LIS used for the realization of the abstract syntactic trees in Figure 5. Note that the LIS word order SOV is formalized in the lexical entry for "superiore" by using the Category $S\backslash NP\backslash NP$, that can be paraphrased as "a sentence is a verb that combines with two nouns (subj and obj) on its left".

DB name	DB ID	#hands	lexical::value	PoS	Category SynSem relations
superiore	3168	2	exceed::meteo-status	verb	S [Y0] \ NP [Y1] \ NP [Y2] @Y0:meteo-status (exceed ^ <SUBJ>Y1:eva-entity ^ <OBJ>Y2:eva-ontity)
oggi	2669	2	today::deictic-day-description	noun	N [Y0] @Y0: deictic-day-description (today)

by using feature structures, which are part of the syntactic categories. Since CCG is a lexicalized formalism, the LIS CCG consists of a set of lexical items plus a number of unary rules, that are used to account for specific syntactic/semantic phenomena. Each lexical item, that corresponds to a single sign, consists of a number of morpho-syntactic and semantic features. In particular a sign entry is composed by: (1) The database name of the sign;[5] (2) The database ID of the sign; (3) The number of hands used to perform the sign; (4) The lexical value of the sign; (5) The Part of Speech (PoS) of the sign (6) The Category of the sign with the syntactic/semantic relations. In table 1, we report an example. The current version of the LIS CCG consists of about 100 lexical entries and 10 unary rules, organized in about 30 lexical families. For instance, by using the first abstract syntactic trees of Figure 5, the realizer produces the CCG derivation shown in Figure 6. We use the CCG derivations together with the syntactic relations of the abstract trees in order to produce a TUT-compliant LIS dependency tree (cf. Fig. 7). Note that at this stage the signs do not have a feature specifying a spatial collocation: this feature is provided by the spatial planner in the next stage.

---

[5] As usual in literature, we use a database name for the sign related to its meaning in the source language, Italian in our case.

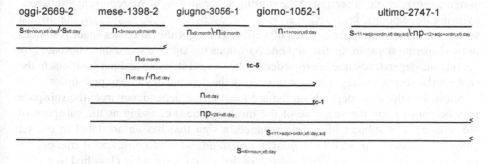

Fig. 6. Syntactic derivation of the LIS sentence "oggi-2669-2 mese-1398-2 giugno-3056-1 giorno-1052-1 ultimo-2747-1" (*sign_name–database_ID*)

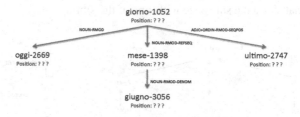

Fig. 7. LIS dependency tree of "Oggi mese giugno giorno ultimo"

4 The Spatial Allocation Planner

In SLs the signing space (namely, a portion of space in front of the signer) has a syntactic function (see Figure 10) [24]. The verb agreement, that is the semantic relation between a verb and its arguments, in SLs is expressed through the use of space. For example, in the LIS sentence "mother child call" (*the mother calls the child*) the sign "call" starts in the position of the sign "mother" and ends in the position of the sign "child". As a consequence, we need to know the position and the hand-movement in the signing space in order to get the "animation" of the virtual character. Broadly speaking, LIS signs can be split in two main classes: relocatable and not relocatable. We modeled this classification by merging it with agreement properties. Agreement, in LIS, may involve one or two arguments; consequently, we introduced three agreement classes: 0-agr (no-agreement and no-relocatable), 1-agr (relocatable, with position depending on a single agreement), 2-agr (relocatable, with position depending on two agreement). Another important morphological feature, that expresses syntactic agreement too, concerns the direction of the hand movement with respect to the other signs. Some signs have a movement that is constrained by the position of the other signs in the space: in this case we assign the value +dir to the sign, otherwise we assign -dir. The Figure 8 provides the reader with a taxonomy that embodies this morpho-syntactic features.

In the spatial planner, we model the total signing space as a 3D box in front of the signer. Each sign has a position in this box expressed by a triple of coordinates $< x, y, z >$. Moreover, to take into account the morpho-syntactic features we need to assign a *subspace*, i.e. a portion of the signing space, to a sign. We represent a subspace by using three points, $[x_{inf}, y_{inf}, z_{inf} \leftarrow x_0, y_0, z_0 \rightarrow x_{sup}, y_{sup}, z_{sup}]$, that lowest vertex, the center. that highest vertex of a cube. The algorithm allocates space resources in two separate steps: in the fist step one partitions the signing space into subspaces by visiting the dependency tree in pre-order; in the second step one assigns positions to the signs in the corresponding subspace in visiting the dependency tree in post-order.

Subspace allocation depends on the arc label in the dependency tree: the subspace may be parasitic on the subspace of the mother node (i.e. as big as the subspace of the mother) or a subpart of it. For instance, a sign that has an arc label in the set {*adjc-rmod, advb-rmod, adjc + qualif-rmod, advb-rmod-time, coord-merge, verb-rmod, conj-arg, coord2nd, adjc + ordin-rmod-seqpos*} is classified as a *Space_reassigner* and shares the same subspace of the mother. In contrast, a sign that has an arc label in the set {*verb + modal-indcompl, verb + modal-predcompl, predcompl, verb-predcompl+subj*} is classified as a $Non\_space\_assigner group$ and identifies signs that share the subspace with one of the sibling.

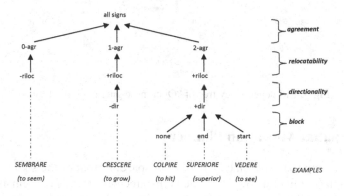

Fig. 8. Taxonomy of phonological and Morphological features

After the pre-order visit, the Spatial Planner algorithm determines in post-order visit the actual position of the sign (hand placement and movement) inside the assigned subspace. This is accomplished by taking into account the agreement properties of the governing node, according to taxonomy based on the phonological and morphological features of a sign (see above, Figure 8). For instance, the sign *superiore* is classified in the taxonomy as $2 - agr$, $+riloc$ and $+dir$, i.e. the sign has two syntactic agreements, it is relocatable and involves a hand movement into the space forward the signs that are in relation with it. Moreover, the final position is *blocked* (see Figure 8), i.e. the initial position corresponds to the position of the first argument and the sign movement has to pass through the second argument.

As example, we trace the execution of the algorithm on the sentence "Oggi ultimo giorno mese di giugno": the reader can refer to Figure 9 to trace the execution.

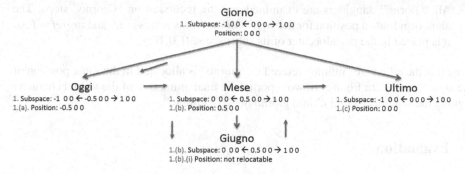

Fig. 9. Execution of the algorithm on the sentence "Oggi ultimo giorno mese di giugno"

1. **Giorno**: it is the root of the tree. It is assigned all the signing space $[-1, -1, -1, \leftarrow 0, 0, 0 \rightarrow 1, 1, 1]$. The root "Giorno" has three daughters, so the algorithm starts the recursion on them. The two daughters "Oggi" and "Mese" are *Space_assigners* (their arc labels are "noun-rmod" and "noun-rmod-refseq"), while the daughter "Ultimo" is a Space_reassigner (its arc label is "adjc+ordin-rmod-seqpos"). The mother subspace is split in two for the space assigners: the subspace $[-1, 0, 0, \leftarrow -0.5, 0, 0 \rightarrow 0, 0, 0$, is assigned to "Oggi" and the subspace $0, 0, 0 \leftarrow 0.5, 0, 0 \rightarrow 1, 0, 0$ is assigned to "Mese"; while the current space of the mother is assigned to "Ultimo".

   (a) **Oggi**: it is a leaf so the algorithm finds a position. It is *riloc+*, *dir-* and *argref* = *0*, so it is placed in the pseudocenter of its subspace $< -0.5, 0, 0 >$.

   (b) **Mese**: it has one daughter, "Giugno", which is a space-assigner. The whole space of the mother "Mese" is assigned to "Giugno".

      i. **Giugno**: It is a leaf so the alghoritm finds a position. It is *riloc-*, (i.e. not relocatable) and its position remains unchanged.

      All "Mese"daughters are examined and the recursion on "Mese" stops. The alghoritm finds a position for the sign "Mese". It is *riloc+*, *dir-* and *argref* = *0*, so it is placed in the pseudocenter of its subspace $< 0.5, 0, 0 >$.

   (c) **Ultimo**: the last daughter of the sign "Giorno". It is a leaf so the algorithm finds a position. It is [riloc+, dir-] and [argref = 1 with agr=subj, loc], so it is placed in the pseudocenter of its subspace $<0, 0, 0>$.

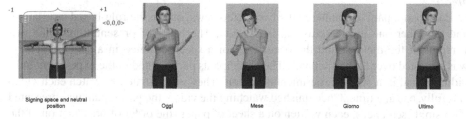

Fig. 10. Virtual character signs the sentence "Oggi ultimo giorno del mese di giugno"

All "Giorno" daughters are examined and the recursion on "Giorno" stops. The alghoritm finds a position for the sign "Giorno". It is *riloc+*, *dir-* and *argref = 1*, so it is placed in the pseudocenter of its subspace $< 0, 0, 0 >$.

Note that the adjective "ultimo" refered to "giorno" is allocated in the same position of the sign "giorno". In Figure 10 we reported the final animation of the virtual character signing the sentence "Oggi ultimo giorno mese di giugno".

5   Evaluation

We test our system following an intrinsic approach: we devised a controlled experiment involving a number of native signers to evaluate the translation. Moreover we compared the results of our rule-based system with the results on the same test of a prototypical statistical translator developed by using the LIS corpus. The rule-based and the statistical systems use the same interface for sign translation. The participants in the experiment were 12 signing deaf (8 females and 4 males, mean age 23 years) adult individuals, with no other disabilities. They voluntarily took part in the experiment. They were deaf individuals with a prelingual and profound hearing deficit (> 90 dB hearing loss), all university students. The experimental materials comprised two meteo news texts, referring to different periods of the year (meteo n. 3 concerns October, meteo n. 19 concerns the period June-July) and of comparable length (259 and 283 words, respectively). For each meteo, we created 9 couples of sentences. Each couple (same length in words) corresponds to a semantic unit in the text, and is either a paraphrase of the original semantic unit or it does not correspond in meaning to the original unit. Consider, as an example, the following couple created from the original sentence "Da dieci giorni più o meno ormai sentiamo parlare di ottobrate, praticamente di bel tempo con temperature decisamente miti, soprattutto nel corso del pomeriggio" (rough translation: *It's more or less ten days that we hear about "ottobrate", i.e. about good weather with particularly mild temperatures especially during the afternoon.*):

Paraphrase: Da dieci giorni temperature molto gradevoli e bel tempo, soprattutto nel corso del pomeriggio. (rough translation: *In the last ten days pleasant temperature and nice weather, especially in the afternoon.*)

Wrong Content: Tempo variabile sull'Italia settentrionale e centrale, molte nubi soprattutto sul versante adriatico e sull'area ionica (rough translation: *Variable weather on north and central Italy. Many clouds especially on the adriatic side and on the ionic area.*).

Each participant encountered the two meteo news, one in the rule-based translation, and the other one in the statistical translation. The order of presentation of the two versions of the meteos, and the occurrence of a specific meteo in a specific version was balanced over all participants. The participants participated in the experiment individually, and in a single experimental session. They were invited to watch each video carefully, one at a time. Once finished watching the video, the participant was presented with single sentences, each written on a sheet of paper (the order of presentation of the sentences was random for each participant). The task of the participant was to decide,

Table 2. Mean of correct recognitions by the participants in the experiment

Type of translation	Paraphrases	Wrong Content	Global performance
Rule based	5.42 (1.51)	4.42 (1.38)	9.83 (2.62)
Statistical based	5,71 (1,75)	4,08 (1,62)	9,25 (2,70)

for each sentence, whether it corresponded to the information provided by the automatic translation (Recognition task). We coded responses of "Yes" to paraphrases and "No" to wrong content sentences as correct. The experimental procedures were exactly the same for the rule based and the statistical systems translations. Table 2 illustrates the mean correct recognition (and standard deviation in parenthesis) with the two types of translations by the participants. A comparison between correct recognitions with the two versions revealed no statistically significant differences (Wilcoxon test: $z = .40$, $p = .69$). The same result holds if we consider separately paraphrases and wrong content sentences (Wilcoxon test: $z = .31$, $p = .75$ and $z = .48$, $p = .63$, respectively). Aim of the experiment was to ascertain whether one of the two automatic translations, i.e., Rule-based or Statistic, was more comprehensible to signing deaf individuals: the numbers support the conclusion that that deaf participants were equally likely to accept paraphrases and to refute wrong content sentences in the recognition task for rule-based and statistical translators.

6   Conclusions

In this paper we have presented a NLP system that translates from Italian into the Italian Signed Language (LIS) by adopting a symbolic approach. After description of the main features of each module, we reported the preliminary results of an evaluation on the meteo domain. In future work we plan to empower the system in order to translate over the rail station domain. We speculate that this domain is more regular w.r.t. to the weather domain and could give better result on evaluation. However, a great effort is required in order to create a new domain lexicon from scratch.

References

1. Bangham, J., Cox, S., Elliott, R., Glauert, J., Marshall, I.: Virtual signing: Capture, animation, storage and transmission – an overview of the VisiCAST project. In: IEE Seminar on Speech and Language (2000)
2. Zhao, L., Kipper, K., Schuler, W., Vogler, C., Badler, N.I., Palmer, M.: A machine translation system from english to american sign language. In: White, J.S. (ed.) AMTA 2000. LNCS (LNAI), vol. 1934, pp. 54–67. Springer, Heidelberg (2000)
3. Huenerfauth, M.: Generating American Sign Language classifier predicates for english-to-asl machine translation. PhD thesis, University of Pennsylvania (2006)
4. Morrissey, S., Way, A., Stein, D., Bungeroth, J., Ney, H.: Combining data-driven mt systems for improved sign language translation. In: Proc. XI Machine Translation Summit (2007)
5. Su, H., Wu, C.: Improving structural statistical machine translation for sign language with small corpus using thematic role templates as translation memory. IEEE Transactions on Audio, Speech and Language Processing 17(7), 1305–1315 (2009)

6. Hutchins, W.J., Somer, H.L.: An Introduction to Machine Translation. Academic Press, London (1992)
7. Volterra, V. (ed.): La lingua dei segni italiana. Il Mulino (2004)
8. Tiotto, G., Prinetto, P., Piccolo, E., Bertoldi, N., Nunnari, F., Lombardo, V., Mazzei, A., Lesmo, L., Principe, A.D.: On the creation and the annotation of a large-scale Italian-LIS parallel corpus. In: 4th Workshop on the Representation and Processing of Sign Languages: Corpora and Sign Language Technologies, Valletta, Malta (2010) ISBN 10: 2-9517408-6-7
9. Lombardo, V., Battaglino, C., Damiano, R., Nunnari, F.: An avatar-based interface for the italian sign language. In: Proc. of CISIS 2011, pp. 589–594. IEEE Computer Society (2011)
10. Lesmo, L., Robaldo, L.: From natural language to databases via ontologies. In: Proceedings of LREC 2006, Genoa, Italy. European Language Resources Association, ELRA (2006)
11. Lesmo, L., Robaldo, L.: Use of Ontologies in Practical NL Query Interpretation. In: Basili, R., Pazienza, M.T. (eds.) AI*IA 2007. LNCS (LNAI), vol. 4733, pp. 182–193. Springer, Heidelberg (2007)
12. Lesmo, L.: The Rule-Based Parser of the NLP Group of the University of Torino. Intelligenza Artificiale 2, 46–47 (2007)
13. Bosco, C., Lombardo, V.: Dependency and relational structure in treebank annotation. In: Proc. of the Workshop Recent Advances in Dependency Grammar (2004)
14. Hudson, R.: Word Grammar. Basil Blackwell, Oxford (1984)
15. Nirenburg, S., Raskin, V.: Ontological Semantics (Language, Speech, and Communication). The MIT Press (2004)
16. Gangemi, A., Guarino, N., Masolo, C., Oltramari, A.: Sweetening WORDNET with DOLCE. AI Magazine 24, 13–24 (2003)
17. Bateman, J.: The place of language within a foundational ontology. In: Press, I. (ed.) Proc. of FOIS 2004 (2004)
18. Buitelaar, P., Cimiano, P., Haase, P., Sintek, M.: Towards linguistically grounded ontologies. In: Aroyo, L., et al. (eds.) ESWC 2009. LNCS, vol. 5554, pp. 111–125. Springer, Heidelberg (2009)
19. Cimiano, P.: Flexible semantic composition with dudes. In: Proc. of IWCS-8 2009, pp. 272–276. Association for Computational Linguistics, Stroudsburg (2009)
20. Reiter, E., Dale, R.: Building natural language generation systems. Cambridge University Press, New York (2000)
21. White, M.: Efficient realization of coordinate structures in combinatory categorial grammar. Research on Language and Computation 2006, 39–75 (2006)
22. Young, D.E.: The Lisa Project (2007), http://lisa.sourceforge.net/
23. Steedman, M.: The syntactic process. MIT Press, Cambridge (2000)
24. Brentani, D. (ed.): Sign Languages. Cambridge University Press (2010)

A Virtual Player for "Who Wants to Be a Millionaire?" based on Question Answering

Piero Molino, Pierpaolo Basile, Ciro Santoro, Pasquale Lops,
Marco de Gemmis, and Giovanni Semeraro

Dept. of Computer Science, University of Bari Aldo Moro
Via E. Orabona, 4 - 70125 Bari, Italy
firstname.lastname@uniba.it,
c.santoro16@studenti.uniba.it

Abstract. This work presents a virtual player for the quiz game "Who Wants to Be a Millionaire?". The virtual player demands linguistic and common sense knowledge and adopts state-of-the-art Natural Language Processing and Question Answering technologies to answer the questions. Wikipedia articles and DBpedia triples are used as knowledge sources and the answers are ranked according to several lexical, syntactic and semantic criteria. Preliminary experiments carried out on the Italian version of the boardgame proves that the virtual player is able to challenge human players.

1 Introduction

Today artificial systems can compete with the best human players in a growing number of games, like chess, checkers, othello and go. These are called *closed world* games since they have a finite number of possible choices and can be solved in a formal way, even though they are hard to play due to the exponential dimension of the search space.

Recently the interest of the researchers shifted to less structured games, like sports, videogames, crosswords, where the states of the game and the actions that the player can take cannot be easily enumerated, making the search through the space of possible solutions impossible.

In particular, language games require a wide linguistic and common sense knowledge and the understanding of the meaning of natural language words. "Who Wants to Be a Millionaire" (WWBM) is a perfect example of a language game. It is a quiz where the player answers questions posed in natural language by selecting the correct answer out of four possible answers. For example, a possible question is: *"Who directed Blade Runner?"* and the four possible answers are *A) Harrison Ford B) Ridley Scott C) Philip Dick D) James Cameron*.

Even though in this game the number of possible answers is limited, the choice is dependent on the player's knowledge, her understanding of the questions and her ability to balance the confidence in the answers and the risk taken in answering.

M. Baldoni et al. (Eds.): AI*IA 2013, LNAI 8249, pp. 205–216, 2013.

This work describes a virtual player for the WWBM game, which leverages Question Answering (QA) techniques and both Wikipedia and DBpedia data-sources to incorporate common sense human knowledge for playing the game.

The WWBM game allows three "lifelines" which provide some form of assistance to the player. In the original game the lifelines are: *50/50* (which randomly removes two wrong answers), *Ask the Audience* (where the audience answers the question and the percentage of people that choice each possible answer is provided to the player), *Phone-a-Friend* (where the player can phone a friend to ask the question having a specific time constraint - 30 or 60 seconds). If the answer given by the player is correct, she earns a certain amount of money and continues to play with questions of increasing difficulty, until she reaches the last question - the 15th - or she decides to stop the game by taking the money earned. If the player gives the wrong answer, she loses everything if she is answering one of the first five questions; she earns 3,000 Euros if she is answering questions from six to ten, 20,000 Euros for questions from eleven to fifteen. The amount of money earned and the lifelines vary from country to country.

In this first attempt to solve the game, we do not manage "lifelines" or the possibility to retire from the game. Our main goal is to evaluate the ability of the virtual player to correctly answer the questions of the game.

The rest of the paper is organized as follows: in section 2 we describe how our system is built. The evaluation of the system is described in section 3, while in section 4 related work are reported. Conclusions and future work close the paper.

2   The Architecture of the Virtual Player

We built a virtual player for WWBM with a layered architecture consisting of three main modules:

1. *Game Manager*: this module allows to manage the game and its specific rules.
2. *Question Answering*: this module queries Wikipedia and DBpedia data-sources and retrieves the most relevant passages of text useful to select the right answer. A detailed description is provided in Section 2.1.
3. *Answer Selection*: this module adopts different criteria to assign a confidence value to each of the four possible answers for a specific question. A detailed description is provided in Section 2.2.

For each question of the game, the *Game Manager* delegates to the *Question Answering* module the selection of the most relevant passages of text, which might contain the correct answer. The *Question Answering* module returns the passages with the highest scores, together with the title of the articles they are extracted from. Those results are processed by the the *Answer Selection* module which computes the confidence of each possible answer using different heuristics. Finally, the *Game Manager* selects and provides the best answer.

2.1   Question Answering Framework

We exploit QuestionCube [9, 10], a multilingual QA framework created using NLP and IR techniques, in order to obtain relevant passages of text. The overall architecture of the framework is shown in Figure 1.

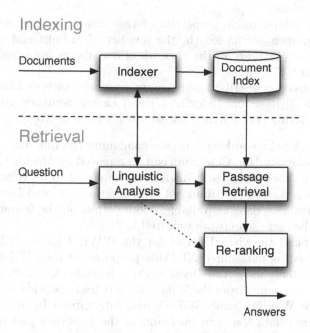

Fig. 1. QuestionCube architecture

The linguistic analysis is carried out by a full-featured NLP pipeline. It includes stopword removal, stemming, part-of-speech tagging, lemmatization, shallow parsing and word sense disambiguation for both English and Italian. Each NLP component adopts state-of-the-art algorithms for the specific task.

The passage retrieval step is carried out by Lucene 4[1], a standard off-the-shelf retrieval framework that allows TF-IDF, Language Modeling and BM25 [13] weighting. Inside the QA process this component is useful to filter passages not related to the question.

The question re-ranking component is designed as a pipeline of different scoring criteria that exploits the different linguistic features obtained from the NLP process (terms, stems, lemmas, chunks, senses). Those criteria include:

– the overlap of specific linguistic features between the question and the answer (or the title of the original document containing the answer)

---

[1] Available at http://lucene.apache.org/

- a density and frequency measure of the question linguistic features inside the answer (or the title of the original document containing the answer). The exact overlapping subsequence, the minimal overlapping span [11] and the number of linguistic features of the question terms in a single sentence of the answer can be considered
- similarity measures based on TF-IDF, Language Modelling and BM25 weighting schemes
- measures based on static properties of the answer and the documents it comes from, such as the length, the number of in-links and out-links (if available), the centrality in the network of documents measured as degree, PageRank or HITS scores
- measures based on distributional semantic models, such as Latent Semantic Analysis [3], Random Indexing [7] and Latent Semantic Analysis over Random Indexing [14]. Further details are available in [10].

We derive a global re-ranking function combining the different scores using the CombSum function [16]. CombSum can be replaced by Machine Learning to Rank algorithms, such as Logistic Regression (reported to be very effective in IBM's DeepQA / Watson [1]), RankNet [2], RankBoost [6] and LambdaMART [17] if enough training data is available. More details on the framework and a description of the main scorers are reported in [9, 10].

In order to build the virtual player for the WWBM game, Wikipedia and DBpedia are used as datasources. To this purpose, we used Wikiedi[2], a specific application built using the QuestionCube framework, which exploits the unstructured data coming from the Wikipedia articles to provide answers to the questions of the WWBM game. Wikiedi also integrates DBpedia as a source for the structured data that can be found in the infoboxes and templates of Wikipedia articles. Search on DBpedia relies on a custom triple searcher built to maximize recall of correct answers. This choice makes the application robust enough to manage both factoid and non-factoid questions. Factoid questions are those whose answers are short excerpts of text, usually named entities, dates or quantities. Non-factoid questions are those whose answers have the form of a passage of text.

Answer re-ranking adopts most of the scorers available in the framework, including those based on the distributional semantics. This enables an approximate matching between question and answers that reduces the impact of question ambiguity.

Table 1 reports the list of five passages returned by Wikiedi for the question "Who directed Blade Runner?". Each passage contains the title of the article it is contained in, and the score computed by Wikiedi.

2.2 Answer Selection

In order to assign a confidence score to each of the four possible answers for a specific question of the WWBM game, we adopt different criteria based on the

---

[2] Available at www.wikiedi.it

Table 1. The list of five passages returned by Wikiedi for the question "Who directed Blade Runner?"

Article Title	Passage Text	Score
Ridley Scott	Sir Ridley Scott (born 30 November 1937) is an English film director and producer. Following his commercial breakthrough with Alien (1979), his best-known works are the sci-fi classic Blade Runner (1982) and the best picture Oscar-winner Gladiator (2000).	5.32
Blade Runner	Blade Runner is a 1982 American dystopian science fiction action film directed by Ridley Scott and starring Harrison Ford, Rutger Hauer, and Sean Young. The screenplay, written by Hampton Fancher and David Peoples, is loosely based on the novel Do Androids Dream of Electric Sheep? by Philip K. Dick.	5.1
Blade Runner	Director Ridley Scott and the film's producers "spent months" meeting and discussing the role with Dustin Hoffman, who eventually departed over differences in vision. Harrison Ford was ultimately chosen for several reasons.	5
Blade Runner	The screenplay by Hampton Fancher was optioned in 1977. Producer Michael Deeley became interested in Fancher's draft and convinced director Ridley Scott to film it.	4.9
Blade Runner	Interest in adapting Philip K. Dick's novel Do Androids Dream of Electric Sheep? developed shortly after its 1968 publication. Director Martin Scorsese was interested in filming the novel, but never optioned it.	1.2

analysis of the passages returned by the QA module. Each individual criterion returns a confidence in the answers, obtained by dividing the score of each answer by the sum of the scores of the four possible answers. Follows a description of each criterion, explained by taking into account the example in Table 1:

- **Title Levenshtein:** this criterion computes the Levenshtein distance (metric for measuring the difference between two sequences of characters) between the candidate answer and the title of the answer returned by Wikiedi. As the Levenshtein distance is a distance measure rather than a similarity measure, we compute $\frac{max(len(a),len(t))-lev(a,t)}{max(len(a),len(t))}$, where $len(a)$ is the length of the candidate answer, $len(t)$ is the length of the title of the Wikipedia page containing the answer provided by Wikiedi, and $lev(a,t)$ is the Levenshtein distance between the candidate answer and the title of the page. This allows to have scores in the $[0,1]$ interval. In our example, the answer B) Ridley Scott occurs in the title of the page containing the passage, so it gets the maximum score of 1 since all the characters are the same, while all the other answers get $\frac{12-11}{12} = 0.083$. The final confidence is 0.8 for answer B) and 0.066 for the others.

- **LCS**: this criterion computes the Longest Common Subsequence between the candidate answer and the passages of text returned by Wikiedi. In our case, answer A) Harrison Ford gets a score equal to 13, answer B) Ridley Scott gets a score equal to 12, answer C) Philip Dick gets a score equal to 11, and answer D) James Cameron gets a score equal to 0 since it does not occur in any of the passages returned by Wikiedi. The final confidence is 0.36 for the candidate answer A), 0.33 for the candidate answer B), 0.31 for the candidate answer C), and 0 for the candidate answer D).
- **Overlap**: this criterion computes the Jaccard index between the set of terms in the candidate answer and the set of terms in the passages of text returned by Wikiedi. In our example, answers A), B) and C) get a score of 0.04651, while the anser D) gets 0. The final confidence is 0.33 for the candidate answer A), B), C), and 0 for the candidate answer D).
- **Exact Substring**: this criterion computes the length in characters of the longest common substring between the candidate answer and the answers from Wikiedi. We normalize the score using the length of the candidate answer. In our example, answer A) Harrison Ford gets a score of $\frac{13}{13} = 1$, answer B) Ridley Scott gets a score of $\frac{12}{12} = 1$, answer C) Philip Dick gets a score of $\frac{6}{11} = 0.54$, and answer D) James Cameron gets score 0. The final confidence is 0.395 for the candidate answer A) and B), 0.21 for the candidate answer C), and 0 for the candidate answer D).
- **Density**: this criterion computes the density of the terms in the candidate answer inside the passages of text returned by Wikiedi, using the minimal overlapping span method described in [11]. In our example, considering only the first passage returned by Wikiedi, answers A) and B) get a score of 1, answer C) gets a score of 0.66 (as the passage reports the full name Philip K. Dick, adding an extra token between the two tokens of the answer), while answer D) gets score 0. The final confidence is 0.375 for the candidate answer A) and B), 0.25 for the candidate answer C) and 0 for the candidate answer C).

Each criterion is parametrized using four parameters: 1) the number of passages returned by Wikiedi; 2) the use of the score of the passages as weights for computing the final value; 3) the level of linguistic analysis to adopt, and 4) the use of the question expansion. Question expansion means that the system asks four different questions obtained by the concatenation of the original question and the four possible candidate answers. For example, the virtual player queries Wikiedi using the following questions: "Who directed Blade Runner? Harrison Ford", "Who directed Blade Runner? Ridley Scott","Who directed Blade Runner? Philip Dick" and "Who directed Blade Runner? James Cameron".

The outcomes of the previous criteria have been also combined using *Majority Vote* and *CombSum*, in order to obtain the final confidence score for each candidate answer. When Majority Vote is used, the confidence of each candidate answer is computed as the ratio between the number of different criteria voting for that candidate answer, and the total number of criteria. When CombSum is used, the scores of the different criteria are standardized and then summed.

3   Experiment

The goal of the evaluation is twofold:

1. to assess the performance of the virtual player
2. to compare the accuracy of the virtual player with that of human players.

   The first experiment aims at evaluating the effectiveness and robustness of the virtual player for different kinds of questions. The second experiment aims at comparing the performance of the system and of the human players by varying the difficulty of the questions.

   The experiments have been carried out using a dataset of 262 questions obtained from the official WWBM Italian boardgame. To the best of our knowledge this is the first attempt to measure the accuracy of a virtual player for the Italian version of the game. This means that we do not have results representing a baseline for our system.

   The metric adopted for the evaluation is *accuracy*, computed as the ratio between the number of correct answers (n_c) and the total number of answers (n), and *c@1*, adopted in the *2010 CLEF QA Competition* [12], computed as $c@1 = \frac{1}{N}\left(n_c + n_n \frac{n_c}{N}\right)$, where N is the number of the questions, n_c is the number of the correct answers provided by the system and n_n is the number of unanswered questions. This measure allows the system to leave the questions unanswered, but the gain in doing so depends on the accuracy, so the metric favors those systems that do not answer the questions they would have answered wrong.

3.1   Results of Experiment 1

Results of the first experiment are reported in Table 2. The first column describes the answer selection criterion and the adopted parameters. Both individual and composite criteria are reported.

   It is worth to note that the composite criteria outperform the individual ones in terms of accuracy and percentage of unanswered questions.

   The best combination is reported in the first row of the table and exploits CombSum of the methods using the following parameters: (1) LCS over 20 passages using lemmas and scores with stopwords removed; (2) Substring over 20 passages adopting keywords and passages score; (3) Overlap over 20 passages, using lemmas and passages score; (4) Density over the first passage, using lemmas and removing stopwords; (5) LCS criterion over 20 passages, using keywords and passages score, with stopwords removed and expanding the question with the four possible answers.

   We have carefully analyzed the questions for which all the criteria failed to provide the correct answer. We found out that some of these questions would require a different kind of knowledge sources to be answered. For example, some of them would require mathematical knowledge, and some others would require English language knowledge. By removing this small subset of questions from the dataset, *accuracy* and *c@1* of the top-ranked criterion increases to 78.43% and 79.66%, respectively.

Table 2. Evaluation results. Criteria: **MV** Majority Vote, **ES** Exact Substring, **LCS** Longest Common Subsequence, **TL** Title Levenshtein. Parameters: the first number is the number of passages, **K** keyword, **L** lemma, **S** uses the score of the passage, **SW** applies stopword removal, **QE** expands the question with the four possible answers.

Criterion	Accuracy	Unansw.	c@1
CombSum: LCS(20,L,S,SW), ES(20,K,S), Overlap(20,L,S), Density(1,L,SW), LCS(20,K,S,SW,QE)	**76.34%**	1.91%	**77.79%**
CombSum: TL(1,K,S,QE), ES(25,K,S), Overlap(25,L,Scored), LCS(25,K,S,SW,QE), LCS(25,L,S,SW,QE)	71.37%	0.38%	71.64%
MV: TL(1,K,S,QE), ES(25,K,S), Overlap(25,L,S), LCS(25,K,S,SW,QE), LCS(25,L,S,SW,QE)	71.76%	0.00%	71.75%
CombSum: TL(1,SW), LCS(25,L,S,SW,QE), LCS(25,K,QE), LCS(25,K,S,SW,QE), Overlap(25,L,S), ES(25,K,S), Overlap(2,K), Overlap(5,K), Overlap(5,K,S), ES(1,K,S), ES(20,L,S)	71.76%	1.14%	72.57%
LCS(20,L,S,SW)	64.89%	16.79%	75.78%
ES(20,L,S)	55.73%	23.66%	68.91%
Overlap(20,L,S)	58.40%	29.01%	75.33%
LCS(20,L,S,SW,QE)	72.90%	1.91%	74.29%
ES(25,K,S)	56.87%	22.14%	69.46%
Overlap(25,K,S)	59.92%	27.48%	76.38%
LCS(25,K,S,SW,QE)	41.60%	20.61%	50.17%
LCS(25,K,L,SW,QE)	71.76%	1.91%	73.12%
Overlap(5,K)	45.80%	44.28%	66.08%
ES(1,K,S)	27.48%	64.50%	45.20%
TL(1,K,SW)	3.44%	96.18%	6.73%
TL(1,K,S,QE)	27.48%	0.00%	27.48%
LCS(25,K,QE)	41.22%	20.61%	49.71%
Overlap(2,K)	36.26%	59.16%	57.71%
Overlap(5,L,S)	45.80%	44.28%	66.08%

The overall outcome of the experiment is consistent with the results presented in [8] (in term of accuracy), even though it is not possible a direct comparison since the experimental settings, the dataset and the language (English) of the questions are different.

3.2   Results of Experiment 2

We involved 60 human players selected using the availability sampling strategy: 85% of the players are graduated, and 10% got a PhD. Each user played the game at least 7 times, and we ensured that each question proposed to the human player was novel so that she never received the same question during different games.

Figure 2 reports the results of the virtual player in terms of *accuracy*, compared with the average results of the human players, for each level of the game.

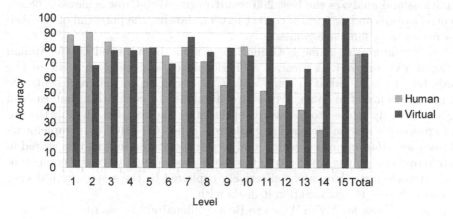

Fig. 2. Accuracy per level

The average accuracy of the virtual player is 76.33%, comparable with the average accuracy reported by humans that is 76.02%.

We performed a further analysis which takes into account the accuracy for each of the fifteen levels of the game separately. Usually lower levels correspond to easier questions, while higher levels correspond to more complex questions. This analysis unveils that human players have higher accuracy on low levels of the game and lower accuracy on higher levels, while the *virtual player* behaves in an opposite way. This means that the virtual player is able to provide the correct answer for the most difficult questions, but it also requires some abilities for providing answers to very simple questions.

This observation is in line from what was previously found in literature [8].

4   Related Work

Teaching a computer how to play games and competing against human players has always been a challenging task since the early days of computing. Games are a good test-bed for Artificial Intelligence as they allow to test the limits of the artificial agents.

Language games are a particular type of open-world games where there is the need to understand the meaning of words and a big amount of knowledge and reasoning are essential to compete at human level. Some examples of language games can be Trivial Pursuit, Punning Riddles, Humoristic Acronyms.

An interesting language game is crosswords, as it relies on linguistic knowledge and requires constraints satisfaction over the possible answers. A system for solving this game is Web-Crow [4], an artificial agent that exploits the Web as a source of information. The problem is subdivided in finding the best words that answer the definitions and in placing them inside the grid. In order to find the best words, the system queries Google with reformulations of the original definitions and analyzes the best 200 result pages. Web-Crow achieves a 68.8% of correct words and a 79.9% of correct letters, showing the potential of the Web as a resource for language games.

Another interesting game is Guillotine, a game broadcasted by the Italian National TV company. It involves a single player, who is given a set of five words (clues), each linked in some way to a specific word that represents the unique solution of the game. Words are unrelated to each other, but each of them is strongly related to the word representing the solution. In [15], the authors propose a system for playing guillotine, called OTTHO, that implements a *knowledge infusion* process which analyzes unstructured information stored in open knowledge sources on the Web to create a memory of linguistic competencies and world facts that can be effectively exploited by the system for a deeper understanding of the information it deals with.

A virtual player for "Who Wants to Be a Millionaire?" is described in [8]. The authors exploit the great amount of knowledge in the Web in order to answer the questions. The query formulation module adopts NLP techniques in order to create different queries. The queries are then sent to Google and the number of obtained results is used to rank the answers, exploiting the redundancy of the information sources. This system reaches an accuracy of 72% showing how useful unstructured data can be for this kind of task, but still fails with questions that require common sense reasoning and access to structured information. The main difference with our work is the adoption of Wikipedia and DBpedia as a selected and reliable source of information rather than the whole Web. Moreover the adoption of a QA framework instead of a search engine allows us to get a more reliable passage filtering.

In February 2011 the IBM Watson supercomputer, adopting technology from the DeepQA [5] project, has beaten the champions of the Jeopardy! TV quiz. In Jeopardy! the classic quiz is reversed, the player is given an ambiguous or ironical piece of the answer and has to find the question for it. To accomplish it, Watson applied several different NLP, IR and ML techniques focusing on the Open-domain QA, answering questions without domain constraint. Watson analyzed 200 million content elements, both structured and unstructured, including the full text of Wikipedia. Watson shows how competitive are actual NLP and ML technologies in managing big amounts of data and exploiting it to compete with humans in a field where they have been considered unbeatable for a long time.

5   Conclusions and Future Work

In this work we propose a virtual player for the game "Who Wants to be Millionaire?". In order to answer the questions from the quiz, our system leverages Natural Language processing and Question Answering techniques and exploits Wikipedia and DBpedia as datasources. A preliminary experiment on the Italian version of the boardgame show that the system is able to provide a correct answer for 76% of questions, and its performance in terms of accuracy is comparable with that of human players.

As future work we plan to add a decision making module able to 1) evaluate whether to provide the answer for a question or to retire from the game; 2) manage the "lifelines" provided by the rules of the game. Furthermore, we shall investigate on the improvement of the answer selection strategy by exploiting learning to rank approaches.

References

1. Agarwal, A., Raghavan, H., Subbian, K., Melville, P., Lawrence, R.D., Gondek, D., Fan, J.: Learning to rank for robust question answering. In: CIKM, pp. 833–842 (2012)
2. Burges, C.J.C., Shaked, T., Renshaw, E., Lazier, A., Deeds, M., Hamilton, N., Hullender, G.N.: Learning to rank using gradient descent. In: ICML, pp. 89–96 (2005)
3. Deerwester, S., Dumais, S.T., Furnas, G.W., Landauer, T.K., Harshman, R.: Indexing by latent semantic analysis. Journal of the American Society for Information Science 41(6), 391–407 (1990)
4. Ernandes, M., Angelini, G., Gori, M.: Webcrow: A web-based system for crossword solving. In: Veloso, M.M., Kambhampati, S. (eds.) AAAI, pp. 1412–1417. AAAI Press/The MIT Press (2005)
5. Ferrucci, D.A., Brown, E.W., Chu-Carroll, J., Fan, J., Gondek, D., Kalyanpur, A., Lally, A., Murdock, J.W., Nyberg, E., Prager, J.M., Schlaefer, N., Welty, C.A.: Building Watson: An Overview of the DeepQA Project. AI Magazine 31(3), 59–79 (2010)
6. Freund, Y., Iyer, R.D., Schapire, R.E., Singer, Y.: An efficient boosting algorithm for combining preferences. Journal of Machine Learning Research 4, 933–969 (2003)
7. Kanerva, P.: Sparse Distributed Memory. MIT Press (1988)
8. Lam, S.K., Pennock, D.M., Cosley, D., Lawrence, S.: 1 billion pages = 1 million dollars? mining the web to play "who wants to be a millionaire?". In: Meek, C., Kjærulff, U. (eds.) UAI, pp. 337–345. Morgan Kaufmann (2003)
9. Molino, P., Basile, P.: Questioncube: a framework for question answering. In: Amati, G., Carpineto, C., Semeraro, G. (eds.) Proceedings of the 3rd Italian Information Retrieval (IIR) Workshop, Bari, Italy, January 26-27. CEUR Workshop Proceedings, vol. 835, pp. 167–178. CEUR-WS.org (2012)
10. Molino, P., Basile, P., Caputo, A., Lops, P., Semeraro, G.: Exploiting distributional semantic models in question answering. In: Sixth IEEE International Conference on Semantic Computing, ICSC 2012, Palermo, Italy, September 19-21, pp. 146–153. IEEE Computer Society (2012)
11. Monz, C.: Minimal span weighting retrieval for question answering. In: Gaizauskas, R., Greenwood, M., Hepple, M. (eds.) Proceedings of the SIGIR Workshop on Information Retrieval for Question Answering, pp. 23–30 (2004)
12. Penas, A., Forner, P., Rodrigo, A., Sutcliffe, R.F.E., Forascu, C., Mota, C.: Overview of ResPubliQA 2010: Question Answering Evaluation over European Legislation. In: Braschler, M., Harman, D., Pianta, E. (eds.) Working Notes of ResPubliQA 2010 Lab at CLEF 2010 (2010)
13. Robertson, S., Zaragoza, H.: The probabilistic relevance framework: Bm25 and beyond. Found. Trends Inf. Retr. 3, 333–389 (2009)
14. Sellberg, L., Jönsson, A.: Using random indexing to improve singular value decomposition for latent semantic analysis. In: LREC (2008)

15. Semeraro, G., de Gemmis, M., Lops, P., Basile, P.: An artificial player for a language game. IEEE Intelligent Systems 27(5), 36–43 (2012)
16. Shaw, J.A., Fox, E.A.: Combination of multiple searches. In: The Second Text REtrieval Conference (TREC-2), pp. 243–252 (1994)
17. Wu, Q., Burges, C.J.C., Svore, K.M., Gao, J.: Adapting boosting for information retrieval measures. Inf. Retr. 13(3), 254–270 (2010)

The Construction of the Relative Distance Fuzzy Values Based on the Questionnaire Experiment

Jedrzej Osiński

Faculty of Mathematics and Computer Science,
Adam Mickiewicz University, Poznan, Poland
josinski@amu.edu.pl

Abstract. The spatio-temporal reasoning is an import field of artificial intelligence. However the qualitative distance (like *near* or *far*) which can be the result of a natural language processing is almost impossible to interpret due to its relative character (connected with a specific language competences of an interlocutor, the context of usage and the perspective of observation). That is why we present a technique for the construction of the fuzzy values associated with such expressions which is based on the questionnaire experiment performed via Internet. We also show how the final linguistic variables can simple became arguments for complex calculations. The presented solution can be treat as a general method applicable in many areas.

Keywords: fuzzy relations, linguistic variables, natural language processing, spatio-temporal reasoning, CDC.

1 Introduction

The natural language processing is one of the crucial areas of artificial intelligence. There are many problems identified in this field, like voice recognition, syntactic analysis, disambiguation solving, etc. However the semantic interpretation is the aspect which moves us from a pure text to a real human-computer interaction. The construction of the semantic map of a described environment, especially in the spatial aspect, is not so complicated as long as we can build it on a precise quantitative sentences, e.g. *Object C is 24 meters to the south of object B and 11 meters to west of object A or Object C is at 52.467 deg North, 16.927 deg East.* But the real problem appears when we have only qualitative knowledge about the position of an analyzed object (what is widely discuss e.g. in [1], [3] and [5]). The interpretation of the qualitative distance relations conceptualized in a natural language like *right next to, close, far* is difficult due to the complex nature of these sentences.

Firstly, the precision of such a expressions is restricted by the natural limitations of the human perception who cannot define the exact distance without the appropriate measuring tools. Even if an observer describes the situation using the quantitative sentences, their reliability is rather low. We have prepared an online questionnaire in which totally 1047 people took part. We asked them to

M. Baldoni et al. (Eds.): AI*IA 2013, LNAI 8249, pp. 217–226, 2013.

judge their frequency of using in an everyday life the particular types of distance expressions - almost 60% pointed at the relative relations (e.g. *near*, *far*), while the interval expressions (e.g. from 20 to 30 meters), rough expressions (e.g. *about 20 meters*) and the exact values (e.g. *21 meters*) were up to few times less popular. What is more when asked about their self-confidence about their ability in precise assessing spatial distance, 49% choose the middle answer (*quite confident*), 31% said there are not sure and 11% directly admitted that their cannot assess a distance. Only 8% said they are very precise and less than 1% described their ability as infallible.

The second aspect generating problems with a correct interpretation is the fact that the relative distance expressions are highly context-dependent. Suppose we analyze the sentence *M is close to N*. The expression close may refer to a totally different quantitative distance - compare:

Peter is close to Kate (about 1 meter),
Washington is close to Baltimore (about 40 miles),
Mercury is close to the sun (almost 58 million kilometers).

Finally, the meaning of the sentence depends on the language competences and life experience. The same distance would be perceived differently by e.g. a trained soldier, a long-distance runner, a helicopter pilot or a small child. All these aspects make the semantic hidden in the relative distance expressions very difficult to analyze. In the next sections we present our solution applied in a well defined context and based on the questionnaire experiment. As a result we get the linguistic variable associated with the three language expressions: *right next to, close, far*. The definition of a linguistic variable was originally introduced in [9] as a quintuple (H, T(H), U, G, M) where H is a name of a variable, T(H) is a collection of its linguistic values (*right next to, close, far* in our example), U is a universe of discourse (context-dependent), G is a syntactic rule which generates the terms in T(H), while M(X) (for each X T(H)) is a semantic rules understood as a fuzzy subset of U. It is worth mentioning that the described idea is widely used as a effective technique to deal with imprecise data, e.g. in systems supporting medical analysis.

In [4] authors use this approach for assessing pulmonary infections by defining the linguistic variable to describe the influence of one concept on another which consists of 12 linguistic values, e.g. *negatively very strong, zero, positively weak, positively medium, positively strong*. Similarly the linguistic values like *very low, medium, high* were defined in [10] for linguistic variables: likelihood level (probability of threat occurrence) and consequence (outcome to the system / asset value).

2   The Problem in a Defined Environment

The problem mentioned above was analyzed during the development of the prototype of the POLINT-112-SMS system (described in details in [7] and [8]). POLINT-112-SMS is a multi-agent expert system designed for improving the quality of monitoring and analysis of complex situations from the point of view of the security of big public events (e.g. soccer matches). The main goal of the system is supporting the security officers by providing the natural language interface combined with the standard SMS communication channel (which is preferably in a noisy or/and hostile environment). This research was partially covered by the Polish Government grant R00 028 02 "Text processing technologies for Polish in application for public security purposes" (2006-2009) within the Polish Platform for Homeland Security. In the first (prototype) stage of the development the analyzed environment was limited to the soccer stadium which is characterized by well-known, general structure. All of such an open areas design for mass sport events has a similar shape and contains the same components: a pitch (being the place where the match is played; its size, material, color are precisely defined by the international sport rules), sectors (located around the pitch and containing seats, entrances and communication routes; their structures are also similar determined by the comfort and safety of fans). As during the corpus analysis the qualitative distance expressions was a popular type of sentences we decided to run the experiment and work out a technique for calculations on such values. There are two assumptions which makes this try interesting and reasonable. Firstly, as we mentioned before the environment being described by users is well-defined, widely-know and perceived similarly by people, even with a different background. Secondly, the system is dedicated to support security officers who has common experience and has been analogically trained. We can assume that the way they see and talk about the distance relations in the same situation would be similar.

3   Experiment

During the corpus analysis (which had been collected during the simulated "Wizard of Oz" experiment when the developers had been manually preparing the answers of the system) we have identified three main spatial distance relations: *right next to, close, far*. Of course one may suggest a farther division by adding other relations like *very close* or *very far*, but the problem is whether such expressions can be reasonable distinguished and similarly understood by different users. Due to these doubts (and the corpus analysis) we have decided to define four basic distance relations associated with the following symbols: D0 (*the same place*), D1 (*right next to*), D2 (*close*), D3 (*far*). D0 corresponds the situation when two objects are located in the same place, in other words the distancebe tween them equals zero. The D1 symbol describes the configuration when objects are externally connected, i.e. their intersection are (and only) the elements of their boundaries. The last two symbols D2 and D3 where defined to introduce two-level scale of the distance between the objects which intersection is an empty set.

The experiment was performed via Internet using the KwikSurveys.com online platform during two weeks in September 2010. The questionnaire was absolutely anonymous and with a free public access. It was also not possible to finish it without answering all the questions. The experiment was promoted by sending the

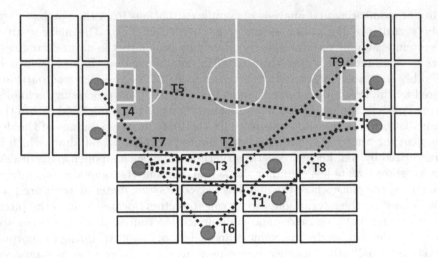

Fig. 1. The compiled diagram of 9 schemas presented to the respondents (tasks T1-T9)

invitation to take part in the research to all the students of the Adam Mickiewicz University in Poznan. The advertise was also placed in the Facebook portal and the popular messengers. Totally 1452 people completed the questionnaire. During the experiment the schema of the soccer stadium has been presented to the respondents. Then they were asked to choose one of the three sentences (*right next to, close, far*) which in their opinion best describes the distance between the objects connected by the dotted line. In next nine tasks the set of the sentences was the same, while the positions of the objects was modified (see fig. 1). That allows us to collect the percentage of the usage of specific relations associated with the presented diagrams (see tab. 1).

Table 1. The percentage results of the questionnaire experiment

Task	T1	T2	T3	T4	T5	T6	T7	T8	T9
D1	1.58%	0.28%	96.01%	0.07%	0.28%	4.89%	2.00%	0.48%	0.41%
D2	89.67%	0.90%	3.86%	3.03%	0.41%	91.46%	69.28%	12.05%	0.14%
D3	8.75%	98.83%	0.14%	96.90%	99.31%	3.65%	28.72%	87.47%	99.45%

4   Results

Having the percentage results mentioned above we have analyzed the presented distances (the dotted lines) normalized by the length of the pitch on the picture which we define as a single unit (denoted by PL). The length of the dotted line (in centimeters) was simple divided by the length (also in centimeters) of the horizontal boundary of the playing pitch - the result was expressed in the PL

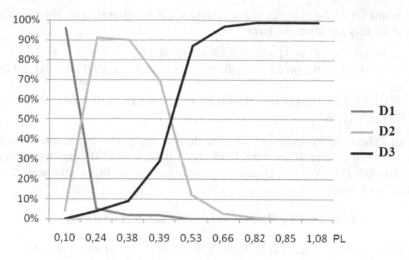

Fig. 2. The results of the experiment in the PL unit

unit. Fig. 2. presents the complete results of the experiment in a chart with a horizontal scale in PL. It is worth mentioning that for each of the three curves we can calculate the abscissa of the center of gravity:

$$\sum_{i=1}^{9}(length(i) \cdot answerPercentage(i))/\sum_{i=1}^{9}(answerPercentage(i)) \qquad (1)$$

These values can treated as a quantitative approximation of the qualitative sentences. Using the above formula we have D1 = 0.11 PL, D2 = 0.34 PL and D3 = 0.76 PL. That is quite interesting as corresponds to some intuivite feellings that if something is *far* from us it means a little more than twice the *close* distance. Similarly *close* is about three times more than the externally connection. These results provide the technique for calculating on the qualitative distances. Of course we cannot recognize it as an absolutely precise mechanizm, however it allows to realize the mathematical operations on the knowledge extracted from the natural language input, e.g. D1 + D2 (e.g. *Object A is just right to the north of the object B + Object C is close to the north of the object A*) = 0.11 + 0.34 = 0.45 PL what we can interpret as follows: *Object C is close (in about 41%) or far (in about 59%) to the north of object B*. Analogically, we can also define subtraction, multiplication, division as well as more complex operation. However the real strength of that solution appears when we look at the results of the experiment as at fuzzy numbers. Let us define the fuzzy set A by the following equation:

$$A = \{(x, \mu_A(x)) : x \in X, \mu_A(x) \in [0,1]\} \qquad (2)$$

where $\mu_A : X \to [0,1]$ is an indicator function describing the membership of the elements from the universe X in the set A. If the universe is finite (as in our example - it contains nine elements), the fuzzy set can be denoted as pairs [membership level, the element of the universe], $\mu|x$, separated by the symbol of sum (+). In this notation we have:

$D1 := 0.96|0.1 + 0.05|0.24 + 0.02|0.38 + 0.02|0.39$.

$D2 := 0.04|0.1 + 0.91|0.24 + 0.9|0.38 + 0.69|0.39 + 0.12|0.53 + 0.03|0.66 + 0.01|0.82$.

$D3 := 0.04|0.24 + 0.09|0.38 + 0.29|0.39 + 0.87|0.53 + 0.97|0.66 + 0.99|0.82 + 0.99|0.85 + 0.99|1.08$.

For the above fuzzy numbers we can simply apply the extension principle. We have X = 0.1, 0.24, 0.38, 0.39, 0.53, 0.66, 0.82, 0.85, 1.08 and let us define $A, B \in D1, D2, D3 : X \rightarrow [0,1]$ and the operation on the fuzzy numbers $A * B :$ $\mathbb{R} \rightarrow [0,1]$ as follows:

$$\forall z \in X : (A * B)(z) = max_{(x,y):x*y=z}\mu_A(x)\mathbf{t}\mu_B(y) \tag{3}$$

where **t** is a t-norm, usually minimum, while $* \in +, -, \cdot$ is an arithmetic operation (addition, subtraction and multiplication respectively).

5 Example of Usage

The presented solution can be successfully applied in many areas in which the value of distance is characterized by low precision (e.g. as a result of a natural language input). We now present how our technique can be combined with a classic models for the qualitative spatio-temporal reasoning. One of the most popular formalism is the Cardinal Direction Calculus (CDC) first presented in [2]. The key idea of that formalism is based on dividing the plane around the reference object (i.e. the object from which the direction relation is determined) into nine regions named after the geographical directions: NW, N, NE, W, O (central region meaning the same location), E, SW, S and SE. These areas, called direction tiles, are closed, unbounded (except for O), their interiors are pairwise disjoint and their union is the whole plane - see fig. 3. Directions between the

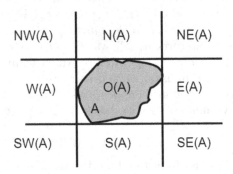

Fig. 3. Division of the plane around the reference object

reference object A and target object B are represented in a 3 x 3 direction-relation matrix denoted by dir(A,B) which we define as follows:

$$dir(A, B) = \begin{bmatrix} f(NW(A) \cap B) & f(N(A) \cap B) & f(NE(A) \cap B) \\ f(W(A) \cap B) & f(O(A) \cap B) & f(E(A) \cap B) \\ f(SW(A) \cap B) & f(S(A) \cap B) & f(SE(A) \cap B) \end{bmatrix},$$

where

$$f(x) = \begin{cases} 0, & if\ Interior(X) \neq \emptyset \\ 1, & if\ Interior(X) = \emptyset \end{cases}.$$

We can also present a relation in an equivalent notation as an expression $R_1 : ... : R_k$ where $1 \leqslant k \leqslant 9$, $R_1, ..., R_k \in NW, N, NE, W, O, E, SW, S, SE$ and $R_i \leqslant R_j$ for every i, j such that $1 \leqslant i, j \leqslant k$ and $i \neq j$. This formalism has been already carefully discussed and used for spatial reasoning including describing spatial relations between objects, computing positions or for assessing similarity between spatial scenes. The algorithm for the composition of the cardinal direction relations was presented in [6] so the operations like $(N\ o\ E)$ or $(N{:}NE\ o\ E\ o\ SW{:}S{:}SE)$ bring no serious difficulties (their calculation time is limited by a constant, i.e. O(1)). However the problem occurs when we try to compose it together with a qualitative distance. Let us analyze to sentences being the source of the knowledge: *Object B is close to the north-east of object A and Object C is far to the east of B*. Fig. 4 presents this situation. We can use the general algorithm to calculate the composition of these distances (denoted by SD_1, SD_2):

1. If (SD_1 is unknown OR SD_2 is unknown) then return UNKNOWN.
2. If SD_1 = D0 then return SD_2.
3. If SD_2 = D1 then return SD_1.
4. If ($O \in dir(A, B)$ AND $O \in dir(B, C)$) then return MAX(SD_1, SD_2).
5. If $O \in dir(A, B)$ then return SD_2.
6. If $O \in dir(B, C)$ then return SD_1.
7. Calculate H1 as a integer number associated with the main direction of the dir(A,B): 0 for N, 1 for NE, , 7 for NW.
8. Calculate H2.
9. $\Delta H := |H1 - H2|$.
10. (Normalization) If $\Delta H > 4$ then $\Delta H := 8 - \Delta H$.
11. $\gamma := 180° - \Delta H \cdot 45°$.
12. $D^2 := SD_1{}^2 + SD_2{}^2 - 2 \cdot SD_1 \cdot SD_2 \cdot cos(\gamma)$ (the law of cosines).
13. Return D.

In the algorithm we have mentioned (step 5.) the main direction of the direction-relation matrix which we understand as the direction tile which determines the most the direction of the distance vector. If we consider only basic relations they immediately generate the main direction (as their matrix representation contains only one 1). The issue is more complex if we analyze all possible

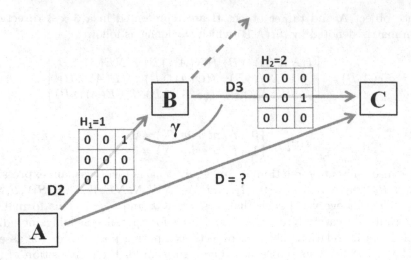

Fig. 4. The example of a spatial situation

relations like *NW:N:NE*, *N:NE* or *NW:N:NE:E*. For a given direction-relation matrix DIR we can calculate H as follows:

1. Consider a square divided into 9 equal tiles.
2. To each tile associate weight 1 if it has non-zero value in DIR and 0 otherwise.
3. Calculate the center of gravity COG of the square.
4. Return the tile containing COG.

Of course we can modify this solution depending on the application and the required level of precision. It is important to notice that the H value do not necessary has to be an integer. If, for example, the COG is close enough to the boundary line between tiles N and NE we can calculate average H1 = (0 + 1)/2 = 0.5 what together with E (H2 = 2) of the second distance gives us $\Delta H := |H1 - H2| = 1.5$ and finally $\gamma := 112.5°$.

In the situation presented on fig. 4. $\gamma := 135°$ what provides all the parameters needed for applying the law of cosines. While using the centers of gravity (D2 = 0.34 PL, D3 = 0.76 PL) we have:

$$D2 = D2^2 + D3^2 - 2 \cdot D2 \cdot D3 \cdot cos(135°) = 0.34^2 + 0.76^2 - 2 \cdot 0.34 \cdot 0.76 \cdot (-0.71) = 1.06$$

so finally D = 1.03 PL.

In the fuzzy version the only difference compared with the classic geometrical calculation is the fact that arguments are fuzzy numbers and the operations +, -, · are to be realize as a consequences of the extension principle. The result of the calculation (the fuzzy D value) is presented on fig. 5. The abscissa of the center of gravity of the area under the curve of D equals 1.10 PL.

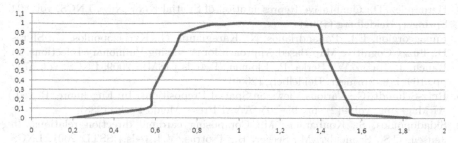

Fig. 5. Fuzzy number D corresponding the composed distance on fig. 3

6   Discussion and Conclusion

We have presented the technique for interpreting and composing qualitative distance relations. Of course the presented solution is useless as long as the knowledge base of an analyzed system contains precise metric (quantitative) values. Our method gives a chance for successful understanding of the spatial relations when the input data are qualitative (e.g. as being the result of natural language processing). In such a situation any other well-defined mechanism may be found insufficient. The key element of the mentioned solution was the experiment performed via Internet. It is important to mention that there were no restrictions on the profile of the respondents (anyone could fill in the online questionnaire) however if necessary the similar research can be repeated for the specific group representing the potential user of the system (if some specific system is dedicated for mines the questions can be asked to miners, etc.). What is also import a similar method can be applied to define fuzzy numbers associated with many other qualitative values popular in a different areas of industry, e.g. the assessment of temperature in a factory or bakery (*too cool, hot, too hot*), the timing in a corporation (*right in time, too late, much too late*), the level of cleanliness (*clean, very clean, dirty*) or the fullness of a storehouse (*almost empty, middle full, almost full*).

Acknowledgements. This research was partially co-financed by European Union under the European Social Fund, within the project "Scholarship support for PH.D. students specializing in majors strategic for Wielkopolska's development" (2010/2011), Sub-measure 8.2.2 Human Capital Operational Program.

References

1. Freksa, C.: Qualitative Spatial Reasoning. In: Frank, D.M., Frank, A.U. (eds.) Cognitive and Linguistics Aspects of Geographic Space. NATO ASI Series, pp. 361–372 (1991)
2. Goyal, R.K., Egenhofer, M.J.: Cardinal directions between extended spatial objects. IEEE Transactions on Knowledge and Data Engineering (2001)

3. Hernández, D.: Qualitative Representation of Spatial Knowledge. LNCS, vol. 804. Springer, Heidelberg (1994)
4. Papageorgiou, E.I., Papandrianos, N., Karagianni, G., Kyriazopoulos, G., Sfyras, D.: Fuzzy Cognitive Map Based Approach for Assessing Pulmonary Infections. In: Rauch, J., Raś, Z.W., Berka, P., Elomaa, T. (eds.) ISMIS 2009. LNCS, vol. 5722, pp. 109–118. Springer, Heidelberg (2009)
5. Retz-Schmidt, G.: Various Views on Spatial Prepositions. In: Engelmore, R. (ed.) AI Magazine, vol. 9(2), pp. 95–105. AAAI Press, Menlo Park (1988)
6. Skiadopoulos, S., Koubarakis, M.: Composing cardinal direction relations. In: Jensen, C.S., Schneider, M., Seeger, B., Tsotras, V.J. (eds.) SSTD 2001. LNCS, vol. 2121, pp. 299–317. Springer, Heidelberg (2001)
7. Vetulani, Z., Marciniak, J., Obrebski, T., Kubis, M., Osinski, J., Walkowska, J., Kubacki, P., Witalewski, K.: POLINT-112-SMS: Beta Prototype. In: Vetulani, Z. (ed.) Proceedings of the 4th Language & Technology Conference, November 6-8, pp. 160–164. Wyd. Poznanskie, Poznan (2009)
8. Vetulani, Z., Marciniak, J., Obrebski, T., Vetulani, G., Dabrowski, A., Kubis, M., Osinski, J., Walkowska, J., Kubacki, P., Witalewski, K.: Zasoby jezykowe i technologie przetwarzania tekstu. POLINT-112-SMS jako przyklad aplikacji z zakresu bezpieczenstwa publicznego (Language resources and text processing Technologies. POLINT-112-SMS as example of public security application with language competence). Wyd. UAM, Poznan (2010) (in Polish)
9. Zadeh, L.: The concept of a linguistic variable and its application to approximate reasoning - I. Information Sciences 8(3), 199–249 (1975)
10. Mohamad Zain, N., Narayana Samy, G., Ahmad, R., Ismail, Z., Abdul Manaf, A.: Fuzzy Based Threat Analysis in Total Hospital Information System. In: Kim, T.-H., Adeli, H. (eds.) AST/UCMA/ISA/ACN 2010. LNCS, vol. 6059, pp. 1–14. Springer, Heidelberg (2010)

Process Fragment Recognition
in Clinical Documents

Camilo Thorne[1], Elena Cardillo[2], Claudio Eccher[2],
Marco Montali[1], and Diego Calvanese[1]

[1] Free University of Bozen-Bolzano, 3 Piazza Domenicani, 39100, Italy
{thorne,montali,calvanese}@inf.unibz.it
[2] Fondazione Bruno Kessler, 18 Via Sommarive, 38123, Italy
{cardillo,eccher}@fbk.edu

Abstract. We describe a first experiment on automated activity and relation identification, and more in general, on the automated identification
and extraction of computer-interpretable guideline fragments from clinical documents. We rely on clinical entity and relation (activities, actors,
artifacts and their relations) recognition techniques and use MetaMap
and the UMLS Metathesaurus to provide lexical information. In particular, we study the impact of clinical document syntax and semantics on
the precision of activity and temporal relation recognition.

Keywords. Clinical entity and relation recognition, UMLS Metathesaurus, natural language processing, process fragment recognition.

1 Introduction

Clinical practice guidelines are systematically developed documents that specify
the activities, resources and personnel required to cure or treat a particular
illness or medical condition, see [6]. The necessity to instantiate them into clinic
and hospital protocols and workflows has given rise to *computer-interpretable
guidelines* (CIGs), see [3], viz., formal representations (constructed typically with
process representation languages) of the care process or plan. On the other hand,
several natural language processing (NLP) techniques have been developed to
fully or partially automate their processing, see [10], until now manual, and
therefore both time and cost consuming.

Clinical NLP approaches leverage on a number of clinical and biomedical annotation resources. Above all, they rely on the crucial US National Library of
Medicine's Unified Medical Language System (UMLS) Metathesaurus[1], which
thanks to the two annotation tools built upon it, MetaMap and SemRel (see
[1]), has become the key lexical semantics resource in this domain. The UMLS
Metathesaurus is a biomedical lexical resource (similar to WordNet) compris-
ing over 1 million biomedical concepts (identified by a CUI – concept unique
identifier) and covering over 5 million terms, which stem from the over 100

---

[1] http://www.nlm.nih.gov/research/umls/

M. Baldoni et al. (Eds.): AI*IA 2013, LNAI 8249, pp. 227–238, 2013.

incorporated controlled vocabularies, nomenclatures and classification systems integrated in it. Concepts in UMLS are structured in a semantic network (or ontology) composed of 150 categories (called "concept types") and 54 semantic relationships.

In this paper we describe some preliminary experiments on how to apply fully-supervised clinical entity recognition techniques inspired by [2] to recognize CIG fragments in medical documents, by leveraging on UMLS concept type and relation annotations. The process dimension of CIGs consists of four pillars: activities, resources, actors, and control flows. We focus on activities, the main building block of CIGs, and their basic temporal relations (before/after). To a lesser extent, we focus also on resources, actors and causal relations. We rely on MetaMap annotations and evaluate our techniques over a small UMLS-annotated clinical corpus, the SemRep corpus. We focus in particular on the issue of feature extraction and selection, to assess which features, be them semantic or (morpho)syntactic, are reasonably good predictors for activity and temporal relation recognition.

This paper is structured as follows. In Section 2 we give an overview of related work on clinical NLP, data mining and unsupervised CIG extraction methodologies. In Section 3 we provide the formal background of CIG fragment recognition. In Section 4 we describe the SemRep corpus, our experimental setting and the goals pursued. In Section 5 we describe and discuss the results of our experiments. Finally, we sum up our conclusions in Section 6, and point out how we intend in the future to study further automated CIG fragment extraction.

2   Related Work

The UMLS Metathesausus has been used for clinical text or data mining purposes in many projects. In the medical domain, early works are MedLEE [8] or MedSyndicate [9]. Also MedIE [21] and SeReMeD [4] apply semantic tagging using the UMLS, but their application is limited to processing radiology reports. Meystre and Haug propose a NLP-based system to extract medical problems from electronic patient records [14]. Clinical NLP frameworks such as cTAKES, proposed by [17], use it for document indexing and retrieval. It has also given rise to automated semi- and fully-supervised annotation techniques and resources: It has inspired the annotation formats used to build clinical annotated corpora such as the CLEF corpus from [16]. Furthermore, as Ben Abacha and Zwiegenbaum in [2] show, it is a key tool for automated clinical entity recognition and for clinical relation recognition and extraction.

The much more complex domain of clinical guidelines and CIGs has been tackled on the other hand, using unsupervised techniques. Serban et al. [18] defined a set of linguistic patterns, which can be used to formally represent the knowledge about medical actions contained in guideline text. By semantic tagging of guidelines, these patterns were identified in the document. After combining them with medical domain knowledge this enables an easier formalization and maintenance of guideline models. A similar approach was pursued by Kaiser et al. [10], who defined syntactic and semantic patterns that are used to develop extraction rules to identify and extract actions and processes out of guidelines. The patterns were based on the UMLS Semantic Network and its semantic relations. But unsupervised techniques are problematic, because the expert knowledge needed to hand-craft such rules is typically much scarce than training corpora.

We believe that UMLS-based supervised clinical entity recognition and annotation techniques and CIG extraction techniques can be successfully combined by following recent advances from the business process modeling community. Friederich et al. [7] and Di Ciccio and Metella [5] have experimented with pipelines combining lexical resources plus supervised (e.g., parsing, entity recognition) and unsupervised (e.g., control flow patterns) NLP techniques to extract, resp., model fragments from formal requirement documents, and process fragments from emails, both with reasonable amounts of success. Following this intuition we experiment in this paper with a UMLS-driven supervised approach that aims at recognizing activities and temporal relations in clinical documents.

3   Process Fragment Recognition

CIG Fragments. There are several ways to formally characterize CIGs. For convenience, we use terminology coming from the Business Process Modeling and Notation (BPMN) standard (see [13]). A CIG is a complex object constituted by the following basic components:

- static components: *(i) activities* (e.g., providing advice, controlling blood glucose levels), representing units of execution in the process; *(ii)* activity agents, viz., the *actors* (e.g., doctors, nurses, patients); *(iii)* artifacts and data used or consumed by activities or *resources* (e.g., metmorfin);
- dynamic components: *(iv) control flows* (e.g., sequence and "if... then... else" control structures) that specify the acceptable orderings among activities.

To extract CIG components the parse trees and MetaMap annotations of guidelines must be mined. Firstly, noun phrases (**NP**s) and verbs must be identified in the parse or constituency trees. At a second step linguistic and domain knowledge in the form of semantic annotations and constituency relations must be considered. In Figure 1 (top) the reader can see a guideline fragment (recommendation 1.4.1 of the NICE diabetes-2 guideline[2]) with its entities highlighted and their candidate annotations. Activities are not only referred to by verbs but also by nouns and noun phrases: to correctly extract the "deep" intended CIG fragment (see Figure 1, bottom left) it is necessary to "filter out" the two wrong "clinical attribute" annotations. Moreover, we need to realize that the verb "continue" introduces a third activity, and rely on syntactic structure to properly order the activities.

We would like to know if the CIG fragment extraction, and in particular the recognition of the activities and control flows intended in clinical documents

---

[2] http://www.nice.org.uk/nicemedia/pdf/CG66NICEGuideline.pdf

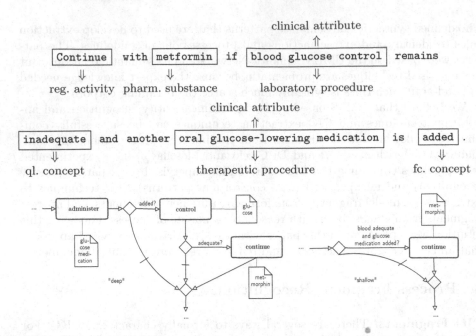

Fig. 1. Top: MetaMap UMLS (automated) annotations of the NICE diabetes guideline fragment; boxes surround entities, annotations are MetaMap's. **Bottom:** Two candidate CIG fragments (represented in BPMN): to the left, the intended "deep" CIG, to the right a "shallow" CIG. Control flows (diamonds) specify the acceptable orderings of the activities (rounded rectangles); activities consume resources (folded-corner rectangles).

can be pursued via supervised clinical entity and relation recognition using an UMLS-annotated corpus for training.

Clinical Entity Recognition. To identify activities and relations in clinical documents, we need to recognize CIG fragments. Let $t_c = (c_1, \ldots, c_n)^T$ denote a vector of n *entity type labels* drawn from a set $\{c_1, \ldots, c_k\}$ of k clinical entities; or, resp., a vector $t_r = (r_1, \ldots, r_n)^T$ of n *relation labels* drawn from a set $\{r_1, \ldots, r_p\}$ of p clinical relations. Let $\alpha = (\alpha_1, \ldots, \alpha_n)^T$ be a vector of n input *noun phrases* (**NP**s) or *entities* (the n **NP**s of a sentence), or resp., a vector $\alpha = ((\alpha_1, \alpha_1), (\alpha_1, \alpha_2), \ldots, (\alpha_n, \alpha_n))^T$ of $n \times n$ input **NP** pairs or relation *arguments* (the $n \times n$ possible pairs of **NP**s in a sentence). The goal of *clinical entity* or, resp., *relation recognition*, see [2], can be formulated as the task of finding the best scoring vector t_β^*:

$$t_\beta^* = \arg\max_{t_\beta} \mu(\rho(\alpha, t_\beta)) \tag{1}$$

where: $\beta \in \{c, r\}$; $\mu(\cdot)$ denotes a *recognizer* built using a classification model (e.g., a logistic regression or neural network algorithm); and $\rho(\cdot, \cdot)$ is a *feature extraction* function, that maps t_β and α into a high-dimensional space of numeric, categorical or ordinal *features* over which the classifier is defined. We study this task w.r.t. the set {activity, resource, actor, other} of entity type labels and the set {temporal, causal, other} of relation labels, and consider *supervised* recognizers, viz., recognizers that can be estimated from a training corpus.

4   Experiments

Features. Our experiments focused in understanding the predictive power of syntax and semantics for recognizing both clinical activities and their temporal relations. Thus, we decided to use linguistically "deep" features extracted from constituency parse trees in addition to semantic annotations. Following strategies similar to the work proposed by [20] we used the Stanford parser (see [12]) to extract syntactic features, and MetaMap to harvest clinical entities and relations via the UMLS concept types they subsume (see Table 1, top), and to compute the lexical semantic features.

By mining parse trees we extracted from **NP**s the following syntactic features: **(1)** depth *nest* of nesting; **(2)** position *pos* in the phrase; and **(3)** occurrence *sub* in a subordinated phrase. The lexical semantic features were extracted by computing several measures of label overlap and frequency. We extracted also the following semantic features: **(4)** the (raw) frequency of the **NP** entity type c in the corpus; **(5)** the degree of *annotation overlap* φ_{hd} between the (possibly repeated) labels *labs* collected using MetaMap from all the constituent nouns of a **NP**, and the (possibly repeated) labels of its head noun *labsh*; **(6)** the *relative frequency* φ_{lf} of the **NP** entity type c w.r.t. *labs*; and **(7)** *label overlap* φ_{ls} that takes into account the taxonomic structure of the UMLS Metathesaurus; viz., respectively,

$$\varphi_{hd} = \frac{||\,labs \cap labsh\,||}{||\,labs\,|| + ||\,labsh\,||} \qquad \varphi_{lf} = \frac{||\,labs \cap \{c\}\,||}{||\,labs\,||} \qquad \varphi_{ls} = \frac{||\,labs \cap sub(c)\,||}{||\,labs\,|| + ||\,sub(c)\,||} \qquad (2)$$

where $||\cdot||$ and $\cap$ denote resp. bag cardinality and intersection, and $sub(c)$ is the bag of all the UMLS concept types that the entity type label c subsumes. In all cases a simple Laplace smoothing was later applied to prevent division by zero errors. See Table 1, bottom.

The SemRep Corpus. Since no UMLS annotated guideline corpora are available for research purposes we ran our experiments over the SemRep corpus (see [11]), a small annotated clinical corpus. It consists of 500 clinical excerpts (MedLine/PubMed) and contains 13, 948 word tokens, manually annotated by clinicians and domain experts, covering the whole clinical domain. UMLS concept types annotate a total of 827 **NP**s (at an average of 2 per sentence). In addition to this, UMLS relations annotate around 200 **NP** pairs.

The domain of SemRep largely overlaps with that of clinical guidelines. Furthermore, they are similar in syntactic structure. Such syntactic structure can be approximated by observing the distribution of function "process-evoking" words (PEWs)[3]. PEWs are tokens belonging to the following word categories:

---

[3] For the part-of-speech tagging we relied on a 3-gram tagger, with 2-gram and unigram backoffs, trained over the Brown corpus; the trained tagger had 0.8 accuracy.

Table 1. Top: Entity types and sample UMLS concept types they subsume; relations and sample UMLS relations they subsume. **Bottom:** Features considered.

activity	actor	resource	other	temporal	causal	other
laboratory procedure	organization	pharmacological substance	qualitative concept	precedes coexists_with	prevents produces	located_in part_of

feature F	description	value f
nest	nesting level in tree	integer $\in \mathbb{N}$
pos	position w.r.t. verb	subject, predicate
sub	occurs in clause?	yes, no
freq	freq. of label in corpus	integer $\in \mathbb{N}$
φ_{lf}	relative frequency of label in **NP**	real $\in [0,1]$
φ_{hd}	head/**NP** overlap	real $\in [0,1]$
φ_{ls}	label/**NP** overlap	real $\in [0,1]$
class	**NP** entity type	activity, actor, resource, other
rel	relation	temporal, causal, other

- conjunctions and prepositions: subordinating prepositions and conjunctions, e.g., "if"; coordinating conjunctions, e.g., "and", "or";
- adverbs: base adverbs, e.g., "after"; comparative adverbs, e.g., "later"; superlative adverbs, e.g., "latest"; and adverbial particles, e.g., "go back".

We thus compared to SemRep: (1) a subset of the NICE diabetes-2 guideline (therapy recommendations, 7,109 words); (2) a subset of the NICE eating disorders guideline[4] (therapy recommendations, 5,078 words); and (3) a subset of the NICE schizophrenia guideline[5] (therapy recommendations, 5,367 words). We also tried to assess whether there is a significant bias in clinical documents towards PEWs. To this end we compared SemRep and the guideline corpora to: (4) a subset of the Brown corpus[6] (A: press articles, 1,391,708 words); (5) Friederich's corpus of business process specifications [7] (3,824 words). See Figure 3. We ran two statistical tests:

1. A t-test with the null hypothesis H_0 that cross-corpora PEW mean relative frequency is $\mu_0 = 0.20$ ($p = 0.01$ significance level). This test showed no (statistically) significant differences in PEW distribution across corpora, where about 17 to 20% of word tokens are PEWs (see Section 3).
2. A χ^2-test of (in)dependence, with the null hypothesis H_0 that PEW relative frequency is correlated to (or depends on) corpus domain ($p = 0.01$ significance level). This test gave no (statistically) significant differences across domains (see Section 3).

Figure 2 seems to indicate that syntax is uniform across domains. This seems to justify at the same time the use of SemRep to experiment with CIG fragment

---

[4] http://www.nice.org.uk/nicemedia/live/10932/29218/29218.pdf
[5] http://www.nice.org.uk/nicemedia/live/11786/43607/43607.pdf
[6] http://nltk.googlecode.com/svn/trunk/nltk_data/index.xml

recognition techniques, as well as the general syntactic features (see Section 3) we extracted. It also suggests that UMLS annotations are independent from syntax. In the following experiments we will thus try to empirically validate the following claim:

$$\text{Semantic features and environment are more significant} \atop \text{for activity and temporal relation recognition than syntactic features.} \tag{3}$$

Feature Vectors. Our main goal was to study feature performance rather than model performance *per se*; we thus relied on standard classification models from the known Weka[7] data mining framework rather than on more sophisticated models. The performance of simple models might be suboptimal, but will nevertheless exhibit recognizable (cross-classifier) trends relatively to the independent features selected for training and prediction. We extracted three sets of observations for our experiments:

corpus tests					
χ^2	p	df.	t-score	p	df.
1.03	0.36	5	43.13	0.00	2

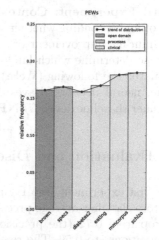

Fig. 2. By "mmcorpus" we mean SemRep

1. a set of **NP** observations: for each **NP** α in SemRep, we extracted the feature vector $(f_1^\alpha, \dots, f_7^\alpha, c^\alpha)^T$;
2. a set of sentence observations: for each vector $(\alpha_1, \dots, \alpha_k)^T$ of (manually annotated) **NP**s in a SemRep sentence, we extracted the feature vectors of the form $(f_1^{\alpha_1}, \dots, f_7^{\alpha_1}, c^{\alpha_1}, \dots, f_1^{\alpha_k}, \dots, f_7^{\alpha_k}, c^{\alpha_k})^T$;
3. a set of relation observations: for each vector $(\alpha, \alpha', r)^T$ of UMLS annotated **NP**s and their UMLS relation r, we extracted the feature vectors $(f_1^\alpha, \dots, f_7^\alpha, c^\alpha, f_1^{\alpha'}, \dots, f_7^{\alpha'}, c^{\alpha'}, r)^T$.

We proceeded to build three parallel sets of training and evaluation observations based on a 2/5 vs. 3/5 split. We considered as feature performance metric, for $\tau \in \{\text{activity}, \text{temporal}\}$, *activity precision* and temporal *relation precision*

$$PR = \frac{|\text{true } \tau \text{s}|}{|\text{true } \tau \text{s}| + |\text{false } \tau \text{s}|}. \tag{4}$$

First Experiment: Feature Significance for Activity Recognition. Our first experiment was designed to measure the significance of each single feature for the activity recognition task. To this end we removed each time a feature F_i from the set $\{F_1, \dots, F_7\}$ of (syntactic and semantic) *independent* features from Table 1, retrained and measured activity precision w.r.t. $\{F_1, \dots, F_{i-1}, F_{i+1}, \dots, F_7\}$.

---

[7] http://www.cs.waikato.ac.nz/~ml/weka/

We considered only **NP** observations. We trained and evaluated the following (Weka) classifiers: logistic classifier, support vector machine, neural network, Bayes classifier, decision tree and a 10-nearest neighbor classifier (baseline).

Second Experiment: Feature Significance for Relation Extraction. Our second experiment was designed to measure the significance of each single feature for the relation recognition task. We proceeded as before, with the difference that we considered an extended set $\{F_1, \ldots, F_8\}$ of 8 independent features including the semantic *class* (i.e., entity type labels) feature, since it arguably describes the *type* of the relation, viz., its domain and range. We considered for this experiment the set of relation observations. We trained and evaluated the following (Weka) classifiers: logistic classifier, neural network, Bayes classifier (baseline) and a decision tree.

Third Experiment: Context Significance. Our third experiment had the aim of understanding whether sentence context as opposed to **NP** context yields a significant improvement of (average) activity recognition precision. Also, we tried to determine which kind of classifier may perform better. We trained and evaluated the following (Weka) classifiers: logistic classifier, support vector machine, neural network, Bayes classifier, decision tree, and 10-nearest neighbor classifier (baseline) over *(i)* **NP** observations and *(ii)* sentence observations.

5   Evaluation and Discussion

The first experiment (see Figure 3, left), shows a statistically significant drop in cross-classifier average precision when label/**NP** overlap is disregarded, viz., a drop from 0.72 (the precision of the decision tree over the full set on **NP** observations) to 0.56. The removal of syntactic features had little to no effect, and the removal of label relative frequency and head/**NP** overlap gave rise to a slight performance decrease. These results suggest that the (lexical) semantic environment of **NP**s is more relevant (on average) for activity recognition than its syntactic environment.

The second experiment shows a statistically significant drop in cross-classifier average precision w.r.t. the best performing classifier, when head/**NP** overlap and label frequency are disregarded, viz., a drop from 0.66 (the precision of the decision tree over the full set on relation observations) to 0.4 and 0.33 resp.. It also showed a statistically significant improvement when subordination is disregarded (viz., from 0.66 to 0.58). This may suggest also that the (lexical) semantic environment of relations is more relevant (on average) for relation recognition. Interestingly, and contrary to our expectation, disregarding the typing of the relation seemed also to boost performance.

In the third experiment (see Figure 3, right) the classifier that performed better w.r.t. **NP** observations was the decision tree classifier (0.69 precision), likely because of its exploiting better the categorical features (position, subordination). It also matched over sentence observations the best performing classifier, the neural network, which showed a statistically significant improvement (from 0.60 to

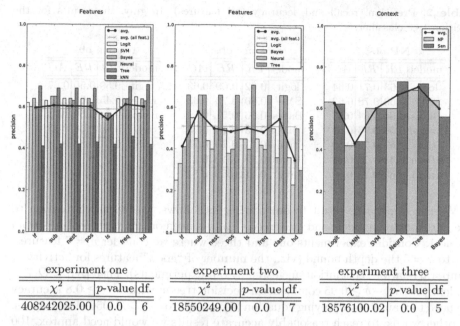

experiment one			experiment two			experiment three		
χ^2	p-value	df.	χ^2	p-value	df.	χ^2	p-value	df.
408242025.00	0.0	6	18550249.00	0.0	7	18576100.02	0.0	5

Fig. 3. Left: Experiment one: impact of removing a feature on activity precision, by classifier. **Center:** Experiment two: impact of removing a feature on relation precision, by classifier. **Right:** Experiment three: impact of context on activity precision. For the significance tests we considered as null hypothesis H_0 the uniform distribution.

0.73). This experiment seems to indicate overall that sentence environment is an important factor for activity recognition.

While the experiments seem to substantiate claim (3), in some cases (e.g., experiment two) we obtained also results that may seem to infirm it: a relation's argument type should be a relevant feature for relation recognition. The SemRep corpus is a small and sparsely annotated corpus – for, e.g., experiment two we extracted only 200 observations. Indeed, the patterns identified in Figure 3 were noisy and error-prone: the average cross-experiment classifier accuracy was in general low, see Table 2. Such behavior contrary to our expectations is likely due to the small size of the dataset.

On way to estimate how much data we actually need to obtain reasonably accurate predictions is to apply the theory of probably approximately correct (PAC) learning [19]. This theory implies that to learn from categorical data with $1 - \delta$ confidence a decision tree classifier m of bounded depth $\leq k$ and n attributes that is is $1 - \epsilon$ accurate[8], we need to train m over

$$N \geq \frac{1}{\epsilon} \times (\ln \frac{1}{\delta} + \ln |\mathcal{M}|) \tag{5}$$

observations, where $\mathcal{M}$ denotes the space of all decision trees m.

---

[8] Parameters ϵ and δ denote classification and learning *error*, resp.

Table 2. Precision, recall and accuracy (all features). In gray, the results for the decision tree classifier.

NP obs.				sentence obs.				relation obs.		
model	PR	RE	AC	model	PR	RE	AC	model	PR	RE	AC
logit	0.63	0.71	0.64	logit	0.62	0.69	0.63	logit	0.50	0.50	0.67
SMV	0.60	0.78	0.62	SMV	0.60	0.73	0.63	Bayes	0.45	0.62	0.71
Bayes	0.64	0.64	0.55	Bayes	0.56	0.51	0.54	neural	0.44	0.50	0.73
neural	0.60	0.76	0.64	neural	0.73	0.74	0.73	tree	0.66	0.50	0.73
tree	0.69	0.75	0.69	tree	0.71	0.79	0.73	(avg)	0.51	0.54	0.71
10-nn	0.43	0.72	0.42	10-nn	0.43	0. 28	0.41				
(avg)	0.60	0.73	0.59	(avg)	0.61	0.62	0.61				

We can apply this result to our setting as follows. We restrict attention to the decision tree models and "discretize" our continuous numeric features from Table 1. Consider experiments one and three where we consider $n = 7$ features. Set to $k = 5$ the depth bound (viz., the number of "good" features for activity or temporal relation recognition suggested by our experiments). Then $|\mathcal{M}| \approx O(7^5)$. To learn with $1 - \delta \geq 0.95$ confidence a decision tree m with $1 - \epsilon \geq 0.8$ accuracy we need to train it, by applying equation (5), on approx. $N \geq 84050$ observations. In other words, to reach reasonably accurate results we would need approx. 100 times more UMLS-annotated **NPs** than the 827 extracted from SemRep.

6   Conclusions and Further Work

We have conducted a preliminary experiment on how to automatically recognize activities and temporal relations using MetaMap and the UMLS Metathesaurus. We used the UMLS-annotated SemRep corpus as our training and evaluation corpus. We focused in the issue of feature selection, seeking to determine if semantics is more relevant than syntax for this task, and hence on feature performance rather than on classifier performance.

Our experiments have shown that in general the lexical semantic environment of an entity is more significant than its syntactic environment for identifying activities. Corpus analysis on SemRep and other clinical and non-clinical corpora showed moreover that the syntax of clinical text is not significantly different both within and across domains. Taking into consideration sentence context gave rise to a slight gain in performance. In all of our experiments the best performing of all the simple annotators used turned out to be the decision tree, better adapted to the categorical features we considered. The small size of the corpus and in particular the small number of relation annotations made our results much less conclusive however regarding temporal relations.

In the future, we plan to consider more powerful techniques, more complex feature sets, and larger corpora to improve our results. Regarding techniques, we intend to use more powerful classification models for NLP such as conditional random fields (CRFs), which can exploit possible dependencies among independent features. Furthermore, such models allow for very complex linguistic features and context models (based on n-grams) that we did not, for the sake of simplicity and scope, consider in this paper, such as the bag of n-words or n-POSs surrounding an entity, or the n-typed dependencies in which it participates, to name three. We intend also to consider a bigger corpus by integrating SemRep with the i2b2 clinical corpus as suggested by [2]. Finally, we will experiment with temporal relation extraction methods (à la TimeML) to tackle CIG control flow extraction. In fact, the current investigation focuses only on before/after temporal relations among tasks, but our final objective is the extraction of complex CIG fragments encompassing also gateways and more elaborated constraints on the process control-flow. Since the nature of the extracted constraints is declarative, we will not only focus on "procedural" specification languages (such as Asbru, Glare, BPMN), but we will also consider, at least as an intermediate format, constraint-based languages such as CigDec [15].

Acknowledgments. The present work has been done within the context of the VERICLIG project supported by a grant from the Free University of Bozen-Bolzano Foundation.

References

1. Aronson, A.R., Lang, F.-M.: An overview of MetaMap: Historical perspective and recent advances. J. of the American Medical Informatics Association 17(3), 229–236 (2010)
2. Ben Abacha, A., Zweigenbaum, P.: Medical entity recognition: A comparison of semantic and statistical methods. In: Proc. of the BioNLP 2011 Work. (2011)
3. De Clercq, P., Kaiser, K., Hasman, A.: Computer interpretable medical guidelines. In: Ten Teije, A., et al. (eds.) Computer-based Medical Guidelines and Protocols: A Primer and Current Trends, ch. 2, pp. 22–43. IOS Press (2008)
4. Denecke, K.: Structuring of and information extraction from medical documents using the UMLS. Methods of Information in Medicine 47(5), 425–434 (2008)
5. di Ciccio, C., Metella, M.: Studies on the discovery of declarative control flows from error-prone data. In: Proc. of the Third International Symposium on Data-Driven Process Discovery and Analysis, SIMPDA 2013 (2013)
6. Field, M.J., Lohr, K.N. (eds.): Clinical Practice Guidelines. Directions for a New Program. National Academy Press (1990)
7. Friedrich, F., Mendling, J., Puhlmann, F.: Process model generation from natural language text. In: Mouratidis, H., Rolland, C. (eds.) CAiSE 2011. LNCS, vol. 6741, pp. 482–496. Springer, Heidelberg (2011)
8. Friedman, C., Hripcsak, G.: Evaluating natural language processors in the clinical domain. In: Proc. of the Conf. on Natural Language and Medical Concept Representation (1997)
9. Hahn, U., Romacker, M., Schulz, S.: MEDSYNDICATE–A natural language system for the extraction of medical information from findings reports. Int. J. of Medical Informatics 67(1-3), 41–52 (2002)
10. Kaiser, K., Akkaya, C., Miksch, S.: How can information extraction ease formalizing of treatment processes in clinical practice guidelines? Artificial Intelligence in Medicine 39(2), 151–163 (2007)

11. Kilicoglu, H., Rosenblat, G., Fiszman, M., Rindfleisch, T.C.: Constructing a semantic predication gold standard from the biomedical literature. BMC Bioinformatics 12(486) (2011)
12. Klein, D., Manning, C.D.: Accurate unlexicalized parsing. In: Proceedings of the 41st Meeting of the Association for Computational Linguistics, ACL 2003 (2003)
13. Ko, R.K.L., Lee, S.S.G., Lee, E.W.: Business process mangament (BPM) standards: A survey. Business Process Management J. 15(5), 744–791 (2009)
14. Meystre, S., Haug, P.: Natural language processing to extract medical problems from electronic clinical documents. J. of Biomedical Informatics 39(6), 589–599 (2006)
15. Mulyar, N., Pesic, M., van der Aalst, W.M.P., Peleg, M.: Declarative and procedural approaches for modelling clinical guidelines: Addressing flexibility issues. In: ter Hofstede, A.H.M., Benatallah, B., Paik, H.-Y. (eds.) BPM 2007 Workshops. LNCS, vol. 4928, pp. 335–346. Springer, Heidelberg (2008)
16. Roberts, A., Gaizaskas, R., Hepple, M., Davis, N., Demetriou, G., Guo, Y., Kola, J., Roberts, I., Setzer, A., Tapuria, A., Wheeldin, B.: The CLEF corpus: Semantic annotation of a clinical text. In: Proc. of the AMIA 2007 Annual Symp. (2007)
17. Savova, G.K., Masanz, J.J., Ogren, P.V., Zheng, J., Sohn, S., Kipper-Schuler, K.C., Chute, C.G.: Mayo clinical text analysis and knowledge extraction system (cTAKES): Architecture, component evaluation and applications. J. of the American Medical Informatics Association 17(5), 507–513 (2010)
18. Serban, R., ten Teije, A., van Harmelen, F., Marcos, M., Polo-Conde, C.: Extraction and use of linguistics patterns for modelling medical guidelines. Artificial Intelligence in Medicine 39(2), 137–149 (2007)
19. Valiant, L.G.: A theory of the learnable. Communications of the ACM 27(11), 1134–1142 (1984)
20. Zhou, D., He, Y.: Semantic parsing for biomedical event extraction. In: Proc. of the Ninth Int. Conf. on Computational Semantics, IWCS 2011 (2011)
21. Zhou, X., Han, H., Chankai, I., Prestud, A., Brooks, A.: Approaches to text mining for clinical medical records. In: Proc. of the 2006 ACM Symposium on Applied Computing (2006)

Offline and Online Plan Library Maintenance in Case-Based Planning

Alfonso E. Gerevini[1], Anna Roubíčková[2], Alessandro Saetti[1], and Ivan Serina[1]

[1] Dept. of Information Engineering, University of Brescia, Brescia, Italy
[2] Faculty of Computer Science, Free University of Bozen-Bolzano, Bolzano, Italy
{gerevini,saetti,serina}@ing.unibs.it, anna.roubickova@stud-inf.unibz.it

Abstract. One of the ways to address the high computational complexity of planning is reusing previously found solutions whenever they are applicable to the new problem with limited adaptation. To do so, a reuse planning system needs to store found solutions in a library of plans, also called a case base. The quality of such a library critically influences the performance of the planner, and therefore it needs to be carefully designed and created. For this reason, it may be also important to update the library during the lifetime of the system, as the type of problems being addressed may evolve or differ from the ones the case base was originally designed for.

In our ongoing research, we address the problem of maintaining the library of plans in a recent case-based planner called OAKPLAN. After having developed offline techniques to reduce an oversized library, we introduce here a complementary online approach that attempts to limit the growth of the library, and we consider the combination of offline and online techniques to ensure the best performance of the case-based planner. The different investigated approaches and techniques are then experimentally evaluated and compared.

1 Introduction

Automated planning is a computationally very hard search problem [1]. One possible approach to improve the performance of a planner is to use an additional knowledge obtained by solving similar problems, to repeat the reasoning that led to a solution or even parts of the solution itself. Case-based planning (e.g., [2–4]) is a family of techniques that implement such a reuse in planning.

A well-designed case-based planner gradually creates a plan library that allows more problems to be solved (or higher quality solutions to be generated) compared to using a classical domain-independent planner, provided that the system frequently encounters problems similar to those already solved and that similar problems have similar solutions. Such a library is a central component of a case-based planning system, which needs a policy to maintain the quality of the library as high as possible in order to be efficient.

In this paper, we investigate two fundamentally different approaches to the plan library maintenance. The first (offline) approach stores all the solution plans

M. Baldoni et al. (Eds.): AI*IA 2013, LNAI 8249, pp. 239–250, 2013.

of the problems the planner encounters and periodically chooses a representative subset of the library elements (also called "cases") to reduce the size of the library while preserving its quality. The second (online) approach decides whether to discard or to include a new solution plan into the library every time the case-based planning system encounters a new problem. This approach, on one hand, slows down the rapid growth of the library, but, on the other hand, it may generate libraries of lower quality, because they are less informative than the reduced libraries obtained by the offline approach. We also propose a way to combine these two approaches into a policy overcoming the relative weaknesses. Finally, we present the results of a preliminary experimental analysis evaluating the proposed maintenance policies.

2 Preliminaries

In this section, we introduce the background and notation of (classical) planning [1], as well as some basic concepts in case-based reasoning and planning [5]. A *planning domain* is a tuple $\mathcal{D} = \langle \mathcal{Q}, \mathcal{O}p \rangle$, where $\mathcal{Q}$ is a finite set of predicates and $\mathcal{O}p$ is a finite set of operators where each operator $o \in \mathcal{O}p$ is defined by three components: a set $\mathcal{V}$ of variable symbols (or parameters of o); a set $pre(o)$ of literals over $\mathcal{Q}$ involving constants or variables in $\mathcal{V}$, forming the preconditions of o; a set $eff(o)$ of literals over $\mathcal{Q}$ involving constants or variables in $\mathcal{V}$, forming the effects of o.

A *planning problem* is obtained by grounding the predicates and operators of a corresponding planning domain over a set of objects. More precisely, given a planning domain $\mathcal{D} = \langle \mathcal{Q}, \mathcal{O}p \rangle$, a planning problem is a tuple $\Pi = \langle \mathcal{O}, \mathcal{F}, \mathcal{I}, \mathcal{G}, \mathcal{A} \rangle$ where: $\mathcal{O}$ is a set of objects (constant symbols); $\mathcal{F}$ is a finite set of ground atomic propositional formulae obtained by instantiating predicates from $\mathcal{Q}$ by objects from $\mathcal{O}$; $\mathcal{I} \subseteq \mathcal{F}$ is the set of atoms that are true in the initial state (and all other propositions are considered to be false in $\mathcal{I}$); $\mathcal{G} \subseteq \mathcal{F}$ is a set of literals over $\mathcal{F}$ defining the problem goals (that is, $\mathcal{G}$ is a partial assignment of truth values to $\mathcal{F}$ where the value of non-assigned propositions is arbitrary); $\mathcal{A}$ is a finite set of actions, where each action $a \in \mathcal{A}$ is an operator in $\mathcal{O}p$ fully instantiated by objects in $\mathcal{O}$. An application of an action a in a state where the action's preconditions are satisfied changes the state according to action's effects; otherwise a cannot be applied.

A *plan* π for a planning problem Π is a partially ordered set of actions from $\mathcal{A}$. A plan π *solves* Π if the application of the actions in π according to their planned order transforms the initial state to a state $\mathcal{S}_g$ where the goals $\mathcal{G}$ of Π are true ($\mathcal{G} \subseteq \mathcal{S}_g$). Classical generative planning is concerned with finding a solution plan for a given planning problem without any additional knowledge about such a plan.

Case-based planning (CBP) is a type of case-based reasoning, exploiting the use of different forms of planning experiences concerning problems previously solved and organized in *cases* forming a case base or *plan library*. The search for a solution plan can be guided by the stored information about previously generated plans in the case base, which may be adapted to become a solution for the new problem. When a CBP system solves a new planning problem, a new case is generated and possibly added to the library for potential reuse in the future. In order to benefit from remembering and reusing past plans, a CBP system needs efficient methods for retrieving analogous cases and for adapting retrieved plans, as well as a case base of sufficient size and coverage to yield useful analogues.

In our work, we focus on planning cases that are planner-independent and consist of a planning problem Π and a solution plan π of Π. A *plan library* is a collection of such cases: $\mathcal{L} = \{c_i \mid c_i = \langle \Pi_i, \pi_i \rangle, i \in \{1, \ldots, n\}\}$. Such a library constitutes the experience of the planner. Usually, the plan library is associated with a specific planning domain, but the problems in this domain can have different sizes and involve different sets of objects.

3 Related Work

The topic of case base maintenance has been of a great interest in the case-based reasoning community for the last two decades. However, the researchers studying case-based planning have not paid much attention to the problem of case base maintenance yet. Therefore, the related work falls mostly in the field of case-based reasoning, where most of the proposed systems handle classification problems.

Leake and Wilson [6] defined the problem of case-based maintenance as "an implementation of a policy to revise the case base to facilitate future reasoning for a particular set of performance objectives". Depending on the evaluation criteria, they distinguish two types of case-base maintenance techniques — the *quantitative criteria* (e.g., time) lead to performance-driven policies, while the *qualitative criteria* (e.g. coverage) lead to competence-driven policies.

The quantitative criteria are usually easier to compute; among these policies belong the very simple random deletion policy [7] and a policy driven by a case utility metric [8], where the utility of a case is increased by its frequent reuse and decreased by costs associated with its maintenance and matching.

The most used qualitative criterion corresponds to the notion of "competence" introduced by Smyth [9]. The notion of competence was used to define footprint deletion and footprint-utility deletion policies [10]. Another extension is the RC-CNN algorithm [11]. Furthermore, Leake and Wilson [12] suggested replacing the relative coverage by relative performance of a case. Zhu and Yang [13] however claim that the competence-driven policies of Smyth and his collaborators do not ensure competence preservation. They propose a case addition policy which mimics a greedy algorithm for set covering, adding always the case that has the biggest coverage until the whole original case base is covered or the limit size is reached.

Reinartz et al. [14] suggest to use quality measure reflecting user requirements, such as correctness, solution's stability and diversity, and combine them into a measure that is easy to compute by a domain independent computation.

Yang and Wu [15] propose a policy that keeps all the cases, but identifies similar problems and gather them into smaller, more focused case bases. The retrieval is then performed in a 2-level hierarchy and is guided by a decision forest. Shiu et al. [16] instead discard the majority of the case base after using the redundant cases to learn new adaptation rules based on the nature of their redundancy.

Muñoz-Avila studied the case retention problem in order to filter redundant cases [2], which is closely related to the problem studied here. However, the policy proposed in [2] can decide only about problems solved by the adopted derivational case-based planner; while the policies studied in this paper are independent from the planner used to generate the solutions of the cases.

4   Plan Library Maintenance

The efficiency of plan reuse is greatly influenced by the "quality" and quantity of stored plans. Higher quality solutions are solutions that allow faster reuse and lead to new better plans. A higher number of stored plans increases the chance of having high-quality plans for reuse; however, the number of stored solutions also significantly increases the computational effort to retrieve such a high-quality plan from the library. In [17, 18], it has been observed that the growing size of the plan library reduces the reuse effort and increases the retrieval effort. At a certain point, the decreased performance of the retrieval outweighs the improvements of the reuse, and the approach becomes ineffective. The goal of the plan library maintenance is to optimize the library size and quality of the cases kept in the library in order to support efficient plan retrieval and reuse.

There are two problems related to the library maintenance. One is concerned with reducing the size of the library when it becomes too big. The other problem concerns bounding the growth of the library by adding only those cases that are beneficial to the overall planning process (plan retrieval and its adaptation for the reuse). More formally:

Definition 1 (Case Base Maintenance Problem [18])

1. *Given a case base* $\mathcal{L} = \{c_i \mid i \in \{1, \ldots, n\}\}$, *decide for each* $i \in \{1, \ldots, n\}$ *whether the case* c_i *should be* removed *from* $\mathcal{L}$.
2. *Given a case base* $\mathcal{L}$ *and a new case* $c = \langle \Pi, \pi \rangle$, $c \notin \mathcal{L}$, *decide whether* c *should be* added *to* $\mathcal{L}$.

The optimal solution to the first maintenance problem provides a reduction of the library without losing (too much) of the knowledge contained within. Unfortunately, a maintenance policy addressing the problem needs to be highly informed in order to select the most suitable subset of the cases from the library, which can incur in a high computational cost. Consequently, such a policy cannot be performed very often or in the periods of high activity of the case-based planning system, and is more suitable for an *offline* use.

On the other hand, if all cases are added and kept in the library, the size of the library may grow very fast, significantly decreasing the performance of the whole system. The second mentioned maintenance problem involves deciding *online* if a solution plan should be added to the library or not according to a criterion measuring its quality relative to the solutions currently stored in the library.

When designing a policy to address the maintenance problem, we need to define a measure of case quality. Aiming at a library that is useful to solve a wide range of problems, the case quality should consider the information redundancy within a case, which is captured by case's diversity (or distance) w.r.t. the other cases in the library.

We define the distance function $d_a(\pi_i, \pi_j)$ between two plans π and π' as the number of actions that are in π and not in π' plus the number of actions that are in π' and not in π, normalized over the total number of actions in π and π' [19], that is, $d_a(\pi, \pi') = \frac{|\pi - \pi'| + |\pi' - \pi|}{|\pi| + |\pi'|}$, unless both plans are empty, in which case their distance is 0. Distance $d_r(\Pi, \Pi')$ between a new problem Π and a library problem Π' is defined as follows [3]: $d_r(\Pi', \Pi) = 1 - \frac{|\mathcal{I}' \cap \mathcal{I}| + |\mathcal{G}' \cap \mathcal{G}|}{|\mathcal{I}'| + |\mathcal{G}|}$, where $\mathcal{I}$ and $\mathcal{I}'$ ($\mathcal{G}$ and $\mathcal{G}'$) are the initial states (sets of goals) of Π and Π', respectively.[1] If $|\mathcal{I}'| + |\mathcal{G}| = 0$, $d_r(\Pi', \Pi) = 0$.

Distance functions d_a and d_r can then be combined in order to define a distance between two cases. We define the distance between two cases as a linear combination of d_r and d_a: $d = \alpha \cdot d_r + (1 - \alpha) \cdot d_a$. Note that d_a only estimates the adaptation effort, and d_r relates to d_a only if the world is regular, i.e., if the similarity of problems determines similarity of the solutions in some way. Other functions or different approximations can be used, especially if some insight in the reuse effort is known due to, e.g., peculiarities of the encountered domains or used algorithms.

4.1   Offline Maintenance

In our previous work [18], we have studied the problem of offline maintenance, devised and experimentally evaluated several policies to address this problem. We implemented a very simple *random policy* [7], which is completely uninformed. This policy is outperformed by a *distance-guided policy* that considers all cases and removes those that are too close to their closest case. The distance-guided policy preserves the knowledge in the case base better than the random policy. However, it may still loose a significant amount of information, as the policy always considers only a pair of cases at a time, and it does not keep track of the removed cases.

In [18], we generalized this approach by considering *all* the cases that may contain redundant information at once. For that, we defined the notion of neighborhood of a case c with respect to a certain similarity threshold δ, denoted

---

[1] The denominator of d_r considers only the initial state of problem Π' and the goals of the new problem Π, because it ignores the presence of additional goals of Π', and the additional initial facts of Π that are not relevant for the executability of the solution π' of Π' stored in the library.

$n_\delta(c)$. The resulting *coverage-guided policy* is briefly described below (the full details can be found in [18]).

There are many possible ways to reduce a case base in accordance with the coverage-guided policy, out of which some are more suitable than others. The work in [18] shows that *maximizing the quality of the coverage of the reduced case base* can significantly influence the performance of a case-based system adopting the coverage-guided policy. The quality of the case base coverage can intuitively be defined as the average distance from the removed cases to the closest kept case (average coverage distance). The quality measure to be optimized is based on the notion of *uncovered neighborhood* defined as $U_\delta(c) = \{c_j \in \mathcal{L} \mid c_j \in \{n_\delta(c) \cap \mathcal{L} \setminus \mathcal{L}'\} \cup \{c\}\}$. The cost of a case c is defined as a real function involving $U_\delta(c)$: $v_\delta(c) = \left(\frac{\Sigma_{c_j \in U_\delta(c)} d(c,c_j)}{|U_\delta(c)|} + p \right)$. The first term of $v_\delta(c)$ within brackets is the average coverage distance of the uncovered neighbors; the second term, $p \in \mathbb{R}$, is a penalization value that is added in order to favor reduced case bases with fewer elements, and to assign a value different from 0 also to isolated cases in the case base.[2] The sum of these costs for all the elements in the maintained case base $\mathcal{L}'$ defines the cost $\mathcal{M}_\delta(\mathcal{L}')$ of $\mathcal{L}'$, i.e., $\mathcal{M}_\delta(\mathcal{L}') = \Sigma_{c \in \mathcal{L}'} v_\delta(c)$. By minimizing $\mathcal{M}_\delta(\mathcal{L}')$ we can defined the following policy:

Definition 2 (Weighted Coverage-Guided Policy [18]). *Given a similarity threshold value $\delta \in \mathbb{R}$ and a case base $\mathcal{L}$, find a reduction $\mathcal{L}'$ of $\mathcal{L}$ that minimizes $\mathcal{M}_\delta(\mathcal{L}')$. We say that $\mathcal{L}'$ reduces $\mathcal{L}$ if $\mathcal{L}' \subseteq \mathcal{L}$ and $\forall c \in \mathcal{L} \; \exists c' \in \mathcal{L}'$ such that $d(c, c') \leq \delta$.*

Computing a reduced case-base according to the weighted coverage-guided policy can be computationally very expensive. Therefore, in [18] we propose to compute an approximation of this reduction policy using a greedy algorithm. In the rest of the paper, such an approximation of the weighted coverage-guided policy is simple called offline maintenance policy.

4.2   Online Maintenance

The challenges of the online maintenance come from the settings in which it is used and may quite differ from the offline ones. The decision whether the new case should be added to the case base or not, and whether some other case should be removed, needs to be done relatively fast, as it is done every time a new case is presented to the system. The goal of the online policy is to maintain the case base in a way that supports efficient queries (case retrieval). To enable a fast retrieval, the case base needs to be small or well structured, and hence the online system should add only new cases of significant "importance" (relatively to the existing cases) to the plan library.

Here, the optimal criterion for deciding whether a new case should be added or not should consider the deterioration of the retrieval step (the cost of maintaining the new case) as well as the improvement of the reuse step (the benefit of the new

---

[2] In our experiments, we use $p = \max d(c_i, c_i^*)$.

case). However, such a measure depends not only on the case itself, but on the rest of the case base as well, and hence it changes over the time as the case base evolves. It is hard to predict the evolution of the case's cost and benefit before the future problems are presented to the system, without any assumption on the planning problems distribution. Therefore, since we are considering domain independent case-based planning, the policy proposed here decides the addition of the case solely based on the current state of the case base, and allowing to remove this element later, if it becomes not beneficial.

As in the offline policy, the insertion of a new case is decided taking into account the diversity of the case with respect to the library. Only if the case contains sufficient amount of new information, it is introduced into the library. More formally, let $c^* \in \mathcal{L}$ denote the element of the case base that is the most similar to the new case c. If $d(c, c^*) > \epsilon$ for some suitable distance threshold ϵ (that is, c is sufficiently diverse from the whole case base), then c is inserted into $\mathcal{L}$.

However, such an approach can still lead to an uncontrollable growth of the plan library if all the solved problems differ from each other more than ϵ. Therefore, we define the bounded online maintenance problem as follows:

Definition 3 (Bounded Online Maintenance). *Given a case base $\mathcal{L}$, a bound $N \in \mathbb{N}$ and a new case $c = \langle \Pi, \pi \rangle$, $c \notin \mathcal{L}$, decide whether c should be added to $\mathcal{L}$. If $|\mathcal{L}| = N$ and c is being included (i.e., $\mathcal{L}' = \mathcal{L} \cup \{c\}$ and $|\mathcal{L}'| > N$), find a suitable case $c_i, i \in \{1, \dots, N\}$, to be removed from $\mathcal{L} \cup \{c\}$.[3]*

The removal of a case is again guided by the cases' diversity. Informally, the policy attempts to identify which case is less diverse. Case diversity is defined by means of the distance function d. Let $c_i^* \in \mathcal{L}$ denote the case that is the most similar to a case c_i, i.e., $d(c_i, c_i^*) \leq d(c_i, c) \ \forall c \in \mathcal{L}$. The case c_i to remove satisfies $d(c_i, c_i^*) \leq d(c_j, c_j^*), \forall j \in \{1, \dots, n\}$. Algorithmically, such an element is selected as follows: for every case $c_i \in \mathcal{L}$, we identify its most similar case c_i^*; we sort the case base elements by the growing distance $d(c_i, c_i^*)$ of their most similar cases, and we remove the first such c_i as described in Figure 1. This step mimics a single iteration of the distance-guided policy described in [18], and requires only polynomial time if the distances of closest elements have been stored during the insertion phase.

4.3   Combined Maintenance

The offline and online approaches defined above have some weaknesses. In case of the offline policies, it is the fast growth of the library between two consecutive maintenance operations. The online maintenance is able to keep the library size limited to a given bound, but its quality is likely to degrade over time. In order to overcome these drawbacks, we propose to combine the two approaches as described below.

---

[3] At this point, also the newly inserted case c is considered for removal.

Input: a plan library $\mathcal{L}$, a case $c \notin \mathcal{L}$, a distance threshold ϵ, a positive integer N
Output: an updated plan library $\mathcal{L}'$

1. find $c^* \in \mathcal{L}$ s.t. $d(c, c^*) \le d(c, c') \, \forall c' \in \mathcal{L}$;
2. if $d(c, c^*) > \epsilon$ then $\mathcal{L}' = \mathcal{L} \cup c$;
3. if $|\mathcal{L}'| > N$ then
   find $c_i \in \mathcal{L}'$ s.t. $\forall c_j \in \mathcal{L}' : d(c_i, c_i^*) \le d(c_j, c_j^*)$;
   remove c_i;
4. **return** $\mathcal{L}'$;

Fig. 1. Algorithmical description of the bounded online maintenance policy

The combined approach stores each new case in the library, but in order to increase the performance during the retrieval, only a subset of the cases are considered; such cases are called *active cases*. The system effectively behaves as if only the active elements were present in the case base, but it can redefine the set of active elements when a certain condition is met. The identification of the active elements is performed using the offline case-base reduction policy, which can be activated when the size of the case base exceeds a certain limit. All and only the elements in the reduced case base are considered active.

In order to implement the combined maintenance policy, each case is marked by an activation flag. When the planner loads the library, it filters the cases by their activation. During the lifetime of the planner, for each new presented problem, the case is stored in the case base and the bounded online maintenance is invoked to decide the activation flag of the new case and possibly modify the activation of another case. This step is realized by an algorithm similar to the one described in Figure 1, with two changes: at line 2 case c is added to the case base regardless the distance from its closes case c^*, and it is marked active if and only if $d(c, c^*) < \epsilon$. When a certain limit condition is met, such as the system has encountered a maximum number of problems, the retrieval time has increased above a specified limit, or the system is in a period of low activity etc., the planner marks all the cases to `active` and computes a suitable reduction of such a full case base by means of the offline maintenance technique.

Intuitively, the resulting combined policy controls the growth of the plan library and consequently also reduces the average retrieval time; moreover the quality of the case base can be in principle better than the one produced by the online maintenance. The main challenge of the combined policy is to find a limit condition at which the offline policy is invoked, which should not happen too often due to its extensive computational costs; although, too rare recomputation of the set of active elements degrades the performance of the planner. This is subject of ongoing investigation; preliminary experimental results suggest that even simple conditions, such as number of encountered problems, lead to some improvement.

5 Experimental Results

The approaches for the library maintenance presented in the previous sections have been implemented in a new version of the CBP system OAKPlan [3].

Table 1. Number of elements, coverage, number of uncovered cases, average minimum distance from the uncovered elements of the case bases and average retrieval time of OAKPlan using the full plan library and the libraries reduced using the online, the offline and the combined maintenance for domains DriverLog, Logistics, and ZenoTravel

Domain	2000 cases encountered		5000 cases encountered			
	Full Library	Online	Full Library	Online	Offline	Combined
DriverLog						
Size	3000	1000	6000	1000	1000	1000
Coverage	1	0.996	1	0.992	0.998	0.997
# Uncovered	0	10	0	43	8	15
Avg. uncov. dist.	0	0.0143	0	0.0165	0.0130	0.0160
Retrieval time	84.81	35.46	137.24	35.44	34.00	34.81
Logistics						
Size	3000	1000	6000	1000	1000	1000
Coverage	1	0.996	1	0.995	1	0.997
# Uncovered	0	9	0	25	0	17
Avg. uncov. dist.	0	0.0384	0	0.0375	0.0262	0.0364
Retrieval time	59.17	44.93	90.40	49.18	22.97	41.15
ZenoTravel						
Size	3000	1000	6000	1000	1000	1000
Coverage	1	0.940	1	0.937	0.979	0.934
# Uncovered	0	179	0	373	124	391
Avg. uncov. dist.	0	0.0464	0	0.0555	0.0454	0.0541
Retrieval time	27.01	23.00	34.14	29.56	11.69	28.86

In our experiments, the plan retrieved by OAKPlan is adapted using planner LPG-Adapt [20]. The benchmark domains considered in the experimental analysis are the available domains DriverLog, Logistics and ZenoTravel from the International Planning Competition [21].

For each considered domain, we generated a plan library with 1000 cases and an additional set of 5000 cases. All these cases create groups of elements similar to each other, that we call case *clusters*. Each cluster c is formed by using either a large-size competition problem or a randomly generated problem $\overline{\Pi}_c$ (with a problem structure similar to the large-size competition problems) plus a random number of cases ranging from 0 to 99 that are obtained by changing $\overline{\Pi}_c$. Problem $\overline{\Pi}_c$ was modified either by randomly changing at most 10% of the literals in its initial state and set of goals, or adding/deleting an object to/from the problem. The solution plans of the planning cases were computed by planner TLPlan [22]. TLPlan exploits domain-specific control knowledge to speedup the search significantly, so that large plan libraries can be constructed by using a relatively small amount of CPU time.

Moreover, for each considered domain, we generated 25 test problems, each of which is derived by (randomly) changing the problem $\overline{\Pi}_c$ of a randomly selected cluster c among those in the case base or in the additional set of 5000 cases. Note that the cases we considered are grouped into clusters, and the test problems were generated from those clusters because the aim of our experimental analysis is studying the effectiveness of the proposed maintenance policies for domains with recurring problems.

In our experimental analysis, we assume that the bounded online maintenance keeps the number of cases in the library equal to 1000, while the offline and the

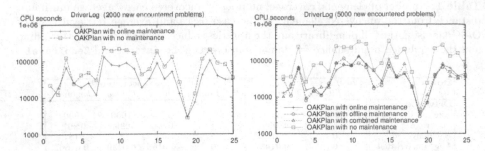

Fig. 2. Total CPU time of OAKPlan with the full library and the reduced libraries for domain DriverLog. On the x-axis, there are the problem names simplified by numbers.

combined maintenance policies reduce the plan library when the number of cases in the library becomes 6000. We considered 4 libraries obtained from the initial plan library by using no maintenance policy, the online, the offline and the combined maintenance policies.

If the number of encountered plans is lower than 5000, the number of cases in the plan library is lower than the bound over which the library is reduced by using the offline and the combined maintenance policies. Therefore, to study the performance of OAKPlan in this case, we generated 2 libraries obtained from the initial plan library by using no maintenance policy and the online maintenance policy after OAKPlan encounters 2000 new solution plans.

Table 1 compares the full library and the libraries reduced by using the considered maintenance policies in terms of size, number of cases of the full library that are not covered by the reduced library[4] and the coverage of the reduced library w.r.t. the full library, average distance from any uncovered case to the closest case in the reduced plan library, and CPU seconds required to retrieve a suitable case from the library. Obviously, the closer the coverage is to 1, or, equally, the lower the number of uncovered cases is, the better the maintenance policy is. Moreover, since a high-quality policy should remove only redundant cases, lower values of the average minimum distance from the uncovered cases indicate better plan libraries.

The results in Table 1 show that with 2000 new encountered plans the coverage of the online plan library is very high (almost equal to 1), and the average coverage distance is quite low. Moreover, for DriverLog and Logistics the number of uncovered cases is very much low. The retrieval of a suitable case from the library reduced by the online maintenance is always faster than from the full library.

Figure 2 shows the total time (retrieval time plus adaptation time) of OAKPlan using the full plan library and the libraries reduced by the analyzed maintenance policies for domain DriverLog. Using the library reduced by the online policy for the DriverLog problems, OAKPlan is almost always faster than using the full

---

[4] We consider a case c to be *covered* if there exists a case c^* in the reduced library such that $d(c, c^*) \leq 0.1$.

library. For `Logistics` and `ZenoTravel`, the performance gap in terms of total CPU time is very similar to `DriverLog`. Moreover, the results in Figure 2 indicate that the performance gap obtained by using the full library and the library reduced by the online policy usually increases with the number of encountered problems. In our future work, we want to confirm this observation considering a much larger number of cases.

As expected, according to the results in Table 1, in terms of coverage, number of uncovered cases and average uncovered distance, the library obtained using the combined maintenance is usually better than using the online maintenance.

Finally, the quality of the library obtained using the offline policy is better than the library derived using the combined maintenance policy, but we observed that the total time of `OAKPlan` with the library obtained using the combined maintenance policy is usually similar to using the offline policy. It is worth noting that, for the results in Figure 2 obtained using the offline and combined policies, the tests were conducted immediately after having performed the library reduction. When a new planning problem is solved using a library that has not been recently reduced, the performance gap can be significative in favor of the combined policy, as using such a policy the size of the case base grows more slowly, and hence retrieving a suitable plan from the library can be much faster.

6   Conclusion and Future Work

In our work, we have studied the problem of maintaining a plan library of a case-based planner at high quality level. We developed an approach complementary to the offline policy introduced earlier, that we called bounded online maintenance. This new approach attempts to control the fast growth of the library over the lifetime of the planner by deciding whether to discard or to include a new solution plan into the library every time the CBP system encounters a new problem. Our experiments show that the average time needed to query the reduced library improves w.r.t. the full library.

To overcome the problems with the quality of the reduced case base in the online manner, we proposed a combined policy that joins benefits from online and offline approaches. It stores each new case in the library considering only certain cases as active — the case base grows slowly and the retrieval time is low, but the quality of the case base is improved by periodical re-evaluation of the case base elements.

In the future, we would like to study other aspects related to the online maintenance, such as different criteria that can guide the policy. In this work, the diversity of a case only estimates the expected costs of reusing the case. However, if the problem is solved locally by the system itself, we have access to additional information about the real effort that was needed to solve the new problem. Hence, the policy could take the observed costs into account, and include in the case base the cases that were hard to solve. Moreover, the benefit of a case can be assessed more accurately if the system keeps track of which elements of the case base are being retrieved and lead to a successful reuse, as the frequency and time of the successful retrievals provide precise information about the real use of the case for the system.

References

1. Ghallab, M., Nau, D.S., Traverso, P.: Automated planning - theory and practice. Elsevier (2004)
2. Muñoz-Avila, H.: Case-base maintenance by integrating case-index revision and case-retention policies in a derivational replay framework. Computational Intelligence 17(2), 280–294 (2001)
3. Serina, I.: Kernel functions for case-based planning. Artificial Intelligence 174(16-17), 1369–1406 (2010)
4. Spalazzi, L.: A survey on case-based planning. AI Review 16(1), 3–36 (2001)
5. Leake, D.B. (ed.): Case-Based Reasoning. The MIT Press, Cambridge (1996)
6. Leake, D.B., Wilson, D.C.: Categorizing case-base maintenance: Dimensions and directions. In: Smyth, B., Cunningham, P. (eds.) EWCBR 1998. LNCS (LNAI), vol. 1488, pp. 196–207. Springer, Heidelberg (1998)
7. Markovitch, S., Scott, P.D., Porter, B.: Information filtering: Selection mechanisms in learning systems. In: 10th Int. Conf. on Machine Learning, pp. 113–151 (1993)
8. Minton, S.: Quantitative results concerning the utility of explanation-based learning. Artificial Intelligence 42(2-3), 363–391 (1990)
9. Smyth, B.: Case-base maintenance. In: Mira, J., Moonis, A., de Pobil, A.P. (eds.) IEA/AIE 1998. LNCS, vol. 1416, pp. 507–516. Springer, Heidelberg (1998)
10. Smyth, B., Keane, M.T.: Adaptation-guided retrieval: Questioning the similarity assumption in reasoning. Artificial Intelligence 102(2), 249–293 (1998)
11. Smyth, B., McKenna, E.: Footprint-based retrieval. In: Althoff, K.-D., Bergmann, R., Branting, L.K. (eds.) ICCBR 1999. LNCS (LNAI), vol. 1650, pp. 343–357. Springer, Heidelberg (1999)
12. Leake, D.B., Wilson, D.C.: Remembering why to remember: Performance-guided case-base maintenance. In: Blanzieri, E., Portinale, L. (eds.) EWCBR 2000. LNCS (LNAI), vol. 1898, pp. 161–172. Springer, Heidelberg (2000)
13. Zhu, J., Yang, Q.: Remembering to add: Competence-preserving case-addition policies for case-base maintenance. In: 16th Int. Joint Conf. on AI, pp. 234–241 (1998)
14. Reinartz, T., Iglezakis, I., Roth-Berghofer, T.: On quality measures for case base maintenance. In: Blanzieri, E., Portinale, L. (eds.) EWCBR 2000. LNCS (LNAI), vol. 1898, pp. 247–259. Springer, Heidelberg (2000)
15. Yang, Q., Wu, J.: Keep it simple: A case-base maintenance policy based on clustering and information theory. In: Hamilton, H.J. (ed.) Canadian AI 2000. LNCS (LNAI), vol. 1822, pp. 102–114. Springer, Heidelberg (2000)
16. Shiu, S.C.K., Yeung, D.S., Sun, C.H., Wang, X.: Transferring case knowledge to adaptation knowledge: An approach for case-base maintenance. Computational Intelligence 17, 295–314 (2001)
17. Veloso, M.M., Carbonell, J.G.: Derivational analogy in PRODIGY: Automating case acquisition, storage, and utilization. Machine Learning 10, 249–278 (1993)
18. Gerevini, A.E., Roubíčková, A., Saetti, A., Serina, I.: On the plan-library maintenance problem in a case-based planner. In: Delany, S.J., Ontañón, S. (eds.) ICCBR 2013. LNCS, vol. 7969, pp. 119–133. Springer, Heidelberg (2013)
19. Srivastava, B., Nguyen, T.A., Gerevini, A., Kambhampati, S., Do, M.B., Serina, I.: Domain independent approaches for finding diverse plans. In: 20th Int. Joint Conf. on Artificial Intelligence, pp. 2016–2022 (2007)
20. Fox, M., Gerevini, A., Long, D., Serina, I.: Plan stability: Replanning versus plan repair. In: 16th Int. Conf. on AI Planning and Scheduling, pp. 212–221 (2006)
21. Koenig, S.: Int. planning competition (2013), http://ipc.icaps-conference.org/
22. Bacchus, F., Kabanza, F.: Using temporal logic to express search control knowledge for planning. Artificial Intelligence 116(1-2), 123–191 (2000)

Integrating Knowledge Engineering for Planning with Validation and Verification Tools

Andrea Orlandini[1], Giulio Bernardi[1], Amedeo Cesta[1], and Alberto Finzi[2]

[1] CNR - Consiglio Nazionale delle Ricerche,
Istituto di Scienze e Tecnologie della Cognizione, Rome, Italy
`name.surname@istc.cnr.it`
[2] Università di Napoli "Federico II",
Dipartimento di Ingegneria Elettrica e Tecnologie dell'Informazione, Naples, Italy
`alberto.finzi@unina.it`

Abstract. Knowledge Engineering environments aim at simplifying direct access to the technology for system designers, and the integration of Validation and Verification (V&V) capabilities in such environments may potentially enhance the users trust in the technology. In particular, V&V techniques may represent a complementary technology with respect to Planning and Scheduling (P&S) contributing to develop richer software environments to synthesize a new generation of robust problem-solving applications. This paper presents the integration of classical knowledge engineering features connected to support design of timeline-based P&S applications taking advantage of services of automated V&V techniques such as domain validation, planner validation, plan verification etc. The result is a Knowledge Engineering ENvironment (called KEEN) that exploits a state-of-the-art verification tool, i.e., UPPAAL-TIGA, as core engine to support the design and development of timeline-based planning and scheduling systems.

Keywords: knowledge engineering, validation and verification, timeline-based planning, domain modeling.

1 Introduction

The deployment of Planning and Scheduling (P&S) tools to implement autonomous control systems in critical environments has been achieving increasing success over the last decade (e.g., [11,8]). Nevertheless, the proposed solutions turn out very often to be neither obvious nor immediately acceptable for users that may have difficulties in validating and verifying them by simple inspection. This is because these tools directly reason on causal, temporal and resource constraints as well as employ resolution processes designed to optimize solutions with respect to non trivial evaluation functions.

Knowledge Engineering environments aim at simplifying direct access to the technology for system designers through the synthesis of software environments that would allow the development of applications to an enlarged users community (i.e., not only "leading edge" specialists). Examples of such environments are ITSIMPLE [22], GIPO [21], and EUROPA [1]. Nevertheless, the full integration of Validation and Verification (V&V) capabilities in such environments seems to be still missing. Considering V&V features may potentially enhance the users trust in the technology as well as also represent a complementary technology with respect to P&S, thus contributing to develop

M. Baldoni et al. (Eds.): AI*IA 2013, LNAI 8249, pp. 251–262, 2013.

richer software environments to synthesize a new generation of robust problem-solving applications.

Over the last year and a half we have been developing a Knowledge Engineering environment built around the APSI-TRF [4], a state of the art framework for P&S with timelines. This paper presents the result of this effort, i.e., a Knowledge Engineering ENvironment (called KEEN) that exploits a state-of-the-art verification tool, i.e., UPPAAL-TIGA [2], as core engine to support the design and development of timeline-based P&S systems. In particular, the perspective we are pursuing is the integration of classical knowledge engineering features – as for example those supporting domain definition, domain refinement, etc. – with services of automated V&V techniques such as those surveyed in [6].

An early paper [18] describes the general idea and formulates an initial plan to develop KEEN as situated within the GOAC project (a robotic project for the European Space Agency [3]). The present paper describes the current KEEN environment: in particular focusing on the new features for Domain Definition and Visualization as well as on some of the V&V tools at work starting from a defined domain. To make the example more concrete, a running example from the GOAC domain is used throughout the paper. Indeed, the experience in a "robotic autonomy" project has fostered us to further investigate problems of plan robustness at execution time. In particular, working on the representation of flexible temporal plans, a key feature of timeline-based plan representation, drives us to synthesize results related to checking the dynamic controllability property [5] as well as to automatically generate robust plan controllers [19]. This paper shows also how such recent results are integrated and exploited in KEEN.

The paper is organized as follows: a section describes basic knowledge on timelines to set the context and shortly introduces the GOAC domain, then the comprehensive idea of the KEEN system is first described in general then two sub-sections are dedicated to the functionalities of domain modeling and V&V respectively. Related works and conclusions end the paper.

2   Timeline-Based Planning

The main modeling assumption underlying the timeline-based approach [17] is inspired by the classical Control Theory: the problem is modeled by identifying a set of *relevant features* whose temporal evolutions need to be controlled to obtain a desired behavior. In this respect, the set of domain features under control are modeled as a set of temporal functions whose values have to be decided over a time horizon. Such functions are synthesized during problem solving by posting planning decisions. The evolution of a single temporal feature over a time horizon is called the *timeline* of that feature.[1] Timeline-based planning has been applied to the solution of several space planning problems – e.g., [11,8].

---

[1] According to Wikipedia, a *timeline* is a way of displaying a list of events in chronological order. It is worth saying that this style of planning synthesizes a timeline for each dynamic feature to be controlled. In this paper, the term "timeline-based planning" is considered because recently it is more widely used, see for instance [9]. Other authors prefer "constraint-based interval planning" [11] following a perspective more connected to the technical way of creating plans.

For the current modeling purpose, we consider multi-valued *state variables* representing time varying features as defined in [17,7]. As done in those papers, the evolution of controlled features are described by causal laws which determine legal temporal evolutions of timelines. For the state variables, such causal laws are encoded in a *Domain Theory* which determines the operational constraints of a given domain. Task of a planner is to find a sequence of control decisions that bring the variables into a final set of desired evolutions (i.e., the *Planning Goals*). A valid plan is composed by a set of completely specified timelines[2] (one for each considered state variable) whose timed intervals are consistent with the domain theory constraints.

2.1 GOAC: A Test Planning Domain

This section introduces the real world planning domain used as a running example. It is derived from the ESA project named GOAC (for Goal Oriented Autonomous Controller), an effort to create a common platform for robotic control software development [3]. The domain considers a planetary rover equipped with a Pan-Tilt Unit (PTU), two stereo cameras (mounted on top of the PTU) and a communication facility. The rover is able to autonomously navigate the environment, move the PTU, take pictures and communicate images to a Remote Orbiter. A safe PTU position is assumed to be (*pan*, *tilt*) = $(0, 0)$. Additionally, during the mission, the Orbiter may be not visible for some periods. The robotic platform can communicate only when the Orbiter is visible. The mission goal is a list of required pictures to be taken in different locations with an associated PTU configuration. A possible mission action sequence is the following: navigate to one of the requested locations, move the PTU pointing at the requested direction, take a picture, then, communicate the image to the orbiter during the next available visibility window, put back the PTU in the safe position and, finally, move to the following requested location. Once all the locations have been visited and all the pictures have been communicated, the mission is considered as successfully completed. The rover should always follow some operative rules to maintain safe and effective configurations. The following conditions must hold during the whole mission: **(C1)** While the robot is moving the PTU must be in the safe position (pan and tilt at 0); **(C2)** The robotic platform can take a picture only if the robot is motionless in one of the requested locations while the PTU is pointing at the related direction; **(C3)** Once a picture has been taken, the rover has to communicate the picture to the base station; **(C4)** While communicating, the rover has to be motionless; **(C5)** While communicating, the orbiter has to be visible.

Timeline-Based Specification. To obtain a timeline-based specification of the robotic domain, we consider two types of state variables: *Planned State Variables* to represent timelines whose values are decided by the planning agent, and *External State Variables* to represent timelines whose values over time can only be observed. Planned state variables are those representing time varying features like the temporal occurrence of navigation, PTU, camera and communication operations. We use four of such state variables, namely the *RobotBase*, *PTU*, *Camera* and *Communication*.

---

[2] A completely specified timeline assumes a unique value for each time instant within the planning horizon.

In Fig. 1, we detail the values that can be assumed by these state variables, their durations and the legal value transitions in accordance with the mission requirements and the robot physics. Additionally, one external state variable represents contingent events, i.e., the communication opportunities. The *Orbiter Visibility* state variable maintains the visibility of the orbiter. The allowed values for this state variable is *Visible* or *Not-Visible* and are set as an external input. The robot can be in a position (*At(x,y)*), moving towards a destination (*GoingTo(x,y)*) or Stuck (*StuckAt(x,y)*).[3] The PTU can assume a

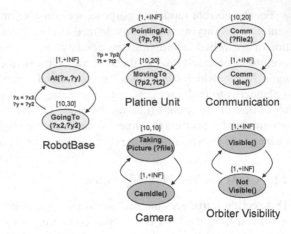

Fig. 1. State variables describing the robotic platform and the orbiter visibility (durations are stated in seconds)

PointingAt(pan,tilt) value if pointing a certain direction, while, when moving, it assumes a *MovingTo(pan,tilt)*. The camera can take a picture of a given object in a position $\langle x, y \rangle$ with the PTU in $\langle pan, tilt \rangle$ and store it as a file in the on-board memory (*TakingPicture(file-id,x,y,pan,tilt)*) or be idle (*CamIdle()*). Similarly, the communication facility can be operative and dumping a given file (*Communicating(file-id)*) or be idle (*ComIdle()*).

Domain operational constraints are described by means of *synchronizations*. A synchronization models the existing temporal and causal constraints among the values taken by different timelines (i.e., patterns of legal occurrences of the operational states across the timelines). The synchronizations considered are: (**C1**) *GoingTo(x,y)* must occur during *PointingAt(0, 0)*; (**C2**) *TakingPicture(pic, x, y, pan, tilt)* must occur during *At(x, y)* and *PointingAt(pan, tilt)*; (**C3**) *TakingPicture(pic, x, y, pan, tilt)* must occur before *Communicating(pic)*; (**C4**) *Communicating(file)* must occur during *At(x,y)*; (**C5**) *Communicating(file)* must occur during *Visible*. In addition to those synchronization constraints, the timelines must respect transition constraints among values and durations for each value specified in the domain (see again Fig. 1).

3   The KEEN System

APSI-TRF [4] is a development environment that provides "a timeline-based support" for modeling P&S domains. Its representation is sketched as the core of Fig. 2, where planning domains and problems can be respectively defined by means of a *Domain Description Language* and a *Problem Description Language* (the timeline-based equivalent of PDDL in classical planning). Then, a *Component-Based Modeling Engine* constitutes the software machinery that produces a data structure, here called *Current Plan*,

---

[3] Sometimes, the robot may be stuck in a certain position and the navigation module should be reset.

i.e., a Decision Network [4], a suitable structure to represent the domain, the current problem, the flaws in the current solution, and the flexible temporal plan obtained at the end of a problem solving session. For the current purpose, APSI-TRF relies on OMPS [12] as the *Problem Solver* that runs the search for a solution.

The Knowledge Engineering ENvironment (KEEN) is built around APSI-TRF as a set of active services to support the Knowledge Engineering (KE) phase (see again the figure).

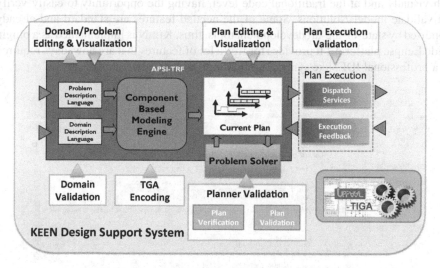

Fig. 2. The Knowledge Engineering ENvironment (KEEN)

KEEN is composed by a *Domain/Problem Editing and Visualization* module, providing a user interaction functionality for creating planning domain models and a set of V&V services taking advantage of the results presented in [5] and follow up papers. Also, a *Plan Execution* block is constituted by a *Dispatch Service* to send actual commands to a controlled system and an *Execution Feedback* module that allows to receive the telemetry from an actual plan execution environment. The pursued idea is that you can connect KEEN to either an accurate simulator of a real environment or a real physical system (e.g., a robot) and have functionalities to monitor with visual tools also the execution phase. Currently, KEEN is composed by a set of "classical KE" tools and the V&V services.

The editing and visualization capabilities have been developed as an *Eclipse* plugin, thus, providing a graphical interface to model, visualize and analyze the P&S domains. The V&V functionalities are based on Timed Game Automata (TGA) model checking and rely on UPPAAL-TIGA [2] as the verification engine.

The *TGA Encoding* module provides the basic TGA automatic translation for P&S specification [5] which constitutes the basis for implementing the KEEN V&V services: the *Domain Validation* module is to support the model building activity allowing to check the P&S model with respect to system requirements; similarly, the *Planner Validation* module is deputed to assess the P&S solver with respect to given requirements. In this regard, two sub-modules are further deployed, i.e., *Plan Verification* to verify the correctness of solution plans and *Plan Validation* to evaluate their adequacy; finally, a *Plan Execution Validation* module is to check whether the proposed solution plans are suitable for actual execution as well as to generate robust plan controllers.

3.1 The Support for Domain Definition

A main goal in KEEN is to provide an integrated environment where users may work both visually and at the traditional code level, having the opportunity to easily verify and validate models/solutions. Some of the needed features are standard and already supported by state-of-the-art development tools, thus, KEEN is implemented as a plugin inside Eclipse platform.[4] It provides for free a lot of features that are nowadays required for a professional IDE, like syntax highlighting, content assist, etc.

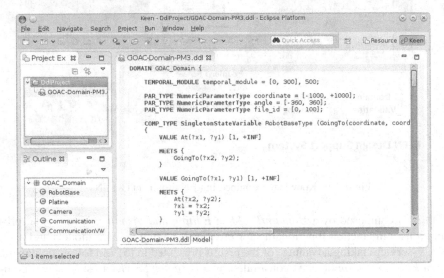

Fig. 3. DDL editor based on Eclipse plugin

KEEN builds on standard Eclipse components to provide traditional code-level functionalities: (i) a syntax highlighter, which uses different colors for different parts of the code to emphasize language keywords, special types, parameters, etc.; (ii) a tree view (Outline view) of relevant code blocks like state variables, providing a fast visualization and navigation inside the source files; (iii) real time syntax checks to easily spot erroneous programming constructs. The *look and feel* for KEEN is then depicted in Figure 3. The central part shows the code editor performing syntax highlighting while in the lower left corner, the Outline view is showing the GOAC Domain state variables.

KEEN is also endowed with a graphical representation of the Domain Model, as shown in Fig. 4, while other graphical representations, like an execution timeline view, are being worked on. In this view, the state variables of the Domain Model are represented on a

---

[4] http://www.eclipse.org – one of the most widespread *Integrated Development Environments* (IDE) for many programming languages.

workbench by blocks, which can be moved around and expanded/collapsed to show/hide their values and constraints. Also, the desired state variables can be selected to show their synchronization relations with other state variables, represented by dotted lines between blocks.

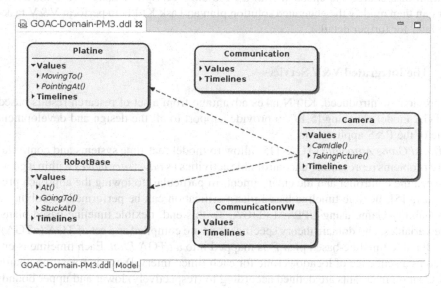

Fig. 4. Detail of the graphical view of the model

In Fig. 4, the five state variables of the GOAC Domain are shown: some of them are expanded (i.e., *Platine, Camera, RobotBase*), while the others are collapsed. The currently selected variable, *Camera*, shows its synchronizations (depicted as arrows) with other two state variables: one from *Camera.TakingPicture()* to *Platine.PointingAt()* and one from *Camera.TakingPicture()* to *RobotBase.At()*.

The user may also define new state variables by interacting with the graphical environment (i.e., right-clicking on an empty workbench area) and add and edit their properties and values. Defining a new synchronization is matter of dragging a line between the two values that should be connected, setting the type of the synchronization (i.e., the temporal constraint) and the needed parameters. This task can be achieved by means of a Synchronization Editor window that allows also for further edit/update.

The environment draws its information from APSI-TRF, and immediately updates the internal representation when the user makes a change. At the end of the chain, when the internal representation of the model is changed, a language-specific component is used to trigger source code modifications, using Eclipse's support for code refactoring. This way the tool allows the user to perform round-trip engineering[5] by synchronizing source code and graphical views. Then, a user can start to define a new model graphically, then switch to the traditional mode and do some text editing, then switch back again to the graphical mode and so on.

---

[5] http://en.wikipedia.org/wiki/Round-trip_engineering

Users can also ask for solution plan generation by means of the KEEN functionalities. Taking advantage of APSI-TRF capabilities, this may be performed by means of the OMPS planner. As for the case of the domain, the plan representation is completely handled by APSI-TRF and a specific language generation component is deputed to the creation of a source file encoded with a *Problem Description Language* syntax. The user can then modify the generated solution plan and ask KEEN to perform V&V tasks (see later for further details).

3.2   The Integrated V&V Services

As previously introduced, KEEN takes advantage from a set of research results based on TGA model checking [5,19] to provide support to all the design and development cycle of the P&S application.

Timed Game Automata (TGA) [15] allow to model real-time systems and controllability problems representing uncontrollable activities as *adversary moves* within a game between the controller and the environment. In particular, following the approach presented in [5], flexible timeline-based plan verification can be performed by solving a Reachability Game using UPPAAL-TIGA. To this end, flexible timeline-based plans, state variables, and domain theory specifications are compiled as a set of TGA (nTGA): (1) a flexible timeline-based plan P is mapped into a nTGA *Plan*. Each timeline is encoded as a sequence of locations (one for each timed interval), while transition guards and location invariants are defined according to (respectively) lower and upper bounds of flexible timed intervals; (2) the set of state variables SV is mapped into a nTGA *StateVar*. Basically, a one-to-one mapping is defined from state variables descriptions to TGA partitioning value transitions as controllable and uncontrollable according to their execution profile. (3) an *Observer* automaton is introduced to check for value constraints and synchronizations violations. In particular, two locations are considered: an Error location, to state constraint/synchronization violations, and a Nominal (OK) location, to state that the plan behavior is correct. The *Observer* is defined as fully uncontrollable. (4) the nTGA PL composed by the set *StateVar* ∪ *Plan* ∪ {*Observer*} encapsulates flexible plan, state variables and domain theory descriptions.

Then, considering a Reachability Game $RG(PL, Init, Safe, Goal)$ where *Init* represents the set of the initial locations of each automaton in PL, *Safe* is the Observer's nominal location, and *Goal* is the set of goal locations (one for each automaton in *Plan*), plan verification can be performed solving/winning the $RG(PL, Init, Safe, Goal)$ defined above. If there is no winning strategy, UPPAAL-TIGA provides a counter strategy for the opponent (i.e., the environment) to make the controller lose. That is, an execution trace showing a faulty evolution of the plan is provided. The encoding PL is considered as the basis for implementing the V&V functionalities discussed in the following.

Domain Validation. Similarly to [14], the TGA encoding PL can be exploited in order to validate planning domains, i.e., checking properties that are useful for ensuring correctness as well as detecting inconsistencies and flaws in the domain specification. For instance, undesired behaviors or safety properties can be checked against the planning model in order to guarantee the validity of the specification. In this regard, the KEEN

Domain Validation module is to support knowledge engineers in the process of eliciting, refining and correcting the domain model with respect to safety- and system-critical
requirements.[6]

Considering the nTGA $Dom = StateVar \cup \{Observer\}$ derived from $\mathcal{PL}$, the allowed
behaviors described by the associated planning domain can be represented. Then, given
a system requirements F, it can be stated as a Computation Tree Logic (CTL) formula
ϕ and checked by means of UPPAAL-TIGA verifying ϕ in Dom, i.e., validating the
planning domain with respect to the requirement F.

Among relevant properties, values reachability is an important aspect that can be
checked. Namely, the reachability of a value stated in the planning domain is checked
starting from one specific initial state (or from each possible initial state). In this regard,
KEEN allows to perform a reachability test for all the values declared in the domain.
In general, finding that a certain value is unreachable may suggest either the presence
of incomplete specifications or that some parts of the model are actually unneeded.
Also, KEEN can check user-defined properties (e.g., additional safety properties). For
instance, a relevant property a user may want to check through KEEN is the violation
of mutual exclusion for timeline's allowed values. In fact, such test is useful for detecting incomplete specification of synchronizations in the planning domain theory. In the
case of a flawed GOAC domain, the property $E\Diamond$ *RobotBase.GoingTo and Communi-
cation.Communicating*, which reads *there exists a possible trace which at some point
in time the rover is moving while communicating*, could be verified, then, providing an
evidence that the (**C4**) domain constraints might be violated.

Planner Validation. In order to validate the planner, we are interested in checking that
the planning solver works properly. In this sense, the application design activity should
be supported by providing effective methods to validate the solver and the generated
solutions, i.e., assessing its capability of generating a correct plan and, in addition, also
the adequacy of the generated solution plans should be checked.

For this purpose, KEEN has been endowed with two important submodules: (1) *Plan
Verification*, which systematically analyzes the solutions proposed by the planner itself,
and (2) *Plan Validation*, which allows to assess the plan adequacy with respect to given
requirements. Errors or negative features possibly found in the generated plans could
help knowledge engineers to revise the model (back to the domain validation step),
the heuristics, or the solver. Furthermore, plan V&V is also to assess the produced
plans with respect to the execution controllability issue. The KEEN Plan Verification
and Plan Validation modules have been implemented exploiting the verification method
presented in [5], i.e., solving the Reachability Game $RG(\mathcal{PL}, Init, Safe, Goal)$ defined
as above.

Plan Verification and Dynamic Controllability Check. The Plan Verification module
is fully relying on UPPAAL-TIGA by *winning* the Reachability Game RG defined as
above. Namely, KEEN invokes UPPAAL-TIGA for checking the CTL formula $\Phi = A$
[Safe U Goal] in $\mathcal{PL}$. In fact, the formula Φ states that along all its possible temporal

---

[6] It is worth underscoring that in [14] only controllable events are considered while, here, also
uncontrollable actions are modeled and taken into account.

evolutions, $\mathcal{PL}$ remains in *Safe* states until *Goal* states are reached. That is, in all the possible temporal evolutions of the timeline-based plan all the constraints are fulfilled and the plan is completed. Thus, if the solver verifies the above property, then the flexible temporal plan is valid. Whenever the flexible plan is not verified, UPPAAL-TIGA produces an execution strategy showing one temporal evolution that leads to a fault. Such a strategy can be analyzed in order to check either for plan weaknesses or for the presence of flaws in the planning model. The feasibility of such method has been shown in [5,19] where the verification methodology has been applied in two real-world planning domains.

Plan Validation. Besides synchronization constraints, users may need also to take into account system requirements to validate the generated solutions. These constraints define a kind of user preferences on the global behavior of the generated plan. Then, to implement the Plan Validation module, it is possible to apply the same verification process as in plan verification, verifying not only plan correctness, but also other domain-dependent constraints, i.e., the requirements. In general, the additional properties to be checked carry a low additional overhead to the verification process. Thus, the verification tool performances are not affected. Examples of such constraints in the robotic case study may be that no unnecessary tasks have been planned (e.g., unneeded robot navigation tasks). This validation task results as an important step in assessing the plan adequacy and, then, planner effectiveness.

Plan Controllers Synthesis. Plans synthesized by temporal P&S systems may be temporally flexible hence they identify an envelope of possible solutions aimed at facing uncertainty during actual execution. A valid plan can be brittle at execution time due to environment conditions that cannot be modeled in advance (e.g., *disturbances*). Previous works have tackled these issues within a Constraint-based Temporal Planning framework deploying specialized techniques based on temporal-constraint networks. Several authors (e.g., [16,20]) proposed a *dispatchable execution* approach where a flexible temporal plan is then used by a plan executive that schedules activities on-line while guaranteeing constraint satisfaction. In a recent work [19] we have addressed aspects of plan execution presenting the formal synthesis of a plan controller associated to a flexible temporal plan. UPPAAL-TIGA is exploited here in a different perspective as it synthesizes robust execution controllers of flexible temporal plans. And the generated execution strategies are embedded as execution policies in the KEEN Dispatch Service.

4   Related Works and Conclusions

This paper describes the current status of KEEN, a knowledge engineering environment for timeline-based planning. In particular, the paper underscores features of the environment that supports domain modeling and the integration of V&V features.

Most of the existing Knowledge Engineering tools have been dedicated to support PDDL as the underlying language (this is the case for ITSIMPLE and to some extent for GIPO). The EUROPA environment is the only timeline-based developing environment endowed with KE features. With respect to both EUROPA and other proposals KEEN contains some distinctive features. The round-trip engineering functionalities in KEEN are rather new. While some of the existing systems can export to PDDL, and sometimes also allow the user to edit the produced PDDL file (as in ITSIMPLE), they do not support an integrated work practice in which the users can seamlessly switch between graphical and code views while maintaining the model consistency. As far as V&V is concerned, it is worth remembering the standalone validation tools VAL [13] for PDDL that is used by integrated environments to perform validation as in ModPlan [10]. Additionally, systems like GIPO and ITSIMPLE support static and dynamic analysis of the domains. The dynamic analysis though is performed by means of manual steppers (for GIPO) or simulation through Petri Nets (for ITSIMPLE). ITSIMPLE also supports plan analysis by simulation. Nevertheless, it is worth underscoring that testing by simulation can only show the presence of errors (i.e., if no error is found there is no guarantee that none exists), whereas formal validation and verification proposed in KEEN can *demonstrate* the absence of errors (i.e., if no error is found we are guaranteed that none exists). Indeed, the integration of V&V techniques within P&S technology in KEEN aims to advance the state of the art of knowledge engineering tools and nicely contribute to the whole picture.

Several directions exist for improving KEEN in future work. The more immediate work will concern three aspects: (a) the capability to easily integrate different planners sharing the timeline-based modeling assumption; (b) the design of an environment for supporting problem definition; (c) the integration of tools for plan visualization and annotation, and the monitoring of execution.

Acknowledgments. CNR authors are partially supported by the Italian Ministry for University and Research (MIUR) and CNR under the GECKO Project (Progetto Bandiera "La Fabbrica del Futuro").

References

1. Barreiro, J., Boyce, M., Do, M., Franky, J., Iatauro, M., Kichkaylo, T., Morris, P., Ong, J., Remolina, E., Smith, T., Smith, D.: EUROPA: A Platform for AI Planning, Scheduling, Constraint Programming, and Optimization. In: The 4th Int. Competition on Knowledge Engineering for Planning and Scheduling, ICKEPS 2012 (2012)
2. Behrmann, G., Cougnard, A., David, A., Fleury, E., Larsen, K.G., Lime, D.: UPPAAL-tiga: Time for playing games! In: Damm, W., Hermanns, H. (eds.) CAV 2007. LNCS, vol. 4590, pp. 121–125. Springer, Heidelberg (2007)
3. Ceballos, A., Bensalem, S., Cesta, A., de Silva, L., Fratini, S., Ingrand, F., Ocon, J., Orlandini, A., Py, F., Rajan, K., Rasconi, R., van Winnendael, M.: A Goal-Oriented Autonomous Controller for Space Exploration. In: Proc. of the 11th Symposium on Advanced Space Technologies in Robotics and Automation, ASTRA 2011 (2011)
4. Cesta, A., Cortellessa, G., Fratini, S., Oddi, A.: Developing an End-to-End Planning Application from a Timeline Representation Framework. In: Proceedings of the 21st Innovative Application of Artificial Intelligence Conference, IAAI 2009, Pasadena, CA, USA (2009)
5. Cesta, A., Finzi, A., Fratini, S., Orlandini, A., Tronci, E.: Analyzing Flexible Timeline Plan. In: Proceedings of the 19th European Conference on Artificial Intelligence, ECAI 2010, vol. 215. IOS Press (2010)

6. Cesta, A., Finzi, A., Fratini, S., Orlandini, A., Tronci, E.: Validation and Verification Issues in a Timeline-Based Planning System. Knowledge Engineering Review 25(3), 299–318 (2010)
7. Cesta, A., Oddi, A.: DDL.1: A Formal Description of a Constraint Representation Language for Physical Domains. In: Ghallab, M., Milani, A. (eds.) New Directions in AI Planning. IOS Press, Amsterdam (1996)
8. Chien, S., Tran, D., Rabideau, G., Schaffer, S., Mandl, D., Frye, S.: Timeline-Based Space Operations Scheduling with External Constraints. In: Proc. of the 20th International Conference on Automated Planning and Scheduling, ICAPS 2010 (2010)
9. Chien, S.A., Johnston, M., Frank, J., Giuliano, M., Kavelaars, A., Lenzen, C., Policella, N.: A Generalized Timeline Representation, Services, and Interface for Automating Space Mission Operations. In: SpaceOps (2012)
10. Edelkamp, S., Mehler, T.: Knowledge acquisition and knowledge engineering in the ModPlan workbench. In: Proceedings of the First International Competition on Knowledge Engineering for AI Planning (2005)
11. Frank, J., Jonsson, A.: Constraint Based Attribute and Interval Planning. Journal of Constraints 8(4), 339–364 (2003)
12. Fratini, S., Pecora, F., Cesta, A.: Unifying Planning and Scheduling as Timelines in a Component-Based Perspective. Archives of Control Sciences 18(2), 231–271 (2008)
13. Howey, R., Long, D., Fox, M.: VAL: automatic plan validation, continuous effects and mixed initiative planning using PDDL. In: 16th IEEE International Conference on Tools with Artificial Intelligence, ICTAI 2004, pp. 294–301 (2004)
14. Khatib, L., Muscettola, N., Havelund, K.: Mapping Temporal Planning Constraints into Timed Automata. In: The Eigth Int. Symposium on Temporal Representation and Reasoning, TIME 2001, pp. 21–27 (2001)
15. Maler, O., Pnueli, A., Sifakis, J.: On the Synthesis of Discrete Controllers for Timed Systems. In: Mayr, E.W., Puech, C. (eds.) STACS 1995. LNCS, vol. 900, pp. 229–242. Springer, Heidelberg (1995)
16. Morris, P.H., Muscettola, N.: Temporal Dynamic Controllability Revisited. In: Proc. of AAAI 2005, pp. 1193–1198 (2005)
17. Muscettola, N.: HSTS: Integrating Planning and Scheduling. In: Zweben, M., Fox, M.S. (eds.) Intelligent Scheduling. Morgan Kauffmann (1994)
18. Orlandini, A., Finzi, A., Cesta, A., Fratini, S., Tronci, E.: Enriching APSI with Validation Capabilities: the KEEN environment and its use in Robotic. In: Proc. of 11th Symposium on Advanced Space Technologies in Robotics and Automation, ASTRA 2011 (2011)
19. Orlandini, A., Finzi, A., Cesta, A., Fratini, S.: TGA-Based Controllers for Flexible Plan Execution. In: Bach, J., Edelkamp, S. (eds.) KI 2011. LNCS, vol. 7006, pp. 233–245. Springer, Heidelberg (2011)
20. Shah, J., Williams, B.C.: Fast Dynamic Scheduling of Disjunctive Temporal Constraint Networks through Incremental Compilation. In: ICAPS 2008, pp. 322–329 (2008)
21. Simpson, R.M., Kitchin, D.E., McCluskey, T.L.: Planning Domain Definition using GIPO. Knowledge Engineering Review 22(2), 117–134 (2007)
22. Vaquero, T.S., Silva, J.R., Tonidandel, F., Beck, J.C.: itSIMPLE: Towards an Integrated Design System for Real Planning Applications. Knowledge Engineering Review 28(2), 215–230 (2013)

Numeric Kernel for Reasoning about Plans Involving Numeric Fluents

Enrico Scala

Dipartimento di Informatica, Universita' di Torino, Torino, Italy
scala@di.unito.it

Abstract. The paper proposes the notion of *numeric kernel* as a means for reasoning about plans involving numeric state variables, i.e. numeric fluents. A *numeric kernel* identifies the sufficient and necessary conditions that allow to directly - without any search and any propagation - assess whether a plan is valid in a specific world state. The notion generalizes the propositional kernels defined for the STRIPS language, to support domains involving numeric information as well. A regression method to build such kernels is reported, and its correctness is theoretically proved. To evaluate the numeric kernels contribution, we report two possible repair strategies that can be employed as a direct application of the numeric kernel properties. Results show the promise of the approach both from the computational point of view and in terms of plan quality.

1 Introduction

In the last decade of research in AI, an increasing amount of work has been devoted in extending the automated planning (and especially the classical paradigm) for dealing with real world problems ([1], [2], [3]). One of the shortcomings of the classical setting is the lack of the management of consumable and continuous resources as well as of the capability of reasoning about quantitative characteristics of the world. To this end, the *numeric fluent* notion and hence the *numeric planning* formalism have been introduced ([4], [2]). As an innovation w.r.t. the classical paradigm, in the numerical planning setting, plans must obey to particular resource profiles. This is achieved by allowing to express conditions and operations over the set of numeric variables of interest.

However, while efforts have been devoted for the problem of off-line plan generation ([1], [5], [2], [3]), few attention has been paid in studying (numeric) plan of actions for the on-line phase, making exception for the works dealing with the temporal dimension. In this context models as STP (Simple Temporal Problem) and DTP (Disjunctive Temporal Problem) have been adopted and some extensions have been proposed ([6] [7] [8] [9] [10]). However, also other continuous resources (e.g. energy, money and so forth) should be managed.

The main contribution of this paper is the notion and the mechanism for the construction of *numeric kernels*. A *numerical kernel* identifies the set of sufficient and necessary conditions allowing, for a given goal G and a plan π, to directly (without performing any search and any propagation) assess whether a

M. Baldoni et al. (Eds.): AI*IA 2013, LNAI 8249, pp. 263–275, 2013.

particular state of the system is consistent with G and π. The *numeric kernel* generalizes the propositional definition ([11]) for the *numeric* setting. Therefore, the extension allows the application of some properties studied in the classical setting for the numeric case, too. Such properties have been exploited (in the classical setting) in the plan execution context for improving the monitoring ([12]) and the repair ([13]).

As a complementary contribution, the paper presents a continual planning agent ([14]) and two effective plan repair strategies for dealing with numeric information. In particular, one of these two strategies shows that is possible to combine *numeric kernels* with the heuristic mechanisms developed for the numeric planning (e.g. [1],[2],[15]).

After a brief introduction on the formal framework of reference, the paper formalizes the notion of *numeric kernels* (section 2.2) and presents how such kernels can be constructed (section 3). Then, the work describes and (experimentally) evaluates the two repair strategies above (section 4 and 5).

2   Formal Framework

This section reports the reference planning formalism and then it formalizes the *numeric kernels* notion. We assume the reader is familiar to the PDDL-like language; for a thorough discussion see [4].

2.1   Basic Definitions

Definition 1 (World State). *A world state is built upon a set F of propositions and a set X of numeric variables. Thus a state s is a pair $< F(s), X(s) >$, where $F(s)$ is the set of atoms that are true in s (Closed World Assumption) and $X(s)$ is an assignment in $\mathbb{Q}$ for each numeric variable in X.*

Definition 2 (Numeric Action[1]). *Given F and X as defined above, a numeric action "a" is a pair $< pre, eff >$ where:*

- *"pre" is the applicability condition for "a"; it consists of:*
  - *a numeric part (pre_{num}), i.e. a set of comparisons of the form $\{exp \; \{<,\leq,=,\geq,>\} \; exp'\}$.*
  - *a propositional part (pre_{prop}), i.e. a set of propositions defined over F.*
- *"eff" is the effects set of a; it consists of:*
  - *a set of numeric operations (eff_{num}) of the form $\{f, op, exp\}$, where $f \in X$ is the numeric fluent affected by the operation, and op is one of $\{+ =, - =, =\}$.*
  - *an "add" and a "delete" list (eff^+ and eff^-), which respectively formulate the propositions produced and deleted after the action execution*

---

[1] For the sake of the explanation we refer to ground actions. However, in our implementation we support action schema as well.

Here, exp and exp' are arithmetic expressions involving variables from X. An expression is recursively defined in terms of (i) a constant in $\mathbb{Q}$ (ii) a numeric fluent (iii) an arithmetical combination among $\{+,*,/,-\}$ of expressions[2].

An action a is said to be applicable in a state s iff its propositional and numeric preconditions are satisfied in s. Meaning that (i) $pre_{prop}(a) \subseteq F(s)$ and (ii) $pre_{num}(a)$ must be satisfied (in the arithmetical sense) by $X(s)$.

Given a state s and a numeric action a, the application of a in s, identified by $s[a]$, (deterministically) produces a new state s' as follows. s' is initialized to be s; each atom present in $eff^+(a)$ is added to $F(s')$ (whether this is not already present); each atom present in $eff^-(a)$ is removed from $F(s')$; each numeric fluent f of the numeric operation $\{f,op,exp\}$ is modified according to the exp and the op involved. The state resulting from a non applicable action is undefined. An undefined world state does not satisfy any condition.

Definition 3 (Numeric plan). *Let I and G be a world state and a goal condition*[3], *respectively, a numeric plan $\pi = \{a_0, a_1, .., a_{n-1}\}$ is a total ordered set of actions such that the execution of these actions (in the order defined by the plan) transforms the state I into a state I' where G is satisfied.*

Given the formulation above, we allow the access to a segment of the plan by subscripting the plan symbol. More precisely, $\pi_{i \to j}$ with $i < j$ identifies the subplan starting from the i-th till the j-th action. Moreover when the right bound is omitted the length of the plan is assumed, i.e. $\pi_i \equiv \pi_{i \to |\pi|}$. Finally, we identify by $s[\pi]$ the state produced executing π starting from s.

2.2 Numeric Kernel

When the plan has to be handled online, the presence of deviations from the nominal state (e.g., unexpected events, wrong assumptions made at planning time or actions that achieve different effects) may prevent the feasibility of the plan, and just checking the next action preconditions could not suffice to establish whether the plan still achieve the goals. Indeed, it is necessary to simulate the whole plan execution to predict if the goal is reachable via such a plan. More formally we can say that:

Definition 4 (Numeric i-th Plan Validity). *Let s be a world state and G a set of goal conditions, the sub-plan π_i is said to be i-th valid w.r.t. s and G iff $s[\pi_i]$ satisfies G.*

Since the PLANEX system ([11]) and more recently in [12], it has been noticed that the simulation step can be avoided by keeping trace of just a subset of (propositional) conditions; such conditions are also referred as the weakest preconditions of a plan.

---

[2] For computational reasons, several numeric planners request such expressions to be linear (e.g. [1],[3]). In our case, such a restriction is not necessary (see section 3).

[3] A goal condition has the same form of the applicability condition of a numeric action.

However, the works in literature discussed so far just focused on the propositional fragment of the planning problem. Hence, to handle the numeric setting, we need to specify conditions even on the numeric part of the problem. That is:

Definition 5 (Numeric Kernel). *Let π be a numeric plan for achieving G, and K a set of (propositional and numeric) conditions built over F and X, K is said to be a numeric kernel of π iff it represents a set of sufficient and necessary conditions for the achievement of the goal G starting from s via π. That is, given a state s, $s[\pi]$ satisfies G iff s_{num} satisfies K_{num} and s_{prop} satisfies K_{prop}.*

A kernel is well defined when all the involved numeric comparisons are satisfiable by at least an assignment of values (e.g., a comparison $4 < 3$ is not allowed). This means that there must exist a state s, satisfying such kernel, for which $X(s)$ gives a value for each numeric variable. An ill-defined kernel comes from a plan π and a goal G in which π cannot be a solution of any planning problem having G as goal.

By considering each suffix of the plan π, i.e. $\pi_0 = \{a_0, ..., a_{n-1}\}$, $\pi_1 = \{a_1, ..., a_{n-1}\}$, $\pi_2 = \{a_2, ..., a_{n-1}\}$,...,$\pi_{n-1} = \{a_{n-1}\}$ till the empty plan $\pi_n = \{\}$, it is possible to individuate an ordered set of *numeric kernels* where the i-th element of the set is the *numeric kernel* of π_i. It is worth noting that, by definition, the goal is a special kind of *numeric kernel* for the empty sub-plan.

Therefore, given a plan of size n we can say that:
- $s^0[\pi_0]$ satisfies G iff s^0 satisfies K^0
- $s^1[\pi_1]$ satisfies G iff s^1 satisfies K^1
- ...
- $K^n = G$ corresponding to the kernel for the empty plan π_n

where the superscript indicates the "time" index of interest. The resulting set of kernels will be denoted with $\mathbb{K}$, i.e. $\mathbb{K} = \{K^0,...,K^n\}$.

Given $\mathbb{K}$ defined as above, and the plan validity notion reported in Definition 4, it is possible to deduce that:

Proposition 1. *A plan π is valid at step i for a goal G iff s^i satisfies K^i, where*

- *s^i is the state observed before the execution of the subplan π_i*
- *K^i is the i-th kernel associated with π_i*

By observing this relation it is also possible to see that:

Proposition 2. *Let $\mathbb{K} = \{K^0,..,K^n\}$ be the kernel set associated with the plan $\pi = \{a_0, ..., a_{n-1}\}$, and s a world state, if there is a plan π' such that $s[\pi']$ satisfies K^i (with $0 \leq i \leq n$), then a plan from s to a state s' satisfying K^j (with $i \leq j \leq n$) exists, too.*

Basically, the proposition above provides a sufficient condition for the reachability of a kernel (included the goal) starting from a state satisfying such a condition. For instance, if it is possible to reach at least one of the kernel K via a given plan π'', a planning task, having s as initial state and G as goal, admits at least a solution, which is the one given by the concatenation of π'' and the suffix of the plan relative to K.

As we will see in section 3, the construction of the *numeric kernels* can be done just once, in a pre-processing phase. Once obtained the kernels set, thanks to Proposition 1, the validity checking process is very easy; it can be performed by simply substituting the numeric values of the state in each comparison appearing in the kernel, as well as, for the propositional part, it suffices to check if each proposition is included in the current world state. For this reason one can infer whether the goal is still supported by the current state without analyzing the plan. Of course, in case the plan undergoes some adjustments, the set of kernels has to be recomputed. Moreover, thanks to Proposition 2, in case some inconsistency is detected, it could be not necessary to replanning from scratch. Indeed, it may suffice, firstly, achieving one of the kernel conditions, and then applying the remaining subplan. Both propositions are exploited in the continual planning agent of section 4.

3   Kernel Construction

Algorithm 1 reports the regression mechanism to build the kernel set.

Algorithm 1. Numeric Kernel Computation (NKC)

Input: $\pi = \{a_0,..,a_{n-1}\}$ - plan ; G - goal
Output: $\mathbb{K}$: an ordered set of numeric Kernels
1   $K^{|\pi|} = G$
2   **for** $i=|\pi|-1$ **to** *0* **do**
3       $K^i_{prop} = \{K^{i+1}_{prop} \setminus eff^+(a_i)\} \cup pre_{prop}(a_i)$
4       $K^i_{num} = \{K^{i+1}_{num} \oplus eff_{num}(a_i)\} \cup pre_{num}(a_i)$

In particular, the procedure starts from the last kernel corresponding to the set of the goals (the propositions that must be achieved at the final state, given the comparisons on the involved numeric fluents, line 1), and produces each i-th kernel by performing two independent steps:
- (propositional fragment) - (i) removing the atoms provided by a_i (i.e. the add-list of the i-th action), (ii) adding the atoms required by a_i (i.e. the propositional preconditions of the i-th action), line 3.
- (numeric fragment) - combining the information involved in (i) the numeric part of the action model and (ii) the next *numeric kernel* (previously computed), keeping trace of the numeric action contribution.

The numeric part construction relies on the operator $\oplus$, which is a function that maps a set of comparisons C and a set of numeric action effects *eff* to a new set of comparisons C'. The operator assures that the new comparisons set will take in consideration the *future* effects of the action.

In particular, for each comparison in C the operator (see algorithm 2) performs a substitution of the numeric fluents involved in c_l and c_r[4], according to the

---
[4] c_l identifies the left part of the comparison while c_r the right one. Thus both c_l and c_r are arithmetical expressions over X.

Algorithm 2. ⊕

Input: eff: numeric effects of a ; C: a set of comparisons
Output: C': a new set of comparisons

1   $C' = \{\}$
2 **foreach** $c \in C$ **do**
3     $c'_l = c_l.\text{substitution}(x_0...x_{m-1}, eff_{x_0}...eff_{x_{m-1}})$
4     $c'_r = c_r.\text{substitution}(x_0...x_{m-1}, eff_{x_0}...eff_{x_{m-1}})$
5     $C' = C' \cup \{c'\}$

---

effects reported in a. For instance, if the numeric effects set of a involves a fluent x with $(x+ = 5)$ and the comparison asserts that $x < 4$, then the outcome C' will be $x + 5 < 4$, i.e. $x < -1$. Of course, the action can affect many fluents involved in the previous comparison; therefore the substitution has to iterate over all the fluents involved in C with the effects described in the action model. In the algorithm, each eff_{x_i} is the numeric effect of an action, having x_i as affected numeric fluent.

As anticipated before, it is evident that the expression exp involved in a numeric effect $\{f, op, exp\}$ does not require to be linear; in fact, the mechanism does not regress to a specific state, but to a set of new conditions that substitute each fluent with the way such a fluent is altered.

For simplicity, let us introduce a small example, focusing on the numeric part of the model. Let us consider a "one action" plan $\pi = \{a_0\}$ and a goal G, where a_0 and G are numerically defined as follows:

$$\text{pre}(a_0) = \begin{cases} f_1 > 5 \\ f_2 < 4 \end{cases} \quad \text{eff}(a_0) = \begin{cases} f_1 = f_1 + 5 \\ f_2 = f_2 + 8 \end{cases} \quad G = \begin{cases} f_1 > 10 \\ f_2 < 4 \end{cases}$$

By following algorithm 1, we have $K^1 = G$ and (the numeric part of) K^0 created by means of $\{K^1_{num} \oplus eff_{num}(a_0)\} \cup pre_{num}(a_0)$. So in the first step, we have to keep trace of the action contribution, that is:

$$K^1 \oplus eff_{num}(a_0) = \begin{cases} f_1 + 5 > 10 \\ f_2 + 8 < 4 \end{cases}$$

Then, the *numeric kernel* is completed by joining the constraints defined by ⊕ with the constraints defined for *pre(a)*, i.e.:

$$(K^1 \oplus eff_{num}(a_0)) \cup pre_{num}(a_0) = \begin{cases} f_1 + 5 > 10 \\ f_2 + 8 < 4 \\ f_1 > 5 \\ f_2 < 4 \end{cases} = \begin{cases} f_1 > 5 \\ f_2 < -4 \end{cases}$$

It is easy to see that if we take an arbitrary (initial) state with f_1 and f_2 satisfying the arising comparisons (e.g. $f_1 = 6$ and $f_2 = $ -5), and, if we apply to such a state the action a_0, we are sure that (i) the action is applicable, since $6 > 5$ and $-5 < -4$, and (ii) we will obtain a state s' where both $f_1 > 10$ and $f_2 < 4$ hold, namely goal conditions will be satisfied.

The example reported above describes a simple scenario where numeric fluents do not depend on each other. But this is not always the case. Theoretically, in fact, an effect for the action can express that $f_1 = f_2$. Also this kind of representation is captured by the substitution performed by the algorithm 2. In general the algorithm will substitute each variable of the comparison with the way in which the variable is modified (line 3 and 4).

Let us conclude this section with the proof of the correctness of the algorithm 1. For the sake of explanation, the correctness proof focuses on the numeric aspect of the problem, so G, s, K and the plan π are analyzed as far as it is concerned by their numeric part.

Theorem 1. *Correctness. Given a plan π and a goal G the algorithm 1 finds a set of numeric kernels for π and G.*

Proof. The proof proceeds by induction on the length of the plan. The base case of our induction is when the plan is empty, i.e. $|\pi| = 0$; in such a case the algorithm terminates after one iteration with the last and the only *numeric kernel* of interest, i.e. $K^n \equiv K^0$ since $n = 0$. This kernel contains the same comparisons present in the goal; hence, given a state s, it follows that $s[\pi]$ satisfies G if and only if s satisfies K^0. In particular $s[\pi]$ corresponds to s and the only way of reaching the goal is to be a state that already satisfies the goal.

Inductive step. For inductive assumption we know that the $i\text{-}1$ steps of the algorithm NKC (the iteration) have computed a set of i *numeric kernels* for a plan of length n; i.e. we have the set of kernels $K = \{K^{n-(i-1)}, K^{n-(i-2)}, ..., K^{n-1}, K^n\}$ which is in relation with each suffix of the plan, i.e. with $\pi_{n-(i-1)}, \pi_{n-(i-2)}, ..., \pi_{n-1}, \pi_n$. At the i-th step, K^{n-i} is computed by combining the preconditions of the first action in π_{n-i}, i.e. a_{n-i} and the comparisons obtained in the previous step. For this reason, a state s satisfying K^{n-i} will be such that (i) the action a_{n-i} is applicable and (ii) the state resulting from the action application, i.e. $s[\{a_{n-i}\}]$, turns out to satisfy the $K^{n-(i-1)}$. This last in fact follows directly from the definition of $\oplus$. Indeed the operation keeps the conditions expressed in $K^{n-(i-1)}$, while considering the effects of a_{n-i} by means of the substitution mechanism. Having both (i) and (ii) we are sure that if s satisfies K^{n-i}, then $s[\pi_{n-i}]$ will satisfy the goal.

For the *necessary condition*, that is if $s[\pi_{n-i}]$ satisfies G then s satisfies K^{n-i}, we can proceed by absurd. Indeed, if s does not satisfy the conditions expressed in K^{n-i}, it means that either the first action of π_{n-i} (a_{n-i}) is not applicable or the state resulting from the application of such an action does not satisfy $K^{n-(i-1)}$. The latter is not possible for inductive assumption as $K^{n-(i-1)}$ is assumed to be a kernel, while for the former it is obviously impossible since if the action is not applicable then $s[\pi_{n-i}]$ will not satisfy the goal. This proves the contradiction.

4   Continual Planning via Numeric Kernels

To validate and evaluate the relevance of the *numeric kernels* formulation, we implemented a continual planning agent able to handle domains with numeric fluents. The continual planning paradigm [14] allows the agent to interleave the execution and the planning all along the task to execute. The idea is that the planning phase cannot be seen as a single one shot task. Instead, the agent should monitor the plan execution and replan or repair the plan when necessary.

In particular, when the current plan becomes inconsistent, it is important that the agent acts in a timely fashion to recover from the unexpected impasse. To this end, in the following, we propose two plan repair strategies which directly exploit the *numeric kernel* properties. For further details on how implementing the overall continual planning loop see [16].

4.1   Greedy Repair

The first implemented repair strategy is a direct application of the Proposition 2. Rather than abandoning the old plan for computing a solution completely from scratch, the greedy repair tries to restore the kernel conditions by performing a patch plan that connects the current state with a state that satisfies the expected kernel condition, at that step. Then the patch can be concatenated with the previous plan, forming a new plan of actions leading the agent to the goal conditions. We expect, in fact, that in some situation (i.e. where the deviation from the nominal trajectory is not so prominent), despite contrasting complexity results on this topic ([17]), it may be actually more convenient to adapt the old solution rather than replanning completely from scratch. However, there may be situations where the greedy nature of this strategy can be a waste of time, and, instead, there may be kernels all along the plan, that are actually easier to reach. To mitigate this behavior we present the kernel heuristic repair.

4.2   Kernel Heuristic Repair

As we have seen, each element of $\mathbb{K}$ identifies the conditions for a suffix of the plan to be valid. This means that, if the agent would have the capability of connecting the current state to one of such conditions, the relative suffix of the plan will bring the agent to reach the final goal.

Therefore, the idea is to search for the kernel that is the one actually *closer* (and hence hopefully simpler) to satisfy. However, the exact estimation of such distance is prohibitive as it would correspond in solving several instances of new planning tasks. For this reason, we combined the kernel set $\mathbb{K}$ with the numeric planning graph heuristic, developed in the context of numeric planning (e.g. in Metric-FF, [1], and Lpg-TD, [2]). In particular, the heuristic is the solution length of a relaxed version of the planning problem; the relaxation consists of considering the actions only in their positive effects[5].

In our context, the heuristic provides us with an estimation of the distance between the current state and the set of kernels $\mathbb{K}$; the resulting strategy is straightforward. Let s be the current world state at step i of the execution violating the condition expressed in the i-th kernel, the *kernel heuristic repair* estimates the distance $d(s, K_j)$ for each $K_j \in \mathbb{K}$ where $j \geq i$. Then the strategy picks the kernel which has the lower distance and will use this kernel as an intermediate point towards which performs the patch-plan. Having computed this new course of actions, as in the greedy repair, it suffices to concatenate the patch plan with the relative plan-suffix to obtain a valid plan from s to the goal G. The procedure is summarized by the algorithm 3.

Algorithm 3. Kernel_Heuristic_Repair

Input: π - numeric plan ; $\mathbb{K}$ - kernels set ; s - world_state
Output: the numeric plan modified
1 K* = best_kernel(s,$\mathbb{K}$);
2 if $d(s, K^*) \neq \infty$ then
3   π' = solve(s,K*);
4   if π' *is not failure* then
5   return $\pi' \cup$ suffix_of(K*)
6 return $\emptyset$

By comparing these two strategies we can see that, on the one hand, the greedy repair immediately commits towards a specific kernel without spending efforts to reason on the previous plan. On the other hand, the kernel heuristic repair has to spend some extra time to estimate what is the "best" kernel to take into account (i.e. the heuristic computation). We expect that, in the long run, the kernel heuristic repair could take advantage from this reasoning and hence outperforming the greedy approach in those cases for which making (or trying to make) a patch towards the current kernel is actually a useless time consumption.

5 Experimental Results

To evaluate the repair mechanisms proposed in the previous section, we tested the greedy repair and heuristic kernel repair on three domains from the third International Planning Competition (IPC3)[6]. The first is the ZenoTravel (ZT), the second is a harder version, where the refuel action is allowed only in presence of the refuel station (Hard ZT), and the third is the Rover domain.

---

[5] The numeric planning graph heuristic extends the original definition for the classical paradigm (where actions are considered without their delete effects) to the domains involving numeric variables. For details see [1] and [2].
[6] http://www.plg.inf.uc3m.es/ipc2011-deterministic.

For each domain, we used 11 problems from the IPC suite (the harder ones), and for each problem, we generated (off-line) a starting plan. To evaluate the plan repair strategies, we simulated the plan execution by injecting discrepancies on the way in which resources are consumed. For each case we injected 5 different amounts of noise; hence each case for our evaluation is identified by the tuple $< domain, problem, plan, noise >$. Totally we have 55 cases for each domain.

The evaluation relies on the metric defined for the IPC2011 and has been focused on the performance of repairing a plan[7]. In particular, we measured the time and the quality score (i.e. the resulting plan length[8]), and the coverage of the strategy (i.e. the percentage of solved problems). For comparison reasons, test cases ran also with a replanning from scratch, and with LPG-ADAPT ([18]).

Each computation is allotted with a maximum of 100 secs of cpu-time. Tests have been executed on Ubuntu 10.04 with an Intel Core Duo@2.53GHz cpu and 4 GB of Ram. The software has been implemented in Java 1.6 and the (re)planner used for the resolution of the arising planning tasks has been *Metric-FF* ([1]).

Table 1. Time Score, Quality Score and Coverage, according to the IPC 2011 metrics, for the three strategies over all the problems and domains

	Time Score				Quality Score				Coverage			
	GR	HKR	REP	LPGA	GR	HKR	REP	LPGA	GR	HKR	REP	LPGA
ZT	**55.0**	27.0	25.9	5.6	53.4	53.4	**54.2**	26.2	**100**	**100**	**100**	70.9
HardZT	**47.0**	24.3	17.7	7.4	46.1	**48.9**	47.4	21.3	85.5	**90.9**	87.3	61.8
Rover	**46.0**	4.7	9.9	22.7	42.2	**41.3**	26.1	46.3	89.1	83.6	49.1	**100**
Total	**148.0**	56.0	53.6	35.7	141.7	**143.6**	127.7	93.8	**91.5**	**91.5**	78.8	77.6

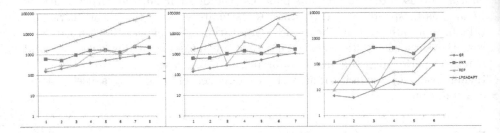

Fig. 1. The y-axis shows the Cpu-Time (in msec) for all the problems (x-axis) in the tested domains: ZenoTravel (left), Hard ZenoTravel (center), Rover (right)

Results are summarized by Table 1 and Figure 1; GR, HKR, REP and LPGA stand for, respectively, the greedy repair, the heuristic kernel repair, the replanning from scratch strategy and LPG-ADAPT. The table reports the scores

---

[7] It is worth noting that the kernel construction can be done in a preprocessing phase so that it can be seen as a form of off-line reasoning which hence does not influence the on-line performances.

[8] For the repair such a length is given by the patch plan and the plan suffix to execute.

obtained, while Figure 1 analyzes the scalability of the three approaches w.r.t. the increase of the difficulty of the problems[9].

Concerning the time score, GR outperformed the other strategies in all the tested domains (Table 1). The difference is large and it is also evident analyzing the gain between the curves reported in Figure 1. As matter of facts, GR remains stably under the 1.5 secs of computational time for all the tested cases. Surprisingly, even for the quality score and the coverage, both the kernel based mechanisms outperformed the other strategies. This has been due to the many timeout situations encountered by running REP and LPGA. HKR is the one with the highest quality score but it is no competitive with GR time score. From our experiments we noticed that a large part of the HKR solving-time has been spent to compute the heuristic function (50%, 44% and 71% respectively for the ZenoTravel, Hard ZenoTravel and Rover). The bottleneck has been probably a not well optimized Java code (differently from others well known C++ implementations of the numeric planning graph heuristic, see for instance [1]); hence, we expect that the performance of HKR could be greatly increased.

6 Conclusions

The paper has proposed the notion of *numeric kernel*, which generalizes the STRIPS kernel for dealing with actions involving changes for continuous variables, i.e. numeric fluents. Concretely a *numeric kernel* is the weakest set of conditions that must be satisfied to make a plan of actions applicable. The paper formally reported a concrete regression mechanism for the construction of such a kernel, and interesting properties arising from its formulation.

Besides this theoretical contribution, the paper provided two effective strategies, which directly apply the kernel facilities for a problem of repair in presence of continuous and consumable resources. As a difference w.r.t. the most of the works appeared in literature ([19], [20], [18], [13]) in the context of on-line planning, the repair mechanisms implemented in this paper can be applied not only for the propositional setting, but also for the numeric one. Moreover, since these strategies do not make any assumption on the planner/heuristic used, they could be easily adapted with existing replanning/plan-adaption tools ([18]).

The performance of the repair strategies have been empirically evaluated on three domains defined by the planning community. Results showed that, in facing unexpected resources consumption, the proposed repair mechanisms are quite efficient, and produce good quality solutions, outperforming in the most of the cases both a replanning mechanism, and the LPG-ADAPT system ([18]). Such results, of course, need further confirmations on a larger set of domains; therefore, as an immediate future work, we are working on extending the experimental phase to assess the generality of the proposed strategies, possibly in combination with others plan-adaption tools.

---

[9] This scalability measures only those cases which have been successfully solved by all the strategies tested.

From a methodological point of view, we are also studying how and when the *numeric kernel* notion can be employed in the context of case based planning with numeric fluents ([21]). Intuitively, we believe in fact that the notion may provide a quite good guidance in the selection/exploitation of past plans, solutions of (hopefully) similar problems.

References

1. Hoffmann, J.: The metric-ff planning system: Translating "ignoring delete lists" to numeric state variables. Journal of Artificial Intelligence Research 20, 291–341 (2003)
2. Gerevini, A., Saetti, I., Serina, A.: An approach to efficient planning with numerical fluents and multi-criteria plan quality. Artificial Intelligence 172(8-9), 899–944 (2008)
3. Coles, A.J., Coles, A., Fox, M., Long, D.: Colin: Planning with continuous linear numeric change. Journal of Artificial Intelligence Research 44, 1–96 (2012)
4. Fox, M., Long, D.: Pddl2.1: An extension to pddl for expressing temporal planning domains. Journal of Artificial Intelligence Research 20, 61–124 (2003)
5. Coles, A.J., Coles, A.I., Fox, M., Long, D.: Forward-chaining partial-order planning. In: Proc. of International Conference on Automated Planning and Scheduling, ICAPS 2010 (2010)
6. Conrad, P.R., Williams, B.C.: Drake: An efficient executive for temporal plans with choice. Journal of Artificial Intelligence Research 42, 607–659 (2011)
7. Dechter, R., Meiri, I., Pearl, J.: Temporal constraint networks. Artificial Intelligence 49(1-3), 61–95 (1991)
8. Kvarnström, J., Heintz, F., Doherty, P.: A temporal logic-based planning and execution monitoring system. In: Proc. of International Conference on Automated Planning and Scheduling (ICAPS 2008), pp. 198–205 (2008)
9. Policella, N., Cesta, A., Oddi, A., Smith, S.: Solve-and-robustify. Journal of Scheduling 12, 299–314 (2009)
10. Stergiou, K., Koubarakis, M.: Backtracking algorithms for disjunctions of temporal constraints. Artificial Intelligence 120(1), 81–117 (2000)
11. Fikes, R., Hart, P.E., Nilsson, N.J.: Learning and executing generalized robot plans. Artificial Intelligence 3(1-3), 251–288 (1972)
12. Fritz, C., McIlraith, S.A.: Monitoring plan optimality during execution. In: Proc. of International Conference on Automated Planning and Scheduling (ICAPS 2007), pp. 144–151 (2007)
13. Garrido, A., Guzman, C., Onaindia, E.: Anytime plan-adaptation for continuous planning. In: Proc. of P&S Special Interest Group Workshop, PLANSIG 2010 (2010)
14. Brenner, M., Nebel, B.: Continual planning and acting in dynamic multiagent environments. Journal of Autonomous Agents and Multiagent Systems 19(3), 297–331 (2009)
15. Chen, Y., Wah, B.W., Hsu, C.-W.: Temporal planning using subgoal partitioning and resolution in sgplan. Journal of Artificial Intelligence Research 26, 369 (2006)
16. Scala, E.: Reconfiguration and Replanning for robust Execution of Plans Involving Continous and Consumable Resources. Phd thesis in computer science, Department of Computer Science - Universita' di Torino (2013)

17. Nebel, B., Koehler, J.: Plan reuse versus plan generation: A theoretical and empirical analysis. Artificial Intelligence 76(1-2), 427–454 (1995)
18. Gerevini, A., Saetti, A., Serina, I.: Case-based planning for problems with real-valued fluents: Kernel functions for effective plan retrieval. In: Proc. of European Conference on AI (ECAI 2012), pp. 348–353 (2012)
19. van der Krogt, R., de Weerdt, M.: Plan repair as an extension of planning. In: Proc. of International Conference on Automated Planning and Scheduling (ICAPS 2005), pp. 161–170 (2005)
20. Fox, M., Gerevini, A., Long, D., Serina, I.: Plan stability: Replanning versus plan repair. In: Proc. of International Conference on Automated Planning and Scheduling (ICAPS 2006), pp. 212–221 (2006)
21. Gerevini, A.E., Roubíčková, A., Saetti, A., Serina, I.: On the plan-library maintenance problem in a case-based planner. In: Delany, S.J., Ontañón, S. (eds.) ICCBR 2013. LNCS, vol. 7969, pp. 119–133. Springer, Heidelberg (2013)

Underestimation vs. Overestimation in SAT-Based Planning

Mauro Vallati, Lukáš Chrpa, and Andrew Crampton

School of Computing and Engineering,
University of Huddersfield, United Kingdom
{m.vallati,l.chrpa,a.crampton}@hud.ac.uk

Abstract. Planning as satisfiability is one of the main approaches to finding parallel optimal solution plans for classical planning problems. Existing high performance SAT-based planners are able to exploit either forward or backward search strategy; starting from an underestimation or overestimation of the optimal plan length, they keep increasing or decreasing the estimated plan length and, for each fixed length, they either find a solution or prove the unsatisfiability of the corresponding SAT instance.

In this paper we will discuss advantages and disadvantages of the underestimating and overestimating techniques, and we will propose an effective online decision system for selecting the most appropriate technique for solving a given planning problem. Finally, we will experimentally show that the exploitation of such a decision system improves the performance of the well known SAT-based planner SatPlan.

1 Introduction

Planning as satisfiability is one of the main approaches to solving the classical planning problem in AI. Planning as satisfiability is commonly used for finding parallel optimal solution plans, and it gives the advantage of reusing the large corpus of efficient SAT algorithms and makes good use of the extensive advancement in SAT solvers. Approaches using this technique compile a classical planning problem into a sequence of SAT instances, with increasing plan length [10–12, 8, 9, 16]. The forward level expansion search keeps increasing the estimated plan length and for each fixed length finds a solution or proves the unsatisfiability of the SAT instance. While exploiting this approach, most of the search time is spent proving unsatisfiability of generated SAT instances, which could be expensive because the entire search space may need to be explored.

A different approach to SAT-based planning is exploited in MaxPlan [21]. This planner estimates an upper bound of the optimal plan length, by using the suboptimal domain-independent planner FF [7], and then exploits a backward search strategy until the first unsatisfiable SAT instance is reached.

Since both the approaches (the forward and the backward search) have advantages and disadvantages, which will be discussed in the paper, we designed a technique for combining them by selecting online the technique to exploit. The resulting system is called SP_{UO} (SatPlan Under-Overestimating).

M. Baldoni et al. (Eds.): AI*IA 2013, LNAI 8249, pp. 276–287, 2013.
© Springer International Publishing Switzerland 2013

We focus our study on SatPlan [10–12], one of the most popular and efficient SAT-based optimal planning systems. Our experimental analysis will show that SP_{UO} is able to efficiently select the most promising strategy, and that it is able to improve the performance of SatPlan.

The rest of the paper is organized as follows. We first provide some background and further information on SatPlan and SAT-based planning. Next, we describe the forward and backward approaches and we present in detail our experimental analysis and results, followed by concluding remarks and a discussion of some avenues for future work.

2   Background

A *classical planning problem* can be described as a tuple $\langle F, G, I, A \rangle$, where F is a set of facts (or state variables) that represent the state of the world, G is a set of facts representing the goal, I is a set of facts describing the initial state and A is a set of actions, that represents the different possibilities of changing the current state, described by their preconditions and effects.

A valid solution plan is a sequence $\langle A_1, \ldots, A_n \rangle$ of sets of actions, where the i-th set A_i represents the actions that can be executed concurrently at the i-th time step of the plan. An optimal (parallel) plan solving a classical planning problem is formed by the shortest sequence of action sets, and hence it can be executed using the minimum number n of time steps.

A well known and efficient optimal parallel planner for classical planning is SatPlan, which essentially computes the optimal plan by encoding the planning problem into a set of SAT problems that are solved by a generic SAT solver.

In particular, SatPlan first estimates an initial bound k on the length of the optimal plan by constructing the planning graph of the problem, as defined in GraphPlan [2]. A planning graph is a directed acyclic levelled graph representing, for each time step until the estimated horizon, the sets of facts that can be achieved, the set of actions that can be planned/executed, including special actions called "no-ops" used to propagate the facts forward in time across the time steps, and sets of mutually exclusive relations (called mutex relations) between facts and actions. Then, SatPlan encodes the planning problem with horizon equal to k into a propositional formula in Conjunctive Normal Form (CNF), and solves the corresponding SAT problem. Finally, SatPlan uses an incorporated SAT solver to solve this SAT problem. If the CNF is satisfiable, the solution to the SAT problem is translated into a plan, otherwise, SatPlan generates another CNF using an increased value for the planning horizon, and so on, until it generates the first satisfiable CNF. The unsolvability of the planning problem can be proved during the construction of the planning graph.

SatPlan has a modular architecture; it can work with different alternative SAT solvers and methods for encoding the planning problem into a CNF. In this work we consider, respectively, PrecoSAT [1] and the SAT-MAX-PLAN encoding [18]. PrecoSAT was the winner of the Application Track of the 2009 SAT Competition[1] and has been successfully exploited in several SAT-based approaches for planning, e.g., [18, 9, 19].

---

[1] http://www.satcompetition.org/

It uses a version of the DPLL algorithm [4] optimized by ignoring the "less important" generated clauses and by using on-the-fly self-subsuming resolution to reduce the number of newly generated clauses. The SAT-MAX-PLAN encoding is a very compact encoding that is directly derived from the planning graph structure. The variables used in the encoding represent time-stamped facts, actions and no-ops; the clauses are: unit clauses representing the initial and final state, clauses representing the link between actions to their preconditions and effects, and clauses expressing a compact representation of the mutex relations between actions and between facts in the planning graph. This encoding has been designed for both reducing the size of the SAT instance, by removing redundant clauses, and promoting unit propagation during the solving process.

3 Forward and Backward Approaches

As described in the previous section, the SatPlan planning system follows an incremental scheme. It starts to solve a planning problem by estimating a lower bound k of the length of the optimal plan, and moves forward until it finds the first satisfiable CNF. In this way SatPlan is able to find the optimal solution or to fail. This forward search approach has three main advantages: (i) it does not require any knowledge on the length of the optimal solution of the planning problem, (ii) the demonstration of optimality of the solution found is given by the planning process; a valid solution shorter than the one found does not exist, and (iii) it avoids the generation of a big CNF corresponding to a longer, w.r.t. the number of time steps encoded, planning graph. The last consideration is especially true while using old encoding strategies (e.g. [11]). On the other hand, new, compact encodings based on the planning graph structure have been proposed (e.g., SAT-MAX-PLAN encoding or the one proposed in [16]), so the size of the CNF is no longer a very constraining element of the SAT-based planning approach. Moreover, there is another fact that should be considered:

– It has been proven that on random SAT instances, unsatisfiable instances tend to be harder, w.r.t. the CPU-time needed, than satisfiable ones [14]. One reason for this behavior is that on satisfiable instances the solver can stop as soon as it encounters a satisfying assignment, whereas for unsatisfiable instances it must prove that no satisfying assignment exists anywhere in the search tree. No studies about the hardness of instances, w.r.t. satisfiability, have been done on structured instances, where propagation and clause learning techniques may have a lot more leverage.

Given this consideration, it seems reasonable that instead of starting from a lower bound of the optimal plan length, it could also be useful to overestimate the optimal plan length and start from a satisfiable SAT instance. This approach, that was first proposed in MaxPlan, has a number of distinct characteristics that make it very interesting. (i) The planning system, even if it does not find the optimal solution, can output suboptimal good quality solution(s), (ii) it is possible to avoid many hard, unsatisfiable instances, that become harder and harder as the system gets closer to the phase transition on solvability, (iii) it is possible to find a preliminary suboptimal solution plan and apply optimisation techniques, as done in MaxPlan, and (iv) it is possible to *jump*; while overestimating it could happen that the SAT-solver finds a satisfying variable assignment that corresponds to a valid solution plan that is shorter than the actual considered plan length. This means that some time steps are composed only by no-ops or unnecessary (w.r.t. the goal facts) actions. In this case it is possible to remove these actions, generate a shorter plan and jump closer to the optimal length.

In order to understand the actual usefulness of the described approaches, we have modified SatPlan. It is now able to exploit both the underestimation and the overestimation approaches. We decided to use SatPlan for two main reasons: it is well known and it has a modular architecture. The modular architecture allows us to suppose that the achieved results could be easily replicated with different combinations of SAT-solver and encodings. Moreover, since we are comparing search strategies and not planners, using the same planner should lead to the fairest possible comparison.

3.1 Related Works

The forward strategy was used in the original GraphPlan algorithm [2], and was then exploited also in SatPlan for finding makespan optimal plans.

The backward strategy, that still preserve the optimality of the solution found, was firstly presented in MaxPlan.

Alternatives to these strategies were investigated in [17]. In this work two algorithms were proposed: Algorithm A runs the first n SAT problems, generated with increasing time horizons, in parallel, each at equal strength. n is a parameter that can be tuned. Algorithm B runs all the SAT problems simultaneously, with the ith problem receiving a fraction of the CPU time proportional to γ^i, where $\gamma \in (0, 1)$ is a parameter.

In [20], Streeter and Smith proposed an approach that exploits binary search for optimising the resolution process. The introduced approaches, Algorithms A and B, and the binary search, were shown to yield great performance improvements, when compared to the original forward strategy.

The aim of these works is to avoid hard SAT problems, i.e. CNFs that require a huge amount of CPU time for deciding about their solvability, for finding a satisficing solution as soon as possible. The solution found by their approaches are not proven to be optimal.

In this paper we are investigating the possibility of improving the performance of SAT-based planners from a different perspective. We are interested in preserving the demonstration of the optimality of solutions found, and we are studying the predictability of the slope of hardness of SAT problems from upper and lower bounds, w.r.t. the first satisfiable CNF of the planning problem.

4 Decision System

We have found that neither forward nor backward approach is always useful in speeding up the resolution process of SatPlan. Intuitively, the forward search is faster in very easy problems, where the CPU time needed for generating bigger CNFs becomes the bottleneck, or on problems that allow a lot of actions, where each new step implies a significant increase in the number of clauses and variables of the SAT instance. The backward search is faster, intuitively, on problems that have many different valid solutions since satisfiable instances tend to be easy to solve.

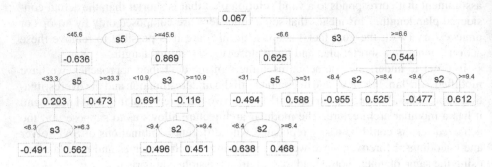

Fig. 1. The domain-independent online decision system represented as an alternating decision tree. Decision nodes are represented as ellipses; sY indicates that the decision is taken on the ratio of the set of clauses Y. Prediction nodes are represented as rectangles. An instance which obtains a positive score is classified as being suitable for the underestimation approach, otherwise the overestimation one is exploited.

For deciding when it is useful to exploit the overestimating approach, and when it is better to use the underestimating one, we designed a decision system, that is shown in Figure 1. Firstly, we tried to extract information on the hardness of structured SAT instances of a given planning problem by evaluating the ratio of clauses to variables, following [13]. However, this ratio is not suitable for taking decisions because its value does not change much between different planning problems (especially from the same domain) and it does not seem to be in a direct relation with the hardness of satisfiable SAT instances generated from a planning problem. Instead, we divided the clauses of the SAT instances into 7 sets, according to information derived from the planning problem that they are encoding: (1) initial and final state, (2) linking actions to their preconditions, (3 & 4) linking actions to their positive and negative effects, (5 & 6) facts are true (false) at time step T if they have been added (removed) before, and (7) mutually exclusive actions and facts. We considered the ratios between the total number of clauses and each set of clauses of the SAT instance. This was done for avoiding too small numbers, that are potentially hard for machine learning techniques and, moreover, are hard to understand for humans. We considered these ratios because we believe they can give some insight about the hardness of structured SAT instances.

First of all, we did not consider the ratios of set 1. The number of these clauses is very small (usually around 20–30 in the considered problems) w.r.t. the size of the CNF (hundreds of thousands), and it does not change significantly across the considered training problems. On the other hand, we noticed that the value of some of the considered ratios remains almost the same throughout the SAT instances generated from the same planning problem, namely ratios of sets 2, 3, 5 and 7. This means that they are related to the planning problems, and moreover this fact allows us to evaluate only a very small SAT instance generated from the given planning problem for taking a decision.

In Table 1 the minimum and maximum ratios of the considered clausal sets per domain are shown. The set 7 seems to be highly related to the structure of the domain, since there is no intersection between minimum and maximum values across the domains.

Table 1. Min – Max values of each considered set of clauses on training problems of the selected domains

Sets	Domains			
	Blocksworld	**Ferry**	**Matching-Bw**	**Satellite**
2	6.4–7.1	6.6–8.1	4.5–5.0	8.7–10.5
3	5.3–7.1	9.0–11.2	5.2–5.8	10.3–13.2
5	21.4–36.2	37.8–48.6	37.2–49.6	49.2–59.3
7	1.8–2.5	1.5–1.7	2.4–2.9	1.4–1.5

Intuitively, one could agree that the information related to the mutexes is related to the domain. From this point of view, this set of clauses is not informative for an instance-specific predictive model.

For automatically finding correlations between the ratios and the effectiveness of the overestimating approach, we exploited an existing machine learning technique: the alternating decision tree (ADTree) [15], which is included in the well known machine learning tool Weka [6], due to its good performance on the training set. The resulting decision system is an alternating decision tree that is composed of decision nodes and prediction nodes. Decision nodes specify a condition on an attribute of the instance that is classified. Prediction nodes contain a single number. A new instance is classified by following all paths for which all decision nodes are true and by summing any prediction nodes that are traversed.

As training examples for the decision system we selected ten planning problems from BlocksWorld, Matching-bw, Sokoban and Ferry domains. For each of the forty problems we generated a SAT instance and extracted the ratios. Each planning problem has been solved using both forward and backward approaches, and classified w.r.t. their usefulness. We selected these domains because in BlocksWorld and Ferry the overestimating approach is very useful, and in the remaining two domains the underestimating approach is often the best choice. It should be noted that the training set is small. This is because we believe that it is easy, especially for an automatic machine learning tool, to extract useful information for classifying the planning problems. Moreover, a small training data set requires a small amount of CPU time for being generated and is easy to evolve in case of future improvements of the decision system.

The generated tree is shown in Figure 1; it is domain-independent and can be used to decide online which approach is better to exploit for the given planning problem. A new instance is classified by the sum of the prediction nodes that are traversed; positive values lead to exploiting the underestimating approach, negative values lead to the overestimating approach. For being used online, the CPU-time needed for taking this decision is critical; given the fact that the ratios are approximately the same on all SAT instances generated from the same planning problem, it is enough to generate the smallest CNF. The decision process usually takes between a tenth and a hundredth of a second.

By analysing the tree, it is possible to have some insight about which sets of clauses are relevant for taking the decision. Intuitively, the sets of clauses that are not changing their value too much through a single planning problem should be considered by the predictive model. In fact, it is easy to note that clauses encoding the connection between actions and facts are relevant (sets 2, 3 and 5), while the sets of clauses encoding the initial and goal states and, surprisingly, clauses encoding the mutually exclusive actions and facts are not significant for the purposes of this classification. This seems counter-intuitive; we are, essentially, trying to predict the hardness of satisfiable instances generated from a planning problem. Intuitively we would imagine some relations between hardness and mutually exclusive actions, but in fact the machine learning technique used, and some others tested, always ignore the ratio of this set of clauses since it does not change much throughout the considered training examples. From a closer look at the ADTree, we can also derive that, most of the times, the bigger is the number of clauses from the considered sets (which corresponds to smaller ratio), the most likely is the decision to overestimate. This lets us suppose that in problems where actions have many effects and require many preconditions, the overestimate approach is useful for achieving better performances.

5 Experimental Evaluation

All experimental tests were conducted using an Intel Xeon(tm) 3 GHz machine, with 2 Gbytes of RAM. Unless otherwise specified, the CPU-time limit for each run was 30 minutes (1800 seconds), after which termination was forced.

Overall, the experiments consider eight well-known planning domains: BlocksWorld, Depots, Ferry, Gripper, Gold-miner, Matching-bw, Sokoban and Satellite. These domains were selected because random instance generators are available for them, and because SatPlan is able to solve not only trivial instances, i.e., that have makespan shorter than 10 timesteps; very short plans are not significant for the scope of our study. Four of them, namely BlocksWorld, Ferry, Matching-bw and Satellite, have been considered for the training of the predictive model. We included them because we are interested in evaluating the performance of the generated model on both training domains and, obviously, new ones. For each selected domain we generated thirty benchmark problems using the available generator. These problems are different from those used for training the predictive model.

The performance of each approach was evaluated using the *speed score*, the performance score function adopted in IPC-7 [3], which is a widely used evaluation criterion in the planning community.

The speed score of a planning system s is defined as the sum of the speed scores assigned to s over all the considered problems. The speed score of s for a planning problem P is defined as:

$$
Score(s, P) = \begin{cases} 0 & \text{if } P \text{ is unsolved} \\ \dfrac{1}{1+\log_{10}(\frac{T_P(s)}{T_P^*})} & \text{otherwise} \end{cases}
$$

where T_P^* is the lowest measured CPU time to solve problem P and $T_P(s)$ denotes the CPU time required by s to solve problem P. Higher values of the speed score indicate better performance. All the CPU times under 0.5 seconds are considered as equal to 0.5 seconds.

Table 2. Number of problems considered per domain (column 2), mean plan length (column 3), IPC score (columns 4-6), mean CPU time (columns 7-9) and percentages of solved problems (columns 10-12) achieved by respectively SatPlan, SP$_O$ and SP$_{UO}$, starting from -10/+10 time steps w.r.t. the optimal plan length

Domain	#	PL	IPC Score			Mean CPU Time			% solved		
			SP$_U$	SP$_O$	SP$_{UO}$	SP$_U$	SP$_O$	SP$_{UO}$	SP$_U$	SP$_O$	SP$_{UO}$
BlocksWorld	30	32.4	25.2	28.5	**28.6**	110.2	**81.5**	81.8	100.0	100.0	100.0
Ferry	30	29.2	11.1	**17.0**	**17.0**	309.4	**144.5**	**144.5**	46.7	**56.7**	**56.7**
Matching-bw	30	29.3	**22.0**	14.4	**22.0**	**261.1**	636.2	**261.1**	**73.3**	70.0	**73.3**
Satellite	30	23.7	**21.2**	19.0	**21.2**	308.0	**272.3**	305.3	73.3	73.3	73.3
Depots	30	16.3	**28.0**	12.7	27.0	**152.1**	724.1	197.9	**93.3**	70.0	**93.3**
Gripper	30	11.0	7.3	**9.9**	**9.9**	28.9	**12.5**	**12.5**	33.3	33.3	33.3
Gold-miner	30	55.2	27.4	**28.3**	27.7	191.3	207.2	**193.1**	100.0	100.0	100.0
Sokoban	30	38.7	28.2	27.6	**28.4**	64.8	64.2	**62.3**	100.0	100.0	100.0
All	240	24.9	170.4	157.4	**181.8**	175.3	267.9	**162.6**	77.5	75.4	**78.8**

In the rest of this experimental analysis, SP$_O$ stands for SatPlan exploiting the overestimation approach and SP$_U$ stands for SatPlan exploiting the underestimation approach. The complete proposed approach is called SP$_{UO}$; it is composed by the modified version of SatPlan, which is able to exploit both over/under-estimation, and the decision system for selecting online whether to overestimate or underestimate. The mean CPU times are always evaluated only on instances solved by all the compared systems.

In Table 2 are shown the results, in terms of IPC score, mean CPU times and percentages of solved problems of a comparison between the original SatPlan underestimating approach, SP$_O$ and SP$_{UO}$ on benchmark instances of selected domains. The overestimating approach starts solving from 10 time steps over the optimal plan length, while the original SatPlan underestimating approach is starting from 10 time steps lower. If the optimal plan is shorter than 10 time steps, the system is starting from the lower bound estimated by SatPlan. We are comparing the CPU times needed for demonstrating the optimality of the solution, i.e. if the optimal plan length is k, it must demonstrate the unsatisfiability of the planning problem using $k - 1$ time steps.

The upper half of Table 2 shows the results on domains that were considered for training the predictive model exploited by SP$_{UO}$, while in the lower half shows results on new domains. As expected, SP$_{UO}$ always achieved the best IPC score in the upper half. But it is worth noting that also on new domains, it is usually able to achieve very good results. It is the best approach on two domains and it is very close to best in the remaining domains.

It is noticeable that in Ferry, the overestimating approach is always the best choice, since the IPC score has exactly the same value of the number of solved problems. In Gripper, it is almost the best approach, since on only one problem the underestimating approach has achieved better results. On the other hand, in planning problems from the Matching-bw and Depots domains underestimating is always the best choice. On the remaining domains, combining the two techniques is the best choice; in all of them there is a significant percentage of planning problems in which either the under or overestimating approach is faster.

Table 3. For each considered domain, the percentages of solved instances in which it is better to apply the underestimating (column 2) or overestimating (column 3) approach, and the percentages of instances in which SP_{UO} exploited the right (fastest) approach (column 4)

Domain	Under	Over	SP_{UO}
BlocksWorld	33.3	66.7	76.7
Ferry	0.0	100.0	100.0
Matching-bw	100.0	0.0	100.0
Satellite	80.0	20.0	72.3
Depots	100.0	0.0	90.0
Gripper	10.0	90.0	90.0
Gold-miner	50.0	50.0	53.3
Sokoban	60.0	40.0	70.0

Although in the presented results we considered only an under/over-estimation of 10 time steps, we experimentally observed that when the best approach is overestimating the optimal plan length, an overestimation of 20 time steps allows the system to achieve better results than an underestimation by only 10 time steps. On the other hand, we also experimentally observed that in the case in which it is better to underestimate, even an underestimation of 20 time steps allows the planning system to achieve better results than through even a small overestimation. Intuitively, we believe that this behavior derives from the fact that the hardness (or the gradient of the hardness) of SAT instances on these problems is very different between satisfiable and unsatisfiable ones.

Looking at the column with mean plan length (PL) over considered problems of each selected domain in Table 2, we can derive that the best strategy to exploit is not really related to the number of steps of the optimal plan. The overestimating approach achieved good results on both problems with short optimal plans (Gripper) and long optimal plans (Gold-miner).

For a better understanding of the relation between the two approaches on each of the selected domains, in Table 3 are shown the percentages of problems in which it is better to apply the underestimating or overestimating approach, and the percentages of instances in which SP_{UO} selected the right approach. The results shown in this table confirm the ones presented in Table 2; in the Matching-bw and Depots domains, underestimating is always the best choice, and in Ferry, overestimating is always better. But it is interesting to note that also in the other domains there is one technique which is better, at least in terms of the percentage of problems in which it is faster; this is the case for Sokoban and Satellite domains, in which underestimating is better, and BlocksWorld and Gripper, in which the overestimating approach allows us to achieve better results. In Gold-miner both the approaches are useful for speeding up the planning process on exactly 50% of the problems. Finally, SP_{UO} is generally able to select the right (fastest) approach for solving a given problem. Only in Gold-miner it shows a precision below 70%. This is probably due to the fact that Gold-miner problems have usually a very similar structure, and that both SP_U and SP_O approaches lead to performances that are not significantly different.

In order to evaluate the accuracy of the decision system developed, we compared SP_{UO} with an Oracle specifying the best approach to exploit for every benchmark

Table 4. For each considered domain, the IPC score (columns 2-4) and the mean CPU time (columns 5-7) of the proposed SP$_{UO}$, an Oracle selecting the best approach on benchmark problems and a random selection

Domain	IPC Score			Mean CPU Time		
	SP$_{UO}$	Oracle	Random	SP$_{UO}$	Oracle	Random
BlocksWorld	28.6	**30.0**	26.5	81.8	**79.7**	110.2
Ferry	17.0	17.0	12.8	144.5	144.5	219.6
Matching-bw	22.0	22.0	14.6	261.1	261.1	512.7
Satellite	21.2	**22.0**	19.2	305.4	**234.1**	314.8
Depots	27.0	**28.0**	19.3	197.9	**152.1**	301.4
Gripper	9.9	**10.0**	8.1	12.5	12.5	28.3
Gold-miner	27.7	**30.0**	27.6	193.1	**166.9**	210.6
Sokoban	28.4	**30.0**	27.6	62.3	**59.2**	66.4
All	181.8	**189.0**	155.7	161.4	**142.0**	220.7

problem and a random approach, that randomly selects the approach to use for the given problem. The results of this comparison are shown in Table 4. The decision system is very accurate on domains used for training the predictive model; on two domains it achieves the same IPC score of an Oracle, and on the remaining it is always very close to it. Also on new domains the decision system is able to achieve results that are very close to the Oracle ones. These are very interesting results, considering that the decision system is instance-based, is very fast, and has been trained on a small training set. The random selection approach usually achieved a low IPC Score; only in domains in which forward and backward approaches have similar behavior on considered problems e.g., Blocksworld, Gold-miner and Sokoban, its performance is close to the SP$_{UO}$ ones.

In terms of mean CPU times SP$_{UO}$ is almost always very close to the Oracle, while the random selected approach usually required a considerable amount of CPU time for solving testing instances. On Satellite the very high mean CPU time of SP$_{UO}$ is mainly due to only two instances (out of 22 solved) in which the selected approach is very slow.

6   Conclusions and Future Work

In this paper we investigated the usefulness of combining two well known approaches to planning as satisfiability: overestimating and underestimating. For combining these approaches, we proposed a machine learning based technique for taking instance-based domain-independent online decisions about the best strategy to exploit for solving a new planning problem. The decision system is represented as an alternating decision tree which evaluates the ratio of three sets of clauses w.r.t. the whole set of clauses of a SAT instance generated from the given planning problem. We have experimentally observed that the ratios are fixed throughout all the SAT instances generated from the same problem, so it is sufficient to generate the smallest one, which corresponds to the encoding of the first level of the planning graph in which all the goals are true, for deciding.

An experimental study, which considered 8 different domains and 240 planning problems, indicates that: (i) the best approach to exploit can be different across planning problems of the same domain; (ii) in planning problems in which overestimating is useful, an inaccurate upper bound performs better than a more accurate lower bound; (iii) the decision system is accurate, even if compared with an Oracle, and helps to effectively combine the underestimating and overestimating approaches; (iv) the SP_{UO} system is very efficient and has outperformed SatPlan. We also observed that the overestimating approach usually is not useful on SAT encoding techniques that do not consider no-ops (see, e.g., [8]). Such encodings make it impossible to exploit jumps while overestimating, and moreover, we experimentally observed that it makes satisfiable instances, longer than the optimal one, harder to solve if compared to different encodings.

A limit of the current approach is that we have considered as known the optimal plan length, however, there already exists works about how to generate an accurate prediction of the optimal plan length (e.g. [5]) which could be exploited. Moreover, for overestimating the optimal plan length, the strategy adopted in [21] could be very useful; using a suboptimal planner to find a suboptimal sequential plan and then parallelize it. Finally, another strategy for overestimating the length of the optimal parallel plan, could be based on extracting information about the level of the fixed point of the planning graph of the given problem.

We see several avenues for future work on SP_{UO}. Concerning the optimal plan length prediction, we are evaluating the possibility of merging the proposed decision system with a predictive model of the optimal plan length. In this way, after deciding the approach to use, it can make a "biased" prediction for starting the planning process from a lower or an upper bound. Moreover, we are interested in evaluating the possibility of applying some sort of parallel solver in the proposed backward search; this has been already done in the SAT-based classical forward search in the well known Mp planner [16], with very interesting results. Furthermore, we are planning to study strategies for reusing part of the knowledge learnt by solving a satisfiable SAT instance, for speeding up the overestimating approach. Finally, we are interested in running additional experiments about the impact of the proposed SP_{UO} on SAT-based planning, and in investigating the relation between the frequency and the length of *jumps* and the performance of the backward search.

Acknowledgements. The authors would like to acknowledge the use of the University of Huddersfield Queensgate Grid in carrying out this work.

References

1. Biere, A.: P{re,i}cosat@sc'09. In: SAT Competition 2009 (2009)
2. Blum, A., Furst, M.L.: Fast planning through planning graph analysis. Artificial Intelligence 90, 281–300 (1997)
3. Celorrio, S.J., Coles, A., Coles, A.: Learning track of the 7th International Planning Competition (2011), http://www.plg.inf.uc3m.es/ipc2011-learning
4. Davis, M., Logemann, G., Loveland, D.: A machine program for theorem-proving. Communications of the ACM 5(7), 394–397 (1962)
5. Gerevini, A.E., Saetti, A., Vallati, M.: Exploiting macro-actions and predicting plan length in planning as satisfiability. In: Pirrone, R., Sorbello, F. (eds.) AI*IA 2011. LNCS, vol. 6934, pp. 189–200. Springer, Heidelberg (2011)

6. Hall, M., Holmes, F.G., Pfahringer, B., Reutemann, P., Witten, I.H.: The WEKA data mining software: An update. SIGKDD Explorations 11(1), 10–18 (2009)
7. Hoffmann, J.: FF: The Fast-Forward Planning System. AI Magazine 22(3), 57–62 (2001)
8. Huang, R., Chen, Y., Zhang, W.: A novel transition based encoding scheme for planning as satisfiability. In: Proceedings of the 24th AAAI Conference on Artificial Intelligence (AAAI 2010), pp. 89–94. AAAI (2010)
9. Huang, R., Chen, Y., Zhang, W.: SAS+ planning as satisfiability. Journal of Artificial Intelligence Research 43, 293–328 (2012)
10. Kautz, H., Selman, B.: Planning as satisfiability. In: Proceedings of the 10th European Conference on Artificial Intelligence (ECAI 1992), pp. 359–363. John Wiley and Sons (1992)
11. Kautz, H., Selman, B.: Unifying sat-based and graph-based planning. In: Proceedings of the 16th International Joint Conference on Artificial Intelligence (IJCAI 1999), pp. 318–325. Morgan Kaufmann (1999)
12. Kautz, H., Selman, B., Hoffmann, J.: SatPlan: Planning as satisfiability. In: Abstract Booklet of the 5th International Planning Competition (2006)
13. Mitchell, D., Selman, B., Levesque, H.: Hard and easy distributions of SAT problems. In: Proceedings of the 10th National Conference on Artificial Intelligence (AAAI 1992), pp. 459–465. AAAI (1992)
14. Nudelman, E., Leyton-Brown, K., Hoos, H.H., Devkar, A., Shoham, Y.: Understanding random SAT: Beyond the clauses-to-variables ratio. In: Wallace, M. (ed.) CP 2004. LNCS, vol. 3258, pp. 438–452. Springer, Heidelberg (2004)
15. Pfahringer, B., Holmes, G., Kirkby, R.: Optimizing the induction of alternating decision trees. In: Cheung, D., Williams, G.J., Li, Q. (eds.) PAKDD 2001. LNCS (LNAI), vol. 2035, pp. 477–487. Springer, Heidelberg (2001)
16. Rintanen, J.: Engineering efficient planners with SAT. In: Proceedings of the 20th European Conference on Artificial Intelligence (ECAI 2012), pp. 684–689. IOS Press (2012)
17. Rintanen, J.: Evaluation strategies for planning as satisfiability. In: Proceedings of the 16th European Conference on Artificial Intelligence (ECAI 2004), pp. 682–687. IOS Press (2004)
18. Sideris, A., Dimopoulos, Y.: Constraint propagation in propositional planning. In: Proceedings of the 20th International Conference on Automated Planning and Scheduling (ICAPS 2010), pp. 153–160. AAAI (2010)
19. Sideris, A., Dimopoulos, Y.: Propositional planning as optimization. In: Proceedings of the 20th European Conference on Artificial Intelligence (ECAI 2012), pp. 732–737. IOS Press (2012)
20. Streeter, M.J., Smith, S.F.: Using decision procedures efficiently for optimization. In: Proceedings of the 17th International Conference on Automated Planning and Scheduling (ICAPS 2007), pp. 312–319. AAAI (2007)
21. Xing, Z., Chen, Y., Zhang, W.: MaxPlan: Optimal planning by decomposed satisfiability and backward reduction. In: Proceedings of the 5th International Planning Competition, IPC-5 (2006)

Social Interactions in Crowds of Pedestrians: An Adaptive Model for Group Cohesion

Stefania Bandini, Luca Crociani, and Giuseppe Vizzari

CSAI - Complex Systems & Artificial Intelligence Research Center,
University of Milano-Bicocca,
Viale Sarca 336, 20126, Milano, Italy
{bandini,luca.crociani,vizzari}@disco.unimib.it

Abstract. The paper introduces an agent-based model for the simulation of crowds of pedestrians whose main innovative aspect is the representation and management of an important type of social interaction
among the pedestrians: members of groups, in fact, carry out of a form
of interaction (by means of verbal or non-verbal communication) that
allows them to preserve the cohesion of the group even in particular conditions, such as counter flows, presence of obstacles or narrow passages.
The paper formally describes the model and presents both qualitative
and quantitative results in sample simulation scenarios.

Keywords: Agent-Based Models, Complex Systems, Crowds, Groups.

1 Introduction

By means of "computer simulation" we generally denote the use of computational
model to gain additional insight on a complex system's working (e.g. biological
or social systems) by envisioning the implications of modelling choices, or the
evaluation of alternative designs and plans for system operation, without actually
bringing them into existence in the real world (e.g. architectural designs, road
networks and traffic lights). The adoption of these *synthetic environments* can
be necessary because the simulated systems cannot be observed (since they are
actually being designed and therefore they still do not exist), or also for ethical
(e.g. the safety of humans would be involved) or practical reasons (e.g. costs of
experiments or data acquisition, systems characterised by a very slow evolution).

Several situations and systems can be considered as being characterised by
the presence of a number of autonomous entities whose behaviours (actions and
interactions) determine (in a non–trivial way) the evolution of the overall system. A crowd of pedestrians is a paradigmatic example of this kind of situation:
pedestrians share the same environment and they generally compete for the
space resource. Nonetheless, they can also exhibit collaborative patterns of interactions, for instance conforming to (written or non-written) shared rules like
giving way to passengers getting off of a train before getting on board, or even
respecting cultural dependant rules (e.g. gallantry). There are evidences of im-
itation among pedestrians whenever they need to cross a road (see, e.g., [8]),

M. Baldoni et al. (Eds.): AI*IA 2013, LNAI 8249, pp. 288–299, 2013.

but basic proxemic considerations indicate that normal behaviour includes a tendency to stay at a distance, to preserve a personal space [6]. Pedestrians continuously adapt their behaviour to the contextual conditions, considering the geometry of the environment but also social aspects. Agent-based models are particularly suited to tackle these situations and they support the study and analysis of topics like decentralised decision making, local-global interactions, self-organisation, emergence, effects of heterogeneity in the simulated system.

Computer models for the simulation of crowds are growingly investigated in the academic context and these efforts, in addition to improving the understanding of the phenomenon, also led to the implementation of commercial simulators employed by firms and decision makers[1]. Pedestrian models can be roughly classified into three main categories that respectively consider pedestrians as *particles subject to forces*, particular *states of cells* in which the environment is subdivided in Cellular Automata (CA) approaches, or *autonomous agents* acting and interacting in an environment. The most widely adopted particle based approach is represented by the *social force model* [7], which implicitly employs fundamental proxemic concepts like the tendency of a pedestrian to stay away from other ones while moving towards his/her goal. *Cellular Automata* based approaches have also been successfully applied in this context, and in particular the floor-field approach (based on gradient-like annotation of cells) is often adopted for representing any kind of environment and pedestrian movement tendency [16]. Finally, works like [9] essentially extend CA approaches, separating the pedestrians from the environment and granting them a behavioural specification that is generally more complex than a simple CA transition rule.

Despite the significant results achieved this area is still lively and we are far from a complete understanding of the complex phenomena related to crowds of pedestrians: one of the least studied and understood aspects of crowds of pedestrians is represented by the implications of the presence of groups [4]. Recently, extensions of the social force model have been defined and employed to simulate small groups in relatively low density situations [13,19]; also CA [14] and agent based [1] approaches can be found in the literature, but results are still not validated against literature and empirical data. Finally, all the above models do not analyse the capability of preserving group cohesion even in relatively high-density situations. The main goal of this paper is to present an agent–based model extending the floor-field approach to endow pedestrians the ability to adapt their behaviour, should they belong to a group, in order to preserve the cohesion of the group even when facing conditions that would hinder their possibility to stay close to other group members, like counter flows, narrow passages, presence of obstacles. The paper will first of all formally introduce the model then some qualitative and quantitative results will be described. Conclusions and future developments will end the paper.

---

[1] See http://www.evacmod.net/?q=node/5 for a significant although not necessarily comprehensive list of simulation platforms.

2   Agent-Based Computational Model

In this section the formalisation of the agent-based computational model will be discussed, by focusing on the definition of its three main elements: *environment*, *update mechanism* and *pedestrian behaviour*.

2.1   Environment

The environment is modelled in a discrete way by representing it as a grid of squared cells with $40 \ cm^2$ size (according to the average area occupied by a pedestrian [18]). Cells have a state indicating the fact that they are vacant or occupied by obstacles or pedestrians:

$$State(c) : Cells \rightarrow \{Free, \ Obstacle, \ OnePed_i, TwoPeds_{ij}\}$$

The last two elements of the definition point out if the cell is occupied by one or two pedestrians respectively, with their own identifier: the second case is allowed only in a controlled way to simulate overcrowded situations, in which the density is higher than $6.25 \ m^2$ (i.e. the maximum density reachable by our discetisation).

The information related to the scenario[2] of the simulation are represented by means of *spatial markers*, special sets of cells that describe relevant elements in the environment. In particular, three kinds of spatial markers are defined:

- *start* areas, that indicate the generation points of agents in the scenario. Agent generation can occur in *block*, all at once, or according to a user defined *frequency*, along with information on type of agent to be generated and its destination and group membership;
- *destination* areas, which define the possible targets of the pedestrians in the environment;
- *obstacles*, that identify all the non-walkable areas as walls and zones where pedestrians can not enter.

Space annotation allows the definition of virtual grids of the environment, as containers of information for agents and their movement. In our model, we adopt the *floor field* approach [2], that is based on the generation of a set of superimposed grids (similar to the grid of the environment) starting from the information derived from spatial markers. Floor field values are spread on the grid as a gradient and they are used to support pedestrians in the navigation of the environment, representing their interactions with static object (i.e., destination areas and obstacles) or with other pedestrians. Moreover, floor fields can be *static* (created at the beginning and not changed during the simulation) or *dynamic* (updated during the simulation). Three kinds of floor fields are defined in our model:

---

[2] It represents both the structure of the environment and all the information required for the realization of a specific simulation, such as crowd management demands (pedestrians generation profile, origin-destination matrices) and spatial constraints.

- *path field*, that indicates for every cell the distance from one destination area, acting as a potential field that drives pedestrians towards it (static). One path field for each destination point is generated in each scenario;
- *obstacles field*, that indicates for every cell the distance from neighbour obstacles or walls (static). Only one obstacles field is generated in each simulation scenario;
- *density field*, that indicates for each cell the pedestrian density in the surroundings at the current time-step (dynamic). Like the previous one, the density field is unique for each scenario.

Chessboard metric with $\sqrt{2}$ variation over corners [12] is used to produce the spreading of the information in the path and obstacle fields. Moreover, pedestrians cause a modification to the density field by adding a value $v = \frac{1}{d^2}$ to cells whose distance d from their current position is below a given threshold. Agents are able to perceive floor fields values in their neighbourhood by means of a function $Val(f, c)$ (f represents the field type and c is the perceived cell). This approach to the definition of the objective part of the perception model moves the burden of its management from agents to the environment, which would need to monitor agents anyway in order to produce some of the simulation results.

2.2 Time and Update Mechanism

Time is also modeled as discrete, employing steps of 1/3 of second. This choice, along with the adoption of side of 40 cm per cell, generates a linear pedestrian speed of about $1.2\ ms^{-1}$, which is in line with the data from the literature representing observations of crowd in normal conditions [18].

Regarding the update mechanism, three different strategies are usually applied in this context [11]: *ordered sequential*, *shuffled sequential* and *parallel* update. The first two strategies are based on a sequential update of agents, respectively managed according to a *static* list of priorities that reflects their order of generation or a *dynamic* one, shuffled at each time step. On the contrary, the parallel update calculates the choice of movement of all the pedestrians at the same time, actuating choices and managing conflicts in a latter stage. The two sequential strategies, instead, imply a simpler operational management, due to an a-priori resolution of conflicts between pedestrians.

In our model we chose to adopt the shuffled sequential strategy, updating the list of agents using a dynamic list of priority that is randomly generated every step. This choice is not totally in line with the current literature, in which the parallel update is considered more realistic and is generally preferred[3], but since we are mainly exploring the potential implications of the presence of groups in the simulated population we accept this simplification to focus on the social interaction mechanisms of the model.

---

[3] In [10] it is suggested that conflicts represent a significant variable of the pedestrian dynamic.

Fig. 1. Examples of simple (a) and structured (b) groups

2.3 Pedestrians and Movement

Formally, our agents are defined by the following triple:

$$Ped : \langle Id, \; Group, \; State \rangle; \quad State : \langle position, \; oldDir, \; Dest \rangle$$

with their own numerical identifier, their group (if any) and their internal state, that defines the current position of the agent, the previous movement and the final destination, associated to the relative path field.

Before describing agent behavioural specification, it is necessary to introduce the formal representation of the nature and structure of the groups they can belong to, since this is an influential factor for movement decisions.

2.3.1 Social Interactions

To represent different types of relationships, two kinds of groups have been defined in the model: a *simple group* indicates a family or a restricted set of friends, or any other small assembly of persons in which there is a strong and simply recognisable cohesion; a *structured group* is generally a large one (e.g. team supporters or tourists in an organised tour), that shows a slight cohesion and a natural fragmentation into subgroups, sometimes simple: Fig. 1 represents an example of simple and structured groups (please consider that all persons in the same waiting box in the right are members of a largely fragmented structured group including smaller simple ones).

Members of a simple group it is possible to identify an apparent tendency to stay close, in order to guarantee the possibility to perform interactions by means of verbal or non-verbal communication [5]. On the contrary, in large groups people are mostly linked by the sharing of a common goal, and the overall group tends to maintain only a weak compactness, with a following behaviour between members. In order to model these two typologies, the formal representation of a group is described by the following:

$$Group : \langle Id, [SubGroup_1, \dots, SubGroup_m], [Ped_1, \dots, Ped_n] \rangle$$

In particular, if the group is simple, it will have an empty set of subgroups, otherwise it will not contain any direct references to pedestrians inside it, which will be stored in the respective leafs of its three structure. Differences on the modelled behavioural mechanism in simple/structured groups will be analysed in the following section, with the description of the utility function.

2.3.2   Agent Behaviour

Agent behaviour in a single simulation turn is organised into four steps: *perception, utility calculation, action choice* and *movement*. The *perception* step provides to the agent all the information needed for choosing its destination cell. In particular, if an agent does not belong to a group (from here called *individual*), in this phase it will only extract values from the floor fields, while in the other case it will perceive also the positions of the other group members within a configurable distance, for the calculation of the *cohesion* parameter. The choice of each action is based on an utility value, that is assigned to every possible movement according to the following function:

$$U(c) = \frac{\kappa_g G(c) + \kappa_{ob} Ob(c) + \kappa_s S(c) + \kappa_c C(c) + \kappa_i I(c) + \kappa_d D(c) + \kappa_{ov} Ov(c)}{d}$$

Function $U(c)$ takes into account the behavioural components considered relevant for pedestrian movement, each one is modelled by means of a function that returns values in range $[-1; +1]$, if it represents an *attractive* element (i.e. its goal), or in range $[-1; 0]$, if it represents a *repulsive* one for the agent. For each function a κ coefficient has been introduced for its calibration: these coefficients, being also able to actually modulate tendencies based on objective information about agent's spatial context, complement the objective part of the perception model allowing agent heterogeneity. The purpose of the function denominator d is to constrain the diagonal movements, in which the agents cover a greater distance $(0.4 * \sqrt{2}$ instead of $0.4)$ and assume higher speed respect with the non-diagonal ones.

The first three functions exploit information derived by local floor fields: $G(c)$ is associated to goal attraction whereas $Ob(c)$ and $S(c)$ respectively to geometric and social repulsion. Functions $C(c)$ and $I(c)$ are linear cobinations of the perceived positions of members of agent group (respectively simple and structured) in an extended neighbourhood; they compute the level of attractiveness of each neighbour cell, relating to group cohesion phenomenon. Finally, $D(c)$ adds a bonus to the utility of the cell next to the agent according to his/her previous direction (a sort of *inertia* factor), while $Ov(c)$ describes the *overlapping* mechanism, a method used to allow two pedestrians to temporarily occupy the same cell at the same step, to manage high-density situations.

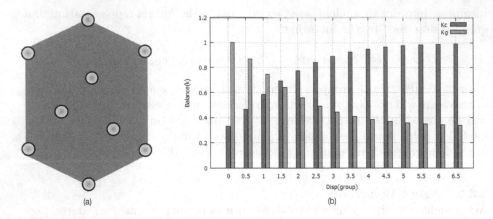

Fig. 2. Graphical representation of a group composed of nine members and the area of the convex polygon that contains all the group members (a) and graph of $Balance(k)$, for $k = 1$ and $\delta = 2.5$ (b)

As we previously said, the main difference between simple and structured groups resides in the cohesion intensity, which in the simple ones is significantly stronger. Functions $C(c)$ and $I(c)$ have been defined to correctly model this difference. Nonetheless, various preliminary tests on benchmark scenarios show us that, used singularly, function $C(c)$ is not able to reproduce realistic simulations. Human behaviour is, in fact, very complex and can react differently even in simple situation, for example by allowing temporary fragmentation of simple groups in front of several constraints (obstacles or opposite flows). Acting statically on the calibration weight, it is not possible to achieve this dynamic behaviour: with a small cohesion parameter several permanent fragmentations have been reproduced, while with an increase of it we obtained no group dispersions, but also an excessive and unrealistic compactness.

In order to face this issue, another function has been introduced in the model, to adaptively balance the calibration weight of the three attractive behavioural elements, depending on the fragmentation level of simple groups:

$$Balance(k) = \begin{cases} \frac{1}{3} \cdot k + (\frac{2}{3} \cdot k \cdot DispBalance) & \text{if } k = k_c \\ \frac{1}{3} \cdot k + (\frac{2}{3} \cdot k \cdot (1 - DispBalance)) & \text{if } k = k_g \vee k = k_i \\ k & \text{otherwise} \end{cases}$$

$$DispBalance = tanh\left(\frac{Disp(Group)}{\delta}\right); \quad Disp(Group) = \frac{Area(Group)}{|Group|}$$

where k_i, k_g and k_c are the weighted parameters of $U(c)$, δ is the calibration parameter of this mechanism and $Area(Group)$ calculates the area of the convex hull defined using positions of the group members. Fig. 2 exemplifies both the group dispersion computation and the effects of the $Balance$ function on parameters. The effective utility computation, therefore, employs calibration weights resulting from this computation, that allows achieving a dynamic and adaptive behaviour of groups: cohesion relaxes if members are sufficiently close to each other and it intensifies with the growth of dispersion.

After the utility evaluation for all the cells in the neighbourhood, the choice of action is stochastic, with the probability to move in each cell c as (N is the normalization factor):

$$P(c) = N \cdot e^{U(c)}$$

On the basis of $P(c)$, agents move in the resulted cell according to their set of possible actions, defined as list of the eight possible movements in the Moore neighbourhood, plus the action to keep the position (indicated as X): $A = \{NW, N, NE, W, X, E, SW, S, SE\}$.

3   Experimental Scenarios and Results

This section explores the effects of the modelled social interactions in pedestrian dynamics, by presenting qualitative and quantitative results in experimental scenarios[4].

3.1   Analysis of Group Dispersion

Tests regarding the simple group cohesion have been performed in two benchmark scenarios, which represent basic situations that can be identified in larger environments: a corridor with a central obstacle (Fig. 3(a) line 1) and a corner with a bidirectional flow (Fig. 3(a) line 2 and 3). The aim is to cause group fragmentation and to verify the capability of the model to preserve cohesion in these situations. As shown in Fig. 3(a), the simple group cohesion mechanism has a strong efficacy: in both the scenarios, different temporary situations of fragmentation can be identified but in all cases the groups managed to reduce the level of dispersion before the exit of the scenario. In particular, while in corridor scenario the obstacle generates lower dispersion, in the corner the counterflow splits the simple group: when the dispersion reaches a high value, a portion of group members waits for the others, to increase cohesion before reaching their goal. Fig. 4(a) illustrates the evolution of group dispersion (calculated with the $Disp(g)$ function described in Sec. 2.3.2) in relationship with the growth of density of pedestrians in the environment: the average group dispersion is not influenced by the increase of pedestrian density. These results confirm the activation of *Balance* function, that concurs in the management of the attractive elements.

In addition to the above results, we have run various simulations in a corridor scenario of size 10.0×4.8 m^2, with a bidirectional pedestrian flow generated by two equally structured groups only including individuals. Fig. 3(b) shows that this mechanism produces a slight attraction between members, which leads to

---

[4] Simulations have been performed with this weights configuration: $\kappa_g = 8.0; \kappa_{ob} = 8.0; \kappa_s = 25.0; \kappa_c = 6.0; \kappa_i = 7.0; \kappa_d = 1.0; \kappa_{ov} = 9.5;$ and $\delta = 2.5$.

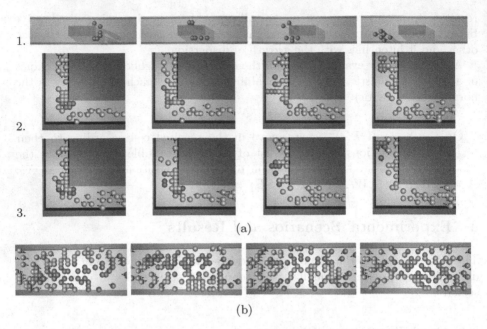

(a)

(b)

Fig. 3. (a) Situations of fragmentations (temporarily ordered from left to right) of a six member simple group in corridor (line 1) and corner (2 and 3). (b) Lane formation with structured groups.

an imitative behaviour and generates a well-known collective phenomena, the so-called *lane formation*[5] (large and stable as well as small and dynamic ones). The capability to reproduce this phenomenon is relevant not only in a qualitative way: formation of lanes makes the overall flow smoother, allowing to preserve velocity at higher density situations.

Despite the fact that cohesion mechanism, both in simple and structured group, reveals its capability and efficacy in the benchmark scenarios, the plausibility of this behaviour cannot be guaranteed due to the lack of empirical data on the phenomenon of groups and group dispersion[6].

3.2   Group Impacts on the Pedestrian Flow

Another relevant result is related to the analysis of the impact of simple groups on the pedestrian flow. In all the previously mentioned scenarios, the presence of groups has generally a negative influence on the flow: Fig. 4(b) illustrates the differences in the tendency of pedestrian flow with and without simple groups (in red and blue, respectively) in case of a 20.0×2.4 m^2 corridor. The diagram

---

[5] It can be detected when groups of people move in opposite directions in a crowded environment and they spontaneously organize themselves into different lanes [15].

[6] This means that the δ parameter in *DispBalance* function cannot be precisely calibrated yet.

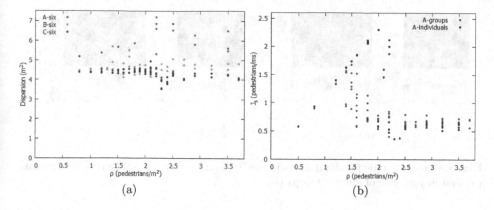

Fig. 4. (a) Dispersion levels of six members groups in corridor scenarios of different width (A = 2.4 m, B = 3.6 m, C = 4.8 m); (b) Flow of pedestrians in corridor A at different densities, with and without simple group.

shows how density influences the overall pedestrian flow and it is in line with empirical data available in the literature [15]: the shape of the graph in the two cases is similar, although in presence of groups the flow starts decreasing at a lower density. This means that estimating the capacity of a corridor without considering the presence of groups might lead to an overestimation at low to moderate densities, and this is in good accordance with [19]. Beyond this *critical density*, however, the presence of groups does not imply any more significant differences in the flow: this is probably due to an increased possibility of lane formation. This result is original but, once again, it calls for additional evidences coming from observations or experiments.

3.3 Space Utilisation

An additional scenario in which the model has been tested is instead associated to a situation in which pedestrians have to perform a turn and two different flows have to merge: geometrically, this environmental structure is a T-Junction in which two branches of a corridor meet and form a unique stream. Also in this kind of situation it is not hard to reach high local densities, especially where the incoming flows meet to turn and merge in the outgoing corridor.

The central result of this scenario is represented by the so-called *cumulative mean density* (CMD) [3], a measure associated to a given position of the environment indicating the average density perceived by pedestrians that passed through that point. We wanted to evaluate the capability of the model to reproduce patterns of spatial utilisation similar to those resulting from the actual observations available in the literature [20]: in these observations the density in the T-junction was not homogeneous, since higher values were observed near the junction, a low density region was observed where branches begin to merge, and generally pedestrians prefered to move along the shorter and smoother path.

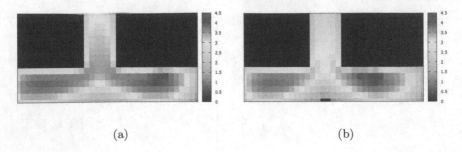

(a)   (b)

Fig. 5. Density profile for simulations in the T-junction scenario without simple groups (a) and in presence of simple groups (b)

Our model is able to reproduce the observed phenomenon: a comparison between the two different density profiles from one run of a simulation with an overall density equal to 2.5 m^{-2} is shown in Fig. 5 (a). The maximum CMD reaches is the same in the two density profile, and equal to 4.5 m^{-2}. Moreover, in Fig. 5 (b), instead, we propose the results of the simulations carried out in the same scenario but introducing some simple groups: while qualitatively the diagram looks similar to the previous one, the highest level of density is actually sensibly lower and in general the CMD is lower in most parts of the environment. This result is original and it requires support by empirical data from observations or experiments for sake of validation.

4   Conclusions and Future Works

The paper has introduced an agent-based model for the modelling and simulation of pedestrians characterised by the possibility to describe and consider the influence of groups in the simulated population. The model encompasses an adaptive mechanism able to preserve the cohesion of a group preserving some desirable overall properties of the model. A more thorough description of the model is available in [17]; results discussed here suggest that although groups reduce the maximum flow (at least in corridor like scenarios) their presence could even have a positive impact in corner-like situations. Future works are aimed, on one hand, at validating the component of the pedestrian behavioural model considering the influence of groups employing empirical data that are being gathered in actual observations. On the other hand, we also plan to extend the model to consider additional social phenomena like, for instance, leader–follower movement schemes.

References

1. Bandini, S., Rubagotti, F., Vizzari, G., Shimura, K.: An agent model of pedestrian and group dynamics: Experiments on group cohesion. In: Pirrone, R., Sorbello, F. (eds.) AI*IA 2011. LNCS, vol. 6934, pp. 104–116. Springer, Heidelberg (2011)

2. Burstedde, C., Klauck, K., Schadschneider, A., Zittartz, J.: Simulation of pedestrian dynamics using a two-dimensional cellular automaton. Physica A: Statistical Mechanics and its Applications 295(3-4), 507–525 (2001)
3. Castle, C., Waterson, N., Pellissier, E., Bail, S.: A comparison of grid-based and continuous space pedestrian modelling software: Analysis of two uk train stations. In: Pedestrian and Evacuation Dynamics, pp. 433–446. Springer US (2011)
4. Challenger, R., Clegg, C.W., Robinson, M.A.: Understanding crowd behaviours: Supporting evidence. Tech. rep., University of Leeds (2009)
5. Costa, M.: Interpersonal distances in group walking. Journal of Nonverbal Behavior 34, 15–26 (2010)
6. Hall, E.T.: The Hidden Dimension. Anchor Books (1966)
7. Helbing, D., Molnár, P.: Social force model for pedestrian dynamics. Phys. Rev. E 51(5), 4282–4286 (1995)
8. Helbing, D., Schweitzer, F., Keltsch, J., Molnar, P.: Active walker model for the formation of human and animal trail systems. Physical Review E 56(3), 2527–2539 (1997)
9. Henein, C.M., White, T.: Agent-based modelling of forces in crowds. In: Davidsson, P., Logan, B., Takadama, K. (eds.) MABS 2004. LNCS (LNAI), vol. 3415, pp. 173–184. Springer, Heidelberg (2005)
10. Kirchner, A., Namazi, A., Nishinari, K., Schadschneider, A.: Role of Conflicts in the Floor Field Cellular Automaton Model for Pedestrian Dynamics (2003)
11. Klüpfel, H.: A Cellular Automaton Model for Crowd Movement and Egress Simulation. Ph.D. thesis, University Duisburg-Essen (2003)
12. Kretz, T., Bönisch, C., Vortisch, P.: Comparison of various methods for the calculation of the distance potential field. In: Pedestrian and Evacuation Dynamics 2008, pp. 335–346. Springer, Heidelberg (2010)
13. Moussaïd, M., Perozo, N., Garnier, S., Helbing, D., Theraulaz, G.: The walking behaviour of pedestrian social groups and its impact on crowd dynamics. PLoS ONE 5(4), e10047 (2010)
14. Sarmady, S., Haron, F., Talib, A.Z.H.: Modeling groups of pedestrians in least effort crowd movements using cellular automata. In: Asia International Conference on Modelling and Simulation, pp. 520–525 (2009)
15. Schadschneider, A., Klingsch, W., Kluepfel, H., Kretz, T., Rogsch, C., Seyfried, A.: Evacuation Dynamics: Empirical Results, Modeling and Applications. ArXiv e-prints (2008)
16. Schadschneider, A., Kirchner, A., Nishinari, K.: Ca approach to collective phenomena in pedestrian dynamics. In: Bandini, S., Chopard, B., Tomassini, M. (eds.) ACRI 2002. LNCS, vol. 2493, pp. 239–248. Springer, Heidelberg (2002)
17. Vizzari, G., Manenti, L., Crociani, L.: Adaptive pedestrian behaviour for the preservation of group cohesion. Complex Adaptive Systems Modeling 1(7) (2013)
18. Weidmann, U.: Transporttechnik der Fussgänger - Transporttechnische Eigenschaftendes Fussgängerverkehrs (literaturauswertung) (1993)
19. Xu, S., Duh, H.B.L.: A simulation of bonding effects and their impacts on pedestrian dynamics. IEEE Transactions on Intelligent Transportation Systems 11(1), 153–161 (2010)
20. Zhang, J., Klingsch, W., Schadschneider, A., Seyfried, A.: Ordering in bidirectional pedestrian flows and its influence on the fundamental diagram. Journal of Statistical Mechanics: Theory and Experiment 2012(02), P02002 (2012)

On the Expressiveness of Attribute Global Types: The Formalization of a Real Multiagent System Protocol

Viviana Mascardi, Daniela Briola, and Davide Ancona

DIBRIS, University of Genova, Italy
{viviana.mascardi,daniela.briola,davide.ancona}@unige.it

Abstract. Attribute global types are a formalism for specifying and dynamically verifying multi-party agent interaction protocols. They allow the multiagent system designer to easily express synchronization constraints among protocol branches and global constraints on sub-sequences of the allowed protocol traces. FYPA (Find Your Path, Agent!) is a multiagent system implemented in Jade currently being used by Ansaldo STS for allocating platforms and tracks to trains inside Italian stations. Since information on the station topology and on the current resource allocation is fully distributed, FYPA involves complex negotiation among agents to find a solution in quasi-real time. In this paper we describe the FYPA protocol using both AUML and attribute global types, showing that the second formalism is more concise than the first, besides being unambiguous and amenable for formal reasoning. Thanks to the Prolog implementation of the transition function defining the attribute global type semantics, we are able to generate a large number of protocol traces and to manually inspect a subset of them to empirically validate that the protocol's formalization is correct. The integration of the Prolog verification mechanism into a Jade monitoring agent is under way.

Keywords: multiagent systems, attribute global types, negotiation, dynamic verification of protocol compliance.

1 Introduction

Artificial Intelligence has been often used in planning and scheduling systems where tasks, or resources, must be assigned to different entities: [10,13,14] describe for example how different techniques have been exploited in different domains, where time constraints must also be taken into account.

Railway dispatching is characterized by being physically distributed and time constrained: classical technologies such as operational research and constraint programming are suitable to model static situations with complete information and a predefined number of actors, while they lack the ability to cope with the dynamics and uncertainty of real railway traffic management. Instead, Multiagent Systems (MASs [16], a modern vision of Distributed Artificial Intelligence)

M. Baldoni et al. (Eds.): AI*IA 2013, LNAI 8249, pp. 300–311, 2013.

suite very well scenarios where coordinating autonomous and heterogeneous entities, possibly in real time, is crucial, and their application to the railway domain is witnessed by many papers [11,19,20]. Since MASs are open and highly dynamic, ensuring that the agents' actual behavior conforms to a given interaction protocol is of paramount importance to guarantee the participants' interoperability and security: many logic-based formalisms have been used to check the compliance to a protocol, such as [3,9,12].

In our recent work, we adopted the global types [8], a behavioral type and process algebra approach, to specify and verify multiparty interactions between distributed components. In [2] the problem of run-time verification of the conformance of a logic-based MAS implementation to a given protocol was tackled by the authors by exploiting "plain" global types: they were represented as cyclic Prolog terms on top of the Jason agent oriented programming language [4], and the transition rules giving the semantics of the conformance verification were translated into Prolog too. An extension with sending action types and constrained shuffle was presented in [1]. In [18] the authors further extended the formalism by adding attributes to both sending action types and constrained global types, thus obtaining "attribute global types". That proposal was mainly inspired by "attribute grammars" [17].

In this paper we assess the expressiveness of attribute global types by formalizing the FYPA protocol, developed with and currently being used by Ansaldo STS, that norms the interaction among agents in a real MAS for allocating platforms and tracks to trains inside stations in Italy.

The paper is organized as follows: Section 2 introduces attribute global types, Section 3 describes the FYPA protocol using AUML and natural language, Section 4 formalizes it using attribute global types and Section 5 concludes.

2   Attribute Global Types

The building blocks of attribute global types are the following:

Sending actions. A sending action a is a communicative event taking place between two agents.

Sending action types. Sending actions types model the message pattern expected at a certain point of the conversation. A sending action type α is a predicate on sending actions.

Producers and consumers. In order to model constraints across different branches of a constrained fork, we introduce two different kinds of sending action types, called *producers* and *consumers*, respectively. Each occurrence of a producer sending action type must correspond to the occurrence of a new sending action; in contrast, consumer sending action types correspond to the same sending action specified by a certain producer sending action type. The purpose of consumer sending action types is to impose constraints on sending action sequences, *without introducing new events*. A consumer is a sending action type, whereas a producer is a sending action type α equipped with a natural superscript n specifying the exact number of consumer sending actions which are expected to be synchronized with it.

Constrained global types. A constrained global type τ represents a set of possibly infinite sequences of sending actions, and is defined on top of the following type constructors:

- λ (empty sequence), representing the singleton set $\{\epsilon\}$ containing the empty sequence ϵ.
- $\alpha^n{:}\tau$ (*seq-prod*), representing the set of all sequences whose first element is a sending action a matching type α ($a \in \alpha$), and the remaining part is a sequence in the set represented by τ. The superscript n specifies the number n of corresponding consumers that coincide with the same sending action type α; hence, n is the least required number of times $a \in \alpha$ has to be "consumed" to allow a transition labeled by a.
- $\alpha{:}\tau$ (*seq-cons*), representing a consumer of sending action a matching type α ($a \in \alpha$).
- $\tau_1 + \tau_2$ (*choice*), representing the union of the sequences of τ_1 and τ_2.
- $\tau_1|\tau_2$ (*fork*), representing the set obtained by shuffling the sequences in τ_1 with the sequences in τ_2.
- $\tau_1 \cdot \tau_2$ (*concat*), representing the set of sequences obtained by concatenating the sequences of τ_1 with those of τ_2.
- The "meta-construct" fc (for *finite composition*) that takes τ, a constructor cn, and a positive natural number n as inputs and generates the "normal" constrained global type $(\tau \ cn \ \tau \ cn \ ... \ cn \ \tau \ cn \ \lambda)$ (n times).

Attribute global types are constrained global types whose sending action types may have attributes, included within round brackets, and that have been extended to provide contextual information by means of further attributes and conditions, included in square brackets. They are regular terms, that is, can be cyclic (recursive), and they can be represented by a finite set of syntactic equations. We limited our investigation to types that have good computational properties, namely *contractiveness* and *determinism*. The attribute global types semantics has been provided in [18].

The formalism without attributes is not comparable with ω-context free grammars, since ω-context free grammars recognize languages that cannot have finite strings, and such languages are not required to be complete metric spaces.

As an example, we show the $ping^n pong^n$ protocol which requires that first Alice sends n (with $n \geq 1$, but also possibly infinite) consecutive ping messages to Bob, and then Bob replies with exactly n pong messages (if Alice decides to send a finite number of messages). The conversation continues forever in this way, but at each iteration Alice is allowed to change the number of sent ping messages:

$$Forever = PingPong \cdot Forever$$
$$PingPong = ping{:}((pong{:}\lambda) + ((PingPong) \cdot (pong{:}\lambda)))$$

We may add attributes to state that, for example, the time between one *ping* message and the next one cannot be greater than 5 time units, thus obtaining

$$Forever = PingPong(0, \_) \cdot Forever$$
$$PingPong(t_1, t_2)[t_2 - t_1 \leq 5] = ping(t_2){:}((pong{:}\lambda) + ((PingPong(t_2, \_)) \cdot (pong{:}\lambda)))$$

where t_1 is the time of the previous sending action with type *ping* (initially set to time 0), t_2 is unified with the time of the current *ping* sending action counted from the protocol start, and the condition to be respected is that $t_2 - t_1 \leq 5$.

3   The FYPA Protocol

The problem and the model. The FYPA (Find Your Path, Agent!) multiagent system [5,6,7] was developed starting from 2009 by the Department of Informatics, Bioengineering, Robotics and System Engineering of Genoa University and Ansaldo STS, the Italian leader in design and construction of signalling and automation systems for conventional and high speed railway lines. It automatically allocates trains moving into a railway station to tracks, in order to allow them to enter the station and reach their destination (either the station's exit or a node where they will stop) considering real time information on the traffic inside the station and on availability of tracks. The problem consists of:
– A set of indivisible resources (tracks in a railway station) that must be assigned to different entities (trains) in different time slots
– A set of entities (trains) with different priorities, each needing to use, always in an exclusive way, some resources for one or more time slots.
– A mixed multi-graph of dependencies among resources: an entity can start using resource R only if it used exactly one resource from $\{R_1, R_2, ..., R_n\}$ in the previous time slot (we represent these dependencies as arcs $R_1 \rightarrow R$, $R_2 \rightarrow R$, ..., $R_n \rightarrow R$ in the graph). There may be many different direct arcs connecting the same two resources.
– A set of couples of conflicting arcs and a set of bidirectional arcs in the graph of dependencies: for presentation purposes, in this paper we discard this kind of arcs.
– A static allocation plan that assigns resources to entities (namely, a path in the graph) for pre-defined time slots, in such a way that no conflicts arise.
   In an ideal world where resources never go out of order and where any entity in the system can always access the resources assigned to it by the static allocation plan, no problems arise. In the real world where entities happen to use resources for longer than planned and where resources can break up, a dynamic re-allocation of resources over time is often required. Thus, the solution of the real world problem is a dynamic re-allocation of the resources to the entities such that: 1. the re-allocation is feasible, namely free of conflicts; in our scenario, conflicts may arise because two or more entities would want to access the same resource in the same time slot; 2. the re-allocation task is completed within a pre-defined amount of time; 3. each entity minimizes the changes between its new plan and its static allocation plan; 4. each entity minimizes the delay in which it reaches the end point of the path with respect to its static allocation plan; 5. the number of entities and resources involved in the re-allocation process is kept to the minimum.
   Considering the requirements of Ansaldo STS's application to find a solution in quasi-real time, the plan resulting from the re-allocation problem may be

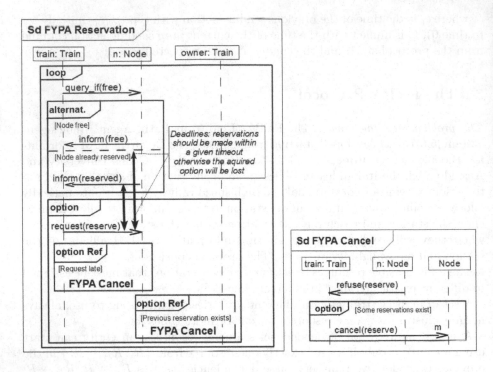

Fig. 1. FYPA Reservation protocol with its FYPA Cancel sub-protocol

sub-optimal. The final solution to the dynamic problem is calculated in a completely distributed and incremental way, and it is the set of the local solutions: each newly reserved path is always compatible with all the others thanks to the negotiation protocol described in the next section.

FYPA Protocol Description. The FYPA Reservation and the Cancel protocols are shown, in the AUML formalism [15], in Figure 1. Messages have been substituted by short labels for readability. This AUML description is an original contribution of this paper, as the previous publications described it in a very informal and shallow way.

The FYPA Reservation protocol involves agents representing trains ("Train" role in the AUML diagram), and agents managing nodes ("Node" role).

Each train knows the paths $\{P_1 = N_{from}...N_{to}; ...; P_k = N_{from}...N_{to}\}$ it could follow to go from the node where it is (N_{from}), to the node where it needs to stop (N_{to}). Such paths are computed by a legacy Ansaldo application[1] which is wrapped by an agent named *PathAgent* and is accessed by the FYPA MAS through web services. We do not model the *PathAgent* here and we limit ourselves to describe the interactions among trains and nodes.

---

[1] The list of possible paths is ordered starting from those nearest to the original one, so the requirement 3 stated in the previous page is fulfilled and the implicit requisite to make connecting trains stop on near platforms is automatically achieved too.

Each train also knows which path it is currently trying to reserve, how many nodes answered to its requests and in which way, and how much delay it can still accept: to reserve a path, the train must obtain a reservation for each node in it. To reserve a node, a train must ask if it is free, wait for the answer from the node (free or already reserved by another train in an overlapping time slot) and then must reserve the resource, which might also mean stealing it to the train that reserved it before. In this case, which usually takes place if the priority of the "thief" is higher than that of the "robbed", the node will inform the robbed train, that will search for another path or for the same path in different time slots. Each node knows the arcs that it manages (those that enter in it). It also knows which trains optioned or reserved the node, in which time slots, from which node they are expected to arrive, and which arc they can traverse.

The protocol control flow is specified by the AUML boxes that represent loops, alternative choices, optional choices, and references to sub-protocols. When a train fires a FYPA Reservation protocol instance by sending a request of information to a node (*query_if(free)* label), a conversation among them starts and a fresh conversation identifier is generated by the train to keep track of it. If the same agent sends another request of information to the same node, the started conversation is considered as a new one and a new fresh id must identify it, whereas when a train contacts more nodes that build up a path, the conversation id remains the same. We discuss the FYPA Reservation protocol from the train perspective, by expanding the message labels into the actual message types, starting from the top.

query_if (free): in order to reserve a path $P_1 = N_{from}...N_{to}$, a train *Train* sends messages of type *msg(Train, N_i, query_if(free(MyPr, T1, T2, From)), cid(CId))* to each node N_i in the path. *MyPr* is *Train*'s priority and can be changed into ∞ if the reservation is not disputable, *T1* and *T2* are the extremes of the time slot when *Train* would reserve the node, and *From* is the node from which it will arrive. A "non-disputable" request is issued only when *Train* has no other chances to move inside the station, and it must necessarily move on N_i.

inform(free): if the node *N* that receives the request is free in the requested time slot, it answers with a message of type *msg(N, Train, inform(free(Arc,From), expires(Timeout)), cid(CId))*, where *Arc* is the arc, among the possibly many ones, that *Train* is allowed to use for moving from *From* to *N*. *N* will not answer other requests for the same or an overlapping time slot until either *Train* reserves definitely the resource or *Timeout* expires, in which case *Train* looses its option.

inform(reserved): if *N* is already allocated to another train in the requested time slot, it answers with a message of type *msg(N, Train, inform(reserved (Owner, OwnerPr, Arc, From, NextT3, NextT4), expires(Timeout)), cid(CId))*, where *Owner* is the train that currently owns the node, *OwnerPr* is its priority, *Arc* is the arc that *Train* could have used to reach *N* from *From*, *NextT3* and *NextT4* are the extremes of the next free slot. *OwnerPr* may assume the value ∞ to mean that the reservation cannot be stolen.

request(reserve): *Train* may decide to reserve the node either if the node is free, or if it is already reserved but the priority of the current owner is lower or equal than its priority (or in other complex situations not described in this paper). In both these cases, *Train* sends a message with type *msg(Train, N, request(reserve(Arc, MyPr, Owner, OwnerPr, T1, T2), expired(AbsTime)), cid(CId))*. *AbsTime* is the absolute time when the option would have expired. We do not enter here in the details of when a train decides to steal a node to another train, that is based on the delay it could undergo and on its priority. If this happens, *N* must start a FYPA Cancel sub-protocol between *N* and *Owner*, to inform it that it lost the resource it reserved. On the other hand, it may also happen that *Train*'s reservation fails because it was issued too late w.r.t. the given timeout, firing a FYPA Cancel sub-protocol between *N* and *Train*. A train can also decide not to confirm the reservation, letting it expire.

In case the initial request of *Train* was not disputable (represented by an infinite priority of *Train*), and the node was already reserved by a train with infinite priority (meaning that the current owner's request was not disputable as well), a human intervention is required. This situation can in fact arise only when the station is undergoing a disaster, with so many failures of tracks and nodes that trains cannot move any longer. Apart from this emergency situation, not shown in the diagram for sake of clarity, the allocation of nodes and arcs to trains is entirely managed by the MAS in an automatic way.

The FYPA Cancel sub-protocol is almost intuitive and we do not discuss it here.

4   Protocol Specification Using Attribute Global Types

Constraints on the FYPA Reservation Protocol's Traces. To design and develop attribute global types, we need to know both the protocol's information flow, and (optionally) the constraints that the protocol traces must respect. The more constraints we are able to formalize, the more complete the verification mechanism will be.

Many of the constraints that the traces of the FYPA protocol must respect could not be expressed in AUML without compromising the readability diagram and involve the content of both single messages and sets of messages. In this section we list some of them. All the traces shown below have been obtained by the automatic trace generation process mentioned in the next paragraph.

"Local" Constraints on Messages. Each message must have the right type. For example, a *request* message must be sent by an agent playing the "Train" role to an agent playing the "Node" role and the arguments of the *reserve* action must be a well formed arc identifier, the priority of the sender, the train to which the resource will be stolen ("none" is the node was free), its priority ("0" if it was free), and the extremes of the reserved time slot. Such a message is correct only if the priority of the train making the reservation is higher than or equal to the priority of the train that already owned the node, if any. These constraints can be easily verified by inspecting the arguments of the message, without needing to exploit attributes. For example this message satisfies them:

msg(t1, n5, request(reserve(arc(4, 5, d), 3, t3, 1, 23, 31), ...))

"Horizontal" Constraints on Queries for Reserving a Path. When a train contacts a sequence of nodes to verify whether they are free in order to optionally issue a reservation request, the arguments of these messages must form a coherent path: the "From" argument in message m_{i+1} must be the same as the receiver of message m_i, the time slot's first extreme in message m_{i+1} must be the same as the time slot's second extreme in message m_i, the conversation id must be the same, and the train cannot change its priority, apart from setting it equal to infinity for non disputable requests. These constraints can be verified without knowing the station's topology. For example, this trace respects them.

    msg(t1, n5, query_if(free(infinity, 10, 15, n4)), cid(c1))
    msg(t1, n6, query_if(free(3, 15, 17, n5)), cid(c1))
    msg(t1, n7, query_if(free(3, 17, 22, n6)), cid(c1))
    msg(t1, n8, query_if(free(infinity, 22, 44, n7)), cid(c1))
    msg(t1, exit, query_if(free(3, 44, 48, n8)), cid(c1))

"Vertical" Constraints on Conversations between a Train and a Node. Apart from the requirement that during a single conversation the train changes neither its distinguishing features (such as its priority) nor the conversation features (the conversation id), we can identify some more constraints:

1. If a node is reserved, it must inform the train of the arc it could have used to reach it and of the time slot when it will be free again. This time slot must start after the time slot's start indicated by the train in its *query_if* message, even if it may overlap with it.

2. When a train decides to reserve a node, it must access it by the arc included in the node's answer.

A trace like the next one respects both constraints:

    msg(t2, n5, query_if(free(3, 22, 44, n4)), cid(c2))
    msg(n5, t2, inform(reserved(t3, 1, arc(4, 5, a), n4, 23, 31), ...))
    msg(t2, n5, request(reserve(arc(4, 5, a), 3, t3, 1, 22, 44), ...))

Since a train can interact with the same node many times, for example because the attempt to reserve a path failed and then the train has to try to reserve a new one, we add another vertical constraint that involves conversation loops: if a train sends more than one *query_if* message to the same node, the conversation id must be different since the messages belong to different conversations. Also, if the node answered that it was not available, in the new *query_if* message the train must ask for a time slot successive to the previous one, unless the train changed its priority to infinity in the meanwhile.

FYPA Reservation Protocol Using Attribute Global Types. As described in [18], the transition function $\delta: \mathcal{CT} \times \mathcal{A} \to \mathcal{P}_{fin}(\mathcal{CT})$, where $\mathcal{CT}$ and $\mathcal{A}$ denote the set of contractive attribute global types and of sending actions, respectively, has been implemented using SWI Prolog[2] and is protocol independent. Thanks to Prolog, this function can be used both to verify traces (in the "on line verification" mode) and to create all the possible traces corresponding to the global type defined in $\mathcal{CT}$.

---

[2] http://www.swi-prolog.org/.

In this section we give the flavor of how the FYPA Reservation attribute global type ($\mathcal{CJ}$) looks like and which features of the formalism were exploited to model the constraints devised at the beginning of Section 4. The correctness of the formalization with respect to the actual protocol has been empirically validated by automatically generating in Prolog a large number of protocol traces that respect the protocol, under different initial conditions (number of trains, station configuration), and by manually inspecting a randomly selected subset of them. The traces are generated by applying the transition function implemented in SWI-Prolog to the initial state of the protocol represented by the global type described in this section. The trace generator and details on its use can be found at http://www.disi.unige.it/person/MascardiV/Software/globalTypes.html.

Core of the Protocol. A single conversation between a train and one node in the FYPA protocol is represented by the following attribute global type:

$$FYPA\_NODE\_RESERVATION = ($$
$$(ND\_QUERY : ((NONDISP\_NONDISP + WT\_DISP) + FREE\_NODE)) +$$
$$(D\_QUERY : ((DISP\_NONDISP + WT\_DISP) + FREE\_NODE)))$$

The train may issue either a non disputable *query_if* (ND_QUERY) or a disputable one (D_QUERY). In the first case, three different situations may take place: either the node has a non disputable reservation ($NONDISP\_NONDISP$), which leads to the involvement of a human operator, or it has a disputable reservation (WT_DISP, for "whatever-disputable"), in which case the train may steal the node, or it is free, in which case the train may reserve the node. In case of a disputable query, the protocol is similar apart from the $DISP\_NONDISP$ branch. The equations defining the sub-types are given below. Despite the complexity of the protocol, whose description using AUML and natural language required many pages of this paper, its representation using attribute global types is rather compact.

$ND\_QUERY = msg(Train, N, nd\_query\_if(free(infinity, T1, T2, From)), cid(CId))^1$
$D\_QUERY = msg(Train, N, d\_query\_if(free(MyPr, T1, T2, From)), cid(CId))^1$
$CONSTR\_ON\_PATH(List, (Train, MyPr, N, T1, T2, From, CId)) =$
$((msg(Train, N, nd\_query\_if(free(MyPr, T1, T2, From)), cid(CId)) :$
$CONSTR\_ON\_PATH([(Train, MyPr, N, T1, T2, From, CId)|List],_{,)} + \lambda)) +$
$(msg(Train, N, d\_query\_if(free(MyPr, T1, T2, From)), cid(CId)) :$
$CONSTR\_ON\_PATH([(Train, MyPr, N, T1, T2, From, CId)|List],_{,)} + \lambda)))$
[the tuple (Train, MyPr, N, T1, T2, From, CId) forms a consistent path with the elements in List]
$ND\_RESERVED(T2, NextT3) = (msg(N, Train, inform(nd\_reserved(Ow, OwPr, Arc, From, NextT3, NextT4), expires(\_TO)), cid(CId))^0 : \lambda)[NextT3 > T2]$
$D\_RESERVED(T2, NextT3) = (msg(N, Train, inform(d\_reserved(Ow, OwPr, Arc, From, NextT3, \_NextT4), expires(\_TO)), cid(CId))^0 : \lambda)[NextT3 > T2]$
$DISASTER = msg(N, \_HumanOperator, inform(disaster(Ow, Train, T1, T2)), cid(CId))^0$
$STEAL\_ON\_TIME = (msg(Train, N, request(reserve\_on\_time(Arc, MyPr, Ow, OwPr, T1, T2), expired(AbsTime)), cid(CId))^0 : REFUSE\_NODE\_ROBBED\_TRAIN)$
$STEAL\_LATE = ((msg(Train, N, request(reserve\_late(Arc, MyPr, Ow, OwPr, T1, T2), expired(AbsTime)), cid(CId))^0 : (REFUSE\_NODE\_SAME\_TRAIN + \lambda)) + \lambda)$
$RESERVE\_ON\_TIME = (msg(Train, N, request(reserve\_on\_time(Arc, MyPr, none, 0, T1, T2), expired(AbsTime)), cid(CId))^0 : \lambda)$
$RESERVE\_LATE = (msg(Train, N, request(reserve\_late(Arc, MyPr, none, 0, T1, T2), expired(AbsTime)), cid(CId))^0 : REFUSE\_NODE\_SAME\_TRAIN)$
$REFUSE\_NODE\_SAME\_TRAIN = (msg(N, Train, refuse(reserve), cid(CId))^0 : \lambda)$

$REFUSE\_NODE\_ROBBED\_TRAIN = (msg(N, Ow, refuse(reserve), cid(\_CIdX))^0 : \lambda)$
$FREE = (msg(N, Train, inform(free(Arc, From), expires(\_TO)), cid(CId))^0 : \lambda)$
$NONDISP\_NONDISP = (ND\_RESERVED \cdot (DISASTER : \lambda))$
$WT\_DISP = (D\_RESERVED(T2, NextT3) \cdot ((STEAL\_ON\_TIME + STEAL\_LATE) + \lambda))$
$DISP\_NONDISP = ND\_RESERVED(T2, NextT3)$
$FREE\_NODE = (FREE \cdot ((RESERVE\_ON\_TIME + RESERVE\_LATE) + \lambda))$

For each message type appearing in the equations, the specification of which actual messages have that type and under which constraints the type is correct, must be provided. To make an example, the message type appearing in $STEAL\_ON\_TIME$ holds on messages $msg(Train, N, request(reserve(Arc, My-Pr, Owner, OwnerPr, T1, T2), expired(AbsTime)), cid(CId))$ if the following conditions are satisfied:

$current\_time(CurrentTime), AbsTime >= CurrentTime, is\_arc(Arc), is\_priority(OwnerPr),$
$is\_priority(MyPr), is\_higher\_priority(MyPr, OwnerPr), role(Train, train), role(N, node),$
$role(Owner, train), different(Owner, Train), is\_conv\_id(CId), is\_time\_range((T1, T2)).$

Note that message types also allow us to distinguish between reservations that arrive on time and reservations that arrive late, just by comparing the actual current time $current\_time(CurrentTime)$ with the time when the reservation expired. This is a very powerful mechanism to model deadlines that may cause different courses of actions if they are not met.

Because of space constraints we only describe the trickiest aspects of the equations above. Given the introduction to the language syntax given in Section 2, most of them should be intuitive. All the messages apart from those in ND_QUERY and D_QUERY have 0 as superscript. This means that 0 consumers for these messages are present in the protocol, and no synchronization among fork branches will take place. This is not true for the messages in ND_QUERY and D_QUERY that will be consumed by the respective consumers in $CONSTR\_ON\_PATH$, in order to synchronize many conversations between the same train and different nodes and to ensure that the $query\_if$ contents form a consistent path. This constraint has been implemented by equipping $CONSTR\_ON\_PATH$ with attributes, namely the contents of all the $query\_if$ sent by train $Train$ to the nodes in the path, accumulated in a list which is passed from one instance of $CONSTR\_ON\_PATH$ to the successive one.

Another constraint that we expressed in Section 4 is that if a node answers to a $query\_if$ that it is reserved, it must also state when it will become available again, and this time slot must be successive to the one requested by the train. This is formalized by adding the two attributes $(T2, NextT3)$ to $D\_RESERVED$, and asking that $[NextT3 > T2]$. Multiple conversations to reserve a path are modeled by the following attribute global type:

$fc(FYPA\_PATH\_RESERVATION, FYPA\_NODE\_RESERVATION, |, NumOfNodes)$
$FYPA = (FYPA\_PATH\_RESERVATION | CONSTR\_ON\_PATH)$

Given $NumOfNodes$ the number of nodes that the train is expected to contact, $fc(FYPA\_PATH\_RESERVATION, FYPA\_NODE\_RESERVATION, |, NumOf-Nodes)$ creates an attribute global type named $FYPA\_PATH\_RESERVATION$, obtained by replicating $FYPA\_NODE\_RESERVATION$ $NumOfNodes$ times, using the fork constructor. From an implementative viewpoint, all these replicas have fresh new variables to ensure that they are independent from one another.

$(FYPA\_PATH\_RESERVATION \,|\, CONSTR\_ON\_PATH)$ defines one single iteration of the $FYPA$ protocol, where the fork is used to synchronize the two branches and to guarantee that inquiries to nodes form a consistent path.

Finally, the external conversation loop modeling more iterations of the $FYPA$ protocol is represented by:

$$LOOP(N, PrevCId, CId, PrevT1, T1) = (FYPA \cdot LOOP(N, CId, \_, T1, \_))$$
$$[\text{different}(PrevCId, CId) \wedge \text{different}(PrevT1, T1)]$$

5   Conclusions and Future Work

The comparison between the informal AUML representation of the protocol, that must be necessarily accompanied by an explanation in natural language, and the representation using attribute global types, which is compact, unambiguous, and amenable for formal reasoning, suggests that the second is definitely better for all those situations where on-line verification of protocol compliance is required, even if it is less intuitive and requires some training to be used.

Nowadays, a Jason version of the monitor agent able to control at runtime a conversation among agents following a specified protocol exists. However, many industrial strength MASs have been implemented in Jade, as the FYPA one. We are thus designing a Jade version of the monitor agent: this agent will integrate a Prolog engine to exploit the Prolog representation of the attribute global type describing a protocol, and the transition function to verify the compliance of actual messages to the protocol. The monitor will be able to check the ongoing execution of the MAS without interfering with the agents, so that no changes are required to the agents' code or behavior. We have already implemented a first prototype of the Jade monitor agent, that acts in a similar way as the Jade sniffer agent: when it starts, it registers itself as a sniffer to the Jade Agent Management System, so that it will be informed of any new agent creation and of all the events concerning messages exchange among the agents that it sniffs. In this way when a new agent is created, if it is in the list of those to be monitored, the monitor will receive a copy of all the messages that involve such agent and will check if the agent is respecting the protocol.

Having such a real time monitor integrated into Jade will help the MAS programmers in testing their MAS in very complex scenarios, for example those of FYPA, where the number of involved entities, exchanged messages and their constraints is large. The monitor can be used to verify the progress of the conversation, which is a very relevant feature in the fault tolerance domain.

Our close future work consists in completing the implementation of the monitor agent in Jade and in using it to monitor the actual FYPA MAS, in order to verify if our approach can be concretely used on top of existing and running MASs, and to assess its performance and scalability.

References

1. Ancona, D., Barbieri, M., Mascardi, V.: Constrained global types for dynamic checking of protocol conformance in multi-agent systems. In: SAC. ACM (2013)
2. Ancona, D., Drossopoulou, S., Mascardi, V.: Automatic generation of self-monitoring MASs from multiparty global session types in Jason. In: Baldoni, M., Dennis, L., Mascardi, V., Vasconcelos, W. (eds.) DALT 2012. LNCS (LNAI), vol. 7784, pp. 76–95. Springer, Heidelberg (2013)
3. Baldoni, M., Baroglio, C., Marengo, E., Patti, V.: Constitutive and regulative specifications of commitment protocols: A decoupled approach. ACM Trans. Intell. Syst. Technol. 4(2), 22:1–22:25 (2013)
4. Bordini, R.H., Hübner, J.F., Wooldridge, M.: Programming multi-agent systems in AgentSpeak using Jason. John Wiley & Sons (2007)
5. Briola, D., Mascardi, V.: Design and implementation of a NetLogo interface for the stand-alone FYPA system. In: WOA 2011, pp. 41–50 (2011)
6. Briola, D., Mascardi, V., Martelli, M.: Intelligent agents that monitor, diagnose and solve problems: Two success stories of industry-university collaboration. Journal of Information Assurance and Security 4, 106–117 (2009)
7. Briola, D., Mascardi, V., Martelli, M., Caccia, R., Milani, C.: Dynamic resource allocation in a MAS: A case study from the industry. In: WOA 2009 (2009)
8. Carbone, M., Honda, K., Yoshida, N.: Structured communication-centred programming for web services. In: De Nicola, R. (ed.) ESOP 2007. LNCS, vol. 4421, pp. 2–17. Springer, Heidelberg (2007)
9. Chesani, F., Mello, P., Montali, M., Torroni, P.: Representing and monitoring social commitments using the event calculus. Autonomous Agents and Multi-Agent Systems 27(1), 85–130 (2013)
10. Elfazziki, A., Nejeoui, A., Sadgal, M.: Advanced internet based systems and applications, pp. 169–179. Springer (2009)
11. Ghosh, S., Dutta, A.: Multi-agent based railway track management system. In: IACC 2013, pp. 1408–1413 (2013)
12. Giordano, L., Martelli, A., Schwind, C.: Specifying and verifying interaction protocols in a temporal action logic. Journal of Applied Logic 5(2), 214–234 (2007)
13. Gomes, C.P.: Artificial intelligence and operations research: challenges and opportunities in planning and scheduling. Knowl. Eng. Rev. 15(1), 1–10 (2000)
14. Hadad, M., Kraus, S., Gal, Y., Lin, R.: Temporal reasoning for a collaborative planning agent in a dynamic environment. Annals of Mathematics and Artificial Intelligence 37(4), 331–379 (2003)
15. Huget, M.-P., Bauer, B., Odell, J., Levy, R., Turci, P., Cervenka, R., Zhu, H.: FIPA modeling: Interaction diagrams. Working Draft Version 2003-07-02 (2003), http://www.auml.org/auml/documents/ID-03-07-02.pdf
16. Jennings, N.R., Sycara, K.P., Wooldridge, M.: A roadmap of agent research and development. Autonomous Agents and Multi-Agent Systems 1(1), 7–38 (1998)
17. Knuth, D.E.: The genesis of Attribute Grammars. In: Deransart, P., Jourdan, M. (eds.) WAGA 1990. LNCS, vol. 461, pp. 1–12. Springer, Heidelberg (1990)
18. Mascardi, V., Ancona, D.: Attribute global types for dynamic checking of protocols in logic-based multiagent systems (technical communication). Theory and Practice of Logic Programming, On-line Supplement (2013)
19. Siahvashi, A., Moaveni, B.: Automatic train control based on the multi-agent control of cooperative systems. TJMCS 1(4), 247–257 (2010)
20. Tsang, C.W., Ho, T.K., Ip, K.H.: Train schedule coordination at an interchange station through agent negotiation. Transportation Science 45(2), 258–270 (2011)

A Distributed Agent-Based Approach for Supporting Group Formation in P2P e-Learning

Fabrizio Messina[1], Giuseppe Pappalardo[1], Domenico Rosaci[2],
Corrado Santoro[1], and Giuseppe M.L. Sarné[2]

[1] Dipartimento di Matematica e Informatica, University of Catania,
V.le Andrea Doria 6, 95125 Catania, Italy
{messina,pappalardo,santoro}@dmi.unict.it
[2] {DIIES, DICEAM}, University "Mediterranea" of Reggio Calabria,
Via Graziella, Feo di Vito 89122 Reggio Calabria (RC), Italy
{domenico.rosaci,sarne}@unirc.it

Abstract. Peer-to-Peer (P2P) technology can be effectively used to implement cooperative e-Learning systems, where the available knowledge for a student is not only from teachers, but also from other students having similar interests and preferences. In such a scenario, a central issue is to form groups of users having similar interests and satisfying personal user's constraints. In this paper we propose a novel approach, called HADEL (Hyperspace Agent-based E-Learning), based on an overlay network of software agents. Our approach preserves user's privacy, allowing to locally maintain sensitive user's data and inferring the properties necessary for determining the groups by using agents acting as personal assistants. The results obtained by some tests performed on simulated e-Learning environments show the efficiency of our approach, that suitably exploits the topology of the overlay network, which exhibits the typical properties of a small-world system.

1 Introduction

Nowadays, due to the rapid development of information technology, important changes have been produced in the fundamental ways people acquire and disseminate knowledge. Several e-Learning systems have been designed to support students and teachers, providing them with the advantages brought by the Internet technology. In particular, the Peer to Peer (P2P) paradigm has been widely recognised as the key driven technology to bring revolutionary impact on the future e-Learning society. P2P technology can be effectively used to implement cooperative e-Learning systems, where the available knowledge for a student is not only from teachers, but also from other students with the same interests. In other words, each actor of an e-Learning community can have both the roles of service client and service provider. In such a scenario, a central issue is to support the formation of groups of users having similar interests [4,5,27] and satisfying personal users' constraints. To this aim, many e-Learning systems construct a user's profile by monitoring his/her choices [3,10,13,26]. Moreover, a lot of approaches are attempting to combine peer-to-peer computing and e-Learning together [2,11,12,14,24,25]. For example, in [12] it is discussed the potential contribution of the P2P technology into

M. Baldoni et al. (Eds.): AI*IA 2013, LNAI 8249, pp. 312–323, 2013.

e-Learning systems and it is introduced a novel P2P based e-Learning environment, called APPLE. As another example, the open source project Edutella is an RDF-based e-Learning P2P network that is aimed to accommodate heterogeneous learning resource metadata repositories in a P2P manner and to further facilitate the exchange of metadata between these repositories based on RDF [25]. Moreover, some approaches for realising personalised support in distributed learning environments have been proposed, as described in [7], based on Semantic Web technologies, or those presented in [9,26] exploiting decentralised recommender systems.

However, any of the aforementioned approaches provides a mechanism to facilitate the formation of groups of nodes in the P2P network, such that the resources of the nodes satisfy the requirements of the e-Learning actors associated with those nodes. Obviously, the involved actors (students and teachers) need to find interlocutors having good affinities, for instance in terms of interest or learning level. Furthermore, in such a context, important privacy concerns arise for users, since the opportunity to find new contacts based on common interests strongly depends on data which users are willing to share in the network[15]. This is a very typical concern, since affinities among users can be found only by analysing sensitive data that they don't want to share. These considerations lead to the necessity of a distributed middleware capable of assisting each user involved in the e-Learning system, acting as a mediator between the P2P network and the user himself. We argue that a software agent, operating on the user's behalf and equipped with a suitable knowledge representation of the user's interests and preferences, could be considered as a suitable solution for this purpose.

In this paper, we propose a mechanism for supporting the actors of a P2P e-Learning system in forming groups of users having similar interests and satisfying specific constraints. The goal of our proposal is to provide each user with homogeneous environments (the groups), suitable for performing e-Learning activities as participating to seminars and lessons, sharing multimedia contents, making discussions, taking tests and exams and so on. Our proposal is based on an overlay network of software agents [8], which allows the e-Learning actors to locally maintain sensitive user's data, and contact interlocutors having a sufficient level of affinity. In particular, software agents are capable of assisting users by analysing local data in order to extract relevant properties which can help to find other users having similar interests. To this purpose, each agent can be delegated by its own user to start the construction of a custom overlay network called *HADEL* (Hyperspace Agent-based Distributed E-Learning) whose topology reflects the distribution of a set of actor's properties. The software agents, in order to build the overlay network, exchange messages by using a gossip protocol, such that each node of the network contacts all the nodes with which it has interacted in the past. The use of the overlay network allows to reach approximately all the agents/nodes with a limited number of steps and messages, as shown in the experiments we provide in Section 4. Once the properties characterising the users of the e-Learning system are mapped into the overlay network, the agents can send their requests over the network in order to find a set of agents reflecting some required properties. Moreover, as we will discuss in the experimental section, the searching of suitable nodes is very efficient. This is also due to the topology of the network, which exhibits the properties of a "small-world" system. It is important to remark that, since properties involved in user aggregation are inferred by local data not shared in the P2P network, our approach preserves the user privacy on sensitive data.

The remaining of the paper is organised as follows. Section 2 describes the details of the basic scenario, while Section 3 introduces the technique for building the HADEL network and finding suitable nodes. Finally, an experimental evaluation of the proposed approach is provided in Section 4 and in Section 5 we draw some conclusions and future works.

2   Scenario

In our framework, we deal with an e-Learning virtual community composed by a set $\mathscr{U}$ of users. Each user $u \in \mathscr{U}$ can be considered in some cases a client and in other cases a provider of e-Learning services (ELSs), where an ELS can be, for instance, an online lesson, an exam, an evaluation report and so on. A P2P e-Learning System is associated with the virtual community, represented by a graph $G = \langle N, A \rangle$, where N is a set of nodes, such that each node $n \in N$ represents a user of $\mathscr{U}$ that we denote by u_n, and A is a set of arcs, such that each arc $a = \langle n_1, n_2 \rangle$ represents the fact that the users u_{n_1} and u_{n_2} mutually interacted in the past. Each user u is assisted by a software agent a_u, resident on his/her user machine. The user u, when assuming the role of e-Learning service provider, makes available his/her services via the agent a_u while, when he assumes the role of client, is assisted by a_u in his/her e-Learning activities. We assume that a software agent is allowed to reside on each user machine. Thus, the e-Learning system is fully distributed except for a unique central component, the *directory facilitator DF* which provides a service of yellow pages (i.e. registration of new users and services, registration of e-Learning resources, search for users, services and resources). As we will discuss in Section 3.1, the *DF* is also actively involved in the process of maintenance of the *Gossip Network* (GN), which is devoted to support the dissemination of the *HADEL* network construction request.

In this perspective, local software agents are allowed to retrieve and analyse the personal (local) data of their users, which are considered "sensitive" by the users themselves. Hence software agents can exploit a detailed profile based not only on data shared in the system, but also on private data. Hence the user profile can be used to create new groups of users with a decentralised technique (see Section 4).

From hereafter, we will use the terms *user* and *node* interchangeably. We also assume that a set *ELP* of *e-Learning properties* is associated with the virtual community, where an e-Learning property can represent, for instance, the set of services that the user provides, the set of services that the user is searching for, the user's interest for an e-Learning topic, the user's expertise in a given topic, the set of e-Learning resources that the user accessed in the past, the set of e-Learning groups which the user joins with, and so on. Moreover, we also assume that each property $p \in ELP$ has a value belonging to a given domain D_p. For instance, if p represents the user's expertise or the user's interest in a given topic, then D_p will be a set of positive integer values. Differently, if p represents the set of the e-Learning resources which the agent accessed in the past, then D_p will be the set of all the possible sets of registered resources available in the e-Learning system. To this aim, note that all the information necessary for determining the property domains are available in the *DF*, where users and e-Learning resources have to be registered. An *e-Learning profile schema ELPS* is also associated with the virtual community, composed of a list of properties belonging to *ELP*, denoted by $p^1, p^2, .., p^l$, where l is a value depending on the given virtual community. In this scenario, a *profile instance* p_n is associated with each node n of N, where p_n is a list $p_n^1, p_n^2, ..p_n^l$, such that each element p_n^i is an instance of the property $p^i \in ELP$.

Each user, in order to evaluate if joining or not with the same group of another user, can express some requirements on the properties of the other user. A *requirement on a property p* is a boolean function that accepts as input a property $p \in ELP$ and returns a boolean value $b \in \{true, false\}$. For instance, a requirement on the expertise on a given topic could state that a user must have an expertise higher than or equal to a given threshold, e.g. having an expertise on the topic *ancient history* higher than or equal to 0.5, if the domain of the expertise values is the real interval $[0, 1]$. The function representing the requirement will return the value *true* if the statement is verified by the property instance of the user, *false* otherwise.

Furthermore, each user u has associated a set of requirements R_u, and u accepts to belong to an e-Learning group if at least the α percent of the members of the group satisfy at least the β percent of the requirements of R_u, where α and β are parameters set by the user u. Moreover, each user u can express his/her preferences regarding the group types he wants to join with, using the parameters *availability, accessibility* and *interests*, where *availability* is a boolean value, *accessibility* is a value belonging to the set $\{public, private, secret\}$ and *interests* is a set of topics. The value *availability* = *false* means that u does not desire to join with any group, while *availability* = *true* means that u is available to join with a new group under the condition that both the topics of the group are contained in the set *interests* and the access type of the group is equal to the value of the parameter *accessibility*.

The software agent a_u associated with the user u is able to perform the following two tasks:

– It can be delegated by its user u to form a group of users, based on a set of requirements ELPR. To this aim, the agent of u sends a joining request to all the nodes belonging to the admissible region S(ELPR), determined by using the *HADEL* protocol (see Section 4). The joining request is a tuple $JR = \langle t, a \rangle$, where t is a set of topics associated with the group and a is the access type of the group.

– It negatively responses to a request of joining $JR = \langle t, a \rangle$ with a group coming from another agent a_m if the parameter *availability* = *false*. Otherwise, if *availability* = *true*, it positively responses if both the topics t of the group are contained in the set *interests* and the access type a of the group is equal to the value of the parameter *accessibility*. After this joining, the agent of u periodically verifies that at least the α percent of the group members' profiles satisfy at least the β percent of the requirements of R_u. If this verification fails, the agent automatically deletes the user u from such a group.

3   The HADEL Overlay Network Construction and Finding Process

According to the scenario reported in the Section above, any agents, on behalf of its own user, can start the process leading the formation of a new group basing on some

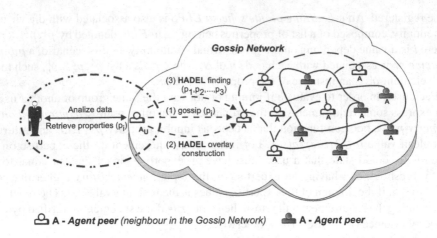

Fig. 1. The HADEL Protocol

properties of interest for the user. For this aim the delegated agent looks for a set of agents representing nodes that satisfy the set of requirements *ELPR*. Since the user data which is analysed by the software agents is distributed over the users machines, therefore to support such a finding process we adopt a *peer-to-peer* approach which is known to be more efficient and scalable than a centralised solution [31].

The finding schema adopted in this work is derived from a decentralised resource finding approach for large scale distributed systems proposed and studied by the authors in the recent years [6,18,17,19,21]. Our approach is based on the construction of an overlay network of software agents which is the key of the *HADEL* protocol. Indeed, as we will discuss in Section 3.2, the *HADEL* topology is built in order to reflect the distribution of the selected properties over the users of the e-learning platform. This characteristic, as we show in Section 5, is the basis for an efficient finding process.

The *HADEL* protocol is structured into the three phases depicted in Figure 1. Supposing that the user data have been already analysed by the agent, whenever the user expresses its intention to start a new e-learning groups, the agent starts with the dissemination (dashed arrow (1) in Figure 1) of a "gossip" message starting from its neighbours. The dissemination phase is detailed in Section 3.1, while the gossip protocol we adopted in our simulation is illustrated in Section 4. All agents receiving the same gossip message start executing the decentralised algorithm of the *HADEL* overlay network construction (dashed arrow (2) in Figure 1). In the proposed approach the *HADEL* overlay construction algorithm is also the basis for the adopted finding process, which is the third phase (dashed arrow (3) in Figure 1). We discuss the second and third phase in Section 3.2 and 3.3.

3.1   *HADEL* Network Construction: Dissemination

In our approach a scale free network [1] is built once and maintained with the aim of supporting the dissemination phase (first phase in Figure 1) within the *HADEL* protocol.

We call this network, for brevity, the "Gossip Network" (GN). Furthermore, we suppose that the *DF* (see Section 2) is able to maintain the *list of the agents*, which is ordered on the basis of their own *node connectivity* [1]. The process leading to the construction and maintenance of the GN is briefly explained below:

- Whenever a user joins the e-Learning virtual community, its agent contacts the *DF*. In this way the agent can select its own neighbours with the connectivity-based preferential attachment proposed in [1].
- Whenever any agents has joined the GN and selectes it own neighbours, it computes its own *connectivity* and sends it to the DF.
- Whenever a user leaves the e-Learning virtual community, its own agent leaves the GN. The agent neighbours contact the DF, which provides the connectivity information in order to add a new neighbour.

The dissemination phase is performed on the GN whenever an agent (say n) is delegated by its own user to start with the construction of the *HADEL* overlay network. It is performed by a gossip message which traverses the GN to trigger the construction of the *HADEL* overlay network. The gossip message contains the list of properties $(p^1, p^2, .., p^l)$ to consider to compute the agents coordinates into the *HADEL* hyperspace (see Section 3.2). This process is explained below:

- the agent n sends a "gossip" message to its own neighbours in the GN.
- Whenever an agent has received the gossip, it analyses user data in order to obtain property values, which are mapped into the *HADEL* hyperspace coordinates.
- The receiver sends the gossip to a subset of its own neighbours, as specified in the gossip algorithm shown in Section 4.
- The receiver connects to the sender in the *HADEL* hyperspace and starts running the *HADEL* overlay construction algorithm as described in Section 3.2.

It should be noted that phase 1 (dissemination) and phase 2 (overlay construction) partially overlap: the gossip messages are spread among the GN, and in the meantime the agents already reached by the gossip are executing the *HADEL* overlay construction algorithm. The benefit coming from the construction and maintenance of the GN is twofold. First of all, the construction process for a new *HADEL* network cannot be managed by a single node, e.g. the DF, which could be easily overloaded when the number of nodes is very high. Moreover the list of properties to be mapped into hyperspace coordinates are carried by the gossip message to the software agents, which reside into the user machines, therefore the list of properties are not collected in a unique node and the privacy over meta-data (i.e. the requested properties and its constraints) is also preserved.

3.2   *HADEL* Overlay Network Construction

The *HADEL* overlay network allows every agent to perform a fast and effective fully decentralised finding process of a set of nodes which exhibit well defined characteristics. The overlay network construction algorithm is fully decentralised and it is performed on the set of agents reached by the gossip message.

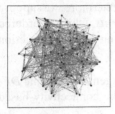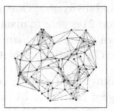

Fig. 2. The overlay network construction

In order to start this phase, property values are mapped into coordinates of a multidimensional space (i.e. the hyperspace); these coordinates are used to compute the Euclidean distance between agents. The overlay construction algorithm is summarised below:

1. Agent a calculates the set of its neighbour at *2-hops*, say N'; it performs this task by asking to its neighbours (say the set N) the lists of their own neighbours.
2. The set N' is ordered basing on the Euclidean distance of each neighbour from a.
3. Agent a selects from N' at least deg_{min} near agents, but at most deg_{trg} agents; in this way a new set of neighbours, say N'', is obtained.
4. The agent connects to the peers belonging to N'' but not already in N (i.e. the original set of neighbour), and the existing links to the peers belonging to N but not in N'' are removed.

Moreover, as detailed in [19], during the last step of the construction algorithm, the so-called *essentially critical* neighbours are preserved to make sure that the overlay network stays connected over time. As proved in [19] the resulting network features a structure quite similar to a *small-world* [32], i.e. a network which shows a high clustering degree and a very low average path length; these characteristics, as shown in Section 4, are decisive to make resource finding effective.

Since the Euclidean distance is a similarity measure for the properties mapped on the coordinates, the clusters of the resulting network are characterised by agents exposing very similar properties. Furthermore, by exploiting short links, a fast navigation within the clusters is possible to e.g. refine the finding process, while, by using long links, the clusters close to the admissible region can be quickly reached.

Some steps in the construction of an overlay network – in this case two properties/attributes have been mapped – are depicted in Figure 2.

3.3   Finding Suitable Nodes with *HADEL*

Once the *HADEL* overlay network is built on the initiative of any agents, several finding requests can traverse the network to find the set of suitable nodes according to a set of requirements (*ELPR*). Moreover, according to the said hyperspace abstraction, not only the single agent can be viewed as a point in the metric space, but also the *finding request* which carries the constraint on the set of properties. In this way the request is a *point* representing a corner of a *region* or *semi-space* whose internal agents are those that expose suitable properties for the given finding request. Such a semi-space is called the *admissible region S(ELPR)*.

The finding process starts with an agent request initially sent to its neighbourhoods. This process is fully decentralised and summarised below:

1. an agent receiving the finding request checks if its own coordinates fall within the admissible region $S(ELPR)$; if this is the case, the finding process continues with step 4;
2. if the coordinates of the agent do not fall within the admissible region $S(ELPR)$, the request is forwarded to a neighbour, which is selected on the basis of appropriate heuristics aiming to approach the admissible region very quickly [21];
3. the set of agents which received the request is collected into the request itself; in this way there are no loops and duplicate messages;
4. when an agent belonging to the admissible region is found, the other peers belonging to the same region can be reached by suitably navigating through links [21]; in this way the needed set for the group formation is collected and sent to the agent which forwarded the request.

We should remark that, when searching suitable nodes with the fully distributed algorithm discussed above, the privacy is still preserved since relevant data (i.e. the properties mapped into *HADEL* coordinates) are distributed into local software agents and not in a central system. Moreover, once the *HADEL* overlay network has been constructed, the finding process can be iterated by several agents according to the additional requirements provided by their own users.

In the next Section we discuss some experimental results collected by simulating the three phases of the *HADEL* protocol.

4   Evaluation

As stated in Section 3.1, the first phase of the protocol consists in a dissemination of the *HADEL* overlay network construction request in the GN (Gossip Network) which is built and maintained through the collaboration of the *DF* and the software agents. In order to evaluate this phase, a GN of 10^6 agents was generated with the ComplexSim simulation platform [20,22,23].

For the dissemination of the overlay network construction request we adopted the probabilistic gossip protocol which is reported in listing 3a. The behaviour of the gossip protocol is basically tuned by the threshold $v \in \mathbb{R}, v \in (0,1)$. When the threshold v assumes a value near to 1, the gossip message is likely to be propagated to the whole agent's neighbourhood; in this case the *hubs* of the network can generate, in average, too many messages, and can involve in an excessive amount of traffic and workload. Conversely, in order to reach most of the nodes, the threshold v should not be too low. A TTL (Time-To-Live) and a message cache – the latter is maintained by each agent – are needed to stop the process in a few steps and to limit the level of redundancy during the gossip.

```
// message msg, threshold  v
if  msg is  NOT gossip
    return FALSE;
if  msg is  in cache OR msg.ttl  is  0
    return FALSE;
else  // send message
    put  msg in cache;
    create  new msg M;
for  all nn in  neighbours
    if  random_uniform(0,1) < v
        send_msg(M, nn)
return TRUE;
```

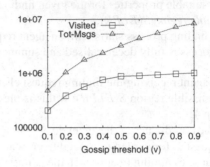

Fig. 3. a) *HADEL* gossip protocol for dissemination. b) Performance of the dissemination phase.

We simulated the dissemination phase and collected the number of peers reached by the gossip and the total number of messages; these simulations were performed for different values of the threshold v and $TTL = 3$. The results are reported in Figure 3b. It can be noted that the gossip can reach about the 80% of the nodes with no more than (about) $3n$ messages (where n is the total number of nodes) with the threshold v in the range $0.4 \div 0.5$.

We report in Figure 4a some characteristics of the resulting *HADEL* overlay network – i.e. the result of the second phase – in terms of maximum and minimum degree (Md and md) and network diameter (l), having set the parameters $deg_{min} = 6$ and $deg_{trg} = 8$ (see Section 3.2). We report only the first 10 steps since we observed that the *HADEL* clustering algorithm leads to a stable network at that time, leading to a network diameter (l) of about 8. At this aim additional details about the clustering algorithm are provided in our previous work[19].

Moreover, the measured clustering coefficient (not shown in Figure 4a) leads to a value around 0.20. These data make evidence that the resulting overlay network exhibits the small world properties (i.e. low geodesic path and high clustering coefficient), which is the basis for an efficient traversal of the network, as we discuss below.

We carried a set of simulations of the finding algorithm (third phase) in the resulting overlay network, and measured the number of time-steps (i.e. the number of hops) to reach the admissible region $S(RSET)$. For this aim the coordinates of the nodes were differently tuned to obtain different ratio for the total number of suitable nodes, i.e. the nodes able to satisfy the constraints of the finding requests. As depicted in Figure 4b, the finding algorithm performs well in terms of time-steps (see 1st quartile, median and 3th quartile) even when the ratio of suitable nodes is rather low (0.1 − 0.2). This result is due to the fact that the agents holding similar properties are connected with high probability, while the diameter of the network is small enough to allow a quick traversal of the network.

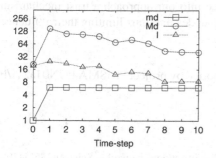 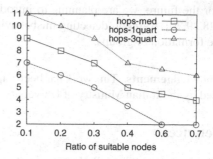

Fig. 4. a) (left) *HADEL* overlay network: md = min. degree, Md = Max degree, l = network diameter. b) (right) Performance of the finding algorithm (ratio of suitable nodes VS time-steps).

5   Conclusions and Future Works

P2P e-Learning systems allow users to collaborate in a virtual e-Learning environment, performing activities as sharing multimedia contents, participating to discussions, lessons and exams, etc. In this context, personalising e-Learning activities is a key issue, and in particular the need of forming groups where the members satisfy personal requirements arises. Although many P2P systems dealing with personalisation issues have been proposed in the past, any of them presents mechanisms to support group formation. In this paper, we have introduced an agent-based P2P approach to support the users of an e-Learning virtual community in the formation of groups composed of users having similar interests and preferences. Our approach is based on an overlay network of software agents, which allows the users to preserve their privacy, locally maintaining sensitive information, and at the same time provides an efficient mechanism for each user to find promising interlocutors. Each personal agent operates on its user's behalf, analysing the local user's information in order to extract relevant properties which are then exploited to select the most appropriate interlocutors. The selected users are then organised into a custom overlay network called *HADEL*, built by exchanging mutual messages between the software agents using a gossip protocol.

An experimental evaluation we have performed on a large network of simulated users shows that our approach is capable of reaching approximately all the agents in a small number of steps and with a limited number of exchanged messages. The experiments also show that the search of suitable nodes is very efficient due to the topology of the network, which exhibits the small-world properties. We highlight that the main advantage of our approach is that the local information about each user involved in group formation are collected by personal agents and used to communicate with each other, without revealing them, and thus preserving the users' privacy with respect to their sensitive data.

It is important to remark that in our framework we have assumed a "honest "behaviour of the users belonging to the e-Learning community when declaring their properties. In other words, in this current version, our framework does not cover the possibility of misleading or fraudulent behaviours, that would compromise the result of the group formation activity.

For the future, we are planning to introduce into our approach a trust mechanism [16,28,29,30], trying to detecting misbehaving nodes and thus limiting their influence in the formation of groups.

Acknowledgements. This work has been supported by project PRISMA PON04a2 A/F funded by the Italian Ministry of University.

References

1. Barabási, A.L., Albert, R.: Emergence of scaling in random networks. Science 286(5439), 509–512 (1999)
2. Chan, E.: An innovative learning approach: Integrate peer-to-peer learning into blended learning. International J. of Global Education 1(1) (2012)
3. Chen, C.M., Lee, H.M., Chen, Y.H.: Personalized e-learning system using item response theory. Computers & Education 44(3), 237–255 (2005)
4. De Meo, P., Nocera, A., Quattrone, G., Rosaci, D., Ursino, D.: Finding reliable users and social networks in a social internetworking system. In: Proc. of the 2009 Int. Database Engineering & Applications Symp., pp. 173–181. ACM (2009)
5. De Meo, P., Nocera, A., Rosaci, D., Ursino, D.: Recommendation of reliable users, social networks and high-quality resources in a social internetworking system. AI Communications 24(1), 31–50 (2011)
6. Di Stefano, A., Messina, F., Pappalardo, G., Santoro, C., Toscano, L.: Evaluating strategies for resource finding in a peer-to-peer grid. In: 16th IEEE International Workshops on WET-ICE 2007, pp. 290–295 (2007)
7. Dolog, P., Henze, N., Nejdl, W., Sintek, M.: Personalization in distributed e-learning environments. In: Proc. of the 13th Int. World Wide Web Conf. on Alternate Track Papers & Posters, pp. 170–179. ACM (2004)
8. Ferber, J.: Multi-agent systems: an introduction to distributed artificial intelligence, vol. 1. Addison-Wesley, Reading (1999)
9. Garruzzo, S., Rosaci, D., Sarné, G.M.L.: ISABEL: A multi agent e-learning system that supports multiple devices. In: Proc. of the 2007 International Conference on Intelligent Agent Technology (IAT 2007), pp. 485–488. IEEE (2007)
10. Garruzzo, S., Rosaci, D., Sarné, G.M.L.: Masha-el: A multi-agent system for supporting adaptive e-learning. In: 19th IEEE International Conference on Tools with Artificial Intelligence, ICTAI 2007, vol. 2, pp. 103–110. IEEE (2007)
11. Huang, L.: e-learning based on semantic P2P networks. In: 2012 3rd Int. Conf. on Networking and Distributed Computing, pp. 105–107. IEEE (2012)
12. Jin, H., Yin, Z., Yang, X., Wang, F., Ma, J., Wang, H., Yin, J.: APPLE: A novel P2P based e-learning environment. In: Sen, A., Das, N., Das, S.K., Sinha, B.P. (eds.) IWDC 2004. LNCS, vol. 3326, pp. 52–62. Springer, Heidelberg (2004)
13. Kritikou, Y., Demestichas, P., Adamopoulou, E., Demestichas, K., Theologou, M., Paradia, M.: User profile modeling in the context of web-based learning management systems. JNCA 31(4), 603–627 (2008)
14. Kuramochi, K., Kawamura, T., Sugahara, K.: Nat traversal for pure P2P e-learning system. In: 3rd Int. Conf. on Internet and Web Applications and Services, pp. 358–363. IEEE (2008)
15. Messina, F., Pappalardo, G., Rosaci, D., Santoro, C., Sarné, G.M.L.: HySoN: A distributed agent-based protocol for group formation in online social networks. In: Klusch, M., Thimm, M., Paprzycki, M. (eds.) MATES 2013. LNCS, vol. 8076, pp. 320–333. Springer, Heidelberg (2013)

16. Messina, F., Pappalardo, G., Rosaci, D., Santoro, C., Sarné, G.M.L.: A trust-based approach for a competitive cloud/Grid computing scenario. In: Fortino, G., Badica, C., Malgeri, M., Unland, R. (eds.) Intelligent Distributed Computing VI. SCI, vol. 446, pp. 129–138. Springer, Heidelberg (2013)

17. Messina, F., Pappalardo, G., Santoro, C.: A self-organising system for resource finding in large-scale computational grids. In: Proc. of WOA (2010)

18. Messina, F., Pappalardo, G., Santoro, C.: Hygra: A decentralized protocol for resource discovery and job allocation in large computational grids. In: Proceedings of ISCC, pp. 817–823. IEEE (2010), doi:10.1109/ISCC.2010.5546559

19. Messina, F., Pappalardo, G., Santoro, C.: Exploiting the small-world effect for resource finding in P2P grids/clouds. In: Proc. 20th IEEE Int. Work. on Enabling Technologies: Infrastructures for Collaborative Enterprises, pp. 122–127 (2011)

20. Messina, F., Pappalardo, G., Santoro, C.: Complexsim: An smp-aware complex network simulation framework. In: 2012 6th Int. Conf. on Complex, Intelligent and Software Intensive Systems (CISIS), pp. 861–866. IEEE (2012)

21. Messina, F., Pappalardo, G., Santoro, C.: Decentralised resource finding in cloud/grid computing environments: A performance evaluation. In: IEEE 21st Int. Work. on Enabling Technologies: Infrastructure for Collaborative Enterprises (WETICE), pp. 143–148. IEEE (2012)

22. Messina, F., Pappalardo, G., Santoro, C.: Complexsim: A flexible simulation platform for complex systems. International J. of Simulation and Process Modelling (2013)

23. Messina, F., Santoro, C., Pappalardo, G.: Exploiting gpus to simulate complex systems. In: 2013 Seventh International Conference on Complex, Intelligent and Software Intensive Systems (CISIS), pp. 535–540 (2013)

24. Navarro-Estepa, Á., Xhafa, F., Caballé, S.: A P2P replication-aware approach for content distribution in e-learning systems. In: 6th Int. Conf. on Complex, Intelligent and Software Intensive Systems, pp. 917–922. IEEE (2012)

25. Nejdl, W., Wolpers, M., Siberski, W., Schmitz, C., Schlosser, M., Brunkhorst, I., Löser, A.: Super-peer-based routing and clustering strategies for rdf-based peer-to-peer networks. In: Proceedings of the 12th Int. Conf. on World Wide Web, pp. 536–543. ACM (2003)

26. Rosaci, D., Sarné, G.M.L.: Efficient personalization of e-learning activities using a multi-device decentralized recommender system. Computational Intelligence 26(2), 121–141 (2010)

27. Rosaci, D., Sarné, G.M.L.: Matching users with groups in social networks. In: Zavoral, F., Jung, J.J., Badica, C. (eds.) Intelligent Distributed Computing VII. SCI, vol. 511, pp. 45–54. Springer, Heidelberg (2014)

28. Rosaci, D., Sarnè, G.M.L., Garruzzo, S.: Integrating trust measures in multiagent systems. International Journal of Intelligent Systems 27(1), 1–15 (2012)

29. Sabater, J., Paolucci, M., Conte, R.: Repage: Reputation and image among limited autonomous partners. J. Artificial Societies and Social Simulation 9(2) (2006)

30. Sabater, J., Sierra, C.: Regret: reputation in gregarious societies. In: Proceedings of the Fifth International Conference on Autonomous Agents, AGENTS 2001, pp. 194–195. ACM, New York (2001)

31. Talia, D., Trunfio, P.: Toward a synergy between P2P and grids. IEEE Internet Computing 7(4), 94–96 (2003)

32. Watts, D., Strogatz, S.J.: Collective dynamics of 'small-world' networks. Nature 393(6684), 440–442 (1998)

Identification of Dynamical Structures in Artificial Brains: An Analysis of Boolean Network Controlled Robots

Andrea Roli[1,3], Marco Villani[2,3], Roberto Serra[2,3],
Lorenzo Garattoni[4], Carlo Pinciroli[4], and Mauro Birattari[4]

[1] Dept. of Computer Science and Engineering (DISI) - Università di Bologna, Italy
[2] Dept. of Physics, Informatics and Mathematics, Università di Modena e Reggio Emilia, Modena, Italy
[3] European Centre for Living Technology, Venezia, Italy
[4] IRIDIA, Université Libre de Bruxelles, Belgium

Abstract. Automatic techniques for the design of artificial computational systems, such as control programs for robots, are currently achieving increasing attention within the AI community. A prominent case is the design of artificial neural network systems by means of search techniques, such as genetic algorithms. Frequently, the search calibrates not only the system parameters, but also its structure. This procedure has the advantage of reducing the bias introduced by the designer and makes it possible to explore new, innovative solutions. The drawback, though, is that the analysis of the resulting system might be extremely difficult and limited to few coarse-grained characteristics. In this paper, we consider the case of robots controlled by Boolean networks that are automatically designed by means of a training process based on local search. We propose to analyse these systems by a method that detects mesolevel dynamical structures. These structures are emerging patterns composed of elements that behave in a coherent way and loosely interact with the rest of the system. In general, this method can be used to detect functional clusters and emerging structures in nonlinear discrete dynamical systems. It is based on an extension of the notion of *cluster index*, which has been previously proposed by Edelman and Tononi to analyse biological neural systems. Our results show that our approach makes it possible to identify the computational core of a Boolean network which controls a robot.

1 Introduction

The design of artificial systems by means of automatic techniques, such as evolutionary computation techniques, is a well-known approach in the AI community since decades. A plethora of studies has been published in the literature with the aim of showing properties, conditions and characteristics of the emergence of intelligent behaviours. A case in point is the study of the emergence of non-trivial cognitive capabilities in robots, such as sensory-motor coordination [9,10].

M. Baldoni et al. (Eds.): AI*IA 2013, LNAI 8249, pp. 324–335, 2013.

Besides these objectives of foundational and investigation flavour, the automatic design of artificial system is currently achieving increasing attention also because it opens the possibility of designing innovative systems or finding solutions for complex problems that are quite hard to solve with classical approaches [3,6]. One of the main advantages of automatic design is that the designer can specify just the minimal set of constraints and objectives on the final system; for example, when designing neural network systems, one can let the design process decide the topology of the network together with the connection weights. This makes it possible to explore larger design spaces than those explored by classical design techniques, but it has the drawback that the resulting system might be very difficult to analyse. In this work, we propose to analyse these systems by a method that detects *mesolevel dynamical structures*, i.e., emerging patterns composed of elements that behave in a coherent way and loosely interact with the rest of the system. The method is based on an extension of the notion of *cluster index*, which has been previously proposed by Edelman and Tononi for analysing biological neural systems [15] and can be used to detect functional clusters and emerging structures in nonlinear discrete dynamical systems. We apply this method to analyse the Boolean network trained to control a robot that must be able to walk along a corridor without collisions. Results show that this makes it possible to identify structures inside the network which perform main information processing jobs. The results we present are preliminary, but, even if at this early stage, we believe they anyway show the potential of this method so that to motivate their diffusion to the AI community. Indeed, the approach is very general as it only requires a collection of states traversed by the system and it does not need information on the topology nor the functions of the elements composing the system. Nevertheless, this method is able to identify structures related to relevant information processing in the actual functioning of the system.

We provide an introduction to the method in Section 2. In Section 3 we illustrate the case study we analyse, namely an application of Boolean networks to robotics. We present the results in Section 4 and we conclude with Section 5.

2 The Cluster Index

In the following, we consider a system U, composed of N elements that can assume values in finite and discrete domains and update their state in discrete time. The value of element i at time $t+1$, $x_i(t+1)$, is a function of the values of a fixed set elements in U at time t. The cluster index is defined with the aim of identifying subsets of U composed of elements that interact much more strongly among themselves than with the rest of the system, i.e., subsets whose elements are characterised by being both *integrated* among themselves and *segregated* w.r.t. the rest of the system. The quantity upon which the cluster index is computed is the entropy of single as well as sets of elements of U. The entropy of element x_i is defined as:

$$H(x_i) = - \sum_{v \in V_i} p(v) \ log \ p(v) \tag{1}$$

where V_i is the set of the possible values of x_i and $p(v)$ the probability of occurrence of symbol v. The entropy of a pair of elements x_i and x_j is defined upon joint probabilities:

$$H(x_i, x_j) = - \sum_{v \in V_i} \sum_{w \in V_j} p(v, w) \ log \ p(v, w). \tag{2}$$

This definition can be extended to sets of k elements by considering the probability of occurrence of vectors of k values. We can estimate the entropy of each element from a long series of states by taking the frequencies of its observed values as proxies for probabilities. Therefore, the sole piece of information we need is a collection of system states, which can be taken by observing the system in different working conditions. For example, the collection can be obtained by composing several trajectories in the state space; however, there are no requirements on the sequence of these states, because the collection is only used to compute frequencies. Once the data have been collected, relevant sets of elements (clusters, from now on) are evaluated by means of the cluster index. A relevant cluster should be composed of elements *(i)* that possess high integration among themselves and *(ii)* that are loosely coupled to other parts of the system. The measure we define provides a value that can be used to rank various candidate clusters (i.e., emergent intermediate-level sets of coordinated elements). Depending on the size of the system, candidate clusters can be exhaustively enumerated, or sampled, or searched by means of suitable heuristics. The cluster index $C(S)$ of a set S of k elements is defined as the ratio of their *integration* $I(S)$ to the *mutual information* between S and the rest of the system $U - S$. The integration is defined as follows:

$$I(S) = \sum_{x \in S} H(x) - H(S) \tag{3}$$

$I(S)$ measures the deviation from statistical independence of the k elements in S, by subtracting the entropy of the whole subset to the sum of the single-node entropies. The mutual information between S and the rest of the system $U - S$ is:

$$M(S; U - S) \equiv H(S) + H(S|U - S) = H(S) + H(U - S) - H(S, U - S) \tag{4}$$

where, as usual, $H(A|B)$ is the conditional entropy and $H(A, B)$ the joint entropy. Finally, the cluster index $C(S)$ is defined by:

$$C(S) = \frac{I(S)}{M(S; U - S)} \tag{5}$$

Special cases are: $I = 0 \wedge M \neq 0 \ \Rightarrow \ C(S) = 0$ and $M = 0 \ \Rightarrow \ C(S)$ not defined. These cases can be diagnosed in advance. The $0/0$ case does not provide any information, whereas $I(S)/0$, with $I(S) \neq 0$, denotes statistical independence of S from the rest of the system and requires a separate analysis.

$C(S)$ scales with the size of S, so a loosely connected subsystem may have a larger index than a more coherent, smaller one. Therefore, to compare the indices of the various candidate clusters, it is necessary to normalise their cluster indices. To this aim, we need a reference system with no clusters, i.e., an homogeneous system, which we define as follows: given a series of states from the system we want to study, we compute the frequency of each symbol and generate a new random series in which each symbol appears with probability equal to that of the original series. The homogeneous system provides us with reference values for the cluster index and makes it possible to compute a set of normalisation constants: for each subsystem size, we compute average integration $\langle I_h \rangle$ and mutual information $\langle M_h \rangle$. We can then normalise the cluster index value of any subsystem S using the appropriate normalisation constants dependent on the size of S:

$$C'(S) = \frac{I(S)}{\langle I_h \rangle} / \frac{M(S; U - S)}{\langle M_h \rangle} \tag{6}$$

In order to compute a statistical significance index (hereinafter referred to as T_c), we apply this normalisation to both the cluster indices in the analysed system and in the homogeneous system:

$$T_c(S) = \frac{C'(S) - \langle C'_h \rangle}{\sigma(C'_h)} \tag{7}$$

where $\langle C'_h \rangle$ and $\sigma(C'_h)$ are respectively the average and the standard deviation of the population of normalised cluster indices with the same size of S from the homogeneous system [2]. Finally, we use T_c to rank the clusters considered.

We have recently applied our method to both artificial test cases and representative natural and artificial systems, such as genetic regulatory networks and catalytic reactions systems [16,17]. In the following, we show that the method can be used also to analyse the network controlling a robot. This case is particularly meaningful because of two main reasons. The first is that it explicitly concerns a system equipped with inputs and outputs, which operates in an environment. Therefore, the relevant structures inside the network are necessarily linked to the interplay between robot behaviour and environment. The second reason is that the networks resulting at the end of the training process are not easily analysable because they have a random topology and the nodes are updated according to Boolean functions: we will show that the method is able to capture relevant structures inside the network without requiring knowledge about network topology and functions.

3   BN-Robot Case Study

We apply our method to analyse the networks trained to control a robot that performs obstacle avoidance. The robot is an *e-puck* [7] and it is controlled by a Boolean network. Boolean networks (BNs) are a model of genetic regulatory networks [5]. BNs have received considerable attention in the community of complex system science. Works in complex systems biology show that BNs provide a powerful model for cellular dynamics [1,13,14]. A BN is a discrete-time discrete-state dynamical system whose state is a N-tuple in $\{0,1\}^N$ and it is updated according to the composition of N Boolean functions, each of which rules the update of one variable of the system. Usually, BNs are subject to a synchronous and parallel node update, but other update schemes are possible. BNs are extremely interesting from an engineering perspective because of their ability to display complex and rich dynamics, despite the compactness of their description. In a previous work, it has been shown that BNs can be used to control robots [12].

In the case study, the robot must navigate along a corridor avoiding any collision with the walls and possible obstacles and finally reach the exit. At the beginning of each experimental run, the robot is placed within the corridor, far from the exit. During the experiment the robot must advance along the corridor avoiding collisions, and finally, within the given total execution time $T = 120$ s, reach the exit. During the execution, if a collision between the robot and the walls of the corridor occurs, the experiment is immediately stopped. Experiments are performed in simulation, by means of the open source simulator ARGoS [11]. The performance measure is the final distance of the robot to the exit (normalised w.r.t. corridor length). The shorter is this distance, the better is the performance of the robot. The robot is equipped with four proximity sensors, placed at positions NE, SE, SW and NW with respect to the heading direction, and with two wheels. At each time step, the readings of the 4 sensors are encoded into the values of the BN input nodes. We use 4 input nodes to encode the readings of the proximity sensors. Values are binarised by introducing a threshold: if the sensor value exceeds the chosen threshold, the corresponding input node value is set to 1. Once the readings of the sensors are encoded in the input nodes, we perform the network state update, and eventually we read and decode the values of the output nodes to set the actuators. Two output nodes are used to set the wheel speeds either to zero or to a predefined, constant value. The robot update frequency is 100 ms. For this case study, we set the network size to 20 nodes. The initial topology of the networks, i.e., the connections among the nodes, is randomly generated with $K = 3$ (i.e., each node has 3 incoming arcs) and no self-connections, and it is kept constant during the training. The initial Boolean functions are generated by setting the 0/1 values in the node Boolean function truth tables uniformly at random. BNs are trained by a local search algorithm which works only on the Boolean functions. At each iteration, the search algorithm flips one bit of a Boolean function. The flip is accepted if the corresponding BN-robot has a performance not worse than the current one.[1] The evaluation of each network is performed on a set of initial conditions, that form the training set. The training set is composed of six different initial orientations of the robot. The six angles are chosen so as to have six equally spaced orientations in the range between $\frac{\pi}{3}$ and $-\frac{\pi}{3}$ (with 0 being the straight direction of the robot towards the exit). In this way, the robot must be able to cope with a wide range of different situations and avoid the walls it detects in any

---

[1] This simple local search algorithm is often called *stochastic descent* or *adaptive walk*.

direction. The final evaluation assigned to the robot is computed as the average of the performance across the 6 trials. We executed 100 independent experiments, each corresponding to a different initial network. For each experiment the local search was run for 1000 iterations. The assessment of the performance of the BN-robots is performed on a test set composed of randomly generated initial positions and orientation. All the final BNs achieved a good performance. For further details, we point the interested reader to a comprehensive report on the experiment [4].

4   Identifying Structures in the Boolean Network Controlling a Robot

We analysed the best performing BNs by means of the method based on the cluster index, with the aim of identifying their core structure, if any. It is important to stress that this method detects dynamical relations among nodes of the system, without requiring information on the topology nor on the function computed by the single nodes. The objective is to identify parts of the system which are dynamically coordinated, rather than to discover topological patterns, such as communities and motifs. Furthermore, this technique captures correlations among sets of nodes and not just between pairs.

In the following, we describe the results of two BN-robots we obtained, which are typical examples. In both cases, the BN-robot learned to walk along the corridor without colliding against the walls and obstacles in the path. The data required by this analysis are simply a collection of states of the system. To this aim, we recorded the states traversed by the BN controlling the robot starting from 200 random initial positions. The length of the trajectories is 1000 steps. We then analysed these states, and searched for clusters of size up to 19 (i.e., $N - 1$) by taking 10^4 random samples for each cluster size.

4.1   Network 1

We analysed the collection of BN states related to the test runs of the BN-robot by looking for the cluster with highest significance T_c. The most significant cluster has size 8, with $T_c \approx 48 \times 10^3$. The following clusters have a much lower significance (about 36×10^3), they are of size 9, and they all contain the first cluster of 8 nodes. Therefore, they are not particularly meaningful. The identified cluster is composed of input, output and internal nodes, namely nodes 1, 3, 5, 7, 8, 10, 12 and 18 of the network depicted in Figure 1. Besides input nodes 1 (NW proximity sensor), and 3 (SE proximity sensor) and node 5 which controls the left wheel, the cluster contains nodes involved in internal dynamics.

The nodes of the cluster should convey relevant information on the overall BN dynamics. A way to reckon the impact of a cluster on the system dynamics is to evaluate its *set-influence*. The notion of influence of a set of nodes is a generalisation of the *node influence*, which is defined as the amount of perturbation induced on a set of nodes by a state change in one node. Informally, the influence

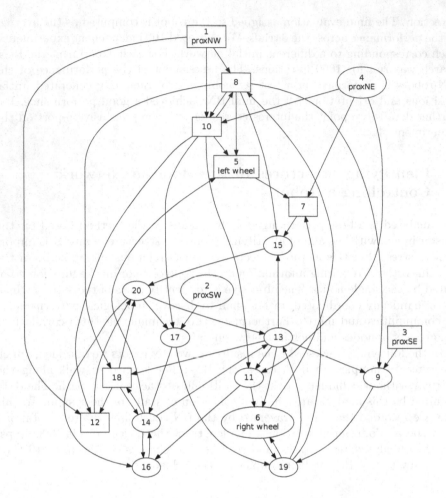

Fig. 1. Structure of BN-1 controlling the robot. Nodes drawn with rectangles are in the most significant cluster.

of a node on the other nodes is the size of the *avalanche* produced by the node perturbation [14]. To estimate the influence of a set S of nodes on the other nodes of the system, $S' = U - S$, we randomly perturb the nodes in S^2 at time t and we compare the state at time $t+l$ in the case with and without perturbation (with time lag $l \in \{1, 2, \ldots, 10\}$). The normalised Hamming distance between the two states is the avalanche produced by perturbing S. We estimated the set-influence of cluster $\{1,3,5,7,8,10,12,18\}$ and of random clusters of the same size by taking the average avalanche over 100 random initial states. Results are shown in Figure 2. As we can observe, cluster $\{1,3,5,8,10,12,18\}$ has a higher set-

---

[2] More precisely, the perturbation is performed by flipping each node state with probability 0.5 .

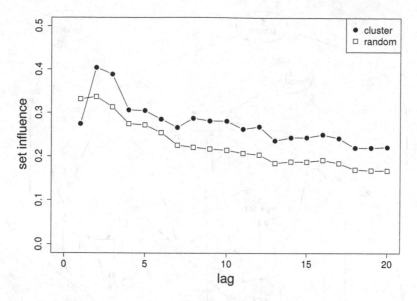

Fig. 2. Set-influence of cluster {1,3,5,7,8,10,12,18} compared with the set-influence of random clusters of the same size (averaged over 100 random clusters)

influence than random sets,[3] hence the functionality of the BN strongly depends on the identified cluster. One might speculate that the most relevant cluster is characterised by a higher set-influence w.r.t. the others, and therefore the search for a significant dynamical structure in the system can be simply performed by inspecting the sets with the highest set-influence. Nevertheless, it should be observed that we are looking for structures relevant for information processing inside the system. These structures are relevant for the *actual functioning* of the system and might not be simply reduced to sets of nodes with high average influence: a node can indeed have high influence by influencing other nodes which are not relevant for the actual dynamics of the system. The next case we present is a prominent example of this phenomenon.

4.2   Network 2

The second BN we analyse has a different topology w.r.t. the first one, because it has been generated with a different random number generator seed. As also the initial BN functions are randomly generated and the search algorithm is stochastic, the training process led to a different solution which anyway achieves the goal.

In this case, the most significant cluster has size 3 and it has a significance value $T_c \approx 1.3 \times 10^6$. The following clusters have a much lower significance (about

---

[3] Except for the case of $l = 1$, which means that the cluster needs some time to spread the information across the network.

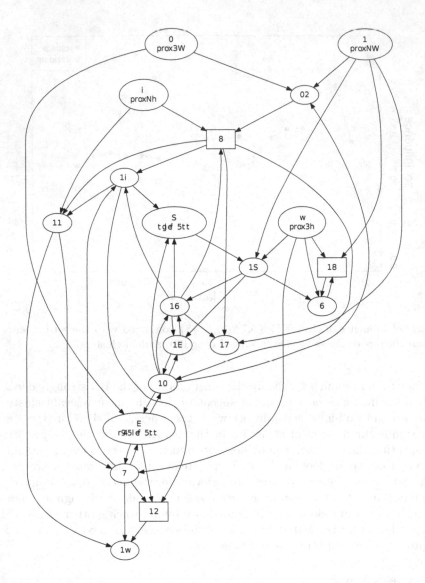

Fig. 3. Structure of BN-2 controlling the robot. Nodes drawn with rectangles are in the most significant cluster.

9.8×10^5) and they are of size 4, all including the first cluster of size 3. The identified cluster is composed of nodes {8,10,18} and it does not contain input nor output nodes (see Figure 3). This result might seem quite surprising, but by looking at the network topology we can observe that (directly or indirectly):

- node 8 collects (and processes) the information coming from sensors SW, NW and NE; moreover, it acts on the right wheel;

- node 10 collects the information coming from node 8 and sensor SE (which produces values obviously correlated with the information coming from the other sensors SW, NW and NE);
- node 18 collects the information coming from sensors SE and NW.

Therefore, these observations help identify sensors SE and NW, and node 8 as the central part of the robot information processing unit, with nodes 10 and 18 playing the role of faithful followers. The set-influence of the cluster is compared against the set-influence of random clusters of the same size in Figure 4. The set-influence of the cluster is lower than the average one; therefore, this sole piece of information would not be sufficient to detect this structure. The reason for low set-influence but high cluster index is very likely that nodes 10 and 18 have low influence on the network, hence low mutual information between the cluster and the remainder of the system. A single node influence analysis is anyway helpful to understand the reason why node 10 is chosen by our method instead of an output node. Node influence is evaluated by computing the *influence matrix* (defined for lag l), in which entry (i, j) is the influence of node i on node j at time l. The influence at lag l is computed as the fraction of times in which the value of node j is affected by a flip in node i occurred l time steps before. If we rank the nodes by influence, among the most five influential nodes on node 10 we find node 6 (left wheel) with the highest influence at lag equal to 1 and nodes 1,2,3 and 4 (i.e., the sensors) for higher time lag values.

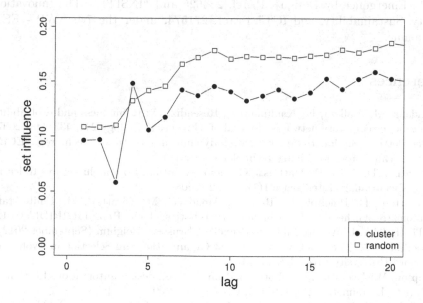

Fig. 4. Set-influence of cluster {8,10,18} compared with the set-influence of random clusters of the same size (averaged over 100 random clusters)

We can conclude that the method based on the cluster index is able to detect subsystems which play a main role in the information processing inside the system: they can be both the causal core of the functioning of the system and a proxy for observing its main dynamical properties.

5 Conclusion

The results we have presented show that the method based on the cluster index can help us detect relevant structures which emerge as the result of the dynamics of the system. This work is at a preliminary stage and further analyses are required to assess the informative power of the cluster index and its possible applications. For example, our method can provide heuristics to reduce the network by pruning irrelevant nodes and links, or it can be used to extract a minimal subset of nodes to observe the system. In addition, whilst in this work we only focused on checking whether our results are meaningful w.r.t. the dynamics of the system, our method can be used as a first step in the identification of *functional modules* of the system [8]. In the future, our method may be applied to detect structures emerging either during the learning process of a single system, or in the evolution of populations of systems, or both.

Acknowledgements. This article has been partially funded by the UE projects "MD – Emergence by Design", Pr.ref. 284625 and "INSITE - The Innovation Society, Sustainability, and ICT" Pr.ref. 271574, under the 7th FWP - FET programme.

References

1. Aldana, M., Balleza, E., Kauffman, S., Resendiz, O.: Robustness and evolvability in genetic regulatory networks. Journal of Theoretical Biology 245, 433–448 (2007)
2. Benedettini, S.: Identifying mesolevel dynamical structures. Tech. rep., ECLT (European Center for Living Technologies) (2013)
3. Floreano, D., Dürr, P., Mattiussi, C.: Neuroevolution: from architectures to learning. Evolutionary Intelligence 1(1), 47–62 (2008)
4. Garattoni, L., Pinciroli, C., Roli, A., Amaducci, M., Birattari, M.: Finite state automata synthesis in boolean network robotics. Tech. Rep. TR/IRIDIA/2012-017, IRIDIA, Université Libre de Bruxelles, Brussels, Belgium (September 2012)
5. Kauffman, S.: The Origins of Order: Self-Organization and Selection in Evolution. Oxford University Press, UK (1993)
6. Lipson, H.: Evolutionary robotics and open-ended design automation. In: Cohen, B. (ed.) Biomimetics, pp. 129–155. CRC Press (2005)
7. Mondada, F., Bonani, M., Raemy, X., Pugh, J., Cianci, C., Klaptocz, A., Magnenat, S., Zufferey, J.C., Floreano, D., Martinoli, A.: The e-puck, a robot designed for education in engineering. In: Gonç alves, P., Torres, P., Alves, C. (eds.) Proceedings of the 9th Conference on Autonomous Robot Systems and Competitions, vol. 1, pp. 59–65 (2009)

8. Müller, G., Wagner, G., Callebaut, W. (eds.): Modularity – Understanding the Development and Evolution of Natural Complex Systems. The Vienna Series in Theoretical Biology. The MIT Press (2005)
9. Nolfi, S., Floreano, D.: Evolutionary robotics. The MIT Press, Cambridge (2000)
10. Pfeifer, R., Scheier, C.: Understanding Intelligence. The MIT Press (2001)
11. Pinciroli, C., Trianni, V., O'Grady, R., Pini, G., Brutschy, A., Brambilla, M., Mathews, N., Ferrante, E., Di Caro, G., Ducatelle, F., Birattari, M., Gambardella, L., Dorigo, M.: ARGoS: A modular, multi-engine simulator for heterogeneous swarm robotics. Swarm Intelligence 6(4), 271–295 (2012)
12. Roli, A., Manfroni, M., Pinciroli, C., Birattari, M.: On the design of Boolean network robots. In: Di Chio, C., et al. (eds.) EvoApplications 2011, Part I. LNCS, vol. 6624, pp. 43–52. Springer, Heidelberg (2011)
13. Serra, R., Villani, M., Barbieri, A., Kauffman, S., Colacci, A.: On the dynamics of random Boolean networks subject to noise: Attractors, ergodic sets and cell types. Journal of Theoretical Biology 265(2), 185–193 (2010)
14. Serra, R., Villani, M., Graudenzi, A., Kauffman, S.: Why a simple model of genetic regulatory networks describes the distribution of avalanches in gene expression data. Journal of Theoretical Biology 246, 449–460 (2007)
15. Tononi, G., McIntosh, A., Russel, D., Edelman, G.: Functional clustering: Identifying strongly interactive brain regions in neuroimaging data. Neuroimage 7, 133–149 (1998)
16. Villani, M., Benedettini, S., Roli, A., Lane, D., Poli, I., Serra, R.: Identifying emergent dynamical structures in network models. In: Proceedings of WIRN 2013 – Italian Workshop on Neural Networks (2013)
17. Villani, M., Filisetti, A., Benedettini, S., Roli, A., Lane, D., Serra, R.: The detection of intermediate-level emergent structures and patterns. In: Liò, P., Miglino, O., Nicosia, G., Nolfi, S., Pavone, M. (eds.) Advances in Artificial Life, ECAL 2013, pp. 372–378. The MIT Press (2013)

Semantic Annotation of Scholarly Documents and Citations

Paolo Ciancarini[1,2], Angelo Di Iorio[1], Andrea Giovanni Nuzzolese[1,2],
Silvio Peroni[1,2], and Fabio Vitali[1]

[1] Department of Computer Science and Engineering, University of Bologna, Italy
[2] STLab-ISTC, Consiglio Nazionale delle Ricerche, Italy
{ciancarini,diiorio,nuzzoles,essepuntato,fabio}@cs.unibo.it

Abstract. Scholarly publishing is in the middle of a revolution based
on the use of Web-related technologies as medium of communication.
In this paper we describe our ongoing study of semantic publishing and
automatic annotation of scholarly documents, presenting several models and tools for the automatic annotation of structural and semantic
components of documents. In particular, we focus on citations and their
automatic classification obtained by CiTalO, a framework that combines
ontology learning techniques with NLP techniques.

Keywords: CiTO, PDF jailbreaking, Semantic Web, citation networks,
citation patterns, semantic annotations, semantic publishing.

1 Introduction

Researchers can access thousands or even millions of scholarly papers. For instance, currently (June 2013) the ACM Digital Library has 382,000 full text
papers and 2,128,000 bibliographic records. There are basically two ways of accessing these sources. First of all, the papers whose identifiers are known (e.g.
DOI, or URI) can be downloaded as PDF files, HTML pages, or other formats.
Second, most papers include bibliographic records (about authors, editors, publication venue and date, etc.) and lists of references to other works that, in
turn, are supplied with similar material. The result is a huge knowledge-base of
references, connections between research works, and valuable pieces of content.

We believe that such knowledge bases are still under exploited. One of the reasons is that a lot of interesting information is not available in a machine-readable
format. While the human readers can access PDF files and harvest information
directly, the software agents can only rely upon a quite limited amount of data for
automatic extraction and interpretation or reasoning. Much more sophisticated
applications can be built if larger and more expressive semantic information were
available.

That is why we are witnessing to an evolution in scholarly publishing driven by
the Semantic Publishing community [16] [4] [14]. The idea is to offer a fully-open
access to both content and metadata, where rich data on the internal structures
of the documents, their components, their (semantic) connections with other

M. Baldoni et al. (Eds.): AI*IA 2013, LNAI 8249, pp. 336–347, 2013.

documents and, in particular, their bibliographic references are available as RDF statements and according to appropriate OWL ontologies. This paves the road to the full integration of existing knowledge bases with other Linked Open Data silos, as well as automatic inferences and information extraction.

This paper presents some highlights of our vision and ongoing research on semantic publishing and automatic annotation of scholarly documents, which include models and tools to semantically annotate scholarly documents. In addition, we introduce in more depth the task of automatic characterisation of citations, exploiting *CiTalO*, a tool that takes as input a sentence containing a citation and infers its nature as modelled in *CiTO*[1] [12], an ontology describing the factual and rhetoric functions of citations (part of a larger ecosystem of ontologies for scholarly publishing called *Semantic Publishing and Referencing – SPAR – Ontologies*[2]). Recent developments of CiTalO are presented, with specific attention to the integration with *PDFX*, an online tool that takes a PDF of a scientific article as input and returns an XML linearisation of it.

These tools are meant to be part of a larger platform on top of which novel services for analysing, collecting and accessing scholarly documents will be built. Our long-term goal is to analyse automatically the pertinence of documents to some research areas, to discover research trends, to discover how research results are propagated, to develop new metrics for evaluating research and to build sophisticated recommenders.

The paper is then structured as follows. In Section 2 we introduce some relevant projects in the context of Semantic Publishing and semantic characterisation of scholarly documents. In Section 3 we give an overview of our vision. CiTalO and its recent developments are presented in Section 4. Section 5 concludes the paper presenting some open issues and future works.

2   Related Works

The semantic annotation and enrichment of scholarly documents, and in particular the adoption of Semantic Web technologies, is a hot research topic. In mid-2010, JISC (the Joint Information Systems Committee, a British funding body) funded two sister projects: the *Open Citation project*[3] and the *Open Bibliography project*[4], held respectively by the University of Oxford and the University of Cambridge. Both projects have the broad goal to study the feasibility, advantages, and applications of using RDF datasets and OWL ontologies when describing and publishing bibliographic data and citations.

The Open Citation project aims at creating a semantic infrastructure that describes articles as bibliographic records and their citations to other related works. One of the outcomes of the project was the development of a suite of ontologies, called *Semantic Publishing And Referencing (SPAR)*, whose aim is to describe

---

[1] CiTO, the Citation Typing Ontology: http://purl.org/spar/cito.

[2] Semantic Publishing and Referencing Ontologies homepage: http://purl.org/spar.

[3] Open Citation project blog: http://opencitations.wordpress.com.

[4] Open Bibliography project blog: http://openbiblio.net.

bibliographic entities such as books and journal articles, reference citations, the organisation of bibliographic records and references into bibliographies, ordered reference lists and library catalogues, the component parts of documents, and publishing roles, publishing statuses and publishing workflows. SPAR has been used in the Open Citation project as reference model to create a corpus of interlinked bibliographic records[5] obtained converting the whole set of reference lists contained in all the PubMed Central Open Access articles[6] into RDF data. The converted RDF data are published as Linked Open Data.

Similarly, the Open Bibliographic project aimed at publishing a large corpus of bibliographic data as Linked Open Data, starting from four different sources: the Cambridge University Library[7], the British Library[8], the International Union of Crystallography[9] and PubMed[10]. The original publishers' models have been modified to natively include the open publication of bibliographic data as Linked Open Data; furthermore, the scholarly community was continuously engaged in the development of both the ontological model and the final data.

The *Lucero project*[11] is another JISC project, held by the Open University, which aims at exploring the use of Linked Data within the academic domain. In particular, it proposes solutions that could take advantages from the Linked Data to connect educational and research content, so as students and researches could benefit from semantic technologies.

Lucero main aims are:

- to promote the publication as Linked Open Data of bibliographic data[12] through a tool to facilitate the creation and use of semantic data;
- to identify a process in order to integrate the Linked Data publication of bibliographic information as part of the University's workflows;
- to demonstrate the benefits derived from exposing and using educational and research data as Open Linked Data, through the development of applications that improve the access to those data.

The automatic analysis of networks of citations and bibliographic data is gaining importance in the research community. Copestake *et al.* [3] present an infrastructure called *SciBorg* that allows one to automatically extract semantic characterisations of scientific texts. The project was a collaboration between the Computer Laboratory of the University of Cambridge, the Unilever Centre for Molecular Informatics and the Cambridge eScience Centre, and was supported by the Royal Society of Chemistry, Nature Publishing Group and the International Union of Crystallography. SciBorg mainly concentrates its extraction efforts on

---

[5] It is available online at http://opencitations.net.
[6] PubMed Central: http://www.ncbi.nlm.nih.gov/pmc/.
[7] Cambridge University Library: http://www.lib.cam.ac.uk.
[8] British Library: http://www.bl.uk.
[9] International Union of Crystallography: http://www.iucr.org.
[10] PubMed: http://www.ncbi.nlm.nih.gov/pubmed/.
[11] Lucero project blog: http://lucero-project.info.
[12] Available at http://data.open.ac.uk.

Chemistry texts in particular, but the techniques available were developed to be domain-independent.

Another recent research [10] on these topics has been presented by Motta and Osborne during the demo session of the 11th International Semantic Web Conference, held in Boston (US). They introduce *Rexplore*, a tool that provides several graphical ways to look at, explore and make sense of bibliographic data, research topics and trends, and so on, which was built upon an algorithm [11] able to identify automatically several relations between research areas.

3   Building Applications on Annotated Scholarly Documents

In this section we describe our long term vision, while in the following ones we will go into details of some models and applications we have already implemented. Our overall goal is to study novel *models* for semantic publishing, to design a *platform* that will offer sophisticated services for collecting and accessing scholarly documents, and to develop algorithmic and visual approaches to *make sense* of bibliographic entities and data, as discussed in the following subsections.

3.1   Semantic Models for Scholarly Publishing

We are currently studying how existing formal models for describing scientific documents – such as the SPAR ontologies and the Semantic Web-based EAR-MARK markup language [8] – can be used to enable the description of scholarly articles according to different semantic facets.

Those facets, we call *semantic lenses* [13], should make it possible to represent document semantics according to different levels of abstraction and to classify all aspects of the publishing production in a more precise, flexible and expressive way (e.g. publication context, authors' affiliations, rhetorical and argumentative organisation of the scientific discourse, citation networks, etc.).

3.2   A Platform for Academic Publishing Data

Our aim is to develop a platform to enhance a document enabling the definition of formal representations of its meaning, its automatic discovery, its linking to related articles, providing access to data within the article in actionable form, and allow integration of data between papers. The platform will support authors, publishing houses, and users at large who "consume" a document for any reasonable goal, for instance evaluating its impact on a scientific community.

In particular, the platform aims at helpings developers and common users to address the following tasks:

– evaluating the pertinence of a document to some scientific fields of its contribution;

- discovering research trends and propagation of research findings;
- tracking of research activities, institutions and disciplines;
- evaluating the social acceptability of the scientific production;
- analysing quantitative aspects of the output of researchers;
- evaluating the multi-disciplinarity of the output of scholars;
- measuring positive/negative citations to a particular work;
- designing and including within the platform efficient algorithms to compute metrics indicators;
- helping final users (e.g. researchers) to find related materials to a particular topic and/or article;
- enabling users to annotate documents through interfaces that facilitate the insertion of related semantic data;
- querying (semantic) bibliographic data.

3.3 Making Sense of Bibliographic Entities and Data

We are developing tools and algorithms to include in the platform, which aim at understanding and making sense of scholarly documents and their research and publishing contexts.

In particular, we are investigating the best ways to apply well-known Semantic Web technologies – e.g. FRED [15] – and other approaches to infer the structural semantics of document components – e.g. the algorithms proposed by Di Iorio *et al.* [7] [6] – to infer automatically the relatedness of different kinds of research, results, and claims, with the intent of automatically linking and recommending set of papers according to a particular topic.

Another work we have done in that direction is the development of *CiTalO*, which we present in detail in Section 4. CiTalO is able to infer the function of citations within a research article, i.e. author's reasons for citing a particular work, and thus can be used to enable the navigation of back and forward citation networks according to a particular dimension, e.g. if the paper is cited positively or negatively.

We believe that the semantically-aware citation networks as well as other semantic characterisations of scholarly papers we are investigating will open new perspectives for evaluating research.

3.4 Scenarios of Use as Evaluation

The evaluation of the whole work must be performed through task-based user testing sessions. Our main aims are:

- to assess the interaction mechanisms the platform provides to people, e.g. scientists, when accessing the semantic data about the documents so as to re-use or link them with/to other relevant and/or external source of data;
- to understand the usefulness of the platform when it is used for measuring the quality of scientific research within the context of a specific institution, like for instance UniBo;

- to evaluate a recommender system built on the top of the repository of bibliographic entities and data provided by the platform, whose aim is to help final users (e.g. researchers) to look for materials concerning a particular topic and/or article.

As examples, we have identified three use-cases addressing different classes of users in three corresponding scenarios:

- a scenario for an author is to build the "related works" section of a paper under preparation by discovering useful citations. In this case, for verification purposes, we could use a published paper, ignore its related work and bibliography sections, and reconstruct them measuring information retrieval indicators with appropriate metrics;
- a scenario for a reader or a research product evaluator is to start from a document and find documents citing it positively or negatively, and discover in which context;
- a scenario useful for a publishing house or a research institution could be annotating pertinent Wikipedia articles automatically with citations of papers published by the house or institution.

4   Discovering the Function of Citations

The scenarios presented in the previous section form a roadmap for the next years and involve several interested research groups. The initial development of CiTalO, we presented in past works, and its extension and integration with other external tools for processing scholarly articles stored in different format (such as XML, PDF, LaTeX, etc.), we introduce herein, represents our initial contribution in this direction, which we introduce in the following subsections.

4.1   CiTalO

CiTalO [5] is a tool that infers the function of citations – which are author's reasons for citing a particular paper – by combining techniques of ontology learning from natural language, sentiment-analysis, word-sense disambiguation, and ontology mapping. The tool is available online at http://wit.istc.cnr.it:8080/tools/citalo and takes as input a sentence containing a citation and returns a set of CiTO properties characterising that citation[13].

The overall CiTalO schema is shown in Fig. 1. Six steps compose the architecture, that are briefly discussed as follows.

Sentiment-Analysis. The first step is optional and consists of running a sentiment-analysis tool on the citation sentence to capture the polarity emerging from the text in which the citation is included. Knowing the polarity of a citation

---

[13] As introduced in Section 1, the *CiTO* is an OWL ontology defining forty different kinds of possible citation functions that may happen in a citation act.

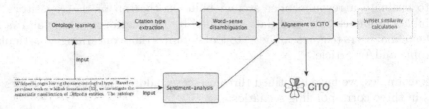

Fig. 1. Pipeline used by CiTalO. The input is the textual context in which the citation appears and the output is a set of properties of the CiTO ontology.

can be eventually exploited to reduce the set of possible target CiTO properties, since all properties are clustered around three different polarities: *positive*, *neuter* and *negative*. The current implementation of CiTalO uses AlchemyAPI[14] as sentiment-analysis module.

Ontology Learning. The next mandatory step of CiTalO consists of deriving a logical representation of the sentence containing the citation. The ontology extraction is performed by using FRED [15], a tool for ontology learning based on discourse representation theory, frames and ontology design patterns. The output of FRED on a sample sentence "We highly recommend X" is shown in Fig. 2 (also available as RDF triples).

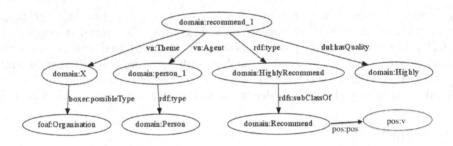

Fig. 2. FRED result for "We highly recommend X"

Citation Type Extraction. The core step of CiTalO consists of processing the output of FRED and extracting candidate terms which will be exploited for characterising the citation. CiTalO recognises some graph patterns and collects the values of some properties expressed in those patterns. We implemented these operations as SPARQL queries where possible; otherwise, we have directly coded them as Java methods. Considering the aforementioned sentence "We highly recommended X", this step returns the terms *domain:HighlyRecommend* and *domain:Recommend* as candidate types for the citation.

Word-Sense Disambiguation. The following step consists of disambiguating candidate types and producing a list of synsets that express their sense.

14 AlchemyAPI: http://www.alchemyapi.com.

To this end, CiTalO uses IMS [17], a word-sense disambiguator, and the disambiguation is performed with respect to OntoWordNet [9]. When running on the last example, IMS provides one disambiguation for the candidate types: *domain:Recommend* disambiguated as *synset:recommend-verb-1*.

Alignment to CiTO. The final step consists of actually assigning CiTO types to citations [12]. We use two ontologies for this purpose: *CiTO2Wordnet*[15] – mapping all the CiTO properties defining citations with the appropriate Wordnet synsets – and *CiTOFunctions*[16] – which classifies each CiTO property according to its factual and positive/neutral/negative rhetorical functions, using the classification proposed by Peroni *et al.* [12]. The final alignment to CiTO is performed through a SPARQL CONSTRUCT query that uses the output of the previous steps, the polarity gathered from the sentiment-analysis phase, OntoWordNet and the two ontologies just introduced. In the case of empty alignments, the CiTO property *citesForInformation* is returned as base case. In the example, the property *citesAsRecommendedReading* is assigned to the citation, derived from *synset:recommend-verb-1*.

Synset Similarity Calculation. It may happen that none of the CiTO properties can be directly aligned. In that case, CiTalO tries to find a new list of synsets that can be mapped into CiTO properties and are close enough to the ones in the current list. This optional step relies on WS4J (WordNet Similarity for Java) a Java API that measures the semantic proximity between synsets[17].

4.2   Extracting Citation Functions from PDF

During the *Jailbreak the PDF Hackathon*[18] we attended in Montpellier in May 2013, the main topic under discussion was to enable the extraction of citations out of scientific articles stored as PDF streams. Processing PDF documents is a complex task that requires the use of some sort of heuristics to fully comprehend the actual organisation of document content. The main assumption PDF makes is that any character (and line, picture, and so on) is assigned to a specific location within the page viewport and this declaration can be placed in any position within the PDF stream. Which means that the positions of three different characters making a word such as "who", each following the others according to a sequential order shown in the visualisation of the PDF, can be declared in different and non-sequential places within the PDF stream.

Of course, there exist tools that try to infer and extract the sequential organisation of document content from a PDF stream. One of the most interesting one is *PDFX*[19] [2], which uses *Utopia Documents* [1]. Basically, PDFX takes a PDF of a scientific article as input and returns an XML linearisation of it based on

---

[15] CiTO2Wordnet ontology: http://www.essepuntato.it/2013/03/cito2wordnet.

[16] CiTOFunctions: http://www.essepuntato.it/2013/03/cito-functions.

[17] WS4J: http://code.google.com/p/ws4j/

[18] Jailbreak the PDF Hackathon homepage: http://scholrev.org/hackathon/

[19] PDFX homepage: http://pdfx.cs.man.ac.uk/

a subset of the *Journal Article Tag Suite* DTD[20], and automatically adds some structural and rhetorical characterisations of blocks of text according to one of the SPAR ontologies, i.e. the *Document Components Ontology (DoCO)*[21].

As a concrete outcome of the hackathon, we started an active collaboration with people (from the Manchester University) responsible for the development of PDFX. From the one hand, they would like to include rhetorical and factual characterisations of citations within the XML document they produce as output of PDFX. On the other hand, we are interested in the development of an automatic process that allows to identify all the citation acts within a scientific document stored as PDF file and then to characterise those citation acts according to the properties in CiTO [12]. We have currently developed such a process and we have also made a Web application called *CiTalO* ^{PDF}, available online[22], to perform the extraction and the semantic characterisation of citations. CiTalO^{PDF} will be also used by the next release of PDFX so as to add citation functions within the appropriate *xref* elements of the XML conversion obtained by processing the PDF. The workflow followed by CiTalO^{PDF} is shown in Fig. 3 and defines three different steps, described as follows.

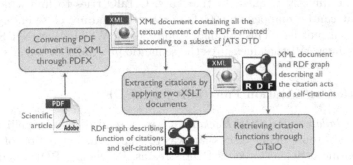

Fig. 3. The workflow followed by CiTalO^{PDF} to extract citation functions from PDF scientific articles

Step 1: from PDF to XML. In this step we simply call the PDFX service to convert our input PDF scientific article into an XML document formatted according to a small subset of the JATS DTD.

Step 2: from XML to Citation Acts. The XML document obtained in the previous step is here processed according to two XSLT documents. The first XSLT document[23] transforms the input XML document into another XML *intermediate* document containing some metadata (element *meta*) of the input document – i.e. the title (element *title*) and the authors (elements *author*) – and

---

[20] Journal Article Tag Suite homepage: http://jats.nlm.nih.gov/

[21] Document Components Ontology: http://purl.org/spar/doco

[22] CiTalO^{PDF} is available at http://wit.istc.cnr.it:8080/tools/citalo/pdf/.

[23] http://www.essepuntato.it/2013/citalo/ExtractRefsFromPDFX.xsl

all the references (elements *ref*) to bibliographic citations that exist within the document itself, which includes an identifier (element *id*), the label (element *label*) used in the text to point to the bibliographic reference (element *description*) contained in the reference section of the article, and the citation sentence[24] (element *context*) for that citation – where the actual label was substituted by an "X". The following excerpt shows an excerpt of the output of the XSLT process:

```
<refs>
  <meta>
    <title>Anhedonia, emotional numbing, and symptom
        overreporting in male veterans with PTSD</title>
    <authors>
      <author>Todd Kashdan</author>
      <author>Jon Elhai</author>
      <author>Christopher Frueh</author></authors></meta>
  <ref>
    <id>16</id><label>Litz (1992)</label>
    <description id="153">Litz, B. T. (1992). Emotional
        numbing in combat-related post-traumatic stress
        disorder: a critical review and reformulation.
        Clinical Psychology Review, 12, 417-432.
    </description>
    <context>Despite an innovative framework proposed by X,
        it is only recently that emotional numbing and
        anhedonia have received empirical attention in
        studies of post-traumatic stress disorder (PTSD).
    </context></ref> ...
</refs>
```

We apply another XSLT document[25] on the previous XML output, so as to describe the all bibliographic entities involved in citations as an RDF graph. In particular, according to the model illustrated in Fig. 4, we produce an instance belonging to the class *c4o:InTextReferencePointer* (representing an inline pointer to a particular bibliographic reference placed in the references section of the article) for each element *ref* defined in the intermediate document, while the entities describing the scientific article in consideration and the cited articles are all defined as individuals of *fabio:Expression*. In addition, this second XSLT process makes self-citations explicit by adding several statements linking the citing article to a cited one through the property *cito:sharesAuthorsWith* every time at least one of the authors of the former co-authored also the latter.

Step 3: from Citation Acts to Citation Functions. In the latter step we actually extend the RDF graph produced before creating an instance of the class *cito:CitationAct* (presented in Fig. 4, representing a citation act) for each element *ref* defined in the intermediate document according to CiTalO's interpretation of the related citation sentence defined in the element *context*. In addition, we

---

[24] A *citation sentence* is the sentence where a particular citation act happens.
[25] Available online at
http://www.essepuntato.it/2013/citalo/SharesAuthorsWith.xsl.

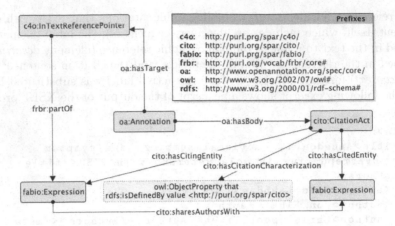

Fig. 4. The diagram defining the model we used to describe the citation acts of scientific documents. Yellow rectangles with solid and dotted border stand for OWL classes (i.e. *owl:Class*) and OWL restrictions (i.e. *owl:Restriction*) respectively, while blue edges define object properties (i.e. *owl:ObjectProperty*) having as domain the class placed under the solid circle and as range the class/restriction indicated by the arrow.

explicitly link the inline pointer reference in consideration (defined previously as individual of *c4o:InTextReferencePointer*) to the related citation act by creating an annotation (i.e. an individual of the class *oa:Annotation*).

5   Conclusions

In this paper we have described our vision of semantically annotating scholarly papers and some methods and technologies we have developed for this goal.

We are still in an early stage of this research. The availability of large repositories of scholarly documents generates new needs and offers new opportunities. Publishers are building and experimenting new services for the social fruition of scholarly literature. Authors and readers are contributing to establish new social uses. Our results are very initial and still far from perfect. We need to extend the capabilities of CiTalO when analysing natural language text from scholarly documents, also improving its ability of recognising rhetorical patterns. In addition, we plan to run experiments with several published scholarly articles written in different languages and end users (e.g. researchers).

Acknowledgments. We would like to thank Alexandru Constantin and Steve Pettifer for their collaboration and support to the CiTalO framework.

References

1. Attwood, T.K., Kell, D.B., McDermott, P., Marsh, J., Pettifer, S., Thorne, D.: Utopia documents: linking scholarly literature with research data. Bioinformatics 26(18), 568–574 (2010), doi:10.1093/bioinformatics/btq383

2. Constantin, A., Pettifer, S., Voronkov, A.: PDFX: fully-automated PDF-to-XML conversion of scientific literature. In: Proceedings of the 2013 ACM Symposium on Document Engineering (DocEng 2013), pp. 181–184. ACM Press, New York (2013), doi:10.1145/2494266.2494271

3. Copestake, A., Corbett, P., Murray-Rust, P., Rupp, C.J., Siddharthan, A., Teufel, S., Waldron, B.: An architecture for language processing for scientific text. In: Proceedings of the UK e-Science All Hands Meeting 2006 (2006)

4. De Waard, A.: From Proteins to Fairytales: Directions in Semantic Publishing. IEEE Intelligent Systems 25(2), 83–88 (2010), doi:10.1109/MIS.2010.49

5. Di Iorio, A., Nuzzolese, A., Peroni, S.: Towards the automatic identification of the nature of citations. In: Proceedings of 3rd Workshop on Semantic Publishing (SePublica 2013), pp. 63–74 (2013), http://ceur-ws.org/Vol-994/paper-06.pdf

6. Di Iorio, A., Peroni, S., Poggi, F., Shotton, D., Vitali, F.: Recognising document components in XML-based academic articles. In: Proceedings of the 2013 ACM Symposium on Document Engineering (DocEng 2013), pp. 177–180. ACM, New York (2013), doi:10.1145/2494266.2494319

7. Di Iorio, A., Peroni, S., Poggi, F., Vitali, F.: Dealing with structural patterns of XML documents. To appear in Journal of the American Society for Information Science and Technology (2013)

8. Di Iorio, A., Peroni, S., Vitali, F.: A Semantic Web Approach To Everyday Overlapping Markup. Journal of the American Society for Information Science and Technology 62(9), 1696–1716 (2011), doi:10.1002/asi.21591

9. Gangemi, A., Navigli, R., Velardi, P.: The OntoWordNet Project: Extension and Axiomatization of Conceptual Relations in WordNet. In: Meersman, R., Schmidt, D.C. (eds.) CoopIS/DOA/ODBASE 2003. LNCS, vol. 2888, pp. 820–838. Springer, Heidelberg (2003)

10. Motta, E., Osborne, F.: Making Sense of Research with Rexplore. In: Proceedings of the ISWC, Posters & Demonstrations Track (2012), http://ceur-ws.org/Vol-914/paper_39.pdf

11. Osborne, F., Motta, E.: Mining Semantic Relations between Research Areas. In: Cudré-Mauroux, P., et al. (eds.) ISWC 2012, Part I. LNCS, vol. 7649, pp. 410–426. Springer, Heidelberg (2012)

12. Peroni, S., Shotton, D.: FaBiO and CiTO: ontologies for describing bibliographic resources and citations. Journal of Web Semantics: Science, Services and Agents on the World Wide Web 17, 33–43 (2012), doi:10.1016/j.websem.2012.08.001

13. Peroni, S., Shotton, D., Vitali, F.: Faceted documents: describing document characteristics using semantic lenses. In: Proceedings of the 2012 ACM Symposium on Document Engineering (DocEng 2012), pp. 191–194 (2012), doi:10.1145/2361354.2361396

14. Pettifer, S., McDermott, P., Marsh, J., Thorne, D., Villéger, A., Attwood, T.K.: Ceci n'est pas un hamburger: modelling and representing the scholarly article. Learned Publishing 24(3), 207–220 (2011), doi:10.1087/20110309

15. Presutti, V., Draicchio, F., Gangemi, A.: Knowledge extraction based on discourse representation theory and linguistic frames. In: ten Teije, A., Völker, J., Handschuh, S., Stuckenschmidt, H., d'Acquin, M., Nikolov, A., Aussenac-Gilles, N., Hernandez, N. (eds.) EKAW 2012. LNCS, vol. 7603, pp. 114–129. Springer, Heidelberg (2012)

16. Shotton, D.: Semantic Publishing: the coming revolution in scientific journal publishing. Learned Publishing 22(2), 85–94 (2009), doi:10.1087/2009202

17. Zhong, Z., Ng, H.T.: It Makes Sense: A wide-coverage word sense disambiguation system for free text. In: Proceedings of the ACL 2010 System Demonstrations, pp. 78–83 (2010)

Common Subsumers in RDF

Simona Colucci[1], Francesco M. Donini[1], and Eugenio Di Sciascio[2]

[1] DISUCOM, Università della Tuscia, Viterbo, Italy
[2] DEI, Politecnico di Bari, Bari, Italy

Abstract. Since their definition in 1992, Least Common Subsumers (LCSs) have been identified as services supporting learning by examples. Nowadays, the Web of Data offers a hypothetically unlimited dataset of interlinked and machine-understandable examples modeled as **RDF** resources. Such an open and continuously evolving information source is then really worth investigation to learn significant facts. In order to support such a process, in this paper we give up to the subsumption minimality requirement of LCSs to meet the peculiarities of the dataset at hand and define Common Subsumers (CSs). We also propose an anytime algorithm to find CSs of pairs of **RDF** resources, according to a selection of such resources, which ensures computability.

1 Introduction

Since its definition and identification as fundamental layer over which building the Semantic Web [4], the Web of Data [22] has led to the availability of a huge amount of perfectly interconnected and machine-understandable data, modeled as **RDF** resources, usually addressed as Linked (Open) Data (LOD). Such an open and free (in most cases) dataset asks for new sorts of information management, which could further support the realization the Semantic Web principles.

One of the most challenging issues in **RDF** resources management is the identification of subsets of resources to some extent related to a common informative content. Finding commonalities in the information conveyed by different **RDF** descriptions may in fact turn out to be useful in several Semantic Web-related tasks. Intuitively, clustering of Web resources is one of these tasks: the need for restricting the search field in a space as wide as the Web has generated several research efforts in the literature.

Since 2001, learning from **RDF** graphs has been investigated [12] with the aim to cluster resources by incrementally constructing new concepts that partially describe all resources of interest, and to classify them according to subsumption. Fanizzi *et al.* [13] perform ontology clustering according to a metric-based approach, with reference to representation languages (namely, OWL-DL[1]) different from **RDF**; lists of so-called medoid are returned as clustering result.

Part of the literature is specifically devoted to the application of clustering algorithms to Semantic Web data. Grimnes *et al.* [16] evaluate, by applying both

---

[1] http://www.w3.org/TR/2004/REC-owl-guide-20040210/

M. Baldoni et al. (Eds.): AI*IA 2013, LNAI 8249, pp. 348–359, 2013.
© Springer International Publishing Switzerland 2013

supervised and unsupervised metrics, different methods for extracting instances from a large **RDF** graph and computing the distance between such instances. The evaluation shows how the behavior of extraction methods and similarity metric strictly depends on the data being clustered, suggesting the need for data-centric flexible solutions.

Mahmoud et al. [20] investigate on clustering for data integration from large numbers of structured data sources, suggesting the need for a pay-as-you-go approach [19] which adopts an initial data integration system (resulting from some fully automatic approximate data integration technique), refined during system use. The approach by Mahmoud et al. clusters data schema into domains only according to attribute names and manages uncertainty in assignment using a probabilistic model.

Also the simpler problem of clustering web pages has met the interest of researchers: a Fuzzy Logic-based representation for HTML document using Self-Organizing Maps has been proposed for clustering [15]. Zeng et al. [24] deal with the re-formalization of the clustering problem in terms of phrase ranking to produce candidate cluster names for organizing web search results.

Proposals taking somehow into account semantics conveyed in resources descriptions while performing clustering of search results have been presented. In particular, Lawrynowicz [18] proposes a method to cluster the results of conjunctive queries submitted to knowledge bases represented in the Web Ontology Language (OWL). Syntactic approaches show some limits in aggregating clustering results when the values instantiating a grouping criterion are all equal or are almost all different. d'Amato et al. . [10] overcome such limits by deductively grouping answers according to the subsumption hierarchy of the underlying knowledge base.

Far from being exhaustive, the above state of the art shows the deep interest for clustering (especially of Web content) and underlines how most of the clustering approaches introduced so far adopt induction to identify clusters according to some—sometimes semantic-based—distance between elements in the same cluster. Our approach instead supports the inference of such clusters, and may provide a description of the informative content associated to the common features of resources belonging to the same cluster.

In particular, we define *Common Subsumers* of pairs of **RDF** resources, in analogy to the *Least Common Subsumer*(LCS) service, well known in Description Logics. Since its proposal in 1992 [5], learning from examples was identified as one of the most important application fields for LCS computation.

We take the same assumption and adapt the original definition to the set of examples we aim at learning from: the Web of Data. The hypothetically unlimited size of the investigated dataset motivates the choice of giving up to the subsumption minimality requirement typical of LCSs and revert to Common Subsumers. Motivated by the chance to compute even rough Common Subsumers, still useful for learning in the Web of Data, we propose an anytime algorithm computing Common Subsumers of pairs of **RDF** resources. The algorithm may work as basis for finding commonalities in collections of **RDF** resources.

The paper is organized as follows: in the next section, we shortly recall related work on common subsumers computation and discuss peculiarities of **RDF** which make it different from other Web languages in terms of such an inference process. Section 3 provides a proof-theoretic definition of Common Subsumers in **RDF**, whose properties are shown in Section 4. The anytime algorithm computing Common Subsumers is given in Section 5, before closing the paper.

2   Common Issues

Coherently with the initial motivation of supporting inductive learning in the DL $\mathcal{LS}$ [5], several algorithms have been developed for computing LCS in different DLs, such as CORECLASSIC [6], $\mathcal{ALN}$ [7], CLASSIC [14].

The idea of reverting to common subsumers which are not "least" has been already investigated to cope with practical applications [2].

The main issues of the problem we investigate here are on the one hand selecting a portion of the knowledge domain in order to ensure computability and, on the other hand, the peculiarities of the language in which resources are modeled: **RDF**. Aimed at learning from the Web of Data, our approach needs in fact to cope with **RDF/RDF-S**. Nevertheless, the **RDF/RDF-S** semantics is not trivial to be investigated, and even less trivial is determining its relationship with DLs [11] and/or other Web languages [21] in which the problem of finding (least or not) common subsumers has been defined and studied. For this reason, we provide in the following section an **RDF**-specific definition of Common Subsumers and clarify the choices about the adopted semantics.

3   Common Subsumers in RDF

We recall that in the Description Logics literature [5,1], a concept L is a Least Common Subsumer of two concepts C_1 and C_2 if: (i) L subsumes both C_1 and C_2 (written as $C_1 \sqsubseteq L$, $C_2 \sqsubseteq L$) and moreover, (ii) L is a $\sqsubseteq$-minimal concept with such property, that is, for every concept D such that both $C_1 \sqsubseteq D$ and $C_2 \sqsubseteq D$ hold, $L \sqsubseteq D$ holds too.

We adopt the semantics based on entailment for **RDF**, that is, the meaning of a set of triples T is the set of triples (theorems) one can derive from T by using **RDF**-entailment rules. The **RDF**-entailment rules we consider are the 18 rules and the axiomatic triples of the official document regarding **RDF** semantics [17], modified according to ter Horst [23] as follows: In his paper, ter Horst proves that the original rules for entailment in **RDF** are incomplete, and that completeness is regained if (i) blank[2] nodes are allowed to stand as predicates in triples (called *generalized RDF triples*), and (ii) Rule rdfs7 is changed accordingly—namely, it can derive triples whose predicate is a blank node. In the following, when

---

[2] Recall that every blank node corresponds to an existentially quantified variable. A blank node without a name is denoted by [], while named blank nodes are prefixed by _:, *e.g.*, _:xxx.

we talk about **RDF** triples and **RDF** entailment, we always assume that the adjustments proposed by ter Horst have been made. To correctly embed **RDF** triples in text we use the notation $<<a\ p\ b>>$ to mean "$a\ p\ b$."

Clearly, the meaning of a resource r changes depending on which triples r is involved in, since different sets of triples derive (in general) different new triples. Hence, we always attach to a resource r the set of triples T_r we consider significant for its meaning and define such a pair as *rooted-graph* of r.

Definition 1 (Rooted Graph(r-graph)). *Let TW_r be the set of all triples with subject r in the Web. A* Rooted Graph(r-graph) *is a pair $\langle r, T_r \rangle$, where*

1. *r is either the URI of an **RDF** resource, or a blank node*
2. *$T_r = \{t \mid t = <<r\ p\ c>>\}$ is a subset of TW_r*

Observe that we implicitly define a set of resources, namely, all resources appearing in some relevant triple of some resource. The idea is to cut out a relevant portion of the Semantic Web, where the resources on the "frontier" of such a portion have no relevant triples—*i.e.*, they are treated as literals[3], even when they are not. In the r-graph $\langle r, \emptyset \rangle$, r has no "meaning" other than a generic resource, for which, *e.g.*, $<<r$ `rdf:type rdfs:Resource`$>>$ is always entailed (Rule rdfs4 in the official document about **RDF** entailment [17]).

We think that it is not realistic to assume that all relevant triples are known at once, and in advance: the usual way in which sets of triples are constructed, is that triples are discovered one at a time, while exploring the Web. In this setting, the question a crawler has to answer while surfing the Web is: "Is this triple t I have just found relevant for r or not?" The answer is provided by the *characteristic function* of T_r, namely, $\sigma_{T_r} : TW_r \to \{false, true\}$, that we will use to determine the set T_r of triples relevant for r (see Algorithm 2 for our current computation).

We clarify that when deciding the entailment of a triple involving r as subject, also triples not involving r at all must be considered. For example, let $\langle r, T_r \rangle$, $\langle a, T_a \rangle$ $\langle b, T_b \rangle$ be three r-graphs, where:

$$T_r = \{<<r\ \texttt{rdf:type}\ a>>\}$$
$$T_a = \{<<a\ \texttt{rdfs:SubClassOf}\ b>>\}$$
$$T_b = \emptyset$$

Observe that the triple $<<r$ `rdf:type` $b>>$, which enriches the "meaning" of r, is not entailed by T_r alone, while it is entailed by $T_r \cup T_a$ (Rule rdfs9 [17]). Hence, we assume that *entailment is always computed with respect to the union of all sets of relevant triples*. We call T such a union set. We are now ready to give our definition of Common Subsumer of two resources.

Definition 2 (Common Subsumer). *Let $\langle a, T_a \rangle$, $\langle b, T_b \rangle$ be two r-graphs and x, w, y be blank nodes. If $\langle a, T_a \rangle = \langle b, T_b \rangle$, then $\langle a, T_a \rangle$ is a Common Subsumer of $\langle a, T_a \rangle$, $\langle b, T_b \rangle$. Otherwise, if $T_a = \emptyset$ or $T_b = \emptyset$, the pair $\langle x, \emptyset \rangle$ is a Common*

---

[3] Recall that literals can never appear as subjects of triples.

Subsumer of $\langle a, T_a \rangle$, $\langle b, T_b \rangle$. *Otherwise, a pair* $\langle x, T \rangle$ *is a Common Subsumer of* $\langle a, T_a \rangle$, $\langle b, T_b \rangle$ *iff:*
$\exists t = <\!<x\ w\ y\!>\!>$ *such that* $(T$ *entails* $t)$

$$\Rightarrow \tag{1}$$

$\exists t_1 = <\!<a\ p\ c\!>\!>, t_2 = <\!<b\ q\ d\!>\!>$ *such that* $(T$ *entails* $t_1) \wedge (T$ *entails* $t_2)$
where $T_a \subseteq T$, $T_b \subseteq T$ *and* $\langle w, T \rangle$ *is a Common Subsumer of* $\langle p, T_p \rangle$ *and* $\langle q, T_q \rangle$,
and $\langle y, T \rangle$ *is a Common Subsumer of* $\langle c, T_c \rangle$ *and* $\langle d, T_d \rangle$.

Observe that if we used "$\Leftrightarrow$" in (1) instead of "$\Rightarrow$", we would define Least Common Subsubers in **RDF**. However, we are more interested here in Common Subsumers, since they are easier to compute, while being still useful enough for our applications. Observe also that the reference to the relevant triples of each resource is crucial for defining the Common Subsumer; in order not to explore the entire Semantic Web, we will restrict the relevant triples either to some specific dataset such as *DBPedia*, or to triples "near" the initial resources—as a distance in the **RDF**-graph—or both.

We now provide an intuition of our definition through an example.

Example 1. Suppose that the following common prefixes have been defined:

```
@prefix dbpedia:     <http://dbpedia.org/resource/> .
@prefix rdf:         <http://www.w3.org/1999/02/22-rdf-syntax-ns#> .
@prefix dbpedia-owl: <http://dbpedia.org/ontology/> .
@prefix dbpprop:     <http://dbpedia.org/property/> .
```

Then, consider the resource `dbpedia:Bicycle_Thieves` (a renowned Italian neorealist movie), along with (what we decide to be) its relevant triples T_{BT} (written in Turtle [3] notation):

```
dbpedia:Bicycle_Thieves
    rdf:type           dbpedia-owl:Film ;
    dbpprop:director   dbpedia:Vittorio_De_Sica ;
    dbpprop:language   dbpedia:Italian_language .
```

and the resource `dbpedia:The_Hawks_and_the_Sparrows` (a post-neorealist Italian movie), along with its relevant triples T_{UU}:

```
dbpedia:The_Hawks_and_the_Sparrows
    rdf:type           dbpedia-owl:Film ;
    dbpprop:director   dbpedia:Pier_Paolo_Pasolini ;
    dbpprop:language   dbpedia:Italian_language .
```

To make the example easy to follow, we consider that there are no relevant triples for all other resources—*i.e.*, `rdf:type`, `dbpedia-owl:Film`, `dbpprop:director`, etc.—involved in the above triples. This corresponds to choose only triples whose subject is either `dbpedia:Bicycle_Thieves` or `dbpedia:The_Hawks_and_the-_Sparrows`.

An intuitive Common Subsumer of $\langle$`dbpedia:Bicycle_Thieves`, $T_{BT}\rangle$ and $\langle$`dbpedia:The_Hawks_and_the_Sparrows`, $T_{UU}\rangle$ is $\langle$`_:cs1`, $T\rangle$, where `_:cs1` is a blank node and T contains, in addition to T_{BT} and T_{UU}, also the new triples

```
_:cs1
    rdf:type              dbpedia-owl:Film ;
    dbpprop:language      dbpedia:Italian_language .
```

A more informative Common Subsumer—the one that would be computed by our algorithm in Sect. 5—is $\langle \_:cs2, T \rangle$ where T contains, in addition to the previous example, also a triple referring to some director:

```
_:cs2
    rdf:type              dbpedia-owl:Film ;
    dbpprop:director      _:cs3 ;
    dbpprop:language      dbpedia:Italian_language .
```

where $\langle \_:cs3, \emptyset \rangle$ is a Common Subsumer of $\langle dbpedia:Vittorio\_De\_Sica, \emptyset \rangle$ and $\langle dbpedia:Pier\_Paolo\_Pasolini, \emptyset \rangle$. Observe that since we did not consider any triple regarding the two directors—*e.g.*, they were both italian—we cannot add to $T_{\_:cs3}$ the triple

$$<<\_:cs3 \text{ dbpprop:director dbpedia:Italian\_director}>>$$

This clearly exemplifies the tradeoff between, on one hand, how large is the portion of the linked data we want to consider, and on the other hand, how many computational resources we are willing to spend. Probably the most intuitive enlargement process is the one based on (**RDF** graph) distance: in that case, considering relevant also the triples whose subject has distance 1,2,... from the initial resources, one obtains progressively more accurate Common Subsumers at the price of a progressively heavier computation.

For sake of completeness, we note that one of the least informative Common Subsumers is $\langle \_:cs4, T_{BT} \cup T_{UU} \rangle$, where $\_:cs4$ is a(nother) blank node. *(End of example).*

In general, a Common Subsumer of a, b is a blank node, except for the case $a = b$, in which the most informative Common Subsumer is the resource itself—see Idempotency in the next section. A Common Subsumer is a blank node because **RDF** can only express—by means of triples—necessary conditions for identifying a resource. For example, even if we state for some resource r exactly the same triples of dbpedia:Bicycle_Thieves, this does not allow us to derive in **RDF** that $r = $ dbpedia:Bicycle_Thieves—one can verify that equality of two resources is never the consequence of **RDF**-entailment Rules.

4   Properties of RDF Common Subsumers

We now explore some properties of Common Subsumers that already hold in Description Logics, and which we prove to hold also—suitably changed—for our definition of Common Subsumers in **RDF**, namely: Idempotency, Commutativity, and Associativity.

Idempotency. Clearly, a Common Subsumer of $\langle a, T_a \rangle$ and $\langle a, T_a \rangle$ is $\langle a, T_a \rangle$ itself. Differently from Description Logics, however, $\langle a, T_a \rangle$ is not the Least Common Subsumer (LCS) of itself—where "LCS" would be defined as in Def. 2, with "$\Leftrightarrow$" in place of "$\Rightarrow$". This is because when the two triples $<<a\ p\ c>>$, $<<a\ q\ d>>$ are in T_a, with $p \neq q$, the Least Common Subsumer should contain the triple $<<a\ []\ []>>$. Such a triple is not derivable from **RDF** rules because of the blank node in the predicate position. Recall that from a triple $<<a\ p\ c>>$, one can derive both $<<[]\ p\ c>>$ (Rule se2 in the W3C document [17]) and $<<a\ p\ []>>$ (Rule se1), but not $<<a\ []\ c>>$. One would need a third rule se3, deriving exactly such a triple. The proposal of such an extension of **RDF** could be a subject of future research.

Commutativity. In Condition (1), it is evident that the role of a and b is completely interchangeable. Hence, a Common Subsumer of $\langle a, T_a \rangle$, $\langle b, T_b \rangle$ is always a Common Subsumer of $\langle b, T_b \rangle$, $\langle a, T_a \rangle$.

For the clustering applications we have in mind, it is necessary to extend the idea of Common Subsumer to a *group* of r-graphs $\langle a_1, T_{a_1} \rangle, \ldots, \langle a_n, T_{a_n} \rangle$. We can adapt the computation of a Common Subsumer from two to n r-graphs, by first computing a Common Subsumer $\langle x_1, T_1 \rangle$ of, say, $\langle a_1, T_{a_1} \rangle$, $\langle a_2, T_{a_2} \rangle$, and then computing a Common Subsumer $\langle x_2, T_2 \rangle$ of $\langle x_1, T_1 \rangle$ and $\langle a_3, T_{a_3} \rangle$, etc., till computing a Common Subsumer $\langle x_{n-1}, T_{n-1} \rangle$ of $\langle x_{n-2}, T_{n-2} \rangle$ and $\langle a_n, T_{a_n} \rangle$.

In this case, it is important to ensure that no matter what pair of r-graphs we start from, and in what order we consider the rest of them, we always compute a Common Subsumer of all r-graphs. We refer to this property as Associativity, for historical reasons, although our computation does not define a proper algebraic operator. We state the property for just three r-graphs, since its inductive extension to n r-graphs is straightforward.

Theorem 1 (Associativity). *Given* $\langle a_1, T_{a_1} \rangle$, $\langle a_2, T_{a_2} \rangle$, $\langle a_3, T_{a_3} \rangle$, *let* $\langle x_1, T_1 \rangle$ *be a Common Subsumer of* $\langle a_1, T_{a_1} \rangle$, $\langle a_2, T_{a_2} \rangle$. *Then, computing the Common Subsumer* $\langle x_2, T_2 \rangle$ *of* $\langle x_1, T_1 \rangle$ *and* $\langle a_3, T_{a_3} \rangle$, *one obtains a Common Subsumer of* $\langle a_1, T_{a_1} \rangle$, $\langle a_2, T_{a_2} \rangle$, $\langle a_3, T_{a_3} \rangle$, *pairwise considered.*

Proof. (Sketch.)We prove that $\langle x_2, T_2 \rangle$ is a Common Subsumer of each of the three pairs of r-graphs $\langle a_1, T_{a_1} \rangle$–$\langle a_2, T_{a_2} \rangle$, $\langle a_2, T_{a_2} \rangle$–$\langle a_3, T_{a_3} \rangle$, $\langle a_1, T_{a_1} \rangle$–$\langle a_3, T_{a_3} \rangle$. Let G be the union of all r-graphs, G_1 be the union of G with the triples involving x_1 and attached resources, and let G_2 be the union of G_1 with the triples involving x_2 and all attached resources.

If $\exists t_2 = <<x_2\ w_2\ y_2>>$ entailed by G_2, then from Def. 2, there exist both a triple $<<x_1\ w_1\ y_1>>$ entailed by G_1, and a triple $<<a_3\ p_3\ c_3>>$ entailed by G, and such that $\langle w_2, W_2 \rangle$ is a Common Subsumer of $\langle w_1, W_1 \rangle$ and $\langle p_3, T_{p_3} \rangle$, and $\langle y_2, Y_2 \rangle$ is a Common Subsumer of $\langle y_1, Y_1 \rangle$ and $\langle c_3, T_{c_3} \rangle$. We now apply Def. 2 to $\langle x_1, T_1 \rangle$: there exist two triples $<<a_1\ p_1\ c_1>>$, $<<a_2\ p_2\ c_2>>$ which are both entailed by G, and such that $\langle w_1, W_1 \rangle$ is a Common Subsumer of $\langle p_1, T_{p_1} \rangle$ and $\langle p_2, T_{p_2} \rangle$, and $\langle y_1, Y_1 \rangle$ is a Common Subsumer of $\langle c_1, T_{c_1} \rangle$ and $\langle c_2, T_{c_2} \rangle$. This proves that $\langle x_2, T_2 \rangle$ is also a Common Subsumer of $\langle a_1, T_{a_1} \rangle$ and $\langle a_2, T_{a_2} \rangle$, and since there exists also the triple $<<a_3\ p_3\ c_3>>$, $\langle x_2, T_2 \rangle$ is also a Common Subsumer of the couples of r-graphs $\langle a_2, T_{a_2} \rangle$–$\langle a_3, T_{a_3} \rangle$, $\langle a_1, T_{a_1} \rangle$–$\langle a_3, T_{a_3} \rangle$. $\qquad\square$

Example 2. If we add to the two r-graphs of Example 1 also the r-graph ⟨The_Big_Sleep_(1946_film), $T_{\text{The\_Big\_Sleep\_(1946\_film)}}$⟩, with the relevant triples:

```
dbpedia:The_Big_Sleep_(1946_film)
     rdf:type            dbpedia-owl:Film;
     dbpprop:director    dbpedia:Howard_Hawks;
     dbpprop:language    "English@en" .
```

one can get as a Common Subsumer ⟨_:cs5, T⟩, where T contains also:

```
_:cs5
     rdf:type            dbpedia-owl:Film;
     dbpprop:director    _:cs6;
     dbpprop:language    _:cs7 .
```

This result can be obtained either by comparing the above r-graph with each of the initial r-graphs of Example 1, or by comparing the above r-graph with the Common Subsumer _:cs2 already computed in Example 1. *(End of example).*

5   Finding Common Subsumers in RDF

In order to find Common Subsumers of a pair of **RDF** resources, we propose Algorithm 1 below, which is an anytime algorithm.

As mentioned in Sect. 3, to define the r-graph of a resource r, we determine the subset $T_r \subseteq TW_r$ of triples relevant for r through its characteristic function $\sigma_{T_r} : TW_r \to \{false, true\}$:

$$\sigma_{T_r}(t) = \begin{cases} true & \text{if } t \in T_r \\ false & \text{if } t \notin T_r \end{cases}$$

Algorithm 1 takes as input two resources, a and b and the related maximum number of investigation calls, n_a and n_b ($n_a \geq 1$, $n_b \geq 1$). At any time, it can return a Common Subsumer x, together with the contextual set T of relevant triples inferred till the moment of its interruption. We notice that T does not only include triples $<<x\ y\ z>>$ involving x as subject, but also triples describing y and z and all resources needed to provide the required representation of x.

Algorithm 1 manages a global data structure S, collecting information about already computed Common Subsumers stored as tuples $[p, q, pq]$, where p and q are **RDF** resources and pq is (a self-evident name for) their Common Subsumer. S can be accessed through the pair p, q.

The adoption of variables n_a and n_b allows for limiting the recursive depth for computing the Common Subsumer of a and b, in presence of a hypothetically unlimited dataset. S supports, instead, the management of cycles in Common Subsumer computation which may occur during investigation.

In Rows 3 and 5, Algorithm 1 asks for the computation of σ_{T_a} and σ_{T_b}, in order to define the sets, T_a and T_b of relevant triples for the input resources.

The rationale for reverting to a subset of relevant triples defining a resource is, as hinted before, the trade-off between the need for a full resource representation and computability of Common Subsumers in a dataset as the Web of Data, too large to explore thoroughly.

Find $CS(a, n_a, b, n_b)$;
Input : a, n_a, b, n_b
Output: x, T
1 Let S be a global data structure collecting computed Common Subsumers ;
2 Let T be a global set of triples describing the Common Subsumer;
3 $\sigma_{T_a} = compute\_\sigma(n_a)$;
4 $T_a = \{t \mid \sigma_{T_a}(t) = true\}$;
5 $\sigma_{T_b} = compute\_\sigma(n_b)$;
6 $T_b = \{t \mid \sigma_{T_b}(t) = true\}$;
7 $T = T_a \cup T_b$;
8 $x = explore(a, \sigma_{T_a}, n_a, b, \sigma_{T_b}, n_b)$;
9 *return* x, T

Algorithm 1. Initialization and start of CS construction

In our current implementation we adopt Algorithm 2, *compute* $\sigma_{T_r}(n_r)$, for computing σ_{T_r} for a generic resource r, to be investigated through up to n_r calls.

Intuitively, Algorithm 2 considers relevant (see Row 5) only triples whose subject belongs to specific datasets of interest (collected in the set D) and returns an empty set of relevant triples when r has not to be further investigated, because the maximum recursive depth has been reached (Rows 2–3). In this way, Algorithm 2 ensures that at least one of the base cases in Definition 2 is reached (see Row 6 in Algorithm 3). We notice that more complex functions could be computed according to some heuristics without affecting Algorithm 1.

$compute\_\sigma(n_r)$;
Input : n_r: maximum recursive depth in the **RDF** graph exploration;
Output: σ_{T_r}
1 let D be set of **RDF** datasets of interest;
2 **if** $n_r = 0$ **then**
3 $\quad \sigma_{T_r} = function\ \sigma_{T_r}(t)\ \{\text{return false}\}$
4 **else**
5 $\quad \sigma_{T_r} = function\ \sigma_{T_r}(t)\ \{\ \textbf{if}\ t \in D\ \textbf{then}\ \text{return true}\ \textbf{else}\ \text{return false}\ \}$;
6 return σ_{T_r};

Algorithm 2. Computation of a simple σ_{T_r}

In order to start searching for a Common Subsumer of a and b, Algorithm 1 calls in Row 8 the recursive function *explore*, defined in Algorithm 3.

Algorithm 3 investigates on the sets (if both not empty – see Row 6) of relevant triples of the input resources a and b (Rows 7–26) and returns a resource x which is a blank node (or one of the input resources, if they are equal to each other – see Rows 3–5).

```
    explore(a, σ_{T_a}, n_a, b, σ_{T_b}, n_b);
    Input  : a, σ_{T_a}, n_a, b, σ_{T_b}, n_b
    Output: x
 1  T_a = {t | σ_{T_a}(t) = true};
 2  T_b = {t | σ_{T_b}(t) = true};
 3  if ⟨a, T_a⟩ = ⟨b, T_b⟩ then
 4  |   add T_a to T ;
 5  |   return a ;
 6  if  T_a = ∅ or T_b = ∅ then return x;
 7  foreach <<a p c>> ∈ T_a do
 8  |   foreach <<b q d>> ∈ T_b do
 9  |   |   if [p, q, pq] ∈ S then
10  |   |   |   y = pq;
11  |   |   else
12  |   |   |   σ_{T_p} = compute_σ(n_a − 1);
13  |   |   |   σ_{T_q} = compute_σ(n_b − 1);
14  |   |   |   y = explore(p, σ_{T_p}, (n_a − 1), q, σ_{T_q}, (n_b − 1));
15  |   |   |   add [p, q, y] to S ;
16  |   |   |   add T_y to T;
17  |   |   if [c, d, cd] ∈ S then
18  |   |   |   z = cd;
19  |   |   else
20  |   |   |   σ_{T_c} = compute_σ(n_a − 1);
21  |   |   |   σ_{T_d} = compute_σ(n_b − 1);
22  |   |   |   z = explore(c, σ_{T_c}, (n_a − 1), d, σ_{T_d}, (n_b − 1));
23  |   |   |   add [c, d, z ] to S ;
24  |   |   |   add T_z to T;
25  |   |   if y ≠ [] or z ≠ [] then  add <<x y z>> to T_x ;
26  |   add T_x to T;
27  return x;
```

Algorithm 3. Investigation on Relevant Graph Portion

For each pair of triples $<<a\ p\ c>> \in T_a$ and $<<b\ q\ d>> \in T_b$, Algorithm 3 performs a recursive call to investigate over the pairs of resources p and q (Row 14), and c and d (Row 22), unless such pairs have been already computed in any previous call (Rows 9–10 and 17–18). All sets of relevant triples inferred during the investigation are added to the global set T (Rows 4, 16, 24, 26).

The maximum number of recursive calls to be still performed, n_a and n_b, is updated at each call (Rows 12 and 22).

Algorithm 1 finally returns the result coming from Algorithm 3, together with the set T of triples required to fully describe x (Row 8 in Algorithm 1).

5.1 Computational Issues

Given $\langle a, T_a \rangle$ and $\langle b, T_b \rangle$ with $|T_a| = |T_b| = n$, we note that, in very artificial cases, a Common Subsumer $\langle x, T_x \rangle$ of them, can grow as large as $|T_x| \in O(n^2)$, and when generalized to k r-graphs, a Common Subsumer of all of them can grow as $O(n^k)$—*i.e.*, exponential in k. This is a very artificial worst case, since it presumes that every triple $<<a\ p\ c>>$ of T_a is "comparable" with every triple $<<b\ q\ d>>$ of T_b—*e.g.*, this case would occur if all triples use the same predicate $p = q$, and every pair of $\langle c, T_c \rangle$, $\langle d, T_d \rangle$ yields a non-trivial Common Subsumer.

A more realistic situation is that for each triple of T_a, only a constant number of triples of T_b, say $\alpha \geqslant 1$, yields a new triple added to T_x. Observe that this was the case for Example 1, with $\alpha = 1$. In this case, $|T_x| = \alpha n$, and for k resources, $O(\alpha^k n)$—still exponential in k for $\alpha > 1$. Even if $\alpha > 1$, a strategy to keep the Common Subsumer "small" could be, for each triple in T_a to choose one among the α triples of T_b which would add a triple to T_x—in practice, to force $\alpha = 1$. The result would be still a Common Subsumer anyway, even if less informative than the "original" one, whose size is $|T_x| = n$, independent of k.

6   Conclusion

Motivated by the need for greedily learning shared informative content in collections of **RDF** resources, we defined Common Subsumers. We decided not to handle Common Subsumers which are also subsumption minimal (known as Least Common Subsumer in DLs) because we need to refer to a selective representation of resources in terms of descriptive triples, in order to ensure computability in the reference dataset (the Web of Data), too large to be explored. The adoption of an anytime algorithm allows for using partial learned informative content for further processing, whenever the search for Common Subsumers is interrupted. Thanks to such distinguishing features, the proposed approach may support the clustering of collections of **RDF** resources, by exploiting associativity of Common Subsumers. Our future work will be devoted to the investigation on **RDF** clustering methods based on Common Subsumers computation, possibly adopting strategies proposed in our past research ([8],[9]).

Acknowledgments. We acknowledge support of project "Semantic Expert Finding for Service Portfolio Creation" funded by HP (Hewlett-Packard) Labs - Innovation Research Program Award.

References

1. Baader, F., Calvanese, D., Mc Guinness, D., Nardi, D., Patel-Schneider, P. (eds.): The Description Logic Handbook, 2nd edn. Cambridge University Press (2007)
2. Baader, F., Sertkaya, B., Turhan, A.Y.: Computing the least common subsumer w.r.t. a background terminology. J. Applied Logic 5(3), 392–420 (2007)
3. Beckett, D., Berners-Lee, T.: Turtle - Terse RDF Triple Language, W3C Team Submission (2011), http://www.w3.org/TeamSubmission/turtle/
4. Berners-Lee, T., Hendler, J., Lassila, O.: The semantic web. Scientific American 248(4), 34–43 (2001)

5. Cohen, W., Borgida, A., Hirsh, H.: Computing least common subsumers in description logics. In: Rosenbloom, P., Szolovits, P. (eds.) Proc. of AAAI 1992, pp. 754–761. AAAI Press (1992)
6. Cohen, W.W., Hirsh, H.: The learnability of description logics with equality constraints. Machine Learning 17(2-3), 169–199 (1994)
7. Cohen, W.W., Hirsh, H.: Learning the CLASSIC description logics: Theoretical and experimental results. In: Doyle, J., Sandewall, E., Torasso, P. (eds.) Proc. of KR 1994, pp. 121–133 (1994)
8. Colucci, S., Di Noia, T., Di Sciascio, E., Donini, F., Piscitelli, G., Coppi, S.: Knowledge Based Approach to Semantic Composition of Teams in an Organization. In: Proc. of SAC 2005, pp. 1314–1319. ACM (2005)
9. Colucci, S., Di Sciascio, E., Donini, F.M., Tinelli, E.: Finding informative commonalities in concept collections. In: Proc. of CIKM 2008, pp. 807–817. ACM (2008)
10. d'Amato, C., Fanizzi, N., Ławrynowicz, A.: Categorize by: Deductive aggregation of semantic web query results. In: Aroyo, L., Antoniou, G., Hyvönen, E., ten Teije, A., Stuckenschmidt, H., Cabral, L., Tudorache, T. (eds.) ESWC 2010, Part I. LNCS, vol. 6088, pp. 91–105. Springer, Heidelberg (2010)
11. De Giacomo, G., Lenzerini, M., Rosati, R.: Higher-order description logics for domain metamodeling. In: Proc. of AAAI 2011 (2011)
12. Delteil, A., Faron-Zucker, C., Dieng, R.: Learning ontologies from RDF annotations. In: Proc. of the Second Workshop on Ontology Learning (2001)
13. Fanizzi, N., d'Amato, C., Esposito, F.: Metric-based stochastic conceptual clustering for ontologies. Information Systems 34(8), 792–806 (2009)
14. Frazier, M., Pitt, L.: CLASSIC learning. In: Proc. of the 7th Annual ACM Conference on Computational Learning Theory, pp. 23–34. ACM (1994)
15. Garcia-Plaza, A., Fresno, V., Martinez, R.: Web page clustering using a fuzzy logic based representation and self-organizing maps. In: Proc. of WI-IAT 2008, vol. 1, pp. 851–854. ACM (2008)
16. Grimnes, G.A., Edwards, P., Preece, A.: Instance based clustering of semantic web resources. In: Bechhofer, S., Hauswirth, M., Hoffmann, J., Koubarakis, M. (eds.) ESWC 2008. LNCS, vol. 5021, pp. 303–317. Springer, Heidelberg (2008)
17. Hayes, P.: RDF semantics, W3C recommendation (2004), http://www.w3.org/TR/2004/REC-rdf-mt-20040210/
18. Ławrynowicz, A.: Grouping results of queries to ontological knowledge bases by conceptual clustering. In: Nguyen, N.T., Kowalczyk, R., Chen, S.-M. (eds.) ICCCI 2009. LNCS, vol. 5796, pp. 504–515. Springer, Heidelberg (2009)
19. Madhavan, J., Cohen, S., Dong, X.L., Halevy, A.Y., Jeffery, S.R., Ko, D., Yu, C.: Web-scale data integration: You can afford to pay as you go. In: Proc. of CIDR 2007, pp. 342–350 (2007)
20. Mahmoud, H.A., Aboulnaga, A.: Schema clustering and retrieval for multi-domain pay-as-you-go data integration systems. In: Proc. of SIGMOD 2010, pp. 411–422. ACM (2010)
21. Pan, J.Z., Horrocks, I.: RDFS(FA): Connecting RDF(S) and OWL DL. IEEE Trans. on Knowl. and Data Eng. 19(2), 192–206 (2007)
22. Shadbolt, N., Hall, W., Berners-Lee, T.: The semantic web revisited. IEEE Intelligent Systems 21(3), 96–101 (2006)
23. ter Horst, H.J.: Completeness, decidability and complexity of entailment for RDF Schema and a semantic extension involving the OWL vocabulary. Journal of Web Semantics 3(2-3), 79–115 (2005)
24. Zeng, H.J., He, Q.C., Chen, Z., Ma, W.Y., Ma, J.: Learning to cluster web search results. In: Proc. of SIGIR 2004, pp. 210–217. ACM (2004)

Personality-Based Active Learning for Collaborative Filtering Recommender Systems

Mehdi Elahi[1], Matthias Braunhofer[1], Francesco Ricci[1], and Marko Tkalcic[2]

[1] Free University of Bozen-Bolzano, Bozen-Bolzano, Italy
{mehdi.elahi,mbraunhofer,fricci}@unibz.it
http://www.unibz.it
[2] Johannes Kepler University, Linz, Austria
marko.tkalcic@jku.at
http://www.jku.at

Abstract. Recommender systems (RSs) suffer from the cold-start or new user/item problem, i.e., the impossibility to provide a new user with accurate recommendations or to recommend new items. Active learning (AL) addresses this problem by actively selecting items to be presented to the user in order to acquire her ratings and hence improve the output of the RS. In this paper, we propose a novel AL approach that exploits the user's personality - using the Five Factor Model (FFM) - in order to identify the items that the user is requested to rate. We have evaluated our approach in a user study by integrating it into a mobile, context-aware RS that provides users with recommendations for places of interest (POIs). We show that the proposed AL approach significantly increases the number of ratings acquired from the user and the recommendation accuracy.

1 Introduction

Recommender systems (RSs) are information and decision support tools providing users with suggestions for items that are likely to be interesting to them or to be relevant to their needs [20]. A common problem of RSs is cold-start; this occurs when a new user or a new item is added to the system, but the system doesn't have enough information (e.g., ratings, purchasing records, browsing history) about them to reliably recommend any item to this new user or the new item to any user. Several approaches have been recently proposed to deal with this problem [20] but the most direct way is to rely on active learning (AL), i.e., to use an initial data acquisition and learning phase. Here the system actively asks the user to rate a set of items, which are identified using a strategy aimed at best revealing the user's interests and consequently at improving the quality of the recommendations [21,7].

In this paper, we present a novel AL rating request strategy that leverages the knowledge of the user's personality in order to predict which items a user will have an opinion about. More specifically, our approach makes use of one of the most influential models in psychology, namely the Five Factor Model (FFM)

M. Baldoni et al. (Eds.): AI*IA 2013, LNAI 8249, pp. 360–371, 2013.

or Big Five dimensions of personality, in which personality is conceptualized in terms of openness, conscientiousness, extraversion, agreeableness and neuroticism [3]. The rationale behind the choice of the FFM for AL is that these factors account for most of the variance among users in terms of trait terms [3]. Previous research has shown that personality influences human behaviours and that there exist direct relations between personality and tastes / interests [19]. Consequently, the incorporation of human personality into AL can help in selecting "good" items to be rated by the user. Moreover, user's personality, as it is shown in this paper, can be acquired with simple and even engaging questionnaires that any user can fill out. The assessed personality can also be illustrated to the user hence making the interaction with the recommender even more rewarding for the user.

We have formulated the following hypotheses: a) the proposed personality-based AL method leads to a higher number of ratings acquired from users compared to a state-of-the-art AL strategy, and b) our proposed AL approach compares favourably to existing state of the art AL methods in terms of achieved recommendation accuracy. To evaluate these hypotheses, we integrated our technique into a context-aware RS that recommends places of interest (POIs) to mobile users [2], and conducted a live user study. The implemented solution takes the personality characteristics of a target user as input to an extended matrix factorization model used for predicting what items the user should be requested to rate. Hence, by improving existing AL strategies, it is able to provide personalized rating requests even to users with few or no past ratings (cold-start).

The remainder of the paper is organized as follows: section 2 reviews the related work. In section 3 we explain the application scenario and in section 4, we present our proposed strategy for AL based on user's personality traits. Section 5 describes the structure of the user study while the obtained results are presented in section 6. Finally, in section 7, conclusions and future work directions are given.

2 Related Work

Active Learning in RSs aims at actively acquiring user preference data to improve the output of the RS [21,7,6]. In [17] six techniques that collaborative filtering systems can use to learn about new users in the sign up process are introduced: *entropy* where items with the largest rating entropy are preferred; *random*; *popularity*; $log(popularity) * entropy$ where items that are both popular and have diverse ratings are preferred; and finally *item-item personalized*, where the items are proposed randomly until one rating is acquired, then a recommender is used to predict the items that the user is likely to have seen. In the considered scenario, the $log(popularity) * entropy$ strategy was found to be the best in terms of accuracy.

In [18] the authors extend their former work using a rating elicitation approach based on decision trees. The proposed technique is called $IGCN$, and builds a tree where each node is labelled by a particular item to be asked to the user to rate. According to the user elicited rating for the asked item a different branch is followed, and a new node, that is labelled with another item to rate is determined. It is worth noting that this type of techniques can be applied only to a recommender system that has already acquired a large dataset of ratings. In fact the ratings are necessary to build the above mentioned decision tree. Hence, they could solve the cold-start problem for a new user but they cannot deal with a situation where the entire system must be bootstrapped, because it has not yet interacted with enough users, as in our study.

In [8] three strategies for rating elicitation in collaborative filtering are proposed. The first method, *GreedyExtend*, selects the items that minimize the root mean square error (RMSE) of the rating prediction (on the training set). The second one, named *Var*, selects the items with the largest $\sqrt{popularity} * variance$, i.e., those with many and diverse ratings in the training set. The third one, called *Coverage*, selects the items with the largest coverage, which is defined as the total number of users who co-rated both the selected item and any other item. They evaluated the performance of these strategies and compared them with previously proposed ones (*popularity*, *entropy*, *entropy0*, *HELF*, and *random*). They show that *GreedyExtend* outperforms the other strategies. However, despite this remarkable achievement, *GreedyExtend* is static, i.e., selects the items without considering the ratings previously entered by the user.

In a more recent work of the same authors [9] an adaptive strategy for rating elicitation in collaborative filtering is proposed. Their strategy is based on decision trees where each node is labelled with an item (movie). The node divides the users into three groups based on their ratings for that movie: lovers, who rated the movie high; haters, who rated the movie low; and unknowns, who did not rate the movie. The proposed strategy has shown a significant reduction of RMSE compared with other strategies. It should be noted that their results are again rather difficult to compare with ours. They simulate a scenario where the system is trained and the decision tree is constructed from a large training dataset with millions of ratings. So they assume a large initial collection of ratings. Then, they focus on completely new users, i.e., those without a single rating in the train set. In contrast, our system had a very limited initial rating dataset with only few hundred ratings. Moreover, we analyze the performance of our system in an online study with real users while they executed only an offline study with simulated users.

Moving now to the topic of personality in RSs, we note that earlier studies conducted on the user personality characteristics support the possibility of using personality information in RSs [13,12,22]. In general, user personality can be enquired either explicitly, i.e., asking users to complete a personality questionnaire using one of the personality evaluation inventories, or implicitly, i.e., by observing users' behavioral patterns [16]. However, a previous study has shown that using explicit personality acquisition interfaces yields to better results in terms of user satisfaction, ease of use and prediction accuracy [4].

For this reason we have decided to adopt (as illustrated in the next section) the explicit approach, and in particular we have used the Ten-Item Personality

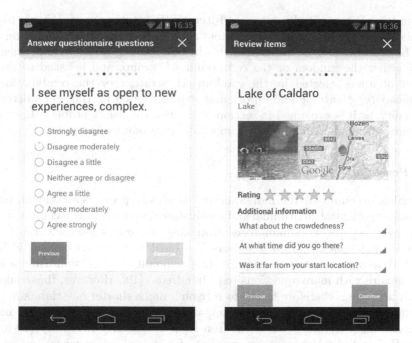

Fig. 1. Sample screenshots of the application

Inventory (10-items TIPI) which takes a few minutes to complete [11]. It represents a sensible option if personality is not the primary topic of research and it is mostly useful when time and space are in short supply (e.g., in a mobile application). We should note here that the personality model we used is a comprehensive model and thus not linked to a specific domain [19].

3   Application Scenario

Our application scenario is a mobile RS for tourists, that recommends interesting places of interests (POIs) in Alto-Adige region in Italy [2]. After the user registration, the system, by using an AL strategy, identifies and presents items to the user in order to collect her ratings (Figure 1, right). The adopted AL strategies analyze the available rating dataset, score the items estimating their ratings usefulness, and the highest scoring items are presented to the user to rate. However, in our application scenario, we encountered a severe cold-start problem, i.e., we needed to bootstrap the RS having just a small dataset with only few hundred ratings. In such a situation, standard AL strategies, as those mentioned in the previous section, fail to select useful and ratable items, and hence they can not elicit any rating from the user.

In order to cope with this problem, we have defined and proposed a novel AL strategy that incorporates an additional source of user information, i.e., user personality information. Hence, in our system before presenting any item to rate, the Ten-Item Personality Inventory (10-items TIPI) statements are shown to the user and she has to indicate to what extent she agrees with each statement on a 7-point Likert scale (Figure 1, left). Then, the user is assigned to an experimental group, either the random or the compound AL group, and is asked to rate a number of items selected by the random AL strategy or the combination of *log(popularity) * entropy* and an original strategy based on the user's detected personality, as it is explained in section 4. Finally, the user's profile is built and the items with highest predicted ratings can be recommended to her.

3.1   Personality Questionnaire

Personality accounts for the individual differences in people's emotional, inter-personal, experiential, attitudinal and motivational styles [14]. It has been shown that personality affects human decision making and user's interests [19]. In order to learn the personality of the users in a ideal way, the system should have enough time and resources to enquire the personality information using a long questionnaire with many questions (e.g. hundreds) [10]. However, this could be difficult and one is therefore forced to rely on a much shorter questionnaire [11]. For example, in our mobile application, we could not ask the users too many questions and therefore we used the Ten-Item Personality Inventory (10-items TIPI). Figure 1 (left) shows a screen shot of our application where one of the questionnaire statements is illustrated. The full questionnaire includes the ten statements that are listed below:

1. I see myself as extraverted, enthusiastic.
2. I see myself as critical, quarrelsome.
3. I see myself as dependable, self-disciplined.
4. I see myself as anxious, easily upset.
5. I see myself as open to new experiences, complex.
6. I see myself as reserved, quiet.
7. I see myself as sympathetic, warm.
8. I see myself as disorganized, careless.
9. I see myself as calm, emotionally stable.
10. I see myself as conventional, uncreative.

For each statements the user has to indicate to what extent she agrees. When the user completes the questionnaire, a brief explanation of her personality, as those used in [1], is shown to her.

4   Active Learning Strategies

In this section we describe the AL strategies that we have used and compared in the experimental study fully described in the next section.

Log(Popularity) * Entropy scores each item i by multiplying the logarithm of the *popularity* of i (i.e., the number of ratings for i in the training set) with the entropy of the ratings for i. Then, the top items according to the computed score are proposed to be rated by the user (4 in our experiments).

This strategy tries to combine the effect of the popularity with a score that favours items with more diverse ratings (larger entropy), which may provide more useful (discriminative) information about the user's preferences [17]. Clearly, more popular items are more likely to be known by the user, and hence it is more likely that a request for such a rating will be fulfilled by the user and will increase the size of the rating database. But many popular items in our dataset had no or only one rating, and rating-based popularity scores cannot distinguish such popular items from less popular items with similar number of ratings. Therefore, this strategy may select items that are unpopular and unknown to the user and thus not rateable.

To cope with that problem, we have designed a second strategy that tries to select the items that the user has most likely experienced, by exploiting the personality information of the users. In that sense, it is similar to the popularity strategy, but it tries to make a better prediction of what items the user can rate by leveraging the knowledge of the items that users with similar personalities have rated in the past.

Personality-Based Binary Prediction first transforms the rating matrix to a matrix with the same number of rows and columns, by mapping null entries to 0, and not null entries to 1. Hence, the new matrix models only whether a user rated an item or not, regardless of its value. Then, the new matrix is used to train an extended version of the popular matrix factorization algorithm. Our model is similar to the one proposed in [15], and profiles users not only in terms of the binary ratings, but also using known user attributes, in our case, gender, age group and the scores for the Big Five personality traits on a scale from 1 to 5. Given a user u, an item i and the set of user attributes $A(u)$, it predicts ratings using the following rule:

$$\hat{r}_{ui} = \bar{i} + b_u + q_i^\top \cdot (p_u + \sum_{a \in A(u)} y_a), \tag{1}$$

where p_u, q_i and y_a are the latent factor vectors associated with the user u, the item i and the user attribute a, respectively. The model parameters are then learnt, as it is common in matrix factorization, by minimizing the associated regularized squared error function through stochastic gradient descent. Finally, the learnt model predicts and assigns a rateable score to each candidate item i (for each user u), with higher scores indicating a higher probability that the target user has consumed the item i, and hence may be able to rate it.

Both strategies have been used in a "compound" AL strategy in order to select the most useful items for the active user to rate. This has been done by getting a short list of items (4) from each of these strategies and merging them together in a way that the final list includes equal numbers of items from the two strategies. Exploiting a combination of two strategies with different characteristics is beneficial. Firstly, such a combination allowed us to compare their performances, in an offline analysis where we built separated training sets with the ratings acquired by each individual strategy. Secondly, in a finally deployed RS one can take advantage of both AL strategies to add some diversity to the system rating requests.

Finally, in order to have a baseline we have also considered a control group that was requested to rate items that were selected at **Random**.

5   User Study Evaluation

In order to evaluate the considered elicitation strategies, we conducted a user study involving 108 participants who were randomly assigned either to the compound AL strategy group (n = 54) or the random item selection strategy group (n = 54). Our goal was to study the influence of the rating elicitation strategies on the evolution of the RS's performance. Given a particular AL strategy, the (training) rating matrix evolves by including all the ratings entered by users on the training items elicited so far. The exact test procedure was as follows. After the user has completed the personality questionnaire (as illustrated in section 3.1) she is asked to rate 13 items (see figure 1, right): 5 of these items are test items, i.e., items selected randomly to test the obtained model accuracy, while 8 are train items, i.e., items selected to train the model. For the random strategy group the 8 train items are selected randomly. For the compound AL strategy group, 4 of the train items are selected by log(pop) * entropy strategy, and 4 by personality-based binary prediction strategy.

This evaluation set-up, i.e., one control group and one AL compound strategies group, allowed us to assign a larger number of users to the considered AL strategies and hence to test and compare their performances more reliably. Then, in order to be fair in the comparison of these strategies vs. the random strategy, i.e., to simulate the same number of rating requests for each individual strategy, when evaluating the accuracy of the RS we randomly sampled with probability 0.5 the train ratings acquired by the random strategy (since 8 items was requested to rate at random, vs. 4 using the two AL strategies). Finally, after having trained the prediction model on all the ratings acquired from the users with a specific AL strategy during the time period of the study, the Mean Absolute Error (MAE) on the ratings of the selected random test items was measured. We have selected the test items randomly in order to avoid any bias that may affect the MAE especially when the rating data set is small.

6   Evaluation Result

6.1   Mean Absolute Error

Figure 2 shows the system's MAE after training our extended matrix factorization model with the users' ratings collected by an individual AL strategy, while at the same time ignoring those collected by the other strategies. After having retrained the model with all the users' ratings elicited by a certain AL strategy, for each user in the user study, we have computed the system MAE on the ratings given to the test items that a user was able to provide and then averaged the MAE of all the users. This allowed us to compute to what extent the prediction accuracy is affected by the addition of new ratings (MAE). Therefore, it indicates the effect of a rating elicitation strategy on the prediction accuracy of the system.

Comparing the results it is clear that our proposed strategy, i.e., personality-based binary prediction, outperforms the other strategies in terms of MAE. As it can be observed in Figure 2, the initial MAE of the system using the 848 training ratings that were available at the beginning of the study was 1.06. The MAE decreased to 0.86 by adding the 125 training ratings acquired by our proposed strategy, whereas it was 0.90 by adding the 112 training ratings collected by the log(popularity) * entropy strategy, and 0.97 by adding the 73 training ratings derived from the random strategy. Hence, the MAE reduction achieved by AL, is 18.8% for personality-based binary prediction, 15.0% for log(popularity) * entropy, and 8.4% for random. Consequently, under application of our proposed strategy, the system's MAE has been reduced the most.

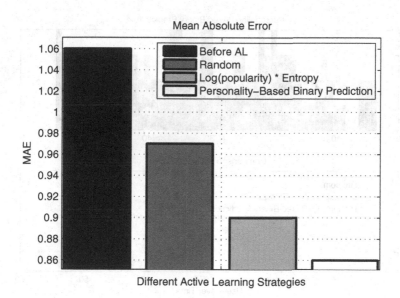

Fig. 2. MAEs of the strategies

6.2 Number of Acquired Ratings

Another important aspect to consider is the number of ratings that are acquired by the considered strategies. As we discussed before, certain strategies can acquire more ratings by better estimating what items are likely to have been experienced or known by the user. Table 1 summarizes our results. Overall,

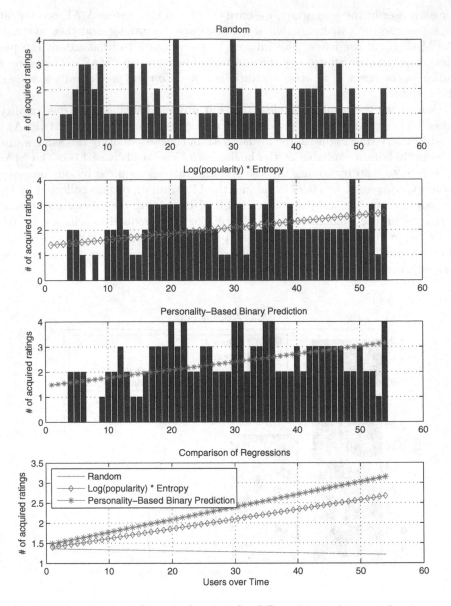

Fig. 3. Number of acquired ratings by different strategies over time

the average number of ratings acquired per user in our train set is 1.9 (out of 4 items that were requested to rate). Ultimately, the personality-based binary prediction strategy has elicited by average 2.31 ratings from each user, whereas the log(popularity) * entropy strategy elicited 2.07 ratings, and random strategy elicited 1.35 ratings. Conducting t-test, we observed that the personality-based binary prediction strategy can elicit significantly more ratings than either log(popularity) * entropy strategy (p-value = 0.0054) or random strategy (p-value<0.0000). Our strategy can obtain more than 57% of the requested ratings. Conversely, log(popularity) * entropy can acquire 51% and random 33% of the requested ratings. Interestingly, these are the results that we conjectured in one of our previous works [5].

In figure 3 we plot the number of ratings that each strategy has acquired from the users over time together with the corresponding regression lines. Observing the results, we can see that over time the average number of ratings acquired by the random strategy is approximately constant. However, this number is increasing when the personality-based binary prediction and log(popularity) * entropy strategies are used. Overall, our proposed strategy, i.e., the personality-based binary prediction, achieves the best result. The number of ratings that this strategy acquires is increasing faster than using the other strategies (its regression line has the largest slope). This means that our proposed strategy is better learning to estimate which items may have been experienced by the users and the users are able to rate. This is an important factor since if a strategy selects the items that are informative but not experienced by the users, it can not acquire their ratings (which is typical for the random strategy). Therefore, a good strategy must focus not only on the quality but also on the quantity of the ratings, as our proposed strategy does.

Finally, when looking at the relationships between user's personality traits and the number of ratings provided, we have observed that people with higher openness to experience can rate significantly more items than those with lower openness to experience (p-value=0.044). This means that the personality of the people can affect their rating behaviors, which is the conjecture that initially motivated this research. This type of relationships opens a new line of research, showing that user's personality can be useful to select and assign a convenient AL strategy to a group of users, when eliciting the largest number of ratings is in order.

Table 1. Pairwise comparison of the number of ratings acquired over 4 requests by different strategies

Pair of strategies	Means	p-value	# of ratings
Random / log(popularity) * entropy	1.35 / 2.07	0.0003	73 / 112
Random / personality-based binary prediction	1.35 / 2.31	0.0000	73 / 125
Personality-based binary prediction / log(popularity) * entropy	2.31 / 2.07	0.0054	112 / 125

7   Conclusions and Future Work

In this paper, we have presented a novel AL technique for addressing the cold-start problem in RSs. The proposed technique uses the Five Factor Model (FFM) of personality traits as its basis in order to provide a user with personalized rating requests, without completely relying on explicit feedback (e.g., ratings) or implicit feedback (e.g., item views or purchases) which is usually more difficult to obtain and are not available in cold-start situations.

We have developed the following two research hypotheses: a) our proposed AL method leads to a higher increase in the number of acquired user ratings in comparison to a state-of-the-art rating elicitation strategy, and b) the prediction accuracy of the recommendation model improves more when utilizing our proposed AL strategy than when using another popular and effective state of the art AL strategy. In a live user study, we successfully verified both hypotheses and we have shown that user personality has an important impact in her rating behaviour.

Our future work includes the further analysis of the data obtained from the study in order to understand potential performance differences among the compared AL strategies that are due to different personality traits. We would like to test our proposed method on larger training and test sets. We would also like to understand the impact of using different rating prediction models, such as context-aware ones, on the performance of our proposed AL approach.

References

1. Short personality quiz - psych central. Based upon the Ten-Item Personality Inventory (TIPI) (February 2013)
2. Baltrunas, L., Ludwig, B., Peer, S., Ricci, F.: Context relevance assessment and exploitation in mobile recommender systems. Personal and Ubiquitous Computing 16(5), 507–526 (2012)
3. Costa, P., McCrae, R.: Toward a new generation of personality theories: Theoretical contexts for the five-factor model. In: The Five-Factor Model of Personality: Theoretical Perspectives, pp. 51–87 (1996)
4. Dunn, G., Wiersema, J., Ham, J., Aroyo, L.: Evaluating interface variants on personality acquisition for recommender systems. In: Houben, G.-J., McCalla, G., Pianesi, F., Zancanaro, M. (eds.) UMAP 2009. LNCS, vol. 5535, pp. 259–270. Springer, Heidelberg (2009)
5. Elahi, M., Repsys, V., Ricci, F.: Rating elicitation strategies for collaborative filtering. In: Huemer, C., Setzer, T. (eds.) EC-Web 2011. LNBIP, vol. 85, pp. 160–171. Springer, Heidelberg (2011)
6. Elahi, M., Ricci, F., Rubens, N.: Adapting to natural rating acquisition with combined active learning strategies. In: Chen, L., Felfernig, A., Liu, J., Raś, Z.W. (eds.) ISMIS 2012. LNCS, vol. 7661, pp. 254–263. Springer, Heidelberg (2012)
7. Elahi, M., Ricci, F., Rubens, N.: Active learning strategies for rating elicitation in collaborative filtering: a system-wide perspective. ACM Transactions on Intelligent Systems and Technology 5(1) (2014)
8. Golbandi, N., Koren, Y., Lempel, R.: On bootstrapping recommender systems. In: Proceedings of the 19th ACM International Conference on Information and Knowledge Management, pp. 1805–1808. ACM (2010)
9. Golbandi, N., Koren, Y., Lempel, R.: Adaptive bootstrapping of recommender systems using decision trees. In: Proceedings of the Fourth ACM International Conference on Web Search and Data Mining, pp. 595–604. ACM (2011)
10. Goldberg, L.R.: The development of markers for the big-five factor structure. Psychological Assessment 4(1), 26–42 (1992)

11. Gosling, S.D., Rentfrow, P.J., Swann, W.B.: A very brief measure of the big-five personality domains. Journal of Research in Personality 37, 504–528 (2003)
12. Hu, R., Pu, P.: A comparative user study on rating vs. personality quiz based preference elicitation methods. In: Proceedings of the 14th International Conference on Intelligent User Interfaces, IUI 2009, pp. 367–372. ACM, New York (2009)
13. Hu, R., Pu, P.: Enhancing collaborative filtering systems with personality information. In: Proceedings of the Fifth ACM Conference on Recommender Systems, RecSys 2011, pp. 197–204. ACM, New York (2011)
14. John, O.P., Srivastava, S.: The big five trait taxonomy: History, measurement, and theoretical perspectives. In: Handbook of Personality: Theory and Research, vol. 2, pp. 102–138 (1999)
15. Koren, Y., Bell, R., Volinsky, C.: Matrix factorization techniques for recommender systems. Computer 42(8), 30–37 (2009)
16. Kosinski, M., Stillwell, D., Graepel, T.: Private traits and attributes are predictable from digital records of human behavior. Proceedings of the National Academy of Sciences, 2–5 (March 2013)
17. Rashid, A.M., Albert, I., Cosley, D., Lam, S.K., Mcnee, S.M., Konstan, J.A., Riedl, J.: Getting to know you: Learning new user preferences in recommender systems. In: Proceedings of the 2002 International Conference on Intelligent User Interfaces, IUI 2002, pp. 127–134. ACM Press (2002)
18. Rashid, A.M., Karypis, G., Riedl, J.: Learning preferences of new users in recommender systems: an information theoretic approach. ACM SIGKDD Explorations Newsletter 10(2), 90–100 (2008)
19. Rentfrow, P.J., Gosling, S.D., et al.: The do re mi's of everyday life: The structure and personality correlates of music preferences. Journal of Personality and Social Psychology 84(6), 1236–1256 (2003)
20. Ricci, F., Rokach, L., Shapira, B., Kantor, P.B.: Recommender Systems Handbook. Springer (2011)
21. Rubens, N., Kaplan, D., Sugiyama, M.: Active learning in recommender systems. In: Ricci, F., Rokach, L., Shapira, B., Kantor, P. (eds.) Recommender Systems Handbook, pp. 735–767. Springer (2011)
22. Tkalcic, M., Kosir, A., Tasic, J.: The ldos-peraff-1 corpus of facial-expression video clips with affective, personality and user-interaction metadata. Journal on Multimodal User Interfaces 7(1-2), 143–155 (2013)

Selection and Ranking of Activities in the Social Web

Ilaria Lombardi, Silvia Likavec, Claudia Picardi, and Elisa Chiabrando

Università di Torino, Dipartimento di Informatica, Torino, Italy
{lombardi,likavec,picardi}@di.unito.it, chiabrando@gmail.com

Abstract. The paper presents a framework for the *selection* and *ranking* of *activities* proposed to users in a social networking service, as for example Facebook's "social events". Our proposal takes into account the peculiarities of this application area, considering a variety of factors including the spatiotemporal relations between activities and between activities and users, and the social opinions weighted by the user's interest in the specific themes and type of an activity.

To this aim, we describe a semantic model for the involved entities, and especially for the available activities. We describe a process that can be easily configured according to user preferences, extended to take into account additional ranking factors, and adapted to include existing recommendation/ranking strategies. Finally, we specifically introduce ranking factors that can be used within the specific model and process we describe. The results of a preliminary evaluation show the effectiveness of our approach.

1   Introduction

In this paper we tackle the problem of selecting and ranking real-life activities proposed to people within a social networking service. By "activity" we intend anything that: (a) has a participatory aspect, i.e. it is offered and people can decide whether to join or not, (b) requires the physical presence of people in the location where the activity takes place, (c) is offered within a given time interval and requires a certain amount of time to be completed. For example, Facebook[1] "events" can be regarded as activities according to this definition. Activities, as "one-and-only items" [4] have some peculiarities:

- Activities have a specific start and end in time; once an activity is finished there is no need to recommend it. Therefore, the opinions of users on past activities cannot be used to recommend them. On the other hand, the recommendations are needed before the users had a chance of participating in the activities.
- Joining an activity is time consuming: we cannot expect a user to join two activities that occur simultaneously.

Many standard ranking strategies used in social networks and mobile applications can be applied to activities, e.g. content-based recommendation [10], where the main topics or themes of activities can be compared to user's interests or to her friends' interests, and context-aware recommendation [1], where the activities can be ranked by proximity, assuming that the user can be geo-localized, or by time of day depending on what is available at what time. Still, many recommender systems fail to combine the

---

[1] http://www.facebook.com

M. Baldoni et al. (Eds.): AI*IA 2013, LNAI 8249, pp. 372–384, 2013.

basic information about the activity with the additional factors which could influence the recommendation process. This is where the motivation for our work comes from. We consider the *themes* of the activity (i.e. the topics the activity is related to, for example "wine" or "english literature") and the activity type (for example "movie" or "concert"), but also the spatial-temporal aspect of the activity, as well as other additional factors which can influence users' decisions to participate. Our goal is to reduce the number of activities shown to the user, and to rank them in a way that is meaningful with respect to the user's context and objective.

To this aim we propose a framework for activity recommendation that takes into account these peculiarities and is composed of (i) a data model giving a semantic description of the actors and their data and (ii) a process for selection and ranking that considers several ranking factors and can consequently be implemented on top of existing ranking strategies. The process we describe can be adapted to the operating context and is flexible to user configuration. We discuss the general framework but we sometimes mention, as examples, the specific configuration choices we selected in our system.

We present the results of a user study, in which our system was tested, showing that:

R1 additional knowledge on activities, besides their themes and type, affects users' choices, so that
R2 recommender systems need to incorporate such knowledge in order to improve the accuracy of their recommendations.

The paper is organized as follows: we describe our data model in Section 2 and our process for selection and ranking in Section 3. Then, we present the results of our experiments in Section 4 and position our work within pertinent research in Section 5. Section 6 concludes the paper.

2   Data Model

In our framework we distinguish three types of entities:

- *spatial-temporal objects* or STOB's for short, the social events which are to be recommended to users;
- *users* which are the entities that use the system and receive the spatial-temporal recommendations or STOB's;
- *activities* that are STOB's selected by the user and inserted in her calendar.

Notice that selection and ranking of activities is strongly influenced by the user's goal when requesting a recommendation. We can distinguish two contexts of usage:

- **daily life context**: the selection and ranking process is aimed at "here and now". In this case the selection and ranking is **proactive**: whenever the user connects to the system, she is presented with a properly ranked selection of activities.
- **journey context**: the user makes an explicit query specifying the spatiotemporal constraints she is interested in (for example, the user is planning a trip somewhere). In this case the selection and ranking process happens **on demand**.

Next we describe STOBs, users and activities and their formal representation.

2.1 STOB

The **STOB**'s are stored in the STOB *Ontology*, that describes the structure of a social event and all its instances. STOB *Ontology* is connected to other ontologies (Domain Products Ontologies, *Time Ontology* and STOB *Type Ontology*). To represent the STOB class we used the Web Ontology Language (OWL 2^2).

The structure of a STOB s can be represented as follows:

$$s = \langle Id(s), T(s), RQT(s), S_t(s), TYP(s), THM(s), RAT(s) \rangle$$

where:

- $Id(s)$ is an **identifier** which univocally identifies s in the system;
- $T(s)$, called **temporal description**, defines the temporal existence of the STOB: $T(s) = I_s$, where I_s is a time interval in which the given STOB is happening; using the *Time Ontology* it is possible to define the *start time* and the *end time* with a specific moment or an interval, and the *duration* of a STOB with a temporal length;
- $RQT(s) = t_{rq}$ is the **required time** to participate at STOB s - notice that in general $RQT(s) \leq T(s)$, for example an art exhibit may be available from 10am to 5pm, but require only an hour to visit;
- $S_t(s)$, called **spatial description**, associates to a STOB s a spatial position in a certain time instance. It is a circular area characterized by its *center* $c(s)$, given with its latitude and longitude coordinates, and its *radius* $r(s)$;
- $TYP(s) = p$ is the **type** of STOB s (concert, seminar, course, birthday, etc.). The available values are described in the STOB *Type Ontology*;
- $THM(s) = \{th_1, \ldots, th_q\}$ are the **themes** that describe more closely the STOB s; they will be matched up with the user model to calculate the similarities between users and STOB's in a classic recommendation system. We rely upon an *ontology-based user model* [17] in which user interests are recorded for each class in the domain ontology. Each user profile is an instance of the domain ontology where every domain entity e has an interest value $i(e)$ associated to it. The themes of the STOB's are a subset of the user model categories, hence the user's interest for a STOB can be computed based on her interest for the themes;
- $RAT(s) = \{r_1, \ldots, r_m\}$, $r_i \in [0, 10]$, is the set of all **ratings** of STOB s. They allow users to express how much they consider a STOB interesting, before it happens.

2.2 User

The **user** is defined in the knowledge base by generic user information (name, surname, gender, age, etc.), some particular spatial-temporal information (spatial-position and area of interest), her social network and a set of activities that make up her *calendar*.

We can denote a user with u and characterize it by:

$$u = \langle Id(u), T(u), S_t(u), SN(u), UM(u), CAL(u) \rangle$$

---

^{2} http://www.w3.org/TR/owl-overview/

where:

- $Id(u)$ is an **identifier** which univocally identifies u in the system;
- $T(u)$, called **temporal description**, that defines t_u as the point in time in which u finds herself and l_u as the time interval for which user u wants to receive the recommendations, obtaining $T(u) = \langle t_u, l_u \rangle$;
- $S(u)$, called **spatial description**, which associates to user u a spatial position in a certain time instance. As for the STOBs, user's spatial description is a circular area characterized by its *center* $c(u)$, given with its latitude and longitude coordinates, and *radius* $r(u)$. But, differently from STOB's, for users it is not always possible to define a fixed center that describes where the "user position" indeed is. Therefore, the area in which the user wants to obtain recommendations can vary, depending on user's preferences. Also, if the user is geo-positioned, $r(u)$ may be very small. If, however, the user is not geo-positioned, for example we know she is in "San Francisco" and nothing more, than $c(u)$ and $r(u)$ are estimated to be the center and radius of San Francisco. Moreover, a user has a propensity to move, expressed as a distance $r_P(u)$ she is willing to cover to reach an event. This is a value that can be frequently changed by the user herself. In the computation of reachability we actually consider $r(u) + r_P(u)$.
- $SN(u)$ is the social network of the user u;
- $UM(u) = \{\langle e_1, i(e_1) \rangle, \ldots, \langle e_n, i(e_n) \rangle\}$ which represents the user model defining a set of entities in which the user u is interested in, together with the interest values $i(e_i) \in [0,1], i = 1, \ldots, n$ for each e_i. The entities in $UM(u)$ can be classes or instances of the domain ontology[3];
- $CAL(u) = \{(s_1, p_1, t_1), \ldots, (s_k, p_k, t_k)\}$, is the *calendar*, which defines a set of activities in which the user u wants to participate, together with the specification if the participation is public or not and the time of participation.

2.3 Activities

We assume that the user can express the intention to participate in a STOB by adding it to her calendar. When a STOB is added to the calendar, it is instantiated with respect to the specific user (so that she may for example change the start/end time) and the STOB becomes an activity. If, for example, an activity is offered regularly (e.g. every Friday, 8-10pm), the user may want to specify she wishes to attend on Friday the 26th.

We distinguish the following activities in the calendar:

- *joined activities*: the user has decided to join these activities; a significant overlapping of a new activity with a joined activity considerably reduces its feasibility;
- *prospective activities*: the user is considering to join these activities but has not decided yet whether to participate. An overlap of a new activity with a prospective activity is shown but has a minor impact, or no impact at all, on its feasibility.

If we denote an activity with a, the basic structure of a can be represented as:

$$a = \langle Id(a), T(a), STOB_{id}(a), USER_{id}(a), PAR(a) \rangle.$$

---

[3] How interest values are calculated is out of the scope of this paper, see for example [17].

where:

- $Id(\mathsf{a})$ is an **identifier** which univocally identifies a in the system;
- $\mathsf{T}(\mathsf{a})$, called **temporal description**, defines the temporal interval when the user wishes to attend the event;
- $\mathsf{STOB}_{id}(\mathsf{a})$ is the **STOB id** which had originated the activity;
- $\mathsf{USER}_{id}(\mathsf{a})$ is the **user id** that inserted this activity in her calendar;
- $\mathsf{PAR}(\mathsf{a}) = \{0, 1\}$ defines the **type of participation**, 1 for joined, 0 for prospective.

3 Reasoning

The reasoning process we propose has two separate phases: selection and ranking. In the selection phase the system discards the STOB's that are not at all interesting for the user, or that do not satisfy her spatiotemporal constraints, creating a *pool* of potentially interesting STOB's. Selection is described in Subsection 3.1. Ranking sorts activities in the pool, by taking into account several factors, detailed in Subsection 3.2.

Selection and ranking are executed asynchronously, communicating by means of data repositories. In particular, when the process is executed in **proactive** mode, as in the *daily life* context, selection phase is performed on a cyclical basis on all the valid (i.e. non expired) STOB's within the system. The ranking phase takes place when the user connects to the system to view its proposals, and it is refreshed whenever the user tunes her settings. When the process is executed **on demand**, as in the *journey* context, selection runs when the user queries the system to plan for her journey, and is immediately followed by the ranking phase. The selection phase however is not run again unless the user changes relevant spatiotemporal constraints, while the ranking phase is again refreshed whenever the user edits her settings.

Figure 1 describes the whole recommendation process, showing how it interleaves with user interaction. In fact, although we neither explicitly address the problem of how the recommendation results are displayed to the user, nor we investigate interaction modalities that allow her to fine tune the recommendation, we believe that a successful recommendation scheme must provide enough information to the UI Module so that it can handle these tasks in a flexible and affordable way.

As we already said, selection creates a *working pool of STOB's* that will be the input for the rest of the process. However, we assume that the user may want either to manually add STOB's she likes to this pool, to be later reminded of them, or to remove once and for all those she discards, in order not to see them anymore. Since we do not want a re-run of the selection process to overwrite these operations, the process saves the add/remove actions performed by the user in a *user operations repository*.

Ranking works on the pool created by selection and it has two additional inputs: *(i)* the *User calendar*, as our spatial-temporal recommendation takes into account the availability of the user in order to suggest other STOB's, and *(ii)* the *Filter configuration*, that allows the user to change the weight of each ranking factor in the final ranking.

The output of ranking goes directly to the UI Module, which provides the means for the user to perform the operations she prefers. Those relevant to our process are: *(i)* manual addition or removal of STOB's into/from the working pool; *(ii)* changing the filters configurations; *(iii)* creating a new activity in the calendar from a STOB.

3.1 Selection

The goal of selection is to build, for each user, a *pool* of potentially interesting activities to recommend, ruling out everything that is definitely out of the user's interest. Of course, an activity that had been previously ruled out, may be added back to the user's pool (depending on the privacy settings for the activity). However, the rationale of the Selection phase is to prevent the user from being flooded with activities proposals and to make recommendation more effective by focusing it on a well-defined group of items.

The pool determined through the selection phase goes through an additional filtering process as more transient user constraints are applied to it. The output of this phase is the *working pool*. To create an efficient *pool* some criteria have been selected:

- all STOB's that have already occurred are removed from the list of STOB's in the case of proactive context, while in on demand context only STOB's in the requested timeframe are considered;
- all STOB's geo-localized outside the maximum radius $r(u) + r_P(u)$ of the user u are not considered;
- all STOB's with a user's interest, that does not exceed a minimum threshold $RAT(s) < \psi$, are eliminated from the list.

In some cases there are STOB's that have a user's interest with a high value. If this interest exceeds a maximum threshold $RAT(s) > \phi$ the criterion of spatial distance could not be considered.

The *working pool* can change dynamically as a consequence of the user's interaction with the system (e.g. she decides to change the recommendation time frame) or it can be edited manually (e.g. the user decides to discard an activity she is not interested in).

3.2 Ranking

The aim of the ranking phase is to sort the STOB's stored in the user's working pool considering different factors. We can further split it in two sub-phases: the computation of the *individual ranking factors* and their *merge into one or more final rankings*.

Let us first discuss the individual ranking factors, which are all computed for a given user u and a given STOB s. We will introduce the following ranking factors: **thematic interest** (THI), **typology interest** (TYI), **average rating** (RAT), **social interest** (SOC), **feasibility** (FSB), **reachability**(RCH). Some are strictly dependent on the user model, some have a social meaning, some depend on spatial-temporal relations.

Each factor $\mathbb{F}$ is computed as a value ranging from 0 to 100, normalizing the result of the computation, if needed[4]. We also compute a qualitative version of the value, that is produced by using two thresholds $0 \leq \theta_{min} < \theta_{max} \leq 100$. The thresholds act as landmarks, thus we obtain the three possible qualitative values *low* ($[0, \theta_{min}]$), *ave* - average (($\theta_{min}, \theta_{max}$)), *exc* - exceptional ($[\theta_{max}, 100]$). We assume that θ_{min} and θ_{max} can be manually adjusted by users, and are initially set to 0 and 100 so as to be rendered ineffective.

Thematic and Typology Interest. The first two ranking factors are the **thematic interest** $THI_u(s)$ of the user u in the themes of the STOB s and **typology interest** $TYI_u(s)$ of

---

[4] We will not include the normalization in the following formulas for the sake of readability.

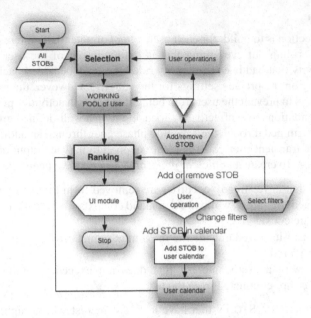

Fig. 1. Selection and Ranking of STOBs

the user u in the type of the STOB s. These two can be derived from the user model and the STOB description by exploiting classic recommendation techniques, the discussion of which is outside the scope of this paper.

Average Rating. We introduce the average rating of a STOB s as:

$$\mathrm{RAT}(\mathsf{s}) = \frac{\sum_{v \in Raters(\mathsf{s})} \omega_v \cdot \mathrm{RAT}_v(\mathsf{s})}{\sum_{v \in Raters(\mathsf{s})} \omega_v}$$

where $Raters(\mathsf{s})$ is a set of users who provided a rating for s, ω_v is equal to $\mathrm{THI}_v(\mathsf{s})$ if $\mathrm{THI}_v(\mathsf{s})$ is *ave* or *exc*, and equal to 0 if $\mathrm{THI}_v(\mathsf{s})$ is *low*. In other words, we weigh the ratings by the thematic interest of the users expressing them, deeming that uninterested people's opinions are less reliable, and we rule out ratings expressed by people whose thematic interests are too low.

Social Interest. The social interest ranks STOB's for a user u considering how much her friends are interested in participating. As explained in Section 2, a user may add an activity into her calendar specifying if she has the firm intention of participating in it (joined activity) or if she is very interested and wants to be reminded of it, but is still deciding whether to join or not (prospective activity). Therefore, the **social interest** $\mathrm{SOC}_u(\mathsf{s})$ for a user u and a STOB s is defined as follows:

$$\mathrm{SOC}_u(\mathsf{s}) = \frac{2|\mathrm{JND}(u, \mathsf{s})| + |\mathrm{PRS}(u, \mathsf{s})|}{2\,|\mathrm{SN}(u)|} \cdot 100 \tag{1}$$

where $\mathrm{JND}(u, s), \mathrm{PRS}(u, s)$ are the following two disjoint sets:

- $\mathrm{JND}(u, s) = \{f \in \mathrm{SN}(u) | f \text{ added } s \text{ as "joined"}\}$ and
- $\mathrm{PRS}(u, s) = \{f \in \mathrm{SN}(u) | f \text{ added } s \text{ as "prospective"}\}$.

In computing social interest, the number of friends who actually joined the STOB weights two times more than the number of friends who simply showed some interest without making any decision.

Feasibility. $\mathrm{FSB}_u(s)$ represents the possibility of a user u to reach the location where the STOB s takes place and to attend it for a relevant portion, given the appointments stored in her calendar. Let us consider the following situation: the user u is adding the STOB s, with the required time to participate $\mathrm{RQT}(s)$, to her calendar which already contains two STOBs, q and p. The scheduling of q, s and p is such that, taking into account the time needed to move from q to s and from s to p[5], the time left to user u for attending s is $\overline{d}_u(s)$. Our measure of feasibility considers s *not* feasible at all if $\overline{d}_u(s)$ is less than half of the required time $\mathrm{RQT}(s)$. Conversely, s has maximum feasibility if $\overline{d}_u(s)$ is at least $\mathrm{RQT}(s)$. Therefore, we define feasibility as follows:

$$\mathrm{FSB}_u(s) = \begin{cases} 0, & \text{if } \overline{d}_u(s) < \frac{\mathrm{RQT}(s)}{2}. \\ 200\, \frac{\overline{d}_u(s)}{\mathrm{RQT}(s)} - 100, & \text{if } \frac{\mathrm{RQT}(s)}{2} \leq \overline{d}_u(s) \leq \mathrm{RQT}(s). \\ 100, & \text{if } \overline{d}_u(s) > \mathrm{RQT}(s). \end{cases} \quad (2)$$

Reachability. The last ranking factor we consider is the **reachability** $\mathrm{RCH}_u(s)$ which measures how inclined the user u is to reach the location of the STOB s. As explained in Section 2, the user's position is a circular area defined by its center $c(u)$ and radius $r(u)$. Besides, an additional radius $r_P(u)$ defines her propensity to move, i.e. the distance she is willing to cover to join the STOB. The reachability is maximum (100) when the user's and STOB's centers coincide and it is $\theta_{\min}$ when the propensity area (whose center is $c(u)$ and whose radius is $r(u) + r_P(u)$) and the STOB area are tangent. Therefore the reachability is calculated as follows:

$$\mathrm{RCH}_u(s) = max\{\frac{\theta_{\min} - 100}{r(u) + r_P(u) + r(s)}\, dist(c(u), c(s)) + 100, 0\}. \quad (3)$$

Let us now discuss how the ranking factors are used to compute the results that ranking provides to the UI Module for visualization and user interaction. Given a (sub)set of ranking factors $\mathbf{RF}_k = \{\mathbb{F}_1, \ldots, \mathbb{F}_n\}$ a *final ranking* $\mathbf{rank}_k(u, s)$ for a user u and a STOB s is defined as:

$$\mathbf{rank}_k(u, s) = \frac{\sum_{i=1}^{n} \phi_i \cdot \mathbb{F}_i(u, s)}{\sum_{i=1}^{n} \phi_i}. \quad (4)$$

The values ϕ_i are called *filters* and determine how much each factor contributes to the final ranking. We assume that the filters are initially set to 1 and can be subsequently changed by the user through the UI Module.

In general we assume that the ranking factors can be combined into one or more final rankings, depending on the kind of visualization that the UI Module proposes to

---

[5] We assume an external service (e.g. Google Maps) provides an estimate of the traveling time.

the user. This kind of flexibility is useful not only to apply the model to several different systems, but also to allow for contextualization within a single system (for example, we can provide different final rankings for the "proactive" mode and for the "on demand" mode). In our prototype, we compute one final ranking combining all the ranking factors but feasibility. The UI we are designing in fact shows the feasibility factor separately; in a sense it provides a second final ranking on its own.

Besides the final rankings, we use the qualitative values of the ranking factors to compute two sets of STOB's that can prove useful when visualizing the information:

- LOW(u) = $\{s \mid \exists \mathbb{F}_i : \tilde{\mathbb{F}}_i(s) = low\}$,
- EXC(u) = $\{s \mid \exists \mathbb{F}_i : \tilde{\mathbb{F}}_i(s) = exc\}$.

LOW(u) includes all STOB's that score a low value in at least one ranking factor. The UI Module may decide to hide these STOB's whenever there are too many of them to show. EXC(u) includes all STOB's that score an exceptionally high value in at least one ranking factor, and may be worth to highlight in the UI.

4 Preliminary Evaluation

In order to provide answers to the research questions raised in the introduction (Section 1), we performed a user study in which we asked users to express their interest in participating in a certain activity given different sets of information[6]. We compared the performance of our algorithm when predicting the users' interest in activity participation for two cases: one when the algorithm takes into account only the themes and type of the activity and the other when it considers also the additional factors. Our evaluation focuses on reachability, social interest and average rating because, as mentioned above, feasibility can be considered as a separate factor. Comparing the recommender predictions with users' answers shows that users do take the additional factors into account, if this additional information is available. Consequently, taking into account additional factors improves recommendation accuracy.

4.1 Experimental Setting

Our experiment involved 200 users recruited according to an availability sampling strategy[7] and a group of 15 events.

The users were asked to provide the following information:

I1 A score $0 - 10$ expressing their potential interest in each event, *knowing only the event description along with its themes and type*. In the following, these will be referred to as the *init* user scores.

I2 A score $0 - 10$ expressing their potential interest in each event, *being aware of all the factors (themes and types, distance, average rating and social interest)*. These will be referred to as the *full* user scores.

---

[6] The available dataset is available on request, sending a mail to one of the authors.

[7] Much research in social science is based on samples obtained through non-random selection, such as the availability sampling, i.e. a sampling of convenience, based on subjects available to the researcher, often used when the population source is not completely defined.

I3 A value $0 - 10$ expressing their personal interest for each theme and type.

I4 A value $0 - 10$ expressing how they think they are affected by distance, average rating and social interest, when deciding if they are interested in an event.

I5 The maximum distance (in kilometers) they would generally travel in order to participate at an event, so that we can compute reachability.

4.2 Results and Discussion

We computed the RMSE (*Root Mean Square Error*) between the predictions given by the algorithm and the values provided by users. Tables 1(a) and 1(b) illustrate our experiment results. Notice that in both tables the last column reports the variation of the RMSE expressed as a percentage with respect to the reference value shown in bold face.

Table 1(a) shows the RMSE which compares system results calculated taking into account only themes and type with (i) *init* user scores (in Line 1) or (ii) *full* user scores (in Line 2). Comparing these two cases, it is evident that ignoring additional factors, while the users are aware of them (Line 2), brings to a worse RMSE (+12.62%). This is our first result [R1]: recommendation accuracy worsens if the system does not take into account the additional factors, since the users are aware of them in real-life.

Table 1. Results of the evaluation

(a) Results of the experiment obtained when the system considers only themes and type.

User scores	System weights					RMSE	$\Delta(\%)$
	THI	TYI	RCH	RAT	SOC		
1 *init*	0.50	0.50	-	-	-	2.583	-
2 *full*	0.50	0.50	-	-	-	2.909	+12.62

(b) Improvement of the RMSE in case the system takes into account additional factors.

User scores	System weights					RMSE	$\Delta(\%)$
	THI	TYI	RCH	RAT	SOC		
1 *full*	0.50	0.50	-	-	-	2.909	-
2 *full*	0.2	0.2	0.2	0.2	0.2	2.850	-2.03
3 *full*	0.2	0.2	user	user	user	2.795	-3.92

Line 2 of Table 1(a) is repeated as Line 1 in Table 1(b) as reference for the RMSE improvement of the following two lines. Lines 2 and 3 in Table 1(b) report the RMSE obtained when the recommender system takes into account also the additional factors. What distinguish those two lines from each other is the adopted set of weights: in Line 2 all the weights are equal while in Line 3 the weights are those provided by the users during the experiment (I4). It is easy to see that the recommender system performs better in these two cases than when it ignores additional factors (Line 1). This is our second result [R2]: the accurate recommender should take the additional factors into account. Moreover, the best result (Line 3) is reached by exploiting user provided weights (-3.92% vs. -2.03%).

5   Related Work

In this section we give a brief overview of the related work in the field of event/activity recommendation, followed by some pointers to works dealing with event ontologies and conclude with some references to works dealing with spatial-temporal representation.

PITTCULT [8] is a collaborative filtering recommender for cultural events, based on trust relations among users for finding similar users. Another collaborative filtering recommender is CUPID [13], which recommends events for the Flemish cultural scene. Similarly to us, they have a pool of events to be recommended, and they apply content and spatial filters after the recommendation process in order to exclude non-compatible events. They use user participation and ratings to find similar users. In [11] the authors propose a collaborative approach to recommendation of future events by starting from a content-based approach, where each user has a parameter vector that relates her preferences to event descriptions and similarity among users is calculated using their past feedback on the events. Only the events within limited distance are considered. A hybrid solution, combining content-based and collaborative filtering techniques, is proposed in [7], which presents an implementation as a Facebook[8] application. Another hybrid approach is described in [3]; in this case an item is recommended to a user if it is similar to the items that this user, or similar users, have liked in the past. In [2] the authors describe *iCity*, an application providing information about cultural resources and events in Torino, Italy. They propose to exploit user's activity to infer her interest about items and therefore to improve content-based recommendations in a Web 2.0 context. Alternatively, location-based recommendation is offered, but it is not combined with interest based recommendation.

All these systems suggest activities ignoring the current user's commitment or her location in the period when the event takes place. Our system offers a relevant improvement, taking into account, in the recommendation phase, the possibility for the user to be present at the event.

In the literature there are already several works on the creation of event ontologies specialized in different domains: [14] used the event ontology for describing music events, [15] is more specialized in the sport domain and [16] considers the events as something that happen. We focus on the general notion of an event intended as a meeting organized earlier, concept applicable across multiple domains and contexts and on which it is possible to think of some recommendations.

Spatial and temporal aspects can be merged in a unique spatial-temporal representation. In this area, there are two main approaches: one considers snapshots of the world at different instants of time [5] and the other uses a 3D or 4D region based representation where time is one of the dimensions [5,6,12]. An example of the application of the second approach can be found in [9], where the aim is to define the spatial-temporal location of facts (e.g. actions performed in real life) described by the users and to define if they are co-located.

6   Conclusions and Future Work

Given the growing use of activities in the social networks, such as the *social events* in Facebook[9], a system being able to recommend activities in sophisticated way is more and more relevant in helping users find interesting suggestions without being overwhelmed by a huge mass of items. The solution proposed by our framework is

---

[8] http://www.facebook.com
[9] http://www.facebook.com

an extension of a classic recommendation system towards spatiotemporal and social aspects. We introduced a semantic model describing three elements (users, STOB's and user's calendar activities) and a process for selecting and ranking new STOB's to be proposed to the user. The final ranking is obtained as a composition of factors, each representing a different aspect of the overall ranking (user's interest, ratings from other users, participation of user's friends, spatiotemporal relationship with other activities and a possibility for the user to reach the location). The opportunity to weight differently the various ranking factors allows for customization of the results according to the taste and needs of the user. Our evaluation shows that recommender systems which take into account additional factors perform better than the ones relying only on type and themes of the events. In addition, allowing users to give weights to these additional factors, brings further improvements into recommendation process. This brings us to a possible future extension of our framework, by adding other ranking factors that can influence the participation. Also, the existing factors can be given various importance in the ranking process and it would be interesting to see how much different factors contribute to the recommendation process. We also plan to investigate how to recommend STOB's to a group of friends, based on their interest in the themes of the STOB or on their participation in similar STOB's.

References

1. Adomavicius, G., Tuzhilin, A.: Context-aware recommender systems. In: Ricci, F., Rokach, L., Shapira, B., Kantor, P.B. (eds.) Recommender Systems Handbook, pp. 217–253. Springer US (2011)
2. Carmagnola, F., Cena, F., Console, L., Cortassa, O., Gena, C., Goy, A., Torre, I., Toso, A., Vernero, F.: Tag-based user modeling for social multi-device adaptive guides. User Modeling and User-Adapted Interaction 18, 497–538 (2008)
3. Cornelis, C., Guo, X., Lu, J., Zhang, G.: A Fuzzy Relational Approach to Event Recommendation. Artificial Intelligence, 2231–2242 (2007)
4. Cornelis, C., Lu, J., Guo, X., Zhang, G.: One-and-only item recommendation with fuzzy logic techniques. Information Sciences 177(22), 4906–4921 (2007)
5. Grenon, P., Smith, B.: SNAP and SPAN: Towards Dynamic Spatial Ontology. Spatial Cognition & Computation 4(1), 69–104 (2004)
6. Hazarika, S., Cohn, A.: Abducing Qualitative Spatio-Temporal Histories from Partial Observations. In: Fensel, D., Giunchiglia, F., McGuinness, D.L., Williams, M.-A. (eds.) Proc. of the 8th International Conference on Principles and Knowledge Representation and Reasoning, KR 2002, pp. 14–25. Morgan Kaufmann (2002)
7. Kayaalp, M., Özyer, T., Özyer, S.T.: A mash-up application utilizing hybridized filtering techniques for recommending events at a social networking site. Social Network Analysis and Mining 1(3), 231–239 (2010)
8. Lee, D.H.: PITTCULT: trust-based cultural event recommender. In: Proc. of the 2008 ACM Conference on Recommender Systems, RecSys 2008, pp. 311–314. ACM (2008)
9. Likavec, S., Lombardi, I., Nantiat, A., Picardi, C., Theseider Dupré, D.: Threading Facts into a Collective Narrative World. In: Aylett, R., Lim, M.Y., Louchart, S., Petta, P., Riedl, M. (eds.) ICIDS 2010. LNCS, vol. 6432, pp. 86–97. Springer, Heidelberg (2010)
10. Lops, P., De Gemmis, M., Semeraro, G.: Content-based Recommender Systems: State of the Art and Trends. In: Ricci, F., Rokach, L., Shapira, B., Kantor, P.B. (eds.) Recommender Systems Handbook, ch. 3, pp. 73–105. Springer US (2011)

11. Minkov, E., Charrow, B., Ledlie, J., Teller, S., Jaakkola, T.: Collaborative future event recommendation. In: Proc. of the 19th ACM International Conference on Information and Knowledge Management, CIKM 2010, pp. 819–828. ACM (2010)
12. Muller, P.: Topological Spatio-Temporal Reasoning and Representation. Computational Intelligence 18(3), 420–450 (2002)
13. Pessemier, T.D., Coppens, S., Geebelen, K., Vleugels, C., Bannier, S., Mannens, E., Vanhecke, K., Martens, L.: Collaborative recommendations with content-based filters for cultural activities via a scalable event distribution platform. Multimedia Tools and Applications 58(1), 167–213 (2012)
14. Raimond, Y., Abdallah, S.A., Sandler, M.B., Giasson, F.: The music ontology. In: Dixon, S., Bainbridge, D., Typke, R. (eds.) Proc. of the 8th International Conference on Music Information Retrieval, ISMIR 2007, pp. 417–422. Austrian Computer Society (2007)
15. Rayfield, J., Wilton, P., Oliver, S.: Sport Ontology (February 2011),
    http://www.bbc.co.uk/ontologies/sport/
16. Shaw, R., Troncy, R., Hardman, L.: LODE: Linking open descriptions of events. In: Gómez-Pérez, A., Yu, Y., Ding, Y. (eds.) ASWC 2009. LNCS, vol. 5926, pp. 153–167. Springer, Heidelberg (2009)
17. Sosnovsky, S., Dicheva, D.: Ontological technologies for user modelling. International Journal of Metadata Semantic Ontologies 5(1), 32–71 (2010)

Granular Semantic User Similarity
in the Presence of Sparse Data

Francesco Osborne, Silvia Likavec, and Federica Cena

Università di Torino, Dipartimento di Informatica, Torino, Italy
{osborne,likavec,cena}@di.unito.it

Abstract. Finding similar users in social communities is often challenging, especially in the presence of sparse data or when working with heterogeneous or specialized domains. When computing semantic similarity among users it is desirable to have a measure which allows to compare users w.r.t. any concept in the domain. We propose such a technique which reduces the problems caused by data sparsity, especially in the cold start phase, and enables granular and context-based adaptive suggestions. It allows referring to a certain set of most similar users in relation to a particular concept when a user needs suggestions about a certain topic (e.g. cultural events) and to a possibly completely different set when the user is interested in another topic (e.g. sport events). Our approach first uses a variation of the spreading activation technique to propagate the users' interests on their corresponding ontology-based user models, and then computes the concept-biased cosine similarity (CBC similarity), a variation of the cosine similarity designed for privileging a particular concept in an ontology. CBC similarity can be used in many adaptation techniques to improve suggestions to users. We include an empirical evaluation on a collaborative filtering algorithm, showing that the CBC similarity works better than the cosine similarity when dealing with sparse data.

Keywords: Recommender systems, ontology, propagation of interests, user similarity, data sparsity, cosine similarity.

1 Introduction

One of the most basic assumptions in social communities is that people who are known to share specific interests are likely to have additional related interests. This assumption is used by many approaches, especially in collaborative-filtering recommender systems, which recommend items that people with tastes and preferences similar to the target user liked in the past [13]. User profiles can be compared using a range of metrics, such as cosine similarity [16], the Pearson Correlation Coefficient [1], Jaccard's index [6], to name just a few.

However, these traditional methods for computing similarity do not come without limitations. First, similarity is difficult to calculate in the presence of data sparsity, a problem which arises when the user ratings are spread over items that seldom overlap. This especially happens in two situations: at the beginning of the interaction, when we do not know much about the user (the *cold start* problem [12]) and when the domain is huge and we do not have the user interest values for all the concepts in the domain.

M. Baldoni et al. (Eds.): AI*IA 2013, LNAI 8249, pp. 385–396, 2013.

Second, these similarity measures usually take into account all the concepts in the domain. This solution may not always be flexible enough, especially in a non-specialized and heterogeneous domain, i.e. the domain covering a range of very different topics and subtopics. *In this case, rarely similar people show to have the same tastes for all the different concepts.* Even when the domain is not so heterogeneous, it frequently happens that people have the same tastes for some portions of the domain, but different ones for others. There are many cases in which it is very important to be able to be more granular and to establish a different set of similar users for different concepts. For example, two users can have a very different tastes for many elements of the domain and still be very similar w.r.t. a certain topic or subculture, such as Rock_Music or Wine. Computing the user similarity over all the elements of the domain does not consider this aspect, wasting useful information, which is a shame especially in the presence of sparse and poor data.

To address these two issues (sparsity problem and computing similarity over the whole domain), we introduce *the concept-biased cosine similarity* (CBC similarity), a novel similarity metric designed to privilege certain concept in an ontology, according to the recommendation context, and overcome data sparsity.

The prerequisites for our approach are the following:

- *semantic representation of the domain knowledge* using an OWL ontology[1] where domain concepts are taxonomically organized;
- *user model defined as an overlay over the domain ontology*;
- a *strategy for propagation of the user interest values in the user model.*
- a methodology to calculate for each user a *set of similar users w.r.t. a certain topic.*

Our approach consists of three steps. First, we use a variation of the spreading activation technique, introduced in [3,4], to propagate the users interests on their ontology based user models (Sect. 2). Then, we pre-compute a matrix in which every pair of users is assigned a similarity score using cosine similarity for each concept in the ontology for which there is enough feedback from both users. Finally, we calculate the concept-biased cosine similarity (CBC similarity), a variation of the cosine similarity designed for privileging a particular concept in an ontology (Sect. 3).

The main contribution of the paper is this novel method for calculating fine grained user similarity for specific sub-portions of the domain (CBC similarity) even in the presence of sparse data. We tested the CBC similarity in a collaborative filtering algorithm and showed that it outperforms the standard cosine similarity when dealing with sparse data. The method is general and applicable in wide variety of domains, provided that the domain is described using an ontology.

The remainder of the paper is organized as follows. In Sect. 2 we provide some details of our ontology-based user model, how to determine the user interest in domain concepts and how to subsequently propagate it to similar concepts in the domain ontology. We describe how to calculate the similarity between users for different specific sub-portions of the domain ontology in Sect. 3 . We present the results of the evaluation of our approach in Sect. 4, followed by related work in Sect. 5. Finally, we conclude and give some directions for future work in Sect. 6.

---

[1] http://www.w3.org/TR/owl-features/

2   User Interest

Our technique for finding similar users in particular contexts as the first step includes the determination and propagation of user interests in order to populate and update the user model. User interests are propagated to similar concepts, using various approaches, in order to reduce data sparsity and reach distant concepts. We start this section with the definition of the user model in Sect. 2.1, followed by a brief overview of how to determine the user interest in Sect. 2.2 and conclude with some possible propagation techniques in Sect. 2.3.

2.1   User Model Definition

In order to satisfy the previously listed requirements, our domain is represented using an OWL ontology[2]. As will be seen later, the way in which this ontology is formulated, influences the choice of propagation technique.In our approach we employ an *ontology-based user model*, defined as an overlay over the domain ontology. More precisely, each ontological user profile is an instance of the domain ontology, and each node in the ontology has an interest value associated to it. This means that each node N in the domain ontology can be seen as a pair $\langle N, I(N) \rangle$, where $I(N)$ is the interest value associated to node N.

2.2   Determining User Interest

We describe here the initial step of our approach which considers determining the user interest in domain concepts and is based on our previous work described in [3,4]. When the user provides feedback for a certain concept in the domain ontology[3], this feedback is implicitly recorded by the system so that it can be further used to calculate user interest values for other domain concepts.

For each concept N in the domain ontology we distinguish two types of interest values: *sensed interest* and *propagated interest*. Given the user feedback for the concept N, we first calculate the user *sensed interest* as follows:

$$I_S(N) = \frac{l(N)}{\text{MAX}(1 + e^{-f(N)})}$$

where $l(N)$ is the level of the node receiving the feedback, MAX is the level of the deepest node in the ontology and $f(N)$ is the user feedback for the node N. The sensed interest depends on the user feedback for the node and the position of the node in the ontology, since the nodes lower down in the ontology represent specific concepts, and as such signal more precise interest than the nodes represented by upper classes in the ontology, expressing more general concepts. Subsequently, the sensed interest value is used in *propagation phase* (see Sect. 2.3) to propagate the user interest to similar objects.

---

[2] http://www.w3.org/TR/owl-ref/
[3] How the feedback is obtained is out of the scope of this paper.

These two interest values, I_S and I_P are kept separated for each concept, and the total interest for each concepts in the ontology is calculated as:

$$I(N) = \phi I_O(N) + \sigma I_S + \Sigma_{i=1}^{n} \pi_i I_P(N_i, N)$$

where $\phi, \sigma, \pi_1, \ldots, \pi_n \in \mathbb{R}$ and $\phi + \sigma + \Sigma_{i=1}^{n} \pi = 1$, I_O is the *old interest value* (initially set to zero), $I_S(N)$ is the sensed interest for the node N and $I_P(N_i, N)$ is the total propagated interest from the node N_i to the node N, n being the number of nodes which propagate their interest values to the node N. By varying the constants ϕ, σ and π_i it is possible to assign different level of importance to either sensed or propagated interest values.

2.3 Propagating User Interest Values

Depending on how the ontology is designed, the concepts to which to propagate the user interest can be determined in different ways. Here we describe two different propagation techniques:

- distance-based propagation,
- property-based propagation.

Distance-Based Propagation. One of the simplest ways to measure the similarity of two concepts in the ontology is to use the ontology graph structure and calculate the distance between nodes by using the number of edges or the number of nodes that need to be traversed in order to reach one node from the other (see [11]).

Using the distance between nodes, the user interest in a certain domain object can be propagated *vertically* in the ontology, upward to the ancestors and downward to the descendants of a given node. In this case the propagated interest is calculated by modulating the sensed interest of the node N that receives the feedback by the exponential factor which describes the attenuation with each step up or down and a weight inversely correlated with the amount of feedback already received by the node as follows:

$$I_P(N, M) = \frac{e^{-kd(N,M)}}{1 + \log(1 + n(M))} I_S(N)$$

where $d(N, M)$ is the distance between the node N receiving the feedback and the node M receiving the propagated interest, $n(M)$ is the number of actions performed in the past on the node M and $k \in \mathbb{R}$ is a constant.

Further improvement of distance-based propagation can be obtained with *conceptual distance* where the main idea is to modify the lengths of the edges in the ontology graph, as initially proposed in [5]. First, the set of relevant concepts is determined. Then for the relevant concepts, the notion of conceptual specificity is introduced, in order to specify their relevance in the given context. Using the conceptual specificity, the edge lengths are modified so that they exponentially decrease as the levels of the nodes increase. Then the propagated interest is calculated using these exponentially decreasing edge lengths to calculate the distances between concepts.

Property-Based Propagation. If the ontology has explicit specification of the concepts' properties, we can use these properties to calculate the similarity between the concepts, the distance between them. To do this, we can adopting the approach described in [4], and decide to which elements to propagate the user interest. This propagation does not have any particular direction and permits propagation of user interests to various nodes in the ontology. The propagated interest value is calculated using the hyperbolic tangent function as:

$$\mathcal{I}_P(N, M) = \frac{e^{2\text{SIM}(N,M)} - 1}{e^{2\text{SIM}(N,M)} + 1} \mathcal{I}_S(N)$$

where $\mathcal{I}_S(N)$ is the sensed interest for the node N which received the feedback and $\text{SIM}(N, M)$ is the *property-based similarity* between the node N which received the feedback and the node M which receives the propagated interest.

Property-based similarity calculates the similarity of classes, starting from Tversky's feature-based model of similarity [17], where similarity between objects is a function of both their common and distinctive characteristics. For two domain elements N_1 and N_2, for each property p, we calculate CF_p, DF_p^1 and DF_p^2, which denote how much p contributes to common features of N_1 and N_2, distinctive features of N_1 and distinctive features of N_2, respectively, depending on how p is defined in N_1 and N_2. We calculate the similarity between N_1 and N_2 as follows:

$$\text{SIM}(N_1, N_2) = \frac{\text{CF}(N_1, N_2)}{\text{DF}(N_1) + \text{DF}(N_2) + \text{CF}(N_1, N_2)}.$$

The property-based similarity of equivalent classes is equal to 1, whereas for instances the values-property pairs declared for each instance are compared.

The choice of one of these propagation methods depends on the ontology structure. If the ontology has a deep taxonomy, (conceptual) distance-based propagation can be used. Our experiments showed that conceptual distance performs better than the standard one. If the ontology does not have a deep hierarchy and the classes are defined by means of properties, property-based propagation is the most suitable one.

3   The Concept-Biased Cosine Similarity

In this section we propose a novel technique for computing granular semantic user similarity w.r.t. any concept formalized in the ontology, called *concept-biased cosine similarity* or *CBC similarity*. The CBC similarity measure introduces a bias in the standard cosine similarity measure, in order to privilege a certain concept in the domain ontology. It is particularly useful when the user is interested in a specific part of the domain and there is a need to find the most similar users concerning that part of the domain. For example, we can refer to a set of the most similar users when a user needs suggestions about one topic (e.g. cultural events) and to a possibly different set when the user is interested in another topic (e.g. sport events). This similarity measure can be used by many kinds of recommendation techniques which build on similarity measures between users, allowing a more granular prospective on the domain. However, the investigation of recommendation strategies is out of the scope of this paper.

The algorithm to compute the CBC similarity exploits a pre-computed tridimensional matrix which contains for each pair of users the similarity values obtained using the standard cosine similarity w.r.t. the significant concepts of the domain ontology. The significant classes may be pre-labeled or can be defined autonomically by setting a threshold on the level or the number of their instances. Table 1 displays a part of this structure using an ontology describing social events. The meaning is intuitive: Ann is similar to Bill w.r.t. some concepts, such as gastronomical events and cooking courses, but different w.r.t. cultural events. Hence, it makes sense to use the information from Bill's user model when recommending gastronomical events to Ann.

To compute this matrix we scan the ontology top-down and for each significant class X and each pair of users we compute the cosine similarity, using as input the interest values for the subclasses and instances of X. To save time and space, it is advisable to ignore all the subclasses of a given class which for a specific pair of users have cosine similarity below a certain threshold (0.5 in our approach) or do not include enough common feedback values to be significant. For example, if Ann and Bill have few feedback values w.r.t. Sport_Event, their similarity w.r.t. the significant classes which are subclasses of Sport_Event (e.g. Race, Open_Day etc.) are not computed.

Table 1. Ann's similarity with Bill, Cindy and Damian w.r.t. different concepts. In bold the highest value for each row, representing the user the most similar to Ann, w.r.t. a certain class.

Similarity with Ann	Bill	Cindy	Damian
Event	0.79	**0.82**	0.75
Gastronomical_Event	**0.89**	0.84	0.65
Cultural_Event	0.54	0.81	**0.94**
Sport_Event	0.72	**0.92**	0.6
Cooking_Course	**0.96**	0.84	0.69
Concert	0.72	0.74	**0.92**
Tasting	**0.83**	0.71	0.63
Soccer_Match	0.56	**0.97**	0.67
...	...	...	...

A naive way to use the matrix given in Table 1 would be to take directly the cosine similarity scores computed for each concept when a suggestion for that concept is needed. This approach, however, may miss the big picture. In fact, while most of the information expressing similarity w.r.t. a certain concept can be derived from the feedback values for the items classified under this concept, by not considering the other feedback values we may miss precious and more subtle information about the users' common preferences. For example, we might compute the cosine similarity using only the interest values for the subconcepts of the class Sport_Event and use this similarity to find similar users with the same interest for sport events. However, if we also consider the user feedback about some Book_Presentation which talks about Yoga, we would reinforce our knowledge about user interest in yoga related sport events.

We address this issue with: (i) *interest propagation* and (ii) *super-class weighing*. Interest propagation, as discussed in Sect. 2, is not only appropriate for reducing sparsity, but it also forces the interest value assigned to each class to take into consideration also the feedback values for any semantically related concepts. However, interest propagation effectiveness depends on how well the ontology is designed. For a well-designed domain ontology which describes all the explicit and subtle relationships among the domain elements, interest propagation would be enough to solve the problem of including the information contained in different classes. In a realistic scenario, instead, the ontology is often far from perfect. For this reason, when computing the similarity w.r.t. concept X it is better not to exclude completely all the other concepts, but rather weigh them differently. This is where CBC similarity comes into picture.

Our approach includes in the computation of the CBC similarity w.r.t. X the similarity scores of the super-classes of X. For example, when the concept Race is taken into consideration we want a similarity metric that does not only award people similar w.r.t. Race but also gives a bonus to people similar with respect to Sport_Event and Event. We compute the CBC similarity for concept X using the following formula:

$$CBC(X, a, b) = w \cdot \cos(v^a(X), v^b(X)) + (1 - w)\frac{\sum_{i \in S} \frac{s_i}{T} \cos(v^a(X_i), v^b(X_i))}{\sum_{i \in S} \frac{s_i}{T}} \tag{1}$$

where

- a, b are the users we want to compare;
- X is the starting class for which we calculate the CBC similarity;
- $\cos(x, y)$ is the cosine similarity for vectors x and y;
- i refers to the element X_i of the set S composed of super classes of X;
- $v^a(X)$ and $v^b(X)$ are the interest value vectors associated to the subclasses/instances of the class X according to the user models of users a and b;
- $v^a(X_i)$ and $v^b(X_i)$ are the interest value vectors associated to the sub-classes/instances of the class X_i according to the user models of users a and b;
- s_i is the number of subclasses of X_i for which both users provided feedback;
- T is the total number of classes in the ontology;
- $0 < w < 1$ weighs the relationship between a class and its super classes.

We believe that w should depend on the sparsity degree of the data for a class: if it is high we will need more support from the super classes, whereas if it is low, it can take care of itself. Thus, in our system we estimate w for concept X as the average of the ratios between the subclasses which received a direct feedback and the total number of subclasses for each ontology-based user model which received at least 10 feedbacks.

It should be noticed that in order for CBC to be meaningful for a user, she/he should have a decent amount of feedback values for class X for which we calculate the CBC similarity. If this is not true, it is better to compute the CBC on the first super class of X that has enough feedback. For example, if the context is Race and the user did not provide enough feedback on Race concept we can check if she/he has the interest values for Sport_Event or for Event, relying to the first usable super class.

We now give an example of the last step of CBC computation, using the values in Table 1. Assume that we need to suggest to Ann a concert for tomorrow night. A standard way would be to exploit the feedback values of the most similar users according to different techniques (e.g. memory-based collaborative filtering). If we adopt the standard cosine similarity we will conclude that Cindy is the most similar user to Ann, followed by Bill and Damian. While this is true taking into consideration the whole domain, the situation may be different if we consider only the items related to the Concert class. Thus, taking into consideration also the context in which the suggestion is needed we can compute CBC relative to class Concert. In the domain ontology the super-classes of Concert are Cultural_Event and Event. For simplicity we assume that Cultural_Event subsumes one third of the items and $w = 0.5$.

$$CBC(\text{Concert}, Ann, Bill) = 0.72 \cdot 0.5 + \frac{0.54 \cdot 1/3 + 0.79}{1/3 + 1} \cdot 0.5 = 0.36 + 0.36 = 0.72$$
$$CBC(\text{Concert}, Ann, Cindy) = 0.74 \cdot 0.5 + \frac{0.81 \cdot 1/3 + 0.82}{1/3 + 1} \cdot 0.5 = 0.37 + 0.41 = 0.78$$
$$CBC(\text{Concert}, Ann, Damian) = 0.92 \cdot 0.5 + \frac{0.54 \cdot 1/3 + 0.79}{1/3 + 1} \cdot 0.5 = 0.46 + 0.40 = 0.86$$

We can see that using the CBC similarity Damian is the user most similar to Ann w.r.t. the class Concert, followed by Cindy and Bill.

4   Evaluation

To evaluate the CBC similarity we implemented a memory-based collaborative filtering recommender that computes the suggested rating for an item as the weighed sum of the feedback values given by the most similar users [2].

Our assumption was that using the CBC similarity would yield more accurate results than adopting the standard cosine similarity, especially in the presence of sparse data. Furthermore, we wanted to compare the conceptual distance-based propagation and the property-based propagation for the CBC similarity computation. We decided to only test conceptual distance-based propagation since it provides better recommendation accuracy than standard distance-based propagation. To this aim we compared three similarity measures:

- Cosine similarity (COS);
- CBC similarity after conceptual distance-based propagation (CBC-C);
- CBC similarity after property-based propagation (CBC-P).

We used as input concept for the CBC metrics the most direct super-class of any item for which the rating had to be suggested.

4.1   Experiment Setup and Sample

We employed a domain ontology describing social events, focusing on gastronomic events (e.g. tastings, food fairs), cultural events (e.g. concerts, movies) and sport events (e.g. races, open days). The ontology was designed for recommending purposes and it includes 16 main classes at 3 levels. In order to collect users preferences regarding the domain items, we used a questionnaire in which 44 events had to be graded on a scale from 1 to 10. The sample was composed of 231 subjects, 19-38 years old, recruited according to an availability sampling strategy. We obtained a total of 10,164 ratings.

To test the accuracy of CBC similarity combined with different propagation techniques we generated an ordered list of rated items starting from a portion of each user's ratings and compared this list with the remaining part of the user ratings. We ran different tests varying the initial percentage of rated items (5%, 10%, 20%), to simulate different degrees of sparsity in the input data.

4.2   Measures

The difference between the estimated ratings and the real ones was computed using the mean absolute error (MAE). The correlation between the list generated by the tested algorithm and the original user's list was estimated using Spearman's rank correlation coefficient ρ, which provides a non-parametric measure of statistical dependence between two ordinal variables and gives an estimate of the relationship between two variables using a monotonic function. In the absence of repeated data values, a perfect Spearman correlation of $\rho = +1$ or $\rho = -1$ is obtained when each of the variables is a prefect monotone function of the other. The performance of the different techniques was compared with the χ-square test.

4.3   Results

The left panel in Fig. 1 shows that the lists produced with the CBC similarity have a higher degree of association ρ with the real ratings than the ones produced by using the standard cosine similarity. CBC similarity is particularly efficient in case of data sparsity, since it allows to use domain knowledge to compensate for the lack of data. Hence, as the percentage of rated items grows the difference between the two approaches becomes less prominent. In all tests the CBC-P approach was more effective than the CBC-C approach confirming the results presented in [4]. The right panel of Fig. 1 displays the mean absolute error of the suggested ratings w.r.t. the real ratings. Also in this case CBC similarity performed better then the cosine similarity for all the tests and CBC-P yielded a slightly better result than CBC-C.

To understand better the aforementioned results we studied the frequency distribution of the Spearman Coefficient for a percentage of rated items equal to 5% and 10%, respectively (Fig. 2). The standard cosine similarity was unable to produce any case with a good association ($\rho > 0.5$) in the 5% test, whereas CBC-P did so for 10% of users. On the 10% test COS obtained a good association in 2% of the cases as opposed to the 15% of CBC-P. This trend continues also for the 20% test yielding 20% for COS and 34% for CBC-P.

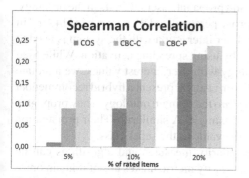
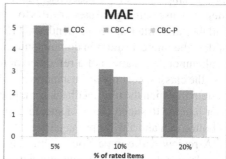

Fig. 1. Left panel: Spearman correlation between the suggested and real preference list in case of 5%, 10% or 20% rated items. Right panel: MAE of the suggested and real rating in case of 5%, 10% or 20% rated items.

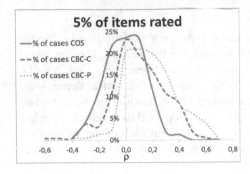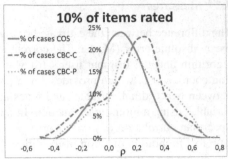

Fig. 2. Distribution of cases for various values of ρ for the 5% (left panel) and 10% (right panel) of rated items

The difference between the cosine similarity and the CBC similarity is statistically significant according to the χ-square test for the 5% and 10% test, but not for the 20% one. More precisely, in the 5% test the χ-square between COS and CBC-C yielded $p = 6 \cdot 10^{-8}$, between COS and CBC-P it yielded $p = 2 \cdot 10^{-100}$ and between CBC-C and CBC-P it yielded $p = 6 \cdot 10^{-19}$. For the 10% it yielded $p = 5 \cdot 10^{-9}$ for COS and CBC-C, $p = 10^{-11}$ for COS and CBC-P and $p = 0.6$ (thus not significant) for CBC-C and CBC-P.

Hence, we can conclude that the CBC similarity approach works significantly better then the standard cosine similarity and can be effective in alleviating the data sparsity and the cold start problem.

5   Related Work

In this section we briefly describe the works in the literature which exhibit similarities with our approach. In particular, we present works that share the cornerstones of our approach: usage of ontology-based user model, and propagation of user interests in an ontology. Then, we present other modalities of calculation of similarity among users.

The most similar works are the ones of Sieg et at. [14,15] and Middleton et al. [8], since they all employ ontology-based user models and propagate user interests in the ontology. Sieg et al. use ontological user profiles to improve personalized Web search in [14] and in collaborative-filtering recommender in [15]. User interaction with the system is used to update the interest values in domain ontology by using spreading activation. While Sieg et al. annotate instances of a reference ontology with user interest values, we annotate also the classes, not only the instances. Middleton et al. [8] present a hybrid recommender system in which the user feedback is mapped onto the domain ontology. They propagate user interest to superclasses of specific topics using IS-A similarity. Their propagation is bottom-up and always propagates 50% of the value to the super class.

A review and comparison of different similarity measures and algorithms can be found in Cacheda et al. [2]. Our alternative similarity metric makes use of domain ontological structure to calculate similarity among users, and in particular uses the interest values of these users' corresponding user models. In a similar fashion to us, other works follow this approach.

For example, Mobasher et al. [9] calculate user similarities based on interests scores across ontology concepts, instead of their ratings on individual items like traditional collaborative filtering. A main difference w.r.t. our approach is the method they use to calculate similarity. They first turn the ontological user profiles into flat vectors of interest scores over the space of concepts, and then compare the user profiles to figure out how distant each users profile is from all other users profiles. The distance between the target user and a neighbor user is calculated using the Euclidean distance.

In Yuan et al. [18] a structured cosine similarity is proposed for supporting text clustering by considering the structure of the documents. Similarly to us they built on the cosine similarity and flatten the values assigned to an ontology classes in a vector to feed to the cosine similarity. However, they simply pass the lower class value to the super-class and do not weigh the relations between different classes.

In Thiagarajan et al. [16] the authors represent user profiles as bags-of-words (BOW) where a weight is assigned to each term describing user interests to create extended user profiles. They then use a spreading activation technique to find and include additional terms in user profiles. The similarity measure is obtained by combining cosine similarity (for overlapping parts) with bipartite graph approaches (for remaining profile terms).

The idea that not all domain concepts should be treated the same is elaborated in [7] which shows that some of the ratings carry more discriminative information than others. They argue that less common ratings for a specific item tend to provide more discriminative information than the most common ratings. Thus, they propose to divide user similarity into two parts: local user similarity (a vector similarity among users) and global user similarity (considers the number of similar neighbors).

6 Conclusions and Future Work

We proposed the concept-biased cosine similarity (CBC similarity), a novel approach to measure the user similarity relative to a specific concept, which is able to alleviate data sparsity problem. Our approach can be adopted for several purposes. For example, it can be used to enhance collaborative filtering recommender in a semantic direction [15], for alleviating data sparsity and for improving of recommendation results. It can also be exploited by social applications in order to suggest new connections to users, based on shared interests.

An empirical evaluation showed that CBC similarity outperforms the cosine similarity in supporting collaborative filtering, especially in the presence of sparse data. In the tests with less then 20% of the items rated the difference between the two techniques was statistically significant.

We are working in several directions to exploit interest propagation and refine ontology-based similarity metrics. We plan on allowing bias in the similarity metric not only for preferring a specific topic, but also for other dimensions such as context and expertise. A more ambitious work would be to even allow this multiple dimensions to influence the structure of the ontology as suggested in [10]. For this reason, we are also working on an approach to compute the CBC similarity across user profiles represented as overlays over different ontologies or personal ontology views. This avenue of work may also be useful to compare user profiles from different systems, making cross-system personalization easier.

References

1. Ahlgren, P., Jarneving, B., Rousseau, R.: Requirements for a cocitation similarity measure, with special reference to Pearson's correlation coefficient. Journal of the American Society for Information Science and Technology 54(6), 550–560 (2003)
2. Cacheda, F., Carneiro, V., Fernandez, D., Formoso, V.: Comparison of collaborative filtering algorithms: Limitations of current techniques and proposals for scalable, high-performance recommender systems. ACM Transactions on the Web, 2 (2011)
3. Cena, F., Likavec, S., Osborne, F.: Propagating user interests in ontology-based user model. In: Pirrone, R., Sorbello, F. (eds.) AI*IA 2011. LNCS, vol. 6934, pp. 299–311. Springer, Heidelberg (2011)
4. Cena, F., Likavec, S., Osborne, F.: Property-based interest propagation in ontology-based user model. In: Masthoff, J., Mobasher, B., Desmarais, M.C., Nkambou, R. (eds.) UMAP 2012. LNCS, vol. 7379, pp. 38–50. Springer, Heidelberg (2012)
5. Chiabrando, E., Likavec, S., Lombardi, I., Picardi, C., Theseider Dupré, D.: Semantic similarity in heterogeneous ontologies. In: 22nd ACM Conference on Hypertext and Hypermedia, Hypertext 2011, pp. 153–160. ACM (2011)
6. Hamers, L., Hemeryck, Y., Herweyers, G., Janssen, M., Keters, H., Rousseau, R., Vanhoutte, A.: Similarity measures in scientometric research: The Jaccard index versus Salton's cosine formula. Information Processing and Management 25(3), 315–318 (1989)
7. Luo, H., Niu, C., Shen, R., Ullrich, C.: A collaborative filtering framework based on both local user similarity and global user similarity. Machine Learning 72(3), 231–245 (2008)
8. Middleton, S.E., Shadbolt, N.R., De Roure, D.C.: Ontological user profiling in recommender systems. ACM Transactions on Information Systems 22, 54–88 (2004)
9. Mobasher, B., Jin, X., Zhou, Y.: Semantically enhanced collaborative filtering on the web. In: Berendt, B., Hotho, A., Mladenič, D., van Someren, M., Spiliopoulou, M., Stumme, G. (eds.) EWMF 2003. LNCS (LNAI), vol. 3209, pp. 57–76. Springer, Heidelberg (2004)
10. Osborne, F., Ruggeri, A.: A prismatic cognitive layout for adapting ontologies. In: Carberry, S., Weibelzahl, S., Micarelli, A., Semeraro, G. (eds.) UMAP 2013. LNCS, vol. 7899, pp. 359–362. Springer, Heidelberg (2013)
11. Rada, R., Mili, H., Bicknell, E., Blettner, M.: Development and application of a metric on semantic nets. IEEE Trans. on Systems Management and Cybernetics 19(1), 17–30 (1989)
12. Salton, G., McGill, M.: Introduction to Modern Information Retrieval. McGraw-Hill Book Company (1984)
13. Schafer, J.B., Frankowski, D., Herlocker, J., Sen, S.: Collaborative filtering recommender systems. In: Brusilovsky, P., Kobsa, A., Nejdl, W. (eds.) Adaptive Web 2007. LNCS, vol. 4321, pp. 291–324. Springer, Heidelberg (2007)
14. Sieg, A., Mobasher, B., Burke, R.: Web search personalization with ontological user profiles. In: 16th ACM Conference on Information and Knowledge Management, CIKM 2007, pp. 525–534. ACM (2007)
15. Sieg, A., Mobasher, B., Burke, R.: Improving the effectiveness of collaborative recommendation with ontology-based user profiles. In: 1st International Workshop on Information Heterogeneity and Fusion in Recommender Systems, HetRec 2010, pp. 39–46. ACM (2010)
16. Thiagarajan, R., Manjunath, G., Stumptner, M.: Finding experts by semantic matching of user profiles. In: 3rd Expert Finder Workshop on Personal Identification and Collaborations: Knowledge Mediation and Extraction, PICKME (2008)
17. Tversky, A.: Features of similarity. Psychological Review 84(4), 327–352 (1977)
18. Yuan, S.-T., Sun, J.: Ontology-based structured cosine similarity in document summarization: with applications to mobile audio-based knowledge management. IEEE Transactions on Systems, Man, and Cybernetics, Part B 35(5), 1028–1040 (2005)

Computing Instantiated Explanations in OWL DL

Fabrizio Riguzzi[1], Elena Bellodi[2], Evelina Lamma[2], and Riccardo Zese[2]

[1] Dipartimento di Matematica e Informatica – University of Ferrara
Via Saragat 1, I-44122, Ferrara, Italy
[2] Dipartimento di Ingegneria – University of Ferrara
Via Saragat 1, I-44122, Ferrara, Italy
{fabrizio.riguzzi,elena.bellodi,evelina.lamma,riccardo.zese}@unife.it

Abstract. Finding explanations for queries to Description Logics (DL) theories is a non-standard reasoning service originally defined for debugging purposes but recently found useful for answering queries to probabilistic theories. In the latter case, besides the axioms that are used to entail the query, it is necessary to record also the individuals to which the axioms are applied. We refer, in this case, to instantiated explanations. The system BUNDLE computes the probability of queries to probabilistic $\mathcal{ALC}$ knowledge bases by first finding instantiated explanations for the query and then applying a dynamic programming algorithm. In order to apply BUNDLE to more expressive DLs, such as $\mathcal{SHOIN}(\mathbf{D})$ that is at the basis of OWL DL, instantiated explanations must be found. In this paper, we discuss how we extended BUNDLE in order to compute instantiated explanations for $\mathcal{SHOIN}(\mathbf{D})$.

1 Introduction

Description Logics (DLs) are at the basis of the Semantic Web and the study of reasoning algorithms for DLs is a very active area of research. The problem of finding explanations for queries has been the subject of various works [22,7,9,6]. An explanation for a query is a set of axioms that is sufficient for entailing the query and that is minimal. Finding all explanations for a query is a non-standard reasoning service that was originally devised for ontology debugging, i.e., for helping the knowledge engineer in spotting modeling errors. However, this reasoning service turned out to be useful also for computing the probability of queries from probabilistic knowledge bases. In [3] we presented the algorithm BUNDLE for "Binary decision diagrams for Uncertain reasoNing on Description Logic thEories", that performs inference from DL knowledge bases under the probabilistic DISPONTE semantics. DISPONTE is based on the distribution semantics for probabilistic logic programming [20] and minimally extends the underlying DL language by annotating axioms with a probability and assuming that each axiom is independent of the others. In [3] BUNDLE computes the probability of queries by first finding all the explanations for the queries using an underlying reasoner (Pellet [23] in particular) and then building Binary Decision Diagrams from which the probability is computed. In [17] we extended

M. Baldoni et al. (Eds.): AI*IA 2013, LNAI 8249, pp. 397–408, 2013.

DISPONTE by including a second type of probabilistic annotation that represents statistical information, while the previous type only represents epistemic information. With these two types of annotations we are able to seamlessly represent assertional and terminological probabilistic knowledge, including degrees of overlap between concepts. Statistical annotations require to consider the individuals to which the axioms are applied. Thus, in order to perform inference, finding explanations is not enough: we need to know also the individuals involved in the application of each axiom.

In [18] we presented a version of BUNDLE that is able to compute the probabilities of queries from DISPONTE $\mathcal{ALC}$ knowledge bases. This required a modification of the underlying DL resoner, Pellet, in order to find *instantiated explanations* for the queries: sets of couples (axiom, substitution) where the substitution replaces (some) of the universally quantified logical variables of the axiom with individuals. This can be seen as a new non-standard reasoning service that, to the best of our knowledge, has not been tackled before. This result was achieved by modifying the rules that Pellet uses for updating the tracing function during the expansion of its tableau.

In this paper, we present an extension of BUNDLE for computing instantiated explanations for the expressive $\mathcal{SHOIN}(\mathbf{D})$ DL, that is at the basis of the OWL DL language. This paves the way for answering queries to DISPONTE OWL DL, thus allowing the introduction of uncertainty in the Semantic Web, an important goal according to many authors [4,24].

To compute instantiated explanations from $\mathcal{SHOIN}(\mathbf{D})$, we will illustrate how we modified the rules for updating the tracing function during the expansion of the tableau. This allows to find a single instantiated explanation. In order to find all of them, the hitting set algorithm of Pellet has been suitably adapted.

The paper is organized as follows. Section 2 briefly introduces the syntax of $\mathcal{SHOIN}(\mathbf{D})$ and its translation into predicate logic. Section 3 defines the problem of finding instantiated explanations while Section 4 illustrates the motivations that led us to consider this problem. Section 5 describes how BUNDLE finds instantiated explanations and, finally, Section 6 concludes the paper.

2 Description Logics

DLs are knowledge representation formalisms particularly useful for representing ontologies and have been adopted as the basis of the Semantic Web [1,2]. In this section, we recall the expressive description logic $\mathcal{SHOIN}(\mathbf{D})$ [12], which is the basis of the web ontology language OWL DL.

DLs syntax is based on concepts and roles: a *concept* corresponds to a set of individuals of the domain while a *role* corresponds to a set of couples of individuals of the domain. Let $\mathbf{A}$, $\mathbf{R}$ and $\mathbf{I}$ be sets of *atomic concepts, roles* and *individuals*, respectively.

A *role* is either an atomic role $R \in \mathbf{R}$ or the inverse R^- of an atomic role $R \in \mathbf{R}$. We use $\mathbf{R}^-$ to denote the set of all inverses of roles in $\mathbf{R}$. An *RBox* $\mathcal{R}$ consists of a finite set of *transitivity axioms* $Trans(R)$, where $R \in \mathbf{R}$, and *role inclusion axioms* $R \sqsubseteq S$, where $R, S \in \mathbf{R} \cup \mathbf{R}^-$.

Concepts are defined by induction as follows. Each $C \in \mathbf{A}$ is a concept, $\bot$ and $\top$ are concepts, and if $a \in \mathbf{I}$, then $\{a\}$ is a concept. If C, C_1 and C_2 are concepts and $R \in \mathbf{R} \cup \mathbf{R}^-$, then $(C_1 \sqcap C_2)$, $(C_1 \sqcup C_2)$, and $\neg C$ are concepts, as well as $\exists R.C$, $\forall R.C$, $\geq nR$ and $\leq nR$ for an integer $n \geq 0$. A *TBox* $\mathcal{T}$ is a finite set of *concept inclusion axioms* $C \sqsubseteq D$, where C and D are concepts. We use $C \equiv D$ to abbreviate $C \sqsubseteq D$ and $D \sqsubseteq C$. An *ABox* $\mathcal{A}$ is a finite set of *concept membership axioms* $a : C$, *role membership axioms* $(a, b) : R$, *equality axioms* $a = b$ and *inequality axioms* $a \neq b$, where C is a concept, $R \in \mathbf{R}$ and $a, b \in \mathbf{I}$.

A *knowledge base* (KB) $\mathcal{K} = (\mathcal{T}, \mathcal{R}, \mathcal{A})$ consists of a TBox $\mathcal{T}$, an RBox $\mathcal{R}$ and an ABox $\mathcal{A}$. A knowledge base $\mathcal{K}$ is usually assigned a semantics in terms of set-theoretic interpretations and models of the form $\mathcal{I} = (\Delta^{\mathcal{I}}, \cdot^{\mathcal{I}})$ where $\Delta^{\mathcal{I}}$ is a non-empty *domain* and $\cdot^{\mathcal{I}}$ is the *interpretation function* that assigns an element in $\Delta^{\mathcal{I}}$ to each $a \in \mathbf{I}$, a subset of $\Delta^{\mathcal{I}}$ to each $C \in \mathbf{A}$ and a subset of $\Delta^{\mathcal{I}} \times \Delta^{\mathcal{I}}$ to each $R \in \mathbf{R}$.

The semantics of DLs can be given equivalently by converting a KB into a predicate logic theory and then using the model-theoretic semantics of the resulting theory. A translation of $\mathcal{SHOIN}$ into First-Order Logic with Counting Quantifiers is given in the following as an extension of the one given in [21]. We assume basic knowledge of logic. The translation uses two functions π_x and π_y that map concept expressions to logical formulas, where π_x is given by

$$\pi_x(A) = A(x) \qquad\qquad \pi_x(\neg C) = \neg \pi_x(C)$$
$$\pi_x(\{a\}) = (x = a) \qquad\qquad \pi_x(C \sqcap D) = \pi_x(C) \wedge \pi_x(D)$$
$$\pi_x(C \sqcup D) = \pi_x(C) \vee \pi_x(D) \qquad \pi_x(\exists R.C) = \exists y.R(x, y) \wedge \pi_y(C)$$
$$\pi_x(\exists R^-.C) = \exists y.R(y, x) \wedge \pi_y(C) \qquad \pi_x(\forall R.C) = \forall y.R(x, y) \to \pi_y(C)$$
$$\pi_x(\forall R^-.C) = \forall y.R(y, x) \to \pi_y(C) \qquad \pi_x(\geq nR) = \exists^{\geq n} y.R(x, y)$$
$$\pi_x(\geq nR^-) = \exists^{\geq n} y.R(y, x) \qquad \pi_x(\leq nR) = \exists^{\leq n} y.R(x, y)$$
$$\pi_x(\leq nR^-) = \exists^{\leq n} y.R(y, x)$$

and π_y is obtained from π_x by replacing x with y and vice-versa. Table 1 shows the translation of each axiom of $\mathcal{SHOIN}$ knowledge bases into predicate logic.

$\mathcal{SHOIN}(\mathbf{D})$ adds to $\mathcal{SHOIN}$ datatype roles, i.e., roles that map an individual to an element of a datatype such as integers, floats, etc. Then new concept definitions involving datatype roles are added that mirror those involving roles introduced above. We also assume that we have predicates over the datatypes.

A query Q over a KB $\mathcal{K}$ is usually an axiom for which we want to test the entailment from the knowledge base, written $\mathcal{K} \models Q$. The entailment test may be reduced to checking the unsatisfiability of a concept in the knowledge base, i.e., the emptiness of the concept. $\mathcal{SHOIN}(\mathbf{D})$ is decidable iff there are no number restrictions on non-simple roles. A role is non-simple iff it is transitive or it has transitive subroles.

Given a predicate logic formula F, a *substitution* θ is a set of pairs x/a, where x is a variable universally quantified in the outermost quantifier in F and $a \in \mathbf{I}$. The application of θ to F, indicated by $F\theta$, is called an *instantiation of* F and is obtained by replacing x with a in F and by removing x from the external quantification for every pair x/a in θ. Formulas not containing variables are called *ground*. A substitution θ is *grounding* for a formula F if $F\theta$ is ground.

Table 1. Translation of $\mathcal{SHOIN}$ axioms into predicate logic

Axiom	Translation
$C \sqsubseteq D$	$\forall x.\pi_x(C) \to \pi_x(D)$
$R \sqsubseteq S$	$\forall x, y.R(x,y) \to S(x,y)$
$Trans(R)$	$\forall x, y, z.R(x,y) \wedge R(y,z) \to S(x,z)$
$a : C$	$\pi_a(C)$
$(a, b) : R$	$R(a, b)$
$a = b$	$a = b$
$a \neq b$	$a \neq b$

Example 1. The following KB is inspired by the ontology `people+pets` [13]:

$\exists hasAnimal.Pet \sqsubseteq NatureLover$

$(kevin, fluffy) : hasAnimal \quad (kevin, tom) : hasAnimal$

$fluffy : Cat \quad tom : Cat \quad Cat \sqsubseteq Pet$

It states that individuals that own an animal which is a pet are nature lovers and that *kevin* owns the animals *fluffy* and *tom*. Moreover, *fluffy* and *tom* are cats and cats are pets. The predicate logic formulas equivalent to the axioms are $F_1 = \forall x.\exists y.hasAnimal(x,y) \wedge Pet(y) \to NatureLover(x)$, $F_2 = hasAnimal(kevin, fluffy)$, $F_3 = hasAnimal(kevin, tom)$, $F_4 = Cat(fluffy)$, $F_5 = Cat(tom)$ and $F_6 = \forall x.Cat(x) \to Pet(x)$. The query $Q = kevin : NatureLover$ is entailed by the KB.

The *Unique Name Assumption* (UNA) [1] states that distinct individual names denote distinct objects, i.e., that $a \neq b$ with $a, b \in \mathbf{I}$ implies $a^{\mathcal{I}} \neq b^{\mathcal{I}}$.

3 Finding Instantiated Explanations

The problem of finding explanations for a query has been investigated by various authors [22,7,9,6]. It was called *axiom pinpointing* in [22] and considered as a non-standard reasoning service useful for debugging ontologies. In particular, in [22] the authors define *minimal axiom sets* or *MinAs* for short.

Definition 1 (MinA). *Let $\mathcal{K}$ be a knowledge base and Q an axiom that follows from it, i.e., $\mathcal{K} \models Q$. We call a set $M \subseteq \mathcal{K}$ a minimal axiom set or MinA for Q in $\mathcal{K}$ if $M \models Q$ and it is minimal w.r.t. set inclusion.*

The problem of enumerating all MinAs is called MIN-A-ENUM. ALL-MINAs$(Q, \mathcal{K})$ is the set of all MinAs for Q in $\mathcal{K}$.

Axiom pinpointing has been thoroughly discussed in [8,10,5,9] for the purpose of tracing derivations and debugging ontologies. The techniques proposed in these papers have been integrated into the Pellet reasoner [23]. Pellet solves MIN-A-ENUM by finding a single MinA using a tableau algorithm and then applying the *hitting set* algorithm to find all the other MinAs.

BUNDLE is based on Pellet and uses it for solving the MIN-A-ENUM problem. However, BUNDLE needs, besides ALL-MINAs$(Q, \mathcal{K})$, also the individuals to which the axiom was applied for each probabilistic axiom appearing in ALL-MINAS$(Q, \mathcal{K})$. We call this problem *instantiated axiom pinpointing*.

In instantiated axiom pinpointing we are interested in instantiated minimal sets of axioms that entail an axiom. We call this type of explanations *Inst-MinA*. An *instantiated axiom set* is a finite set $\mathcal{F} = \{(F_1, \theta_1), \ldots, (F_n, \theta_n)\}$ where $F_1, \ldots, F_n$ are axioms and $\theta_1, \ldots, \theta_n$ are substitutions. Given two instantiated axiom sets $\mathcal{F} = \{(F_1, \theta_1), \ldots, (F_n, \theta_n)\}$ and $\mathcal{E} = \{(E_1, \delta_1), \ldots, (E_m, \delta_m)\}$, we say that $\mathcal{F}$ *precedes* $\mathcal{E}$, written $\mathcal{F} \preceq \mathcal{E}$, iff, for each $(F_i, \theta_i) \subset \mathcal{F}$, there exists an $(E_j, \delta_j) \in \mathcal{E}$ and a substitution η such that $F_j \theta_j = E_i \delta_i \eta$.

Definition 2 (InstMinA). *Let $\mathcal{K}$ be a knowledge base and Q an axiom that follows from it, i.e., $\mathcal{K} \models Q$. We call $\mathcal{F} = \{(F_1, \theta_1), \ldots, (F_n, \theta_n)\}$ an* instantiated minimal axiom set *or* InstMinA *for Q in $\mathcal{K}$ if $\{F_1 \theta_1, \ldots, F_n \theta_n\} \models Q$ and $\mathcal{F}$ is minimal w.r.t. precedence.*

Minimality w.r.t. precedence means that axioms in a InstMinA are as instantiated as possible. We call INST-MIN-A-ENUM the problem of enumerating all InstMinAs. ALL-INSTMINAS$(Q, \mathcal{K})$ is the set of all InstMinAs for Q in $\mathcal{K}$.

Example 2. The query $Q = kevin : NatureLover$ of Example 1 has two MinAs, the first of which is: $\{\ hasAnimal(kevin, fluffy),\ Cat(fluffy),\ \forall x.Cat(x) \to Pet(x),\ \forall x.\ \exists y.hasAnimal(x, y) \land Pet(y) \to NatureLover(x)\ \}$ while the other is: $\{\ hasAnimal(kevin, tom),\ Cat(tom),\ \forall x.Cat(x) \to Pet(x),\ \forall x.\exists y.hasAnimal(x, y) \land Pet(y) \to NatureLover(x)\ \}$, where axioms are represented as first order logic formulas.

The corresponding InstMinAs are $\{\ hasAnimal(kevin, fluffy),\ Cat(fluffy),\ Cat(fluffy) \to Pet(fluffy),\ hasAnimal(kevin,\ fluffy) \land Pet(fluffy) \to NatureLover(kevin)\ \}$ and $\{\ hasAnimal(kevin, tom),\ Cat(tom),\ Cat(tom) \to Pet(tom),\ hasAnimal(kevin, tom) \land Pet(tom) \to NatureLover(kevin)\ \}$ respectively, where instantiated axioms are represented directly in their first order version.

4   Motivation

INST-MIN-A-ENUM is required to answer queries to knowledge bases following the DISPONTE probabilistic semantics. DISPONTE applies the distribution semantics [20] to probabilistic DL theories. In DISPONTE, a *probabilistic knowledge base* $\mathcal{K}$ is a set of certain axioms or probabilistic axioms. *Certain axioms* take the form of regular DL axioms. *Probabilistic axioms* take the form $p ::_{Var} E$ where p is a real number in $[0, 1]$, Var is a set of variables from $\{x, y, z\}$ and E is a DL axiom. Var is usually written as a string, so xy indicates the subset $\{x, y\}$. If Var is empty, then the $::$ symbol has no subscript. The variables in Var must appear in the predicate logic version of E shown in Table 1.

In order to give a semantics to such probabilistic knowledge bases, we consider their translation into predicate logic and we make the UNA. The idea of DISPONTE [17,16] is to associate independent Boolean random variables to (instantiations of) the formulas in predicate logic that are obtained from the translation of the axioms. By assigning values to every random variable we define a *world*, the set of predicate logic formulas whose random variable takes value 1. The purpose of the UNA is to ensure that the instantiations of formulas are really distinct.

To obtain a world w, we include every formula from a certain axiom. For each probabilistic axiom, we generate all the substitutions for the variables of the equivalent predicate logic formula that are indicated in the subscript. The variables are replaced with elements of $\mathbf{I}$. For each instantiated formula, we decide whether or not to include it in w. In this way we obtain a predicate logic theory which can be assigned a model-theoretic semantics. A query is entailed by a world if it is true in every model of the world. Here we assume that $\mathbf{I}$ is countably infinite and, together with the UNA, this entails that the elements of $\mathbf{I}$ are in bijection with the elements of the domain. These are called *standard names* according to [11].

If *Var* is empty, the probability p can be interpreted as an *epistemic probability*, i.e., as the degree of our belief in axiom E, while if *Var* is equal to the set of all allowed variables, p can be interpreted as a *statistical probability*, i.e., as information regarding random individuals from the domain.

For example, the statement that birds fly with 90% probability can be expressed by the epistemic axiom $0.9 :: Bird \sqsubseteq Flies$. Instead the statistical axiom $0.9 ::_x Bird \sqsubseteq Flies$ means that a random bird has 90% probability of flying. Thus, 90% of birds fly. If we query the probability of a bird flying, both axioms give the same result, 0.9. For two birds, the probability of both flying will be $0.9 \cdot 0.9 = 0.81$ with the second axiom and still 0.9 with the first one.

In order to compute the probability of queries, rather than generating all possible worlds, we look for a covering set of explanations, i.e., the set of all InstMinAs for the query. The query is true if all its instantiated axioms in an InstMinA are chosen, according to the values of the random variables. To compute the probability, the explanations must be made mutually exclusive, so that the probability of each individual explanation is computed and summed with the others. This is done by means of Binary Decision Diagrams [18].

Instantiated axiom pinpointing is also useful for a more fine-grained debugging of the ontology: by highlighting the individuals to which the axiom is being applied, it may point to parts of the ABox to be modified for repairing the KB.

5   An Algorithm for Computing Instantiated Explanations

Pellet, on which BUNDLE is based, finds explanations by using a tableau algorithm [7]. A *tableau* is a graph where each node represents an individual a and is labeled with the set of concepts $\mathcal{L}(a)$ it belongs to. Each edge $\langle a, b \rangle$ in the graph is labeled with the set of roles $\mathcal{L}(\langle a, b \rangle)$ to which the couple (a, b) belongs. Pellet repeatedly applies a set of consistency preserving *tableau expansion rules* until a clash (i.e., a contradiction) is detected or a clash-free graph is found to which no more rules are applicable. A clash is, for example, a concept C and a node a where C and $\neg C$ are present in the label of a, i.e. $\{C, \neg C\} \subseteq \mathcal{L}(a)$.

Each expansion rule updates as well a *tracing function* τ, which associates sets of axioms with labels of nodes and edges. It maps couples (concept, individual) or (role, couple of individuals) to a fragment of the knowledge base $\mathcal{K}$. τ is initialized as the empty set for all the elements of its domain except for $\tau(C, a)$ and $\tau(R, \langle a, b \rangle)$ to which the values $\{a : C\}$ and $\{(a, b) : R\}$ are assigned if $a : C$ and $(a, b) : R$ are in the ABox respectively. The output of the tableau algorithm is a set S of axioms that is a fragment of $\mathcal{K}$ from which the query is entailed.

In BUNDLE we are interested in solving the INST-MIN $\wedge$ ENUM problem. In order to find all the instantiated explanations, we modified the tableau expansion rules of Pellet, reported in [7], to return a set of pairs (axiom, substitution) instead of a set of axioms. The tracing function τ now stores, together with information regarding concepts and roles, also information concerning individuals involved in the expansion rules, which will be returned at the end of the derivation process together with the axioms. Fig. 1 shows the tableau expansion rules of BUNDLE for OWL DL, where $(A \sqsubseteq D, a)$ is the abbreviation of $(A \sqsubseteq D, \{x/a\})$, $(R \sqsubseteq S, a)$ of $(R \sqsubseteq S, \{x/a\})$, $(R \sqsubseteq S, a, b)$ of $(R \sqsubseteq S, \{x/a, y/b\})$, $(Trans(R), a, b)$ of $(Trans(R), \{x/a, y/b\})$ and $(Trans(R), a, b, c)$ of $(Trans(R), \{x/a, y/b, z/c\})$, with a, b, c individuals and x, y, z variables (see Table 1). In [18] we presented BUNDLE for the $\mathcal{ALC}$ DL and only the rules $\rightarrow$ *unfold* , $\rightarrow CE$ and $\rightarrow \forall$ were modified with respect to Pellet's ones.

Example 3. Let us consider the knowledge base presented in Example 1 and the query $Q = kevin : NatureLover$.

After the initialization of the tableau, BUNDLE can apply the $\rightarrow$ *unfold* rule to the individuals *tom* or *fluffy*. Suppose it selects *tom*. The tracing function τ becomes (in predicate logic)

$\tau(Pet, tom) = \{ Cat(tom), Cat(tom) \rightarrow Pet(tom) \}$

At this point BUNDLE applies the $\rightarrow CE$ rule to *kevin*, adding $\neg(\exists hasAnimal.Pet) \sqcup NatureLover = \forall hasAnimal.(\neg Pet) \sqcup NatureLover$ to $\mathcal{L}(kevin)$ with the following tracing function:

$\tau(\forall hasAnimal.(\neg Pet) \sqcup NatureLover, kevin) = \{$
$\quad \exists y.hasAnimal(kevin, y) \wedge Pet(y) \rightarrow NatureLover(kevin) \}$

Then it applies the $\rightarrow \sqcup$ rule on *kevin* generating two tableaux. In the first one it adds $\forall hasAnimal.(\neg Pet)$ to the label of *kevin* with the tracing function

$\tau(\forall hasAnimal.(\neg Pet), kevin) = \{$
$\quad \exists y.hasAnimal(kevin, y) \wedge Pet(y) \rightarrow NatureLover(kevin)\}$

Now it can apply the $\rightarrow \forall$ rule to *kevin*. In this step it can use either *tom* or *fluffy*, supposing it selects *tom* the tracing function will be:

$\tau(\neg(Pet), tom) = \{ hasAnimal(kevin, tom),$
$\quad hasAnimal(kevin, tom) \wedge Pet(tom) \rightarrow NatureLover(kevin) \}$

At this point this first tableau contains a clash for the individual *tom*, thus BUNDLE moves to the second tableau and tries to expand it. The second tableau was found by applying the $\rightarrow CE$ rule that added $NatureLover$ to the label of *kevin*, so also the second tableau contains a clash. At this point, BUNDLE joins the tracing functions of the two clashes to find the following InstMinA:

$\{hasAnimal(kevin, tom) \wedge Pet(tom) \rightarrow NatureLover(kevin),$
$\quad hasAnimal(kevin, tom), Cat(tom), Cat(tom) \rightarrow Pet(tom) \}.$

$\rightarrow$ *unfold*: **if** $A \in \mathcal{L}(a)$, A atomic and $(A \sqsubseteq D) \in K$, **then**
  if $D \notin \mathcal{L}(a)$, **then**
    $\mathcal{L}(a) = \mathcal{L}(a) \cup \{D\}$
    $\tau(D, a) := \tau(A, a) \cup \{(A \sqsubseteq D, a)\}$
$\rightarrow$ *CE*: **if** $(C \sqsubseteq D) \in K$, **then**
  if $(\neg C \sqcup D) \notin \mathcal{L}(a)$, **then**
    $\mathcal{L}(a) = \mathcal{L}(a) \cup \{\neg C \sqcup D\}$
    $\tau(\neg C \sqcup D, a) := \{(C \sqsubseteq D, a)\}$
$\rightarrow$ $\sqcap$: **if** $(C_1 \sqcap C_2) \in \mathcal{L}(a)$, **then**
  if $\{C_1, C_2\} \not\subseteq \mathcal{L}(a)$, **then**
    $\mathcal{L}(a) = \mathcal{L}(a) \cup \{C_1, C_2\}$
    $\tau(C_i, a) := \tau((C_1 \sqcap C_2), a)$
$\rightarrow$ $\sqcup$: **if** $(C_1 \sqcup C_2) \in \mathcal{L}(a)$, **then**
  if $\{C_1, C_2\} \cap \mathcal{L}(a) = \emptyset$, **then**
    Generate graphs $G_i := G$ for each $i \in \{1, 2\}$, $\mathcal{L}(a) = \mathcal{L}(a) \cup \{C_i\}$ for each $i \in \{1, 2\}$
    $\tau(C_i, a) := \tau((C_1 \sqcup C_2), a)$
$\rightarrow$ $\exists$: **if** $\exists S.C \in \mathcal{L}(a)$, **then**
  if a has no S-neighbor b with $C \in \mathcal{L}(b)$, **then**
    create new node b, $\mathcal{L}(b) = \{C\}$, $\mathcal{L}(\langle a, b \rangle) = \{S\}$,
    $\tau(C, b) := \tau((\exists S.C), a)$, $\tau(S, \langle a, b \rangle) := \tau((\exists S.C), a)$
$\rightarrow$ $\forall$: **if** $\forall(S_1, C) \in \mathcal{L}(a_1)$, a_1 is not indirectly blocked and there is an S_1-neighbor b of a_1, **then**
  if $C \notin \mathcal{L}(b)$, **then** $\mathcal{L}(b) = \mathcal{L}(a) \cup \{C\}$
  if there is a chain of individuals $a_2, \ldots, a_n$ and roles $S_2, \ldots, S_n$ such that
    $\bigcup_{i=2}^{n}\{(Trans(S_{i-1}), a_i, a_{i-1}), (S_{i-1} \sqsubseteq S_i, a_i)\} \subseteq \tau(\forall S_1.C, a_1)$
    and $\neg \exists a_{n+1} : \{(Trans(S_n), a_{n+1}, a_n), (S_n \sqsubseteq S_{n+1}, a_{n+1})\} \subseteq \tau(\forall S_1.C, a_1)$, **then**
    $\tau(C, b) := \tau(\forall S_1.C, a_1) \setminus \bigcup_{i=2}^{n}\{(Trans(S_{i-1}), a_i, a_{i-1}), (S_{i-1} \sqsubseteq S_i, a_i)\} \cup$
    $\bigcup_{i=2}^{n}\{(Trans(S_{i-1}), a_i, a_{i-1}, b), (S_{i-1} \sqsubseteq S_i, a_i, b)\} \cup \tau(S_1, \langle a_1, b \rangle)$
  else
    $\tau(C, b) := \tau(\forall S_1.C, a_1) \cup \tau(S_1, \langle a_1, b \rangle)$
$\rightarrow$ $\forall^+$: **if** $\forall(S.C) \in \mathcal{L}(a)$, a is not indirectly blocked
    and there is an R-neighbor b of a, $Trans(R)$ and $R \sqsubseteq S$, **then**
  if $\forall R.C \notin \mathcal{L}(b)$, **then** $\mathcal{L}(b) = \mathcal{L}(b) \cup \{\forall R.C\}$
    $\tau(\forall R.C, b) := \tau(\forall S.C, a) \cup \tau(R, \langle a, b \rangle) \cup \{(Trans(R), a, b), (R \sqsubseteq S, a)\}$
$\rightarrow$ $\geq$: **if** $(\geq nS) \in \mathcal{L}(a)$, a is not blocked, **then**
  if there are no n safe S-neighbors $b_1, \ldots, b_n$ of a with $b_i \neq b_j$, **then**
    create n new nodes $b_1, \ldots, b_n$; $\mathcal{L}(\langle a, b_i \rangle) = \mathcal{L}(\langle a, b_i \rangle) \cup \{S\}$; $\dot{\neq}(b_i, b_j)$
    $\tau(S, \langle a, b_i \rangle) := \tau((\geq nS), a)$
    $\tau(\dot{\neq}(b_i, b_j)) := \tau((\geq nS), a)$
$\rightarrow$ $\leq$: **if** $(\leq nS) \in \mathcal{L}(a)$, a is not indirectly blocked
    and there are m S-neighbors $b_1, \ldots, b_m$ of a with $m > n$, **then**
  For each possible pair b_i, b_j, $1 \leq i, j \leq m; i \neq j$ **then**
    Generate a graph G'
    $\tau(Merge(b_i, b_j)) := (\tau((\leq nS), a) \cup \tau(S, \langle a, b_1 \rangle) \ldots \cup \tau(S, \langle a, b_m \rangle))$
    if b_j is a nominal node, **then** $Merge(b_i, b_j)$ in G',
    else if b_i is a nominal node or ancestor of b_j, **then** $Merge(b_j, b_i)$
    else $Merge(b_i, b_j)$ in G'
    if b_i is merged into b_j, **then** for each concept C_i in $\mathcal{L}(b_i)$,
      $\tau(C_i, b_j) := \tau(C_i, b_i) \cup \tau(Merge(b_i, b_j))$
      (similarly for roles merged, and correspondingly for concepts in b_j if merged into b_i)
$\rightarrow$ *O*: **if**, $\{o\} \in \mathcal{L}(a) \cap \mathcal{L}(b)$ and not $a \dot{\neq} b$, **then**
  $Merge(a, b)$
  $\tau(Merge(a, b)) := \tau(\{o\}, a) \cup \tau(\{o\}, b)$
  For each concept C_i in $\mathcal{L}(a)$, $\tau(C_i, b) := \tau(C_i, a) \cup \tau(Merge(a, b))$
  (similarly for roles merged, and correspondingly for concepts in $\mathcal{L}(b)$)
$\rightarrow$ *NN*: **if** $(\leq nS) \in \mathcal{L}(a)$, a nominal node, b blockable S-predecessor of a
    and there is no m s.t. $1 \leq m \leq n$, $(\leq mS) \in \mathcal{L}(a)$
    and there exist m nominal S-neighbors $c_1, \ldots, c_m$ of a s.t. $c_i \dot{\neq} c_j$, $1 \leq j \leq m$, **then**
  generate new G_m for each m, $1 \leq m \leq n$
  and do the following in each G_m:
    $\mathcal{L}(a) = \mathcal{L}(a) \cup \{\leq mS\}$, $\tau((\leq mS), a) := \tau((\leq nS), a) \cup (\tau(S, \langle b, a \rangle))$
    create $b_1, \ldots, b_m$; add $b_i \dot{\neq} b_j$ for $1 \leq i \leq j \leq m$. $\tau(\dot{\neq}(b_i, b_j) := \tau((\leq nS), a) \cup \tau(S, \langle b, a \rangle)$
    $\mathcal{L}(\langle a, b_i \rangle) = \mathcal{L}(\langle a, b_i \rangle) \cup \{S\}$; $\mathcal{L}(b_i) = \mathcal{L}(b_i) \cup \{\{o_i\}\}$;
    $\tau(S, \langle a, b_i \rangle) := \tau((\leq nS), a) \cup \tau(S, \langle b, a \rangle)$; $\tau(\{o_i\}, b_i) := \tau((\leq nS), a) \cup \tau(S, \langle b, a \rangle)$

Fig. 1. BUNDLE tableau expansion rules for OWL DL

For applying BUNDLE to $\mathcal{SHOIN}(\mathbf{D})$, we further modified the rules $\rightarrow \forall^+$ and $\rightarrow \forall$. For the rule $\rightarrow \forall^+$, we record in the explanation a transitivity axiom for the role R in which only two individuals, those connected by the super role S, are involved. For the rule $\rightarrow \forall$ we make a distinction between the case in which $\forall S_1.C$ was added to $\mathcal{L}(a_1)$ by a chain of applications of $\rightarrow \forall^+$ or not. In the first case, we fully instantiate the transitivity and subrole axioms. In the latter case, we simply obtain $\tau(C, b)$ by combining the explanation of $\forall S_1.C(a_1)$ with that of $(a_1, b) : S_1$.

Example 4. Let us consider the query $Q = eva : Person$ with the following knowledge base:

$Trans(friend)$ $kevin : \forall friend.Person$ $(kevin, lara) : friend$ $(lara, eva) : friend$

BUNDLE first applies the $\rightarrow \forall^+$ rule to $kevin$, adding $\forall friend.Person$ to the label of $lara$. In this case $friend$ is considered as a subrole of itself. The tracing function τ is updated as (in predicate logic):

$\tau(\forall friend.Person, lara) = \{ \forall y.friend(kevin, y) \rightarrow Person(y),$
$friend(kevin, lara), \forall z.friend(kevin, lara) \wedge friend(lara, z) \rightarrow friend(kevin, z)\}$

Note that the transitivity axiom is not fully instantiated, the variable z is still present. Then BUNDLE applies the $\rightarrow \forall$ rule to $lara$ adding $Person$ to eva. The tracing function τ is modified as (in predicate logic):

$\tau(Person, eva) = \{ \forall y.friend(kevin, y) \rightarrow Person(y), friend(kevin, lara),$
$friend(lara, eva), friend(kevin, lara) \wedge friend(lara, eva) \rightarrow friend(kevin, eva)\}$

Here the transitivity axiom has become ground: all variables have been instantiated. At this point the tableau contains a clash so the algorithm stops and returns the explanation given by $\tau(Person, eva)$.

Example 5. Let us consider the knowledge base

$kevin : \forall kin.Person$ $(kevin, lara) : relative$ $(lara, eva) : ancestor$
$(eva, ann) : ancestor$ $Trans(ancestor)$ $Trans(relative)$
$relative \sqsubseteq kin$ $ancestor \sqsubseteq relative$

The query $Q = ann : Person$ has the explanation (in predicate logic):

$\tau(Person, ann) = \{ \forall y.kin(kevin, y) \rightarrow Person(y),$
$relative(kevin, lara), ancestor(lara, eva), ancestor(eva, ann),$
$relative(kevin, lara) \wedge relative(lara, ann) \rightarrow relative(kevin, ann),$
$ancestor(lara, eva) \wedge ancestor(eva, ann) \rightarrow ancestor(lara, ann),$
$relative(kevin, ann) \rightarrow kin(kevin, ann),$
$ancestor(lara, ann) \rightarrow relative(lara, ann) \}$

It is easy to see that the explanation entails the axiom represented by the arguments of τ. In general, the following theorem holds.

Theorem 1. *Let Q be an axiom entailed by $\mathcal{K}$ and let S be the output of BUNDLE with the tableau expansion rules of Figure 1 with input Q and $\mathcal{K}$. Then $S \in$ ALL-INSTMINAS$(Q, \mathcal{K})$.*

Proof. The proof, whose full details are given in Theorem 5 of [15], proceeds by induction on the number of rule applications following the proof of Theorem 2 of [7].

The complexity of BUNDLE is similar to the one of Pellet due to the fact that only the $\rightarrow \forall$ rule requires a non-constant number of additional operations. BUNDLE has to find chains of individuals in the current explanation, thus the complexity increment depends on the size of the current explanation.

The tableau algorithm returns a single InstMinA. To solve the problem ALL-MINAs$(Q, \mathcal{K})$, Pellet uses the *hitting set algorithm* [14]. It starts from a MinA S and initializes a labeled tree called *Hitting Set Tree* (HST) with S as the label of its root v [7]. Then it selects an arbitrary axiom F in S, it removes it from $\mathcal{K}$, generating a new knowledge base $\mathcal{K}' = \mathcal{K} - \{F\}$, and tests the entailment of Q w.r.t. $\mathcal{K}'$. If Q is still entailed, we obtain a new explanation for Q. The algorithm adds a new node w in the tree and a new edge $\langle v, w \rangle$, then it assigns this new explanation to the label of w and the axiom F to the label of the edge. The algorithm repeats this process until the entailment test returns negative: in that case the current node becomes a leaf, the algorithm backtracks to a previous node and repeats these operations until the HST is fully built. The distinct non-leaf nodes of the tree collectively represent the set of all MinAs for the query E.

As in Pellet, to compute ALL-INSTMINAs$(E, \mathcal{K})$ we use the hitting set algorithm that calls BUNDLE's tableau algorithm for computing a single InstMinA. However in BUNDLE we need to eliminate instantiated axioms from the KB. This cannot be done by modifying the KB, since it contains axioms in their general form. Therefore we modified BUNDLE's tableau algorithm to take as input a set of banned instantiated axioms, *BannedInstAxioms*, as well. BUNDLE, before applying a tableau rule, checks whether one of the instantiated axioms to be added to τ is in *BannedInstAxioms*. If so, it does not apply the rule. In this way we can simulate the removal of an instantiated axiom from the theory and apply Pellet's hitting set algorithm to find ALL-INSTMINAs$(E, \mathcal{K})$.

Example 6. Let us consider Example 3. Once an InstMinA is found, BUNDLE applies the hitting set algorithm to it. Thus BUNDLE selects an axiom from the InstMinA and removes it from the knowledge base. Suppose it selects the axiom $Cat(tom)$. At this point, a new run of the tableau algorithm w.r.t. the KB without the axiom $Cat(tom)$ can be executed. BUNDLE can apply the $\rightarrow$ *unfold* rule to the individual *fluffy* and following the same steps used in Example 3 it finds a new InstMinA:

$\{hasAnimal(kevin, fluffy) \land Pet(fluffy) \rightarrow NatureLover(kevin),$
$hasAnimal(kevin, fluffy), Cat(fluffy), Cat(fluffy) \rightarrow Pet(fluffy) \}$

Now all the tableaux contain a clash so the hitting set algorithm adds a new node to the HST and selects a new axiom from this second InstMinA to be removed from the knowledge base. At this point, whatever axiom is removed from the KB, the query $Q = kevin : NatureLover$ will be no longer entailed w.r.t. the KB.

Once the HST is fully built, BUNDLE returns the set of all instantiated explanations which is:

$$\text{ALL-INSTMINAS}(kevin : NatureLover, \mathcal{K}) = \{$$

$$\{ \; hasAnimal(kevin, tom) \wedge Pet(tom) \rightarrow NatureLover(kevin),$$
$$hasAnimal(kevin, tom), \; Cat(tom), \; Cat(tom) \rightarrow Pet(tom) \; \},$$
$$\{ \; hasAnimal(kevin, fluffy) \wedge Pet(fluffy) \rightarrow NatureLover(kevin),$$
$$hasAnimal(kevin, fluffy), \; Cat(fluffy), \; Cat(fluffy) \rightarrow Pet(fluffy) \; \} \; \}$$

Theorem 2. *Let Q be an axiom entailed by $\mathcal{K}$ and let* $\text{INSTEXPHST}(Q, \mathcal{K})$ *be the set of instantiated explanations returned by BUNDLE's hitting set algorithm. Then* $\text{INSTEXPHST}(Q, \mathcal{K}) = \text{ALL-INSTMINAS}(Q, \mathcal{K})$.

Proof. See Theorem 6 in [15].

6   Conclusions

We have presented an approach for finding instantiated explanations for expressive DLs such as $\mathcal{SHOIN}(\mathbf{D})$. The approach is implemented in the system BUNDLE for performing inference on probabilistic knowledge bases following the DISPONTE semantics. BUNDLE is available for download from `http://sites.unife.it/ml/bundle`. BUNDLE is thus now able to compute the probability of queries from DISPONTE OWL DL theories. We plan to perform an extensive test of BUNDLE performance on ontologies of various sizes. Moreover, we plan to perform parameter learning of DISPONTE OWL DL KBs by extending EDGE [19], an algorithm that learns DISPONTE $\mathcal{ALC}$ KBs' parameters.

References

1. Baader, F., Calvanese, D., McGuinness, D.L., Nardi, D., Patel-Schneider, P.F. (eds.): The Description Logic Handbook: Theory, Implementation, and Applications. Cambridge University Press (2003)
2. Baader, F., Horrocks, I., Sattler, U.: Description logics. In: Handbook of Knowledge Representation, ch. 3, pp. 135–179. Elsevier (2008)
3. Bellodi, E., Lamma, E., Riguzzi, F., Albani, S.: A distribution semantics for probabilistic ontologies. In: Bobillo, F., et al. (eds.) URSW 2011. CEUR Workshop Proceedings, vol. 778, pp. 75–86. Sun SITE Central Europe (2011)
4. Carvalho, R.N., Laskey, K.B., Costa, P.C.G.: PR-OWL 2.0 - bridging the gap to OWL semantics. In: Bobillo, F., et al. (eds.) URSW 2010. CEUR Workshop Proceedings, vol. 654. Sun SITE Central Europe (2010)
5. Halaschek-Wiener, C., Kalyanpur, A., Parsia, B.: Extending tableau tracing for ABox updates. Tech. rep., University of Maryland (2006)
6. Horridge, M., Parsia, B., Sattler, U.: Explaining inconsistencies in OWL ontologies. In: Godo, L., Pugliese, A. (eds.) SUM 2009. LNCS, vol. 5785, pp. 124–137. Springer, Heidelberg (2009)
7. Kalyanpur, A.: Debugging and Repair of OWL Ontologies. Ph.D. thesis, The Graduate School of the University of Maryland (2006)

8. Kalyanpur, A., Parsia, B., Cuenca-Grau, B., Sirin, E.: Tableaux tracing in SHOIN. Tech. Rep. 2005-66, University of Maryland (2005)
9. Kalyanpur, A., Parsia, B., Horridge, M., Sirin, E.: Finding all justifications of OWL DL entailments. In: Aberer, K., et al. (eds.) ISWC/ASWC 2007. LNCS, vol. 4825, pp. 267–280. Springer, Heidelberg (2007)
10. Kalyanpur, A., Parsia, B., Sirin, E., Hendler, J.A.: Debugging unsatisfiable classes in OWL ontologies. J. Web Sem. 3(4), 268–293 (2005)
11. Levesque, H., Lakemeyer, G.: The Logic of Knowledge Bases. MIT Press (2000)
12. Lukasiewicz, T., Straccia, U.: Managing uncertainty and vagueness in description logics for the semantic web. J. Web Sem. 6(4), 291–308 (2008)
13. Patel-Schneider, P.F., Horrocks, I., Bechhofer, S.: Tutorial on OWL (2003)
14. Reiter, R.: A theory of diagnosis from first principles. Artif. Intell. 32(1), 57–95 (1987)
15. Riguzzi, F., Bellodi, E., Lamma, E., Zese, R.: Probabilistic description logics under the distribution semantics. Tech. Rep. ML-01, University of Ferrara (2013), http://sites.unife.it/ml/bundle
16. Riguzzi, F., Bellodi, E., Lamma, E.: Probabilistic Datalog+/- under the distribution semantics. In: Kazakov, Y., Lembo, D., Wolter, F. (eds.) DL 2012. CEUR Workshop Proceedings, vol. 846, pp. 519–529. Sun SITE Central Europe (2012)
17. Riguzzi, F., Bellodi, E., Lamma, E., Zese, R.: Epistemic and statistical probabilistic ontologies. In: Bobillo, F., et al. (eds.) URSW 2012. CEUR Workshop Proceedings, vol. 900, pp. 3–14. Sun SITE Central Europe (2012)
18. Riguzzi, F., Bellodi, E., Lamma, E., Zese, R.: BUNDLE: A reasoner for probabilistic ontologies. In: Faber, W., Lembo, D. (eds.) RR 2013. LNCS, vol. 7994, pp. 183–197. Springer, Heidelberg (2013)
19. Riguzzi, F., Bellodi, E., Lamma, E., Zese, R.: Parameter learning for probabilistic ontologies. In: Faber, W., Lembo, D. (eds.) RR 2013. LNCS, vol. 7994, pp. 265–270. Springer, Heidelberg (2013)
20. Sato, T.: A statistical learning method for logic programs with distribution semantics. In: ICLP 1995. pp. 715–729. MIT Press (1995)
21. Sattler, U., Calvanese, D., Molitor, R.: Relationships with other formalisms. In: Description Logic Handbook, ch. 4, pp. 137–177. Cambridge University Press (2003)
22. Schlobach, S., Cornet, R.: Non-standard reasoning services for the debugging of description logic terminologies. In: Gottlob, G., Walsh, T. (eds.) IJCAI 2003, pp. 355–362. Morgan Kaufmann (2003)
23. Sirin, E., Parsia, B., Cuenca-Grau, B., Kalyanpur, A., Katz, Y.: Pellet: A practical OWL-DL reasoner. J. Web Sem. 5(2), 51–53 (2007)
24. URW3-XG: Uncertainty reasoning for the World Wide Web, final report (2005)

Using Agents for Generating Personalized Recommendations of Multimedia Contents

Domenico Rosaci[1] and Giuseppe M.L. Sarné[2]

[1] DIIES, University "Mediterranea" of Reggio Calabria,
Via Graziella, Feo di Vito 89122 Reggio Calabria (RC), Italy
[2] DICEAM, University "Mediterranea" of Reggio Calabria, Via
Graziella, Feo di Vito 89122 Reggio Calabria (RC), Italy
{domenico.rosaci,sarne}@unirc.it

Abstract. Agent-based recommender systems assist users based on their preferences and those of similar users. However, when dealing with multimedia contents they need of: (*i*) selecting as recommenders those users that have similar profiles and that are reliable in providing suggestions and (*ii*) considering the effects of the device currently exploited. To address these issues, we propose a multi-agent architecture, called MART, conceived to this aim and based on a particular trust model. Some experimental results are presented to evaluate our proposal, that show MART is more effective, in terms of suggestion quality, than other agent-based recommenders.

1 Introduction

Currently, Web sites offer an increasing number of multimedia contents (MCs) and their exploration could require many time. Therefore, a Web site should present to its visitors those MCs compatible with their preferences. To this aim, many adaptive systems support users in their Web sessions by means of recommender systems [29] that can adopt (*i*) *Content-based* (based on past users' interests [21]), (*ii*) *Collaborative Filtering* (using knowledge-sharing techniques to find people similar for interests [4]), or (*iii*) (*Hybrid*) a combination of these two types of approaches [6].

In this scenario, an emerging issues consists of providing different personalization levels also considering the device exploited in accessing MCs [36]. For instance, a user accesses to a full HD video only if his/her device (e.g. a mobile device) correctly displays it and/or if its download is not expensive for him/her (e.g. by means of a wireless connection). Therefore, both in generating content-based and collaborative filtering suggestions seems necessary to consider, behind similar preferences, also the exploited devices, Internet access costs and network bandwidth.

Currently, many systems associate a *software agent* with each user's device to construct a profile by monitoring his/her choices performed on that device. Agents use such profiles to interact with the site that should compute personalized suggestions. Differently, agents can generate suggestions and/or adapt the site presentation by using a user's global profile obtained aggregating the different device profiles [22,32,34].

A promising solution for improving the recommender performance is represented by exploiting trust-based systems [37,30,42].

M. Baldoni et al. (Eds.): AI*IA 2013, LNAI 8249, pp. 409–420, 2013.
© Springer International Publishing Switzerland 2013

Trust systems can exploit information derived by direct experiences (i.e., reliability) and/or opinions of others (i.e., reputation) to trust potential partners and can integrate such measures into a single synthetic score [12,14,40]. Reputation is essential in absence of a sufficient direct knowledge to make credible a direct trust measure and its accuracy increases with the number of cooperating members but, as in a real context, malicious manipulations can happen. For such reason many reputation systems consider information coming from multiple sources in order to obtain a strength reputation mechanism as, for instance, in [33,35,41]. This issue is particularly relevant in distributed and/or hostile environments [17,20].

In this context, we propose a recommender system for dealing with MCs, called **M**ultimedia **A**gent-based **R**ecommender using **T**rust (**MART**), able to suggest MCs contacting those users that appear the most promising for (*i*) similar preferences, (*ii*) type of device used in accessing MCs and (*iii*) trustworthiness in providing suggestions. In MART each user's device is provided with a *device agent* to build a local profile storing the information about his/her behaviour on that device. A *user agent* builds a user's global profile exploited to associate the user with the partitions, grouping users mutually similar for interests. Each partition is managed by a *friend agent* that generates personalized recommendations sent to the *site agent* of the visited MC site, to build a personalized presentation with the suggested MCs without the necessity for the site of computing suggestions.

To generate collaborative filtering suggestions the friend agent chooses those users that are both the most similar to the current user and the most effective in providing suggestions. To this aim the *friend agent* adopts the Trust-Reputation-Reliability (*TRR*) model proposed in [39,40] that represents into a global trust measure, both reliability and reputation of an agent, suitably weighted on the basis of the number of performed interactions with it, taking into account the interdependencies among all the users' trust measures, differently from the other trust models. This feature is important in a MC context, since the trustworthiness of an agent can be determined only after a sufficient observation time. Moreover, MART is a distributed system where the suggestions are pre-computed off-line by the friend agents for reducing the computation costs for the MC site. As a consequence, also the partitions have to be periodically recomputed to adapt the system to the evolution of the users' behaviors. We have experimentally evaluated MART by comparing it with other recommender systems, and a significant improvement in terms of effectiveness and global efficiency has been observed.

2   The MART Architecture: Technical Details

The proposed architecture supports both users and MC sites in their MC activities. In particular, each MC belongs to a *category c* (e.g., *movies*, *songs*, *news*, etc) presents the platform, belonging to the *catalogue 𝒞* and common to all the agents. For each user a global profile is built and updated, where for each accessed MC an *interest value* is stored, representing a rough measure of the user's interest in the category which the MC belongs to, and depending on the characteristics of the used device.

More in detail, a *device agent* is associated with each user's device to monitor and build his/her profile for that device. A *user agent*, running on a server machine, collects all the user's device profiles to obtain his/her global profile. Based on the global profiles, user agents are associated with one or more partitions, each one managed by a *friend agent* that provides to: (*i*) compute similarities and trustworthiness of its users; (*ii*) store the categories offered by each affiliated site and those accessed by its users by collaborating with the *site agent*; (*iii*) pre-compute suggestions for its users based both on their preferences and type of device and those of the other affiliated (trusted) users that are similar to him/her and exploiting the same type of device.

In particular (see Figure 1), when a user *u* visits the Web site *S*, its site agent *s* receives the *Working Profile* (see below) storing *u*'s MC preferences by using that device by the *u*'s device agent. To propose attractive suggestions for *u*, the site agent interacts with all the friend agents of the partitions whose *u* belongs to and sends to them the received *u*'s profile. The friend agents, that pre-computed content-based and collaborative filtering suggestions, based on the *Working Profile*, select those most suitable for *u* and send them to *s* for being presented to *u*.

The catalogue $\mathscr{C}$ and all the Web sites are XML based and contain MCs belonging to the categories stored in $\mathscr{C}$. Furthermore, each MC can be associated, by one or more *hyperlinks*, with other MCs stored in the same site. A MC hyperlink between *a* and *b* is represented by a pair (a, b) that can be clicked by a user for accessing *b* coming from *a*. Below, all the agents will be described in detail.

2.1 The Device Agent

Each device exploited by a user *u* is associated with a device agent monitoring *u*'s Web sessions in order to identify his/her MC interests and preferences. The device agent builds and updates its *Device Profile* consisting of the *Working Profile* (*WP*) and the *Local User Profile* (*LUP*). In turn, the *Working Profile* consists of:

– *Constraint Data* (*CD*), storing device parameters and user's restrictions in accessing (i.e, downloading) MCs (e.g., video codec unsupported). Based on his/her preferences, the user *u* sets the following parameters: (*i*) *x*, max. number of categories belonging to the site considered in a Web session; (*ii*) *y*, max. number of MCs for

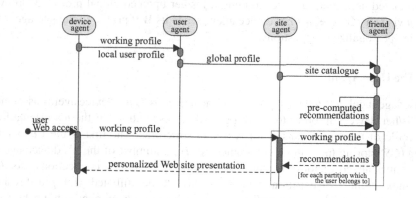

Fig. 1. The behaviour of MART

category of interest considered in the suggestions; (iii) z, max. number of agents for each partition which u belongs to that have to consider in collaborative filtering suggestions; (iv) k, a coefficient, ranging in $[0..1]$, weighting the number of similar vs trust agents. Note that only similar (resp. trusted) users will be considered in generating collaborative filtering suggestions if $k = 1$ (resp. 0).

– *Preference Data (PD)*, a set of preference weights, one for each category $c \in \mathscr{C}$, referred to the device and stored as a tuple $\langle c, w_c \rangle$, where c is a category and w_c is a real value, ranging in $[0..1]$, initially set to 1 for default. When the user desires to "a priori" address his/her agent in accessing to a specific category, he/she can assign a different value to the associated weight. In particular, if u sets w_c to 0, it means that u does not want to access a MC belonging to c on the current device.

– *System Data (SD)*, storing some parameters, used to build the Local User Profile LUP and to generate suggestions, among which the most significant are: (i) T_1, T_2, T_3: three system time thresholds (in seconds, with $T_1 \geq T_2 \geq T_3$) related to video, audio, text MC, respectively (see below); (ii) Π_u: the set of the friend agents associated with the partitions which u belongs to.

The *Local User Profile LUP* is updated based on all the hyperlinks that u clicked using the device d for accessing/downloading MCs. More in detail, LUP is a set of tuples $\langle c, IW_c, LUC_c \rangle$, where c is a category, IW_c (*Interest Weight*) is a measure (a real value set in $[0..1]$) of the u's interest in c by using d and LUC_c (*Last Update Category*) is the date of its last update. The measure IW_c of the u's interest in a MC (i.e., video, audio, text) is maximum when: (i) the time spent in accessing a MC resource is higher than T_i [28]; (ii) when the MC is downloaded. More formally, $IW_c = (IW_c + w_c \cdot \phi)/2$ for each accessed or downloaded MC belonging to c, where w_c is the preference weight associated with c, while ϕ is set to 1 if a MC belonging to c has been downloaded or accessed for a time $t > T_i$ (where T_i is the threshold time associated with the MC typology), otherwise $\phi = t/T_i$. Note that in computing the new IW_c, the current value is decreased based on the time distance (in days) between the current access/download and LUC_c by using the coefficient $(365 - \Delta)/365$ that is 0 when the last access to c is older than a year. The behavior of the device agent consists in building the u's LUP profile, by monitoring the accessed/downloaded categories, and periodically sends it to the associated user agent in order to build a his/her updated global profile. Moreover, when u visits a MC site, his/her device agent sends its WP profile to the site agent for receiving personalized suggestions.

2.2 The User Agent

The user agent collects the local profiles of the user u from the device agents associated with his/her devices in order to build a global representations of the u's orientations. The profile of this agent consists of two data structures. The first is the *Connection Setting (CS)* storing the following parameters: (i) D, number of the u's device agents; (ii) F, a normalized vector of D elements (f_i) storing the device connection costs; (iii) n, the max. number of partitions which u desires to be affiliated; (iv) q, a threshold value, ranging to $[0..1]$, used to affiliate u to a partition; (v) h, number of hours between two consecutive computation of the users' similarities. The second data structure is the

Global User Profile (GUP) storing a global representation of *u*'s interests as a list of pairs $\langle c, GIW_c \rangle$, where $c \in \mathscr{C}$ identifies an accessed/downloaded category and GIW_c is its *Global Interest Weight* computed as $GIW_c = \frac{1}{D} \cdot \sum_{g=1}^{D} f_g \cdot IW_{c,g}$. In particular, $IW_{c,g}$ is the interest weight computed for the given category *c* by the *g*-th device, $r = 1,..,D$, and f_g is the normalized device cost of the *g*-th device.

The behaviour of the User Agent of the user *u* consist of the following activities:

1. It periodically receives from each partition *i* which *u* belongs to, the global profile of each other user $j \in i$ to store them into the list L_i.
2. It periodically computes a *similarity measure* m_i, a real value ranging in $[0..1]$, as the mean of all the similarity measure s_{uj} between the global user profile of its user *u* and those of each other user *j* belonging to the same partition *i*. Each s_{uj} is computed by using the Jaccard similarity measure [13], a real number ranging in $[0..1]$, defined as the number of elements common to *u* and *j*, divided by the total number of unique elements in the profiles (i.e., $s_{uj} = |GUP_u \cap GUP_j| / |GUP_u \cup GUP_j|$). As a result, the mean similarity m_i with respect to the partition *i* is computed as $m_i = (\sum_{j \in i} s_{uj}) / |L_i|$.
3. Each *h* hours, the user agent determines *n* partitions having the highest m_i value, and for which $m_i \geq q$, in order to require the membership of *u* for such partitions by sending the global user profile *GUP* to the friend agents of the *u*'s partitions (see Figure 1). The parameters *n*, *q* and *h* are arbitrarily set by *u* and stored into his/her user profile.

In conclusion, this computation is performed in off-line mode without affecting the other users' activities. This implies that recommendations are generated on a configuration of the partitions that could not consider the last modifications.

2.3   The Friend and the Site Agents

A *site agent s* is associated with a MC site *S* and its data structure only consists of the catalogue of the categories present on the site. A *friend agent p*, associated with a partition *i* involving users similar for interests, manages the *Site Catalogue Set (SCS)*, the *Global User Profile Set (GUPS)* and the *Interest Collector (IC)*. The *SCS* stores all the site catalogues. The *GUPS* stores the global profiles of all the affiliated users. For each site *s* and user *u* that visited *S*, by using the device *d*, *IC* stores in its data section *(DS_S)* a list $DS_S[u,d]$, where each its element of $DS_S[u,d]$ represents (*i*) a category $c \in \mathscr{C}$ present in *S* and interesting for *u* and (*ii*) the interest weight IW_c. Each time that the user *u* visits a site, the associated site agent sends the *u*'s working profile *WP* (received from the *u*'s device agent) to all the agents of the partitions which *u* belongs to and obtains from them the suggestions they pre-computed for *u*.

Each *u*'s friend agent *p* computes for *u* suggestions both content-based and collaborative filtering. For the first, it builds the list *CB* of the *x* categories in the *u*'s *GUP* having the highest *GIW* measures. For the other, it compares the *u*'s profile stored in its $DS_S[u,d] \in DS_S$ with each profile $DS_S[j,d]$ of each other user *j* that visited *S* using the same device *d* exploited by *u*. The similarities between such users *u*, denoted by $D(u,j,d)$, are computed by considering all the categories $c \in C$ common to their profiles as $D(u,j,d) = \sum_{c \in DS_S[u,d] \cap DS_S[j,d]} |IW_u(c) - IW_j(c)|$, where $DS_S[u,d]$ and $DS_S[j,d]$

(i.e., $IW_u(c)$ and $IW_j(c)$) are the data sections stored in their IC (i.e., the interest weights assigned to c), respectively. Then p normalizes among them all the similarity measures and assigns a score to each agent j based on both similarity and trust, computed as $k \cdot D(u, j, d) + (1 - k) \cdot \tau$, where τ_{uj} is the trust of j assigned by u (see Section 3). Finally, the friend agent selects and stores in the list CF the first x categories most relevant of the z agents having the highest score. When a u's Web session ends, the site agent informs the u's friend agents about his/her performed choices for updating u's profile stored in $GUPS$ and the trust values of its agents for u also based on the CF list.

Then the site agent s exploits its copy of the working profile WP and the list CB (resp., CF) pre-computed from each u's friend agent to select the first y MCs (resp., the MCs not present in CB) for each of the most relevant x categories of the list. Note that the MC relevance takes into account the number of visits and downloads performed by all the site visitors. The agent s uses the selected MCs to build on-the-fly a site presentation for u compatible with the characteristics of the exploited device as shown in Figure 2, where the smartphone presentation is lighter than that of the personal computer, due to the preferences stored in the u's WP profile.

3   The Trust Model

This section deals with the trust model used in MART derived by [40]. More in detail, for each partition i, each user (resp. agent) $u \in i$ can request a suggestion r for a category $c \in \mathscr{C}$ to each other user $t \in i$. The relevance of this suggestion depends on the trust (τ) assigned by u to t, based on the reliability (ρ) due to the u's past direct interactions with t, and on the reputation (σ) that the whole agent community assigns to t.

Reliability - The reliability of t perceived by u and dealing with the category $c \in \mathscr{C}$ is assumed to take into account the level of knowledge that u has about t in the context of c. It is represented by $\rho_{ut}^c = \langle i_{ut}^c, rel_{ut}^c \rangle$, where i_{ut}^c is the number of interactions already performed by u and t and rel_{ut}^c is the *reliability value*, a real number ranging in $[0..1]$, such that 0 (resp. 1) means that t is totally unreliable (resp. reliable), computed as the number of times that u has accepted a suggestion r from t, normalized with respect to the total number of recommendations provided by t.

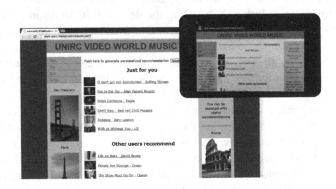

Fig. 2. A personalized site representation on a desktop PC and on a smartphone

Reputation - In the partition i, the reputation of t perceived by u for the category $c \in \mathcal{C}$, denoted by σ_{ut}^c, is computed by assuming that u requires to each user $j \neq t \in i$ the "opinion" (represented by the trust that j has in t) about the capability of t to provide good suggestions in c, suitably weighted by the trust measure τ_{uj}^c that u has of j. Both reputation and opinions are real values ranging in $[0..1]$. In this way, σ_{ut}^c is the *weighted mean* of all the trust measures τ_{jt}^c that each user j (with $j \neq u,t$) assigns to t. Formally, σ_{ut}^c is computed as $\sigma_{ut}^c = (\sum_{j \in i-\{u,t\}} \tau_{jt}^c \cdot \tau_{uj}^c)/(\sum_{j \in i-\{u,t\}} \tau_{uj}^c)$.

Trust - The trust τ_{ut}^c that a user u assigns to another user t in a given category $c \in \mathcal{C}$ combines the corresponding reliability (ρ_{ut}^c) and the reputation (σ_{ut}^c) measures, arbitrarily weighted by using a real coefficient (α_{ut}^c) ranging in $[0..1]$, in the form $\tau_{ut}^c = \alpha_{ut}^c \cdot \rho_{ut}^c + (1 - \alpha_{ut}^c) \cdot \sigma_{ut}^c$, where the coefficient α_{ut}^c should increase with the the number of interactions. To make explicit this linear dependence of α_{ut}^c from i_{ut}^c, we set $\alpha_{ut}^c = i_{ut}^c/N$, if $i_{ut}^c < N$ and 1 otherwise, where the integer N is a system threshold. This way, τ_{ut}^c is computed as:

$$\tau_{ut}^c = \alpha_{ut}^c \cdot r_{ut}^c + (1 - \alpha_{ut}^c) \cdot \frac{\sum_{j \in i-\{u,t\}} \tau_{jt}^c \cdot \tau_{uj}^c}{\sum_{j \in i-\{u,t\}} \tau_{uj}^c} \tag{1}$$

This equation, written for all the agents $u,t \in \mathcal{A}$ and for all the categories c with which i deals with, leads to a system of $m \cdot n \cdot (n-1)$ linear equations, where m is the number of available categories and n is those of the agents of i and admits only one solution (see [5], where this statement is proved).

4   Related Work

Currently a large number of recommender systems (RSs), centralized (CRSs) or distributed (DRSs), have been proposed to help users in their Web activities, often combining content-based (CB) and collaborative filtering (CF) techniques [6]. CRSs are easy to implement but limited for efficiency, fault tolerance, scalability and privacy. DRSs imply computational and data storage distribution providing RSs of scalability, fault tolerance and privacy but design and performances optimization are more complex than CRSs [25], while their complexity increase with the distribution degree [16,38]. Commonly RSs represent users' past habits in profiles [1] based on Web usage and behavioral information [19] that could be influenced by the device exploited in accessing Web resources. Besides, users can use more device typologies giving rise to different behaviors and implying the adaptivity of the RSs also with respect to the devices. Personalized CB and CF suggestions about goods, including MCs, are provided by Amazon [3], exploiting data about goods related to their profiles or those of other users, similarly for eBay [11], also using users and sellers feedbacks, and in CDNOW [8] and Dandang [9]. A CF-like CRS, designed to recommend music files, is Push!Music [15] that deals with the devices of interacting social network. Similarly to MART, the DRSs EC-XAMAS [10] and MASHA [31] consider the device in generating suggestions on the client and on the server side, respectively. MWsuggest [36] is an evolution of MASHA designed to suggest multimedia web services but without exploiting trust information as in MART. A multi-tiered distributed approach is adopted in DIRECT

[26,27] where each personal agent is specialized and runs on a different tier. DIRECT introduces advantages in terms of effectiveness, openness, privacy and security.

P2P infrastructure are also used to implement CF recommenders as (*i*) in [45], where based on similarity ranks between multimedia files locally stored in buddy tables to generate suggestions about those files that can be easily located, or (*ii*) in PEOR [18], where to suggest the most suitable MCs the personal agents use recent peers' ratings, search for nearest peers with similar preferences in order to minimizes time and computational costs. Personalized suggestions about videos are also proposed in [43] by monitoring users by means of an interactive interface and exploiting data mining and CB reasoning techniques for generating suggestions.

The Content-Boosted CF suggests unknown movies to see analysing the users' rates stored in their profiles. The CB component adopts a Bayesian classifier, while CF contributions are due to a neighbourhood-based algorithm using the Pearson correlation. To provide personalize suggestions about MCs, in [2] the user's behaviour and the exploited device are taken into account. A hybrid RS ranking music items by users and acoustic features of audio signals is presented in [46] in which a probabilistic generative model unifies CF and CB data by using the probability theory.

Some RSs exploit trust operating at the level of the profile or of the items. In [24] the authors proposed two computational models of trust easily incorporating into standard collaborative filtering frameworks realizing both these schemas and estimating trust by monitoring the accuracy of a profile at making predictions.

In synthesis, all the presented systems exploit some type of user profile in order to catch user's interests and preferences and to generate personalized suggestions of MCs potentially interesting for him/her. Moreover, some systems, similarly to MART, take into account the device currently exploited in accessing MCs and/or adapt their Web presentation to it and/or exploit trust information.

5   Experiments

This section describes some experiments devoted to measure the contribution on the suggestions quality due to the TRR trust system by comparing MART with a CF trust based system (i.e. CFT1 [24], a hybrid RS (i.e. CBCF [23]) and a hybrid RS taking into account the device in generating suggestions (i.e. MWSuggest [36]).

The experiments involved 20 MC XML-based Web sites, each one with 50 MCs belonging to a common dictionary of the categories $\mathscr{C}$ implemented by a unique XML Schema. The first 10 of such Web sites have been used to build users profiles and the other sites have been used to measure the RSs performances.

To consider a wide user population, we have exploited simulated users visiting the sites and generated their choices about the available MCs and their behaviors in requesting suggestions and opinions to other users. For each simulated user, his/her first 200 simulated choices for different MCs have been stored in a personal file as a list of tuple $\langle s,d,t \rangle$ consisting of source (s) and destination (d) MCs on the site and the associated timestamp (t). The agents exploited in the RSs have been realized under JADE (Java Agent Development Framework) [44] and JADE/LEAP [7]. In particular, the MART platform has been completely implemented and, in detail, 3 device agents (associated

Table 1. Setting of the MART parameters of the device agents

device	T_1	T_2	T_3	Π	x	y	z	k	€ per Mb
PC	180	120	120	3	2	4	3	0.5	0.01
tablet	120	90	60	3	2	4	3	0.5	0.2
smartphone	60	60	45	3	2	4	3	0.5	1.0

with a desktop PC, a tablet and a smartphone) have been exploited and their parameters are set as in Table 1. Remember that the interest for a category is assumed as "saturated" if a MC of that category is either "used" for more than T_i seconds (where i specifies the device typology) or downloaded, and such interest is linearly decreased based on its age in days from its last access. Note that the coefficient k is set to 0.5, so that CF suggestions are due in equal parts to the most similar and to the most trusted users.

The implementation of the recommender agents of the other systems is conform to their algorithms [24,23,36]. In particular, the systems CFT1 and CBCF have been adapted to work under the MC classification provided by the XML Schema. Users' agents have built their initial profiles on 10 Web sites used as training-set. For the other 10 sites, 200 triplets have been collected for each user and exploited as test-set. The experiment involved three sets $S200$, $S400$, and $S800$ containing 200, 400 and 800 users, respectively. For each user u of each given set, when u visits a Web page p, each of the four recommender systems proposes to u a set $R(p)$ of suggestions in the form of links to recommended MCs, ordered based on their computed relevance. If r is clicked (resp. not clicked) by u, than it is considered as a *true positive* (resp. *false positive*) and inserted in the set TP_u (resp. FP_u) of all the true (resp. false) positives generated for u. Finally, if u selects a link not belonging to $R(p)$, this link is considered as a *false negative* (i.e., a link that is not suggested but that is considered as relevant by the user), and it is inserted in a set FN_u. Then, we compute the standard measures called *Precision* π_u and *Recall* ρ_u for the set of the suggestions produced for u. Such measures can be interpreted by the probability that a recommended link is considered as relevant by u and that a relevant link is suggested, respectively. Formally:

$$\pi_u = \frac{|TP_u|}{|TP_u \cup FP_u|} \qquad \rho_u = \frac{|TP_u|}{|TP_u \cup FN_u|}$$

The Average Precision $\overline{\pi}$ (resp. the Average Recall $\overline{\rho}$) of each system, defined as the average of the π_u (resp. ρ_u) values for all the user u, have computed for each of the tested recommender systems and for each of the sets S_1, S_2 and S_3

The experimental results show that MART performs better than the other competitors, in terms of average precision (Figure 3 A) and average recall (Figure 3 B) for all the agent sets, and increases with the size of the sets. The advantage in term of precision (resp. recall) on the second best performer (MW-Suggest) is equal to 10,25% (resp. 12,5%) for a size of 200 users, and becomes equal to about 16% (resp. 19%) for a size of 800 users. Moreover, these results highlight that the MART performances decreases very slowly when the population size increases.

The advantage of MART with respect to CFT1 can be attributed to have considered the device in generating the suggestions, but with respect to MW-Suggest and CBCF its advantage is due to the use of the trust model in computing CF suggestions. It is show in

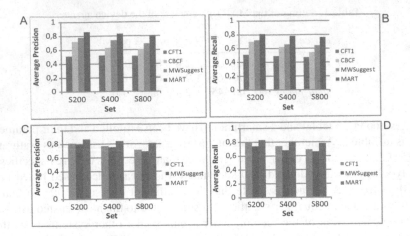

Fig. 3. A) Average precision; B)Average recall; C) CF component of the average precision; D) CF component of the average recall

Figures 3 C and D, where the only CF component of the average precision and average recall of MART, CFT1 and MW-Suggest is plotted. We can note that the advantage of MART vs MW-Suggest is almost exactly equal to the advantage that was achieved in terms of overall average precision and overall average recall, while the content-based component of the two systems perform in an almost identical way.

6 Conclusions

In this paper, we have presented MART, a fully distributed agent recommender system designed to provide users with personalized suggestions about potentially interesting MCs. To this aim, in MART, as a relevant feature for generating both content-based and collaborative filtering recommendations, it is taken into account the particular device exploited when accessing MCs. Moreover, MART also uses a trust approach when computing collaborative filtering suggestions, considering those users that are both the most similar to the current user and are resulted as the most effective in providing suggestions. Some experimental results have shown that the MART features increase the performances in terms of effectiveness of the generated collaborative filtering component of the recommendations with respect to the other tested systems. For the future, we plan to analyse the effect of the tuning parameters in the trust model, and in particular to weight the relevance of the reliability versus the reputation.

References

1. Adomaviciu, G., Tuzhilin, A.: Using Data Mining Methods to Build Customer Profiles. Computer 34, 74–82 (2001)
2. Aldua, M., Sanchez, F., Alvarez, F., Jimenez, D., Menendez, J.M., Cebrecos, C.: System Architecture for Enriched Semantic Personalized Media Search and Retrieval in the Future Media Internet. Communication Magazine 49(3), 144–151 (2011)

3. Amazon (2013), http://www.amazon.com
4. Breese, J., Heckerman, D., Kadie, C.: Empirical Analysis of Predictive Algorithms for Collaborative Filtering. In: Proc. 14th Int. Conf. on Uncertainty in Artificial Intel., pp. 43–52. Morgan Kaufmann (1998)
5. Buccafurri, F., Rosaci, D., Sarné, G.M.L., Palopoli, L.: Modeling Cooperation in Multi-Agent Communities. Cognitive Systems Research 5(3), 171–190 (2004)
6. Burke, R.D.: Hybrid Recommender Systems: Survey and Experiments. User Modeling and User-Adaptivity Interaction 12(4), 331–370 (2002)
7. Caire, G.: LEAP 3.0: User Guide, TLAB
8. CDNOW (2010), http://www.cdnow.com
9. Dandang (2013), http://www.dandang.com
10. De Meo, P., Rosaci, D., Sarné, G.M.L., Terracina, G., Ursino, D.: EC-XAMAS: Supporting e-Commerce Activities by an XML-Based Adaptive Multi-Agent System. Applied Artificial Intelligence 21(6), 529–562 (2007)
11. eBay (2013), http://www.ebay.com
12. Garruzzo, S., Rosaci, D.: The Roles of Reliability and Reputation in Competitive Multi Agent Systems. In: Meersman, R., Dillon, T.S., Herrero, P. (eds.) OTM 2010, Part I. LNCS, vol. 6426, pp. 326–339. Springer, Heidelberg (2010)
13. Greenstette, G.: Explorations in Authomatic Thesaurus Construction. Kluwer (1994)
14. Huynh, T.D., Jennings, N.R., Shadbolt, N.R.: An Integrated Trust and Reputation Model for Open Multi-Agent System. Autonmous Agent and Multi Agent Systems 13 (2006)
15. Jacobsson, M., Rost, M., Holmquist, L.H.: When Media Gets Wise: Collaborative Filtering with Mobile Media Agents. In: Proc. 11th Int. Conf. on Intel. User Interfaces, pp. 291–293. ACM (2006)
16. Jogalekar, P., Woodside, M.: Evaluating the Scalability of Distributed Systems. IEEE Trans. on Parallel Distributed Sys. 11(6), 589–603 (2000)
17. Kamvar, S., Schlosser, M., Garcia-Molina, H.: The Eigentrust Algorithm for Reputation Management in P2P Networks. In: Proc. 12th Conf. WWW, pp. 640–651. ACM (2003)
18. Kim, J.K., Kim, H.K., Cho, Y.H.: A User-Oriented Contents Recommendation System in Peer-to-Peer Architecture. Expert Sys. with App. 34(1), 300–312 (2008)
19. Kim, Y.S., Yum, B.J., Song, J., Kim, S.M.: Development of a Recommender System Based on Navigational and Behavioral Patterns of Customers in e-Commerce Sites. Expert Sys. with App. 28, 381–393 (2005)
20. Lax, G., Sarné, G.M.L.: CellTrust: A Reputation Model for C2C Commerce. Electronic Commerce Research 8(4), 193–216 (2006)
21. Lops, P., Gemmis, M., Semeraro, G.: Content-based Recommender Systems: State of the Art and Trends. In: Recommender Systems Handbook, pp. 73–105. Springer (2011)
22. Macskassy, S.A., Dayanik, A.A., Hirsh, H.: Information Valets for Intelligent Information Access. In: Proc. AAAI Spring Symp. Series on Adaptive User Interfaces. AAAI (2000)
23. Melville, P., Mooney, R.J., Nagarajan, R.: Content-boosted Collaborative Filtering for Improved Recommendations. In: Proc. 18th National Conf. on AI, pp. 187–192. AAAI (2002)
24. O'Donovan, J., Smyth, B.: Trust in Recommender Systems. In: Proc of 10th Int. Conf. on Intelligent User Interfaces, pp. 167–174. ACM (2005)
25. Olson, T.: Bootstrapping and Decentralizing Recommender Systems. Ph.D. Thesis, Dept. of Information Technology, Uppsala Univ. (2003)
26. Palopoli, L., Rosaci, D., Sarné, G.M.L.: A Multi-Tiered Recommender System Architecture for Supporting E-Commerce. In: Fortino, G., Badica, C., Malgeri, M., Unland, R. (eds.) Intelligent Distributed Computing VI. SCI, vol. 446, pp. 71–80. Springer, Heidelberg (2013)
27. Palopoli, L., Rosaci, D., Sarné, G.M.L.: Introducing Specialization in e-Commerce Recommender Systems. Concurrent Engineering: Research and Applications 21(3), 187–196 (2013)

28. Parsons, J., Ralph, P., Gallagher, K.: Using Viewing Time to Infer User Preference in Recommender Systems. In: AAAI Work. Semantic Web Personalization, pp. 52–64. AAAI (2004)

29. Pu, P., Chen, L., Hu, R.: Evaluating Recommender Systems from the User's Perspective: Survey of the State of the Art. UMUAI 22(4), 317–355 (2012)

30. Ramchurn, S.D., Huynh, D., Jennings, N.R.: Trust in Multi-Agent Systems. Knowledge Engeenering Review 19(1), 1–25 (2004)

31. Rosaci, D., Sarnè, G.M.L.: MASHA: A Multi-Agent System Handling User and Device Adaptivity of Web Sites. User Modeling User-Adaptivity Interaction 16(5), 435–462 (2006)

32. Rosaci, D., Sarnè, G.M.L.: Efficient Personalization of e-Learning Activities Using a Multi-Device Decentralized Recommender System. Computational Intelligence 26(2), 121–141 (2010)

33. Rosaci, D., Sarnè, G.M.L.: EVA: An Evolutionary Approach to Mutual Monitoring of Learning Information Agents. Applied Artificial Intelligence 25(5), 341–361 (2011)

34. Rosaci, D., Sarnè, G.M.L.: A Multi-Agent Recommender System for Supporting Device Adaptivity in e-Commerce. Journal of Intelligent Information System 38(2), 393–418 (2012)

35. Rosaci, D., Sarnè, G.M.L.: Cloning Mechanisms to Improve Agent Performances. Journal of Network and Computer Applications 36(1), 402–408 (2012)

36. Rosaci, D., Sarnè, G.M.L.: Recommending Multimedia Web Services in a Multi-Device Environment. Information Systems 38(2) (2013)

37. Rosaci, D., Sarné, G.M.L.: REBECCA: A Trust-Based Filtering to Improve Recommendations for B2C e-Commerce. In: Zavoral, F., Jung, J.J., Badica, C. (eds.) Intelligent Distributed Computing VII. SCI, vol. 511, pp. 31–36. Springer, Heidelberg (2014)

38. Rosaci, D., Sarnè, G.M.L., Garruzzo, S.: MUADDIB: A Distributed Recommender System Supporting Device Adaptivity. ACM Transansacion on Information Systems 27(4) (2009)

39. Rosaci, D., Sarné, G.M.L., Garruzzo, S.T.: An integrated Reliability-Reputation Model for Agent Societies. In: Proceedings of 12th Workshop dagli Oggetti agli Agenti, WOA 2011. CEUR Workshop Proceedings, vol. 741. CEUR-WS.org (2011)

40. Rosaci, D., Sarnè, G.M.L., Garruzzo, S.: Integrating Trust Measures in Multiagent Systems. International Journal of Intelligent Systems 27(1), 1–15 (2012)

41. Sabater, J., Sierra, C.: Reputation in Gregarious Societies. In: Proc. 5th Int. Conf. on Autonomous Agents, pp. 194–195. ACM (2001)

42. Sabater-Mir, J., Paoulucci, M.: On Open Representation and Aggregation of Social Evaluationa in Computational Trust and Reputation Models. International Journal of Approximate Reasoning 46(3), 458–483 (2007)

43. Sullivan, D.O., Smyth, B., Wilson, D.C., McDonald, K., Smeaton, A.: Improving the Quality of the Personalized Electronic Program Guide. UMUAI 14, 5–36 (2004)

44. JADE (2012), http://www.jade.tilab.org

45. Wang, J., Pouwelse, J., Lagendijk, R.L., Reinders, M.J.T.: Distributed Collaborative Filtering for Peer-to-Peer File Sharing Systems. In: Proc. 2006 ACM Symp. on Applied Computing, pp. 1026–1030. ACM (2006)

46. Yoshii, K., Goto, M., Komatani, K., Ogata, T., Okuno, H.G.: An Efficient Hybrid Music Recommender System Using an Incrementally Trainable Probabilistic Generative Model. IEEE Trans. on Audio, Speech, and Language Processing 16(2), 435–447 (2008)

Outlier Detection
with Arbitrary Probability Functions

Fabrizio Angiulli and Fabio Fassetti

DIMES Dept., University of Calabria
87036 Rende (CS), Italy
{f.angiulli,f.fassetti}@dimes.unical.it

Abstract. We consider the problem of unsupervised outlier detection in large collections of data objects when objects are modeled by means of arbitrary multidimensional probability density functions. Specifically, we present a novel definition of outlier in the context of uncertain data under the attribute level uncertainty model, according to which an uncertain object is an object that always exists but its actual value is modeled by a multivariate pdf. The notion of outlier provided is distance-based, in that an uncertain object is declared to be an outlier on the basis of the expected number of its neighbors in the data set. To the best of our knowledge this is the first work that considers the unsupervised outlier detection problem on the full feature space on data objects modeled by means of arbitrarily shaped multidimensional distribution functions. Properties that allow to reduce the number of probability distance computations are presented, together with an efficient algorithm for determining the outliers in an input uncertain data set.

1 Introduction

Traditional knowledge discovery techniques deal with feature vectors having *deterministic* values. Thus, data *uncertainty* is usually ignored in the analysis problem formulation.

However, it must be noted that *uncertainty* arises in real data in many ways, since the data may contain errors or may be only partially complete [1]. The uncertainty may result from the limitations of the equipment, indeed physical devices are often imprecise due to *measurement errors*. Another source of uncertainty are *repeated measurements*, e.g. sea surface temperature could be recorded multiple times during a day. Also, in some applications data values are *continuously changing*, as positions of mobile devices or observations associated with natural phenomena, and these quantities can be approximated by using an uncertain model.

Simply disregarding uncertainty may lead to less accurate conclusions or even inexact ones. This has raised the need for uncertain data management techniques [2], that are techniques managing data records typically represented by probability distributions [3–8]. In this work it is assumed that an *uncertain object* is an object that always exists but its actual value is uncertain and modeled by a

M. Baldoni et al. (Eds.): AI*IA 2013, LNAI 8249, pp. 421–432, 2013.

multivariate probability density function [9]. This notion of uncertain object has been extensively adopted in the literature and corresponds to the *attribute level uncertainty model* viewpoint [9].

In particular, we deal with the problem of *detecting outliers in uncertain data*. An *outlier* is an observation that differs so much from other observations as to arouse suspicion that it was generated by a different mechanism [10]. As a major contribution, we introduce a novel definition of uncertain outlier representing the generalization of the classic distance-based outlier definition [11–13] to the management of uncertain data modeled as arbitrary pdfs.

There exists several approaches to detect outliers in the certain setting [14], namely, statistical-based [15], distance-based [16], density-based [17, 18], MDEF-based [19], and others [14]. However, as far as the uncertain setting is concerned, the investigation of the problem of detecting outliers is still in its infancy. Indeed, only recently some approaches to outlier detection in uncertain data have been proposed [8, 20, 21].

The method described in [8] is a density based approach designed for uncertain objects which aims at selecting outliers in subspaces. The underlying idea of the method is to approximate the density of the data set by means of kernel density estimation and then to declare an uncertain object as an outlier if there exists a subspace such that the probability that the object lies in a sufficiently dense region of the data set is negligible. We note that, differently from our approach, in [8] the density estimate does not take directly into account the form of the pdfs associated with uncertain objects, since it is performed by using equi-bandwidth Gaussian kernels centered in the means of the object distributions. Pdfs are then taken into account to determine the objects lying in regions of low density, where the density is computed as before mentioned.

In [20] authors present a distance-based approach to detect outliers which adopts a completely different model of uncertainty than ours, that is the existential uncertainty model, according to which an uncertain object x assumes a specific value v_x with a fixed probability p_x and does not exist with probability $1 - p_x$. According to this approach, uncertain objects are not modeled by means of distribution functions, but are deterministic values that may either occur or not occur in an outcome of the dataset. Hence, although [20] deals with distance-based outliers, the scenario there considered is completely different from that considered here, and the two methods are not comparable at all.

In [21] authors assume that the space of attributes is partitioned in a space of conditioning attributes and a space of dependent attributes. An uncertain object consists of a pair (l, r), where l is a tuple on a set of conditioning attributes and r is a set of tuples on a dependent attributes, also called instances. To each instance $r_j \in r$ a measure of normality is assigned, consisting in the probability of observing r_j given that both r and l have been observed. The normality of an object is then obtained as the geometric mean of the normality of all its instances. We notice that the approach presented in [21] essentially aims at detecting the abnormal instances, that, loosely speaking, are the abnormal outcomes of the uncertain objects. Thus, the task on interest in [21] is not comparable to that considered here. Moreover, uncertain objects are modeled in a way which is completely different from that considered here.

The contributions of this work are summarized next:

- To the best of our knowledge, this is the first work that considers unsupervised outlier detection on the full feature space on data objects modeled by means of arbitrarily shaped multidimensional distribution functions;
- We introduce a novel definition of uncertain outlier representing the generalization of the classic distance-based outlier definition [11–13] to the management of uncertain data modeled as pdfs;
- Specifically, our approach consists in declaring an object as an outlier if the probability that it has at least k neighbors sufficiently close is low. Hence, it corresponds to perform a nearest neighbor density estimate on all the possible outcomes of the dataset. As such, its semantics is completely different from previously introduced unsupervised approaches for outlier detection on uncertain data;
- We provide an efficient algorithm for the computation of the uncertain distance-based outliers, which works on any domain and with any distance function.

The rest of the paper is organized as follows. Section 2 introduces the definition of uncertain outlier and some other preliminary definitions and properties. Section 3 details how to compute the outlier probability. Section 4 presents the outlier detection method. Section 5 illustrates experimental results. Finally, Section 6 concludes the work.

2 Preliminaries

2.1 Uncertain Objects

Let $(\mathbb{D}, d)$ denote a metric space, where $\mathbb{D}$ is a set, also called *domain*, and d is a *metric distance* on $\mathbb{D}$. (e.g., $\mathbb{D}$ is the d-dimensional real space $\mathbb{R}^d$ equipped with the Euclidean distance d).

A *certain object* v is an element of $\mathbb{D}$. An *uncertain object* x is a random variable having domain $\mathbb{D}$ with associated probability density function f^x, where $f^x(v)$ denotes the density of x in v.

We note that a certain object v can be regarded as an uncertain one whose associated pdf f^v is $\delta_v(t)$, where $\delta_v(t) = \delta(0)$, for $t = v$, and $\delta_v(t) = 0$, otherwise, with $\delta(t)$ denoting the Dirac delta function.

Given a set $S = \{x_1, \ldots, x_N\}$ of uncertain objects, an *outcome* I_S of S is a set $\{v_1, \ldots, v_N\}$ of certain objects such that $f^{x_i}(v_i) > 0$ $(1 \leq i \leq N)$. The pdf f^S associated with S is

$$f^S(v_1, \ldots, v_N) = \prod_{i=1}^{N} f^{x_i}(v_i).$$

Given two uncertain objects x and y, $d(x, y)$ denotes the continuous random variable representing the *distance* between x and y.

2.2   Uncertain Outliers

Given an uncertain data set **DS**, $D_k(x, \mathbf{DS} \setminus \{x\})$ (or $D_k(x)$, for short) denotes the continuous random variable representing the distance between x and its k-th nearest neighbor in **DS**.

We are now in the position of providing the definition of uncertain distance-based outlier.

Definition 1. Given an uncertain data set **DS**, an *uncertain distance-based outlier* in **DS** according to parameters k, R and $\delta \in (0, 1)$, is an uncertain object x of **DS** such that the following relationship holds:

$$Pr(D_k(x, \mathbf{DS} \setminus \{x\}) \leq R) \leq 1 - \delta.$$

That is to say, an uncertain distance-based outlier is a data set object for which the probability of having k data set objects besides itself within distance R is smaller than $1 - \delta$.

Let N be the number of objects in **DS**. In order to determine the probability $D_k(x)$, the following multi-dimensional integral has to be computed, where **DS**′ denotes the data set $\mathbf{DS} \setminus \{x\}$:

$$\int_{\mathbb{D}^N} f^x(v) \cdot f^{\mathbf{DS}'}(I_{\mathbf{DS}'}) \cdot \mathbf{I}[D_k(v, I_{\mathbf{DS}'}) \leq R] \, dI_{\mathbf{DS}'} \, dv,$$

where the function $\mathbf{I}(\cdot)$ outputs 1 if the probability of its argument is 1, and 0 otherwise.

It is clear that deciding if an object is an uncertain distance-based outlier is a difficult task, since it requires to compute an integral involving all the outcomes of the data set. However, in the following sections we will show that this challenging task can be efficiently addressed.

2.3   Further Definitions and Properties

W.l.o.g. it is assumed that each uncertain object x is associated with a finite region $\mathrm{SUP}(x)$, containing the support of x, namely the region such that $Pr(x \notin \mathrm{SUP}(x)) = 0$ holds. For example, SUP could be defined as an hyper-ball or an hyper-rectangle (e.g. the minimum bounding rectangle or MBR).

If the support of x is infinite, then $\mathrm{SUP}(x)$ is such that $Pr(x \notin \mathrm{SUP}(x)) \leq \pi$, for a fixed small value π, and the probability for x to exist outside $\mathrm{SUP}(x)$ is considered negligible. In this case the error involved in the calculation of the probability $Pr(\mathrm{d}(x, y) \leq R)$ is the square of π.

For example, assume that the data set objects x are normally distributed with mean μ_x and standard deviation σ_x. If the region $\mathrm{SUP}(x)$ is defined as $[\mu_x - 4\sigma_x, \mu_x + 4\sigma_x]$ then the probability $\pi = Pr(x \notin \mathrm{SUP}(x))$ is $\pi = 2 \cdot \Phi(-4) \approx 0.00006$ and the maximum error is $\pi^2 \approx 4 \cdot 10^{-9}$.

The *minimum distance* $mindist(x, y)$ between x and y is defined as $\min\{\mathrm{d}(u, v) : u \in \mathrm{SUP}(x) \wedge v \in \mathrm{SUP}(y)\}$, while the *maximum distance* $maxdist(x, y)$ between x and y is defined as $\max\{\mathrm{d}(u, v) : u \in \mathrm{SUP}(x) \wedge v \in \mathrm{SUP}(y)\}$.

Consider the two following definitions.

Definition 2. Let $D_k^m(x)$ denote the smallest distance for which there exists exactly k objects y of **DS** such that $maxdist(x, y) \leq D_k^m(x)$.

Definition 3. Let $d_k^m(x)$ denote the smallest distance for which there exists exactly k objects y of **DS** such that $mindist(x, y) \leq d_k^m(x)$.

The following two properties hold.

Property 1. Let x be an uncertain object for which $d_k^M(x)$ is less or equal than R. Then x is not an outlier.

As a matter of fact, if the condition of the statement of Property 1 is verified, then each outcome of x has certainly k neighbors within radius R in every outcome of the dataset.

Property 2. Let x be an uncertain object for which $d_k^m(x)$ is greater than R. Then x is an outlier.

3   Outlier Probability

In this section we show how the value of $Pr(D_k(x) \leq R)$ can be computed, for x a generic uncertain object of **DS**.

Given a certain object v and an uncertain object y, let $p_v^y(R) = Pr(d(v, y) \leq R)$ denote the cumulative density function representing the relative likelihood for the distance between objects v and y to assume value less or equal than R, that is

$$p_v^y(R) = Pr(d(v, y) \leq R) = \int_{\mathcal{B}_R(v)} f^y(u) \, du, \tag{1}$$

where $\mathcal{B}_R(v)$ denotes the hyper-ball having radius R and centered in v.

Let v be an outcome of the uncertain object x. For $k \geq 1$, it holds

$$Pr(D_k(v, \mathbf{DS} \setminus \{x\}) \leq R) =$$

$$= 1 - \left(\sum_{S \subseteq \mathbf{DS}: |S| < k} \left(\prod_{z \in S} p_v^z(R) \cdot \prod_{z \in \mathbf{DS} \setminus S} (1 - p_v^z(R)) \right) \right), \tag{2}$$

that is one minus the probability that less than k data set objects lie within distance R from v.

Thus, the probability $Pr(D_k(x) \leq R)$ can be eventually obtained as follows:

$$Pr(D_k(x) \leq R) = \int_{\mathbb{D}} f^x(v) \cdot Pr(D_k(v, \mathbf{DS} \setminus \{x\}) \leq R) \, dv. \tag{3}$$

Probability values $p_v^y(R)$ depend on the objects v and y, and on the real value R, and involve the computation of one integral with domain of integration $\mathbb{D}$ (more precisely, the hyper-ball in $\mathbb{D}$ of center v and radius R).

Algorithm 1. *UncertainDBOutlierDetector*

// *Candidate selection phase*
1: Determine the set *OutCands* of candidate outliers by detecting objects x such that $d_k^M(x) > R$
// *Candidate filtering phase*
2: Set *Outliers* to the empty set
3: **foreach** x *in OutCands* **do**
4:     **if** $d_k^m(x) > R$ **then**
5:         Insert x into *Outliers*;
6:     **else**
7:         **if** $Pr(D_k(x) \leq R) \leq 1 - \delta$ **then**
8:             Insert x into *Outliers*;
9: **return** the set *Outliers*

It is known [22] that given a function g, if m points w_1, w_2, ..., w_m are randomly selected according to a given pdf f, then the following approximation holds:

$$\int g(u)\,\mathrm{d}u \approx \frac{1}{m} \sum_{i=1}^{m} \frac{g(w_i)}{f(w_i)}. \tag{4}$$

Thus, in order to compute the value $p_v^y(R)$ reported in Equation (1), the function $g_v^y(u)$ such that $g_v^y(u) = f^y(u)$ if $d(v, u) \leq R$, and $g_v^y(u) = 0$ otherwise, can be integrated by evaluating the formula in Equation (4) with the points w_i randomly selected according to the pdf f^y. This procedure reduces to compute the relative number of sample points w_i lying at distance not greater than R from v, that is

$$p_v^y(R) = \frac{|\{w_i : \mathrm{d}(v, w_i) \leq R\}|}{m}. \tag{5}$$

Let $\mathbf{DS}_v$ be the subset of $\mathbf{DS} \setminus \{x\}$ such that $\mathbf{DS}_v = \{y \in \mathbf{DS} \setminus \{x\} : mindist(v, y) \leq R\}$. Probability $Pr(D_k(v, \mathbf{DS} \setminus \{x\}) \leq R)$ depends on probabilities $p_v^y(R)$ for the objects y belonging to the set $\mathbf{DS}_v$.

Equation (2) can be computed on the objects in the set $\mathbf{DS}_v$ by means of a dynamic programming procedure, as that reported in [23], in time $O(k \cdot |\mathbf{DS}_v|)$, that is linear both in k and in the size $|\mathbf{DS}_v|$ of $\mathbf{DS}_v$.

4   Algorithm

In this section we describe the algorithm *UncertainDBOutlierDetector* that mines the distance-based outliers in an uncertain data set $\mathbf{DS}$ of N objects.

The algorithm is reported in figure. It basically consists of two phases: the candidate selection phase, and the candidate filtering phase.

The *candidate selection phase* (see step 1 in the figure) is described next. As stated in Section 2 the uncertain objects x of $\mathbf{DS}$ satisfying $d_k^M(x) > R$ are a superset of the outliers in **DS**, a suitable set $OutCands$ of uncertain *candidate outliers* can be obtained by considering **DS** as a set of certain objects equipped with the certain distance $maxdist$. Indeed, in this case the certain outliers in **DS** are precisely the objects x of **DS** such that $d_k^M(x) > R$. Thus, an efficient certain outlier detection algorithm can be exploited to select the candidate outliers. In particular, in step 2 the state of the art certain distance-based outlier detection algorithm DOLPHIN is employed [24].

After having determined the set of candidate outliers $OutCands$, the *candidate filtering phase* (see steps 3-9 in the figure) determines the set $Outliers$ of uncertain outliers in **DS**. The objects x of $OutCands$ such that $d_k^m(x) > R$ can be safely inserted into $Outliers$ since, as stated in Section 2, they are outliers for sure. We call these objects *ready outliers*. As for the non-ready outliers x, it has to be decided whether $Pr(D_k(x) \leq R) \leq 1 - \delta$ or not. We note that the sets $\mathbf{DS}_x$ associated with the object x in $OutCands$, which are needed in order to compute the probability $Pr(D_k(x) \leq R)$, are computed during the candidate selection phase.

Next we analyze the temporal cost of the algorithm. Let d denote the cost of computing the distance d between two certain objects of $\mathbb{D}$ and also the distances $maxdist$ and $mindist$ between two uncertain objects of $\mathbb{D}$. Let N_c denote the number of outlier candidates, let m denote the number of samples employed to evaluate integrals by means of the formula in Equation (5), let $m_c \leq m$ denote the mean number of samples needed to decide if $Pr(D_k(x) \leq R) \leq 1 - \delta$, and let N_n denote the mean size of the sets $\mathbf{DS}_x$, for x a candidate outlier. The worst case cost of the candidate selection phase (step 1) corresponds to $O(\frac{k}{p}Nd)$, where $p \in (0,1]$ is an intrinsic parameter of the data set at hand (for details, we refer the reader to [24], where it is shown that the method has linear time performance with respect to the data set size). As for step 7, for each outcome v of x, computing the probability $Pr(\mathrm{d}(v,y) \leq R)$, with $y \in \mathbf{DS}_v$, costs $O(md)$, while computing the probability $Pr(D_k(v, \mathbf{DS} \setminus \{x\}) \leq R)$ costs kN_n. Thus, deciding for $Pr(D_k(x) \leq R) \leq 1 - \delta$ costs $m_cN_n(md + k)$. As a whole, the candidate filtering phase costs $O(N_cm_cN_n(md+k))$. Thus, the last phase of the algorithm is the potentially heaviest one, since it involves integral calculations needed to compute the outlier probability. In order to be practical, the algorithm must be able to select a number of outlier candidates N_c close to the value αN of expected outliers ($\alpha \in [0,1]$) and to keep as lower as possible the number $m_c \ll m$ of integral computations associated with the terms $Pr(D_k(v, \mathbf{DS} \setminus \{x\}) \leq R)$.

5   Experimental Results

In this section, we describe experimental results carried out by using the *UncertainDBOutlierDetector* algorithm. In all the experiments, we employ the following parameters: the number of neighbors was set to $k = \lfloor \varrho N \rfloor$, with $\varrho = 0.001$, the probability threshold δ is set to 0.9, m_0 to 30 and m to 100. The experiments are conducted on a Intel Xeon 2.33 GHz based machine under the Linux operating system.

5.1   Data Sets Employed

In order to evaluate the performance of the introduced method, we performed two sets of experiments.

Firstly, we considered a family of synthetic data in order to show the scalability of the approach when the number of objects and the number of dimensions of the data set increases. Secondly, we considered two families of real data sets in order to study how parameters influence the number of candidate outliers and of ready outliers.

Each data set is characterized by a parameter γ (also called *spread*) used to set the degree of uncertainty associated with data set objects.

As for the *Synthetic* data sets, a family differing for the number N of uncertain objects and the number D of attributes is generated according to the following strategy. The uncertain objects in each data set form two normally distributed separated clusters with mean $(-10, 0, \ldots, 0)$ and $(10, 0, \ldots, 0)$, respectively. Moreover, the 3‰ of the data set objects are uniformly distributed in a region lying on the hyper-plane $x = 0$ (that is to say, their first coordinate is always zero). Uncertain objects are randomly generated and may use a normal, an exponential or a uniform distribution whose spread is related to the standard deviation of the overall data by means of the parameter γ.

As far as the real data sets are concerned, we employed two 2-dimensional data sets from the R-Tree Portal[1], that are *Cities*, containing 5,922 city and village locations in Greece, and *US Point*, containing 15,206 points of populated places in USA. For both data sets, a family of uncertain data sets has been obtained as follows. An uncertain object x_i has been associated with each certain object v_i in the original data set, whose pdf $f^{x_i}(u)$ is a two-dimensional normal, uniform or exponential randomly selected distribution centered in x_i and whose spread is, again, related to the standard deviation of the overall data by means of the parameter γ.

5.2   Scalability Analysis

These experiments are intended to study the scalability of the method when the size of the dataset increases both in terms of the number of objects and in terms of the number of dimensions.

All the experiments are conducted with three different values of the parameter γ, namely 0.02, 0.05 and 0.1, in order to show how the method behaves for different levels of uncertainty.

Figure 2 on the left shows the scalability of the method with respect to the number N of data set objects. In this experiment, N has been varied between 10,000 and 1,000,000, while the number of dimensions D has been held fixed to 3. These curves show that the method has very good performances for different values of spread. In particular, the execution time is below 1,000 seconds even for one million of objects, confirming that the method is able to manage large data sets.

---

[1] See http://www.rtreeportal.org.

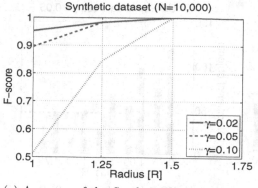

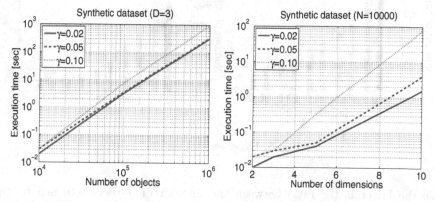

γ\R	1.0	1.25	1.5	1.75
0.02	3.3‰	3.1‰	3.0‰	3.0‰
0.05	3.7‰	3.1‰	3.0‰	3.0‰
0.1	8.7‰	4.1‰	3.0‰	3.0‰

(a) Accuracy of the *Synthetic* data set family.   (b) Outlier detected.

Fig. 1. *Synthetic* data set family

Fig. 2. Scalability with respect to the data set size and the number of dimensions for the *Synthetic* data set family

Conversely, Figure 2 on the right shows the scalability of the method with respect to the number of dimensions D of the data set. In this case the number of objects has been held fixed to 10,000. Also in this case, time performances are good. The execution time clearly increases with the dimensionality, due to the increasing cost of evaluating outcomes of the distributions, but in this experiments remained below 100 seconds even for 10-dimensional data sets, confirming that the method can be profitably employed to analyze multidimensional data.

We studied also the accuracy of the method. Figure 1a reports the F-score as a function of the radius R. The F-score is a well known measure used to evaluate the accuracy of a method. Specifically, the F-score is a combined measure of precision and recall, where the former is the ratio between the number of outliers returned by the method and the total number of objects returned by the method,

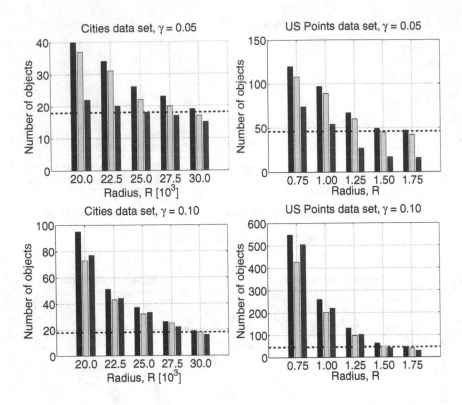

Fig. 3. Experimental results

while the latter is the ratio between the number of outliers returned by the method and the total number of outliers in the dataset. In order to compute such measures, it is assumed that the actual outliers are all and only the objects lying in the hyper-plane $x = 0$.

In the figure three curves are reported, each referred to a different value of spread. Such curves highlight the efficiency of the approach. Indeed, for values of radius above 1.5 the F-score is equal to 1 for every considered spread, and for spread equal to 0.02 and 0.05 the F-score is almost always above 0.9 for every radii considered. For the highest spread and the lowest radius considered, the F-score lowers. This situation can be understood by considering table in Figure 1b which reports the number of outliers returned by the method.

It can be seen that for spread equal to 0.1 and radius set to 1, the number of outliers returned by the method is notably larger than the actual number of outliers. However, all the clear outliers (those lying in the hyper-plane) are correctly retrieved by the method but the method start to consider as outliers the objects lying in the tails of the distributions associated with the clusters (that we have not considered as outliers).

5.3   Sensitivity Analysis

In this section, we study the behavior of the algorithm on the data sets *Cities* and *US Points* in order to study how parameters influence the number of candidate outliers and of ready outliers. Different values for the radius R and for the spread γ ($\gamma \in \{0.05, 0.1\}$) have been considered.

Figure 3 shows the result of these experiments. The first column shows results on the *Cities* data set, while the second column shows results on the *US Points* data set. The first row concerns experiments for $\gamma = 0.05$, while the second one for $\gamma = 0.10$. The diagrams report the number of candidate outliers detected at the end of the candidate selection phase (bar on the left), the actual number of outliers detected (middle bar), and the number of non-ready outliers (bar on the right). From the figure it is clear the effect of the radius on the efficiency of the method and on the number of actual outliers. The dashed line represents the number αN, with $\alpha = 0.003$. It is clear that a proper value for the radius has to be selected in order to control the actual number of outliers and, moreover, that if the radius is properly determined, the computational effort of the method results negligible (see the first two rows of the following table).

The following table shows the execution times (in seconds) of the candidate filtering phase of the algorithm.

R	Cities		R	US Points	
	$s = 0.05$	$s = 0.10$		$s = 0.05$	$s = 0.10$
30,000	0.01	0.05	1.75	0.07	0.29
27,500	0.02	0.09	1.50	0.08	0.97
25,000	0.03	0.17	1.25	0.14	9.17
22,500	0.03	0.47	1.00	0.30	18.77
20,000	0.04	2.04	0.75	0.41	64.07

Since the time sensibly increases only when the number of candidate outliers is very different from the desired one, the table confirms that by properly tuning the value of the radius the *UncertainDBOutlierDetector* algorithm is able to solve very efficiently the computationally heavy uncertain distance-based outlier detection problem.

6   Conclusions

In this work, a novel definition of uncertain outlier has been introduced to deal with multidimensional arbitrary shaped probability density functions and representing the generalization of the classic distance-based outlier definition.

Specifically, our approach corresponds to perform a nearest neighbor density estimate on all the possible outcomes of the dataset and, to the best of our knowledge, has no counterpart in the literature.

Moreover, it has been presented a method to efficiently compute the uncertain outliers, thus overcoming the difficulties raised by the introduced definition. Experiments have confirmed the effectiveness and the efficiency of the approach.

References

1. Lindley, D.: Understanding Uncertainty. Wiley-Interscience (2006)
2. Aggarwal, C., Yu, P.: A survey of uncertain data algorithms and applications. IEEE Trans. Knowl. Data Eng. 21(5), 609–623 (2009)
3. Mohri, M.: Learning from uncertain data. In: Schölkopf, B., Warmuth, M.K. (eds.) COLT/Kernel 2003. LNCS (LNAI), vol. 2777, pp. 656–670. Springer, Heidelberg (2003)
4. Ngai, W., Kao, B., Chui, C., Cheng, R., Chau, M., Yip, K.: Efficient clustering of uncertain data. In: Proc. Int. Conf. on Data Mining (ICDM), pp. 436–445 (2006)
5. Kriegel, H.P., Pfeifle, M.: Density-based clustering of uncertain data. In: Proc. Int. Conf. on Knowledge Discovery and Data Mining (KDD), pp. 672–677 (2005)
6. Ren, J., Lee, S., Chen, X., Kao, B., Cheng, R., Cheung, D.: Naive bayes classification of uncertain data. In: Proc. Int. Conf. on Data Mining (ICDM), pp. 944–949 (2009)
7. Bi, J., Zhang, T.: Support vector classification with input data uncertainty. In: Proc. Conf. on Neural Information Processing Systems (NIPS), pp. 161–168 (2004)
8. Aggarwal, C., Yu, P.: Outlier detection with uncertain data. In: Proc. Int. Conf. on Data Mining (SDM), pp. 483–493 (2008)
9. Green, T., Tannen, V.: Models for incomplete and probabilistic information. IEEE Data Eng. Bull. 29(1), 17–24 (2006)
10. Hawkins, D.: Identification of Outliers. Monographs on Applied Probability and Statistics. Chapman & Hall (May 1980)
11. Knorr, E., Ng, R., Tucakov, V.: Distance-based outlier: algorithms and applications. VLDB Journal 8(3-4), 237–253 (2000)
12. Ramaswamy, S., Rastogi, R., Shim, K.: Efficient algorithms for mining outliers from large data sets. In: Proc. Int. Conf. on Management of Data (SIGMOD), pp. 427–438 (2000)
13. Angiulli, F., Pizzuti, C.: Outlier mining in large high-dimensional data sets. IEEE Trans. Knowl. Data Eng. 2(17), 203–215 (2005)
14. Chandola, V., Banerjee, A., Kumar, V.: Anomaly detection: A survey. ACM Comput. Surv. 41(3) (2009)
15. Barnett, V., Lewis, T.: Outliers in Statistical Data. John Wiley & Sons (1994)
16. Knorr, E., Ng, R.: Algorithms for mining distance-based outliers in large datasets. In: Proc. Int. Conf. on Very Large Databases (VLDB 1998), pp. 392–403 (1998)
17. Breunig, M.M., Kriegel, H., Ng, R., Sander, J.: Lof: Identifying density-based local outliers. In: Proc. Int. Conf. on Managment of Data, SIGMOD (2000)
18. Jin, W., Tung, A., Han, J.: Mining top-n local outliers in large databases. In: Proc. ACM Int. Conf. on Knowledge Discovery and Data Mining, KDD (2001)
19. Papadimitriou, S., Kitagawa, H., Gibbons, P., Faloutsos, C.: Loci: Fast outlier detection using the local correlation integral. In: Proc. Int. Conf. on Data Enginnering (ICDE), pp. 315–326 (2003)
20. Wang, B., Xiao, G., Yu, H., Yang, X.: Distance-based outlier detection on uncertain data. In: Proc. Computer and Information Technology (CIT), pp. 293–298 (2009)
21. Jiang, B., Pei, J.: Outlier detection on uncertain data: Objects, instances, and inference. In: Proc. Int. Conf. on Data Engineering, ICDE (2011)
22. Lepage, G.: A new algorithm for adaptive multidimensional integration. Journal of Computational Physics 27 (1978)
23. Rushdi, A.M., Al-Qasimi, A.: Efficient computation of the p.m.f. and the c.d.f. of the generalized binomial distribution. Microeletron. Reliab. 34(9), 1489–1499 (1994)
24. Angiulli, F., Fassetti, F.: Dolphin: An efficient algorithm for mining distance-based outliers in very large datasets. ACM Trans. Knowl. Disc. Data 3(1), Art. 4 (2009)

Enhancing Regression Models with Spatio-temporal Indicator Additions

Annalisa Appice[1], Sonja Pravilovic[1,2],
Donato Malerba[1], and Antonietta Lanza[1]

[1] Dipartimento di Informatica, Università degli Studi di Bari Aldo Moro
via Orabona, 4 - 70126 Bari - Italy
[2] Montenegro Business School, Mediterranean University Vaka Djurovica
b.b. Podgorica - Montenegro
{annalisa.appice,sonja.pravilovic,donato.malerba,
antonietta.lanza}@uniba.it

Abstract. The task being addressed in this paper consists of trying to forecast the future value of a time series variable on a certain geographical location, based on historical data of this variable collected on both this and other locations. In general, this time series forecasting task can be performed by using machine learning models, which transform the original problem into a regression task. The target variable is the future value of the series, while the predictors are previous past values of the series up to a certain p-length time window. In this paper, we convey information on both the spatial and temporal historical data to the predictive models, with the goal of improving their forecasting ability. We build technical indicators, which are summaries of certain properties of the spatio-temporal data, grouped in the spatio-temporal clusters and use them to enhance the forecasting ability of regression models. A case study with air temperature data is presented.

1 Introduction

Spatio-temporal systems are systems that evolve over both space and time. They can be studied by means of spatio-temporal data mining techniques which discover patterns and models from data characterized by both spatial and temporal dimensions. The focus of this paper is on models for forecasting data generated by a spatio-temporal system based on past observations of the system. The most frequent approach to forecast data by using machine learning consists in transforming the original problem into a regression task, where the target variable is the future value of the series, while the predictors are previous values of the series up to a certain p-length time window. This transformation technique is usually known as "time delay embedding" [10] (also called sliding window model in data stream mining [2]). The limitation of this approach is that the location of data is overlooked, although ignoring space may lead to false conclusions about spatial relationships.

The main issue of mining spatial data is the presence of spatial autocorrelation. This autocorrelation is the property of a variable taking values at pairs

M. Baldoni et al. (Eds.): AI*IA 2013, LNAI 8249, pp. 433–444, 2013.

of locations, which are a certain distance apart (neighborhood), to be more similar (positive autocorrelation) or less similar (negative autocorrelation) than expected for randomly associated pairs of observations [1]. Goodchild [4] noted that positive autocorrelation is common in the majority of geo-physical variables, while negative autocorrelation is an artifact of poorly devised spatial units. This justifies the attention given to positive forms of spatial autocorrelation.

Spatial autocorrelation violates a basic assumption made by traditional data mining methods (included regression methods), namely the independence of observations in the training sample. The inappropriate treatment of data with spatial dependencies, where the spatial autocorrelation is ignored, can obfuscate important insights into the forecasting models [6]. On the other hand, a data mining process, which takes into account the property of spatial autocorrelation, offers several advantages which are well synthesized in [3]. It determines the strength of influence of spatial structure on the data, accounts for stationarity and heterogeneity of the autocorrelation property across space, focuses on the spatial neighborhood to exploit its effect on other neighborhoods (and vice-versa) during the data mining process. In this paper, we decide to take into account the property of positive spatial autocorrelation when forecasting geo-referenced data by assuming that the future data at any location depends not only on the recent data at the same location, but also on data at neighboring locations. In this context, our proposal may be described as an attempt to convey information on both spatial and temporal dimensions of historical data to the predictive models, with the goals of both describing prominent spatio-temporal dynamics of data and improving the forecasting ability of the predictive model. Similarly to [7], we investigate an improvement over the traditional regression strategy through the addition of technical indicators as predictor variables. These variables are summaries of data of the time series falling in a spatio-temporal neighborhood. Differently from [7], summaries are computed for spatio-temporal clusters, which are automatically discovered in data.

The paper is organized as follows. In Section 2, we describe the background of this work. In Section 3, we present the spatio-temporal clustering algorithm for the discovery of the aggregating neighborhoods. In Section 4, we describe spatio-temporal indicators added to the regression model. In Section 5, we illustrate a case study. Finally conclusions are drawn in Section 6.

2   Problem Definition and Basic Concepts

Machine learning usually transforms the problem of forecasting data of a spatio-temporal system into a regression task. We add a phase of spatio-temporal clustering to this transformation. This permits us to convey information on spatial autocorrelation, in addition to temporal information, to the predictive task.

Let (Z, K, T) be the spatio-temporal system with Z a geo-physical numeric variable, K a set of stations and T a discrete, linear time line. Each station $k \in K$ is geo-referenced with a 2D point of an Euclidean space. Each measure of Z is taken by a station of K at a discrete time point of T. A data snapshot

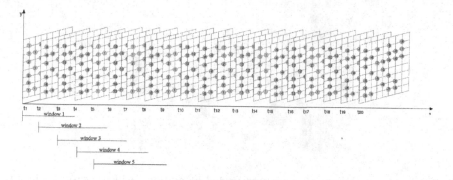

Fig. 1. Sliding window model of a geodata stream with window size $w = 4$

$data(Z, K, t)$ is the set of measures, which are collected for Z by K, at the certain time point $t \in T$. $z(t, k)$ denotes the value observed for the variable Z at the time point t in $data(Z, K, t)$. A spatio-temporal stream is an open-ended sequence of data snapshots, $data(Z, K, 1), data(Z, K, 2), \ldots, data(Z, K, t), \ldots$, with $t \in T$, which are continuously produced, at consecutive time points, by a spatio-temporal system.

The sliding window model [2] is a common model used in stream data mining to address a data mining task over the data of the recent past only. Let t be a certain time point of T. When the data snapshot, $data(Z, K, t)$, is produced in the stream, it is inserted in a window with size p, while the latest data snapshot, $data(Z, K, t - p)$, is discarded (see Figure 1). We use the sliding window model to transform the problem of forecasting data of a spatio-temporal stream into a regression task, whose input and output are reported below.

Input: (1) a spatio-temporal system (Z, K, T), a sliding window model with size p, a certain time point t; (2) a descriptive space $\mathbf{Z}$ spanned by p independent (or predictor) variables $Z(t - p + 1), Z(t - p + 2), \ldots,$ and $Z(t)$, which are variables associated to the measures of Z at the time points $t - p + 1, t - p + 2, \ldots,$ and t, respectively; (3) a target variable $Z(t + 1)$, which is associated to measures of Z at the time point $t + 1$; (4) a set D of training examples, with one example, $z(t - p + 1, k), z(t - p + 2, k), \ldots, z(t, k), z(t + 1, k)$, for each station $k \in K$, such that $(z(t - p + 1, k), z(t - p + 2, k), \ldots, z(t, k)) \in \mathbf{Z}$ and $z(t + 1, k) \in Z(t + 1)$; (5) the projection $D(\mathbf{Z})$ of D on the descriptive space $\mathbf{Z}$.

Output: (1) a spatio-temporal clustering $\mathcal{P}(C)$ of $D(\mathbf{Z})$ and (2) a spatio-temporal predictive function $f \colon \mathbf{Z} \times \mathcal{P}(C)\mathbf{Z} \mapsto Z(t + 1)$.

$\mathcal{P}(C)$ is a partition set of $D(\mathbf{Z})$. Each $C \subseteq D(\mathbf{Z})$ is a spatio-temporal cluster, which groups tuples of $D(\mathbf{Z})$ (or equivalently stations of K) such that, for each predictor variable $Z(t - i)$ (with $i = p - 1, p - 2 \ldots, 0$), there exists an overall pattern of positive autocorrelation for the values of $Z(t - i)$ measured across the set of stations grouped in C. The pattern reveals that each station of C turns out

Fig. 2. Clusters of stations that are surrounded by stations, which measure values in a close range at the specific time point

to be *surrounded* by stations of C, which measure values for Z that are *similar* to one another at the time point $t - i$ (see Figure 2).

$\mathcal{P}(C)\mathbf{Z}$ denotes an additional descriptive space with new predictor variables to convey information on the spatial autocorrelation to the predictive function. It is spanned by predictor data aggregating indicators, which are defined on the spatio-temporal clusters of $\mathcal{P}(C)$.

Several autocorrelation statistics are defined in literature (see [3] for a survey) to quantify the degree of spatial clustering (positive autocorrelation) or dispersion (negative autocorrelation) in a data snapshot. They can be eitherglobal and local measures. Measures of global autocorrelation permit the evaluation of the spatial autocorrelation and identify whether a data snapshot exhibits (or not) an "overall pattern" of regional clustering. They may suffer in looking for a pattern of spatial dependence, if the underlying process is not stationary in space, and particularly if the size of region under examination is large. Measures of local spatial autocorrelation overcome this limit by looking for "local patterns" of spatial dependence. Unlike global measures, which return one value for the entire data sample, local measures return one value for each sampled location of a variable. This value is the degree to which that location is part of a cluster.

In particular, spatial cluster analysis [9,5] broadly uses the standardized Getis and Ord local Gi*. Let k be a station, t be a certain time point, $Gi^*(t, k)$ is defined as follows:

$$Gi^*(t,k) = \left(\sum_{h \in K, h \neq k} \lambda_{hk} z(t,k) - \overline{Z}(t)\ \overline{\Lambda_k} \right) \Bigg/ \sqrt{\frac{S^2}{|K|-1} \left(|K| \sum_{h \in K, h \neq k} \lambda_{hk}^2 - \Lambda_k^2 \right)},$$

$$(1)$$

where $\overline{\Lambda_k} = \sum_{h \in K, h \neq k} \lambda_{hk}$, $\overline{Z}(t) = \dfrac{\sum_{h \in K} z(t,h)}{|K|}$ and $S^2 = \frac{1}{|K|} \left(\sum_{h \in K} z(t,h)^2 \right) - \overline{Z}(t)^2$.

λ_{hk} is a power function of the inverse distance-based weight between h and k.

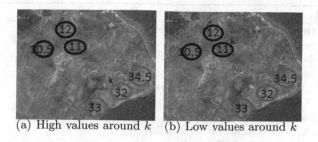

(a) High values around k (b) Low values around k

Fig. 3. Spatial-aware clusters of high/low values

The interpretation of Gi* is straightforward: significant positive values indicate clusters of high values around k (see Figure 3(a)), while significant negative values indicate clusters of low values around k (see Figure 3(b)).

It is noteworthy that the statistic Gi* is computed by using only the spatial dimension, so it ignores the time axis (i.e. t is fixed in Equation 1). For each station $k \in K$, Gi* is computed at each time point $t \in T$ and used for the spatial clustering phase. The temporal clustering phase is performed subsequently, by grouping stations which are repeatedly clustered together along a time horizon.

3 Spatio-temporal Clustering

Spatio-temporal clusters are discovered over sliding window data produced by the spatio-temporal system. A buffer consumes data snapshots as they arrive (line 2, Algorithm 1), spatial clusters are discovered and used to classify each station of the snapshot (lines 3-5, Algorithm 1).

In this phase, we divide each data snapshot into regions, which group similar values. Similarly to [9], we account for positive spatial autocorrelation of data by computing, for each station $k \in K$, $Gi^*(t,k)$, according to Equation 1 (line 3, Algorithm 1). Let $data^*(Z, K, t)$ be the snapshot of Gi* data, computed for the variable Z, the stations of K, and the time point t. We apply the K-means algorithm to $data^*(Z, K, t)$ (line 4, Algorithm 1) with $K = 2$.

The choice $K = 2$ is motivated by the traditional interpretation formulated in the spatial analysis for the Gi* statistic. In fact, Gi* permits us to discriminate between high values surrounded by high values and low values surrounded by low values. Initial centroids for the cluster search are the $25th$ percentile and the $75th$ percentile of the box plot of the data collected in the snapshot.

This spatial-aware use of K-means allows us to cluster data, based on the Gi* spatial profile, which contains both location and attribute information. Operatively, the addition of the Gi* analysis to the traditional K-means looks for the optimal spatial partitioning, which minimizes the following objective function:

$$\sum_{h \in C_1} \|Gi^*(t,h) - \overline{Gi^*(C_1)}\| + \sum_{h \in C_2} \|Gi^*(t,h) - \overline{Gi^*(C_2)}\|, \qquad (2)$$

Algorithm 1. Spatio-Temporal Clustering Procedure

Require: W, the sliding window buffer of size p
Require: Z, the geo-physic variable
Require: K, the set of measuring stations
Require: T, the time line
Ensure: $\mathcal{P}(C)$, the spatio-temporal clusters for the sliding window
1: **for all** $t \in T$ **do**
2:     acquire($data(Z, K, t)$)
3:     $data^*(Z, K, t) \leftarrow$ computeGi*($data(Z, K, t)$)
4:     $\mathcal{P}(C)^* \leftarrow$ kmeans($data^*(Z, K, t)$)
5:     classify($data(Z, K, t), \mathcal{P}(C)^*$)
6:     **if** W is FULL **then**
7:         removeFirst(W)
8:         addLast($W, data(Z, K, t)$)
9:         $\mathcal{P}(C) \leftarrow$ groupby(W)
10:    **else**
11:        addLast($W, data(Z, K, t)$)
12:    **end if**
13: **end for**

---

where $\overline{Gi^*(C)}$ denotes the mean of Gi* measures for data clustered in C. According to the interpretation of Gi* reported in literature, this permits us to identify two clusters. Each cluster groups near values, which are similar to one another throughout the map. In particular, the former groups high values of Z surrounded by other high values. The latter groups low values of Z surrounded by other low values.

When a data snapshot is completely classified in the two clusters (high-valued cluster and low-valued cluster) (line 5, Algorithm 1), the mean $\mu(C, t)$ of the real data, originally measured for Z over the snapshot and aggregated though clustering, is computed for both clusters. $\mu(C, t)$ is the class assigned to each sensor of C for the current time point t. The classified snapshot is then added to the sliding window buffer with this clustering information (lines 7-8, Algorithm 1).

Spatio-temporal clusters can be now obtained with a group-by query, which outputs stations, which are classified in the same sequence of spatial clusters over the window time horizon (line 9, Algorithm 1). It is noteworthy that the number of spatio-temporal clusters is automatically defined. It is equal to the number of distinct sequences of spatial clusters over the window (see Figure 4).

4   Spatio-temporal Indicators

Spatio-temporal clusters discovered over sliding windows are used to derive aggregating indicators, which are added as new predictors to the task of developing forecasting models. These indicators reflect different properties of data over the spatio-temporal clusters.

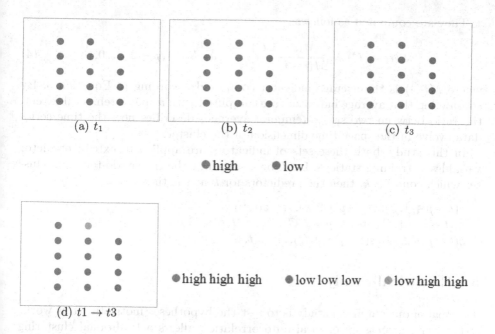

(a) t_1 (b) t_2 (c) t_3

● high   ● low

● high high high   ● low low low   ● low high high

(d) $t1 \to t3$

Fig. 4. From spatial clusters (4(a)-4(c)) to spatio-temporal clusters (4(d)). Three spatio-temporal clusters are built. The former groups the stations labeled as part of the high spatial cluster along the entire time window. The second groups the stations labeled as part of the low spatial cluster along the entire time window. Finally, the latter groups the only station which is part of the low cluster at the starting time point of the window and of the high cluster at the ending time points of the window.

Formally, let C be a spatio-temporal cluster and $\mu(C_{t-p+1}, t-p+1), \mu(C_{t-p+2}, t-p+2), \ldots, \mu(C_t, t)$ be the sequence of mean labels which are assigned to the stations of C along the window time horizon $[t - p + 1, t]$.

We calculate p average indicators, which summarize the information on the typical value of the time series within C. These indicators are calculated to provide information on the spatial dispersion of values to encompass sizes growing over time. They are computed as follows:

$$M(t - i, C) = \frac{\sum_{j=0}^{i} c \times \mu(C_{t-j}, t - j)}{c \times (i + 1)}, \quad \text{with } i = p - 1, \ldots, 0, \qquad (3)$$

where c is the number of stations grouped in C (cluster size).

We calculate $p - 1$ speed indicators, which are the ratio between two averages calculated by using two different encompassed sizes. They provide indications on the tendency of the data over time. If the value of the average with the shorter encompassing surpasses the longer average, we know that the time series is on an upwards tendency, while the opposite indicates a downwards direction.

They are computed as follows:

$$S(t-i,C) = \frac{M(t-i,C)}{M(t-i-1,C)}, \quad with \; i =, p-2 \ldots, 0, \tag{4}$$

where $M(\cdot,\cdot)$ is the average indicator computed according to Equation 3. By considering that average indicators are computed with a spatio-temporal extent, the ratio between two spatio-temporal averages describes how the time series data evolve in the space-time dimension of the cluster.

In this study, both these sets of indicators are applied to extend predictor variables of training stations. Let k be a station, C be the spatio-temporal cluster which contains k, then the predictors for k at the time t are:

$z(t-p+1,k), z(t-p+2,k), \ldots, z(t,k),$
$m(t-p+1,C), m(t-p+2,C), \ldots, m(t,C),$
$s(t-p+2,C), s(t-p+3,C), \ldots, s(t,C).$

5   Empirical Study

The goal of our empirical study is to test the hypotheses motivating this work:

(1) using a measure of local autocorrelation allows a traditional clustering algorithm to look for spatial-aware clusters of data;

(2) aggregating data over temporal-aware clusters of stations, whose data are similarly clustered in recent time, improves the predictive accuracy of regression models when forecasting the future data of a certain station.

With the goal of collecting experimental evidence to support this hypothesis, we have designed an experiment where we have used a well-known model tree learner like M5' [11] to tackle the forecasting task. We compare model trees learned by adding a spatio-temporal clustering phase and including spatio-temporal indicators as predictors, against model trees learned without spatio-temporal indicators as predictors.

5.1   Data Description

We use two spatio-temporal data collections, which record temperature measurements. In both cases, spatial dimension is defined by the latitude and longitude of measuring stations.

The TexasTemperature data collection[1] contains outdoor temperature measures hourly collected by the Texas Commission On Environment Quality in the period May 5-10, 2009. Data are obtained from 26 stations installed in Texas.

The SouthAmericaTemperature data set[2] collects monthly-mean air temperature measurements, recorded between 1960 and 1990 and interpolated over a 0.5 degree by 0.5 degree latitude/longitude grid in South America. The grid nodes are centered on 0.25 degrees for a total of 6477 stations.

---

[1] Available at http://www.tceq.state.tx.us/
[2] Available at http://climate.geog.udel.edu/~climate/html_pages/archive.html

5.2 Experimental Methodology

Spatial-aware clusters are discovered in data snapshots, spatio-temporal clusters are identified over sliding windows by considering a time delay $p = 12$ and model trees are learned with predictors defined over sliding windows. Model trees are used in the short-term forecast of the next data snapshot.

The clustering phase is validated by using Silhouette Index. This index, defined in [8], is a measure of how appropriately data have been clustered. Formally,

$$s(C) = \sum_{k \in C} \frac{b(k) - a(k)}{\max\{a(k), b(k)\}}, \tag{5}$$

where $a(k)$ is the average dissimilarity of the station k with all other stations within the same cluster and $b(k)$ is the lowest average dissimilarity of the station k with stations of any cluster different from C. The index assumes values between -1 and +1. Silhouette Index close to one means that data are appropriately clustered. Silhouette Index close to minus one means that it would be more appropriate if data were clustered in a different cluster. In this study, Silhouette Index is computed twice: for the measures of temperature, which are grouped in a cluster, as well as for the spatial coordinates of stations, which are grouped in a cluster. In both cases, the Euclidean distance is computed as a dissimilarity measure.

The regression phase is evaluated by using the root mean squared error. The lower the error, the better the forecasting ability of the regression model.

5.3 Experimental Results and Discussion

Tables 1 and 2 summarize the results of the clustering phase. Table 1 describes spatial clusters computed snapshot by snapshot. It shows: *i)* Silhouette Index, computed for validating how well the temperature measure of each station lies within its proper spatial cluster (*Silhouette(Temperature)*); *ii)* Silhouette Index, computed for validating how well the location (i.e. spatial coordinates) of each station lies within its proper spatial cluster (*Silhouette(Location)*); *iii)* the time (in msecs) spent to complete the clustering of a data snapshot. The results are averaged per snapshot. Table 2 describes spatio-temporal clusters determined window by window. It presents Silhouette Index computed for validating how well each station's location lies within its proper spatio-temporal cluster (Silhouette(Location)), the time (in msecs) spent to group by stations over the window, as well as the number of spatio-temporal clusters output by the temporal analysis. The results are averaged per window. The analysis of these results leads to several considerations.

Silhouette Index (columns 2 and 3 of Table 1, column 2 of Table 2) always assumes positive values. This validates the hypothesis that clustering data by accounting for a measure of local spatial autocorrelation appropriately clusters similar data throughout the map by picturing regions of spatial variation of data.

Table 1. Spatial clustering analysis: Silhouette Index and computation time are collected snapshot-by-snapshot and averaged on the entire stream

Data	Silhouette(Temperature)	Silhouette(Location)	time (msecs)
TexasTemperature	0.45	0.21	0.8
SouthAmericaTemperature	0.67	0.32	120.3

Table 2. Spatio-temporal clustering analysis: Silhouette Index, computation time and number of clusters are collected window-by-window and averaged on the entire stream

Data	Silhouette(Location)	time (msecs)	n. clusters
TexasTemperature	0.14	0.5	7.14
SouthAmericaTemperature	0.31	12	132

This is shown in Figure 5, where clusters, detected in the first data snapshot of SouthAmericaTemperature, are visualized (Figure 5(a)). We observe that these clusters really picture a positive spatial variation of temperature in the map by correctly identifying regions of neighboring high/low data (Figure 5(b)).

The computation time (column 4 of Table 1, column 3 of Table 2) depends on the number of stations. It should be noted that that measuring local autocorrelation, as well as running a K-means algorithm, has a cost complexity which is quadratic in the number of stations. The higher the number of stations, the more time consuming the spatial clustering phase. This explains differences observed in computation times reported for TexasTemperature (26 stations) and SouthAmericaTemperature (6477 stations), respectively.

The number of spatio-temporal clusters (column 4 of Table 2), which is built by performing the temporal analysis over spatial clustering results, is mainly lower than the number of stations. This reveals that the aggregation schema, detected by processing the local spatial autocorrelation throughout data snapshots, holds over time.

Figure 6 summarizes the results of the regression phase. We recall that the main goal of these experiments is to compare model trees, learned with the addition of spatio-temporal indicators (M5'+indicators) as predictors, with model trees, learned by using data from the same location only (M5'). This comparison is made by plotting the difference of the root mean squared error (rmse) of both solutions. By considering that the learning algorithm is M5' in both cases, differences in rmse are attributed to the presence/absence of spatio-temporal indicators. The results with TexasTemperature (Figure 6(a)) and SouthAmericaTemperature (Figure 6(b)) show that there are several snapshots where M5' uses the spatio-temporal indicators to really improve the forecasting accuracy.

Although there are cases where the addition of indicators worsens the forecasting ability of model trees, we observe that higher peaks of errors are more frequent without the addition of indicators. In general, we can say that forecasting ability has superior performances with spatio-temporal indicators.

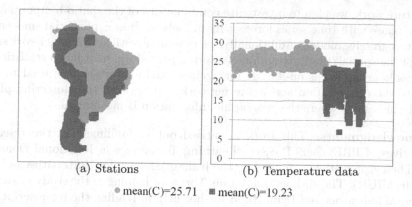

(a) Stations   (b) Temperature data

● mean(C)=25.71   ■ mean(C)=19.23

Fig. 5. SouthAmericaTemperature: spatial-aware clusters of stations, which are discovered by processing measures of local autocorrelation Gi* computed for data produced on January 1960. Blue colored values are grouped in the high cluster, purple colored values are grouped in low clusters.

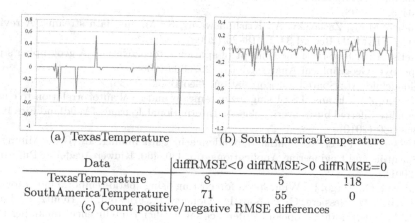

(a) TexasTemperature   (b) SouthAmericaTemperature

Data	diffRMSE<0	diffRMSE>0	diffRMSE=0
TexasTemperature	8	5	118
SouthAmericaTemperature	71	55	0

(c) Count positive/negative RMSE differences

Fig. 6. Forecasting accuracy: snapshots are represented on axis X, the difference diffRMSE (with diffRMSE=rmse(M5'+indicators)-rmse(M5)) is represented on axis Y. Negative diffRMSE denotes that M5'+indicators outperforms M5', while positive diffRMSE denotes that M5' outperforms M5'+indicators

6   Conclusions

This paper describes a two-phased methodology for forecasting data of a spatio-temporal system. A clustering phase determines spatio-temporal neighborhoods used in the computation of spatio-temporal indicators. A regression phase learns regression models by using both data from the same location and spatio-temporal indicators as predictors. We tested our proposal in the forecasting of temperature data in two distinct environmental settings. Experimental results showed that spatio-temporal information can be meaningful in this prediction task.

As future work, we plan to investigate combinations of the spatio-temporal clustering phase with time series forecasting models, such as exponential smoothing or autoregressive models. Additionally, we plan to investigate an in-network solution of the spatial/spatio-temporal clustering procedure, in order to realistically process large networks on-line. A hierarchical architecture can be used to distribute the computation across the network and perform the clustering phase locally in regions where the geographic information is meaningful.

Acknowledgments. This work is carried out in fulfillment of the research objectives of PRIN 2009 Project "Learning Techniques in Relational Domains and Their Applications", funded by the Italian Ministry for Universities and Research (MIUR). The authors thank Luis Torgo for helping in the study of spatio-temporal indicators and Lynn Rudd for her help in reading the manuscript.

References

1. Dubin, R.A.: Spatial autocorrelation: A primer. Journal of Housing Economics 7, 304–327 (1998)
2. Gaber, M.M., Zaslavsky, A., Krishnaswamy, S.: Mining data streams: A review. ACM SIGMOD Record 34(2), 18–26 (2005)
3. Getis, A.: A history of the concept of spatial autocorrelation: A geographer's perspective. Geographical Analysis 40(3), 297–309 (2008)
4. Goodchild, M.: Spatial autocorrelation. Geo Books (1986)
5. Holden, Z.A., Evans, J.S.: Using fuzzy c-means and local autocorrelation to cluster satellite-inferred burn severity classes. International Journal of Wildland Fire 19(7), 853–860 (2010)
6. LeSage, J., Pace, K.: Spatial dependence in data mining. In: Data Mining for Scientific and Engineering Applications, pp. 439–460. Kluwer Academic Publishing (2001)
7. Ohashi, O., Torgo, L.: Wind speed forecasting using spatio-temporal indicators. In: Raedt, L.D., Bessière, C., Dubois, D., Doherty, P., Frasconi, P., Heintz, F., Lucas, P.J.F. (eds.) 20th European Conference on Artificial Intelligence. Including Prestigious Applications of Artificial Intelligence (PAIS 2012), System Demonstrations Track, vol. 242, pp. 975–980. IOS Press (2012)
8. Rousseeuw, P.: Silhouettes: A graphical aid to the interpretation and validation of cluster analysis. J. Comput. Appl. Math. 20(1), 53–65 (1987)
9. Scrucca, L.: Clustering multivariate spatial data based on local measures of spatial autocorrelation. Technical Report 20, Quaderni del Dipartimento di Economia, Finanza e Statistica, Università di Perugia (2005)
10. Takens, F.: Detecting strange attractors in turbulence. Dynamical Systems and Turbulence Warwick 898(1), 366–381 (1981)
11. Wang, Y., Witten, I.: Induction of model trees for predicting continuous classes. In: Proc. Poster Papers of the European Conference on Machine Learning, pp. 128–137. Faculty of Informatics and Statistics, Prague (1997)

A Reduction-Based Approach for Solving Disjunctive Temporal Problems with Preferences

Jean-Rémi Bourguet[1], Marco Maratea[2], and Luca Pulina[1]

[1] POLCOMING - University of Sassari, Viale Mancini 5, 07100 Sassari, Italy
{boremi,lpulina}@uniss.it
[2] DIBRIS - University of Genova, Viale F. Causa 15, Genova, Italy
marco@dibris.unige.it

Abstract. Disjunctive Temporal Problems with Preferences (DTPPs) extend DTPs with piece-wise constant preference functions associated to each constraint of the form $l \leq x - y \leq u$, where x, y are (real or integer) variables, and l, u are numeric constants. The goal is to find an assignment to the variables of the problem that maximizes the sum of the preference values of satisfied DTP constraints, where such values are obtained by aggregating the preference functions of the satisfied constraints in it under a "max" semantic. The state-of-the-art approach in the field, implemented in the DTPP solver MAXILITIS, extends the approach of the DTP solver EPILITIS.

In this paper we present an alternative approach that reduces DTPPs to Maximum Satisfiability of a set of Boolean combination of constraints of the form $l \bowtie x - y \bowtie u$, $\bowtie \in \{<, \leq\}$, that extends previous work that dealt with constant preference functions only. Results obtained with the Satisfiability Modulo Theories (SMT) solver YICES on randomly generated DTPPs show that our approach is competitive to, and can be faster than, MAXILITIS.

1 Introduction

Temporal constraint networks [1] provide a convenient formal framework for representing and processing temporal knowledge. Along the years, a number of extensions to the framework have been presented to deal with, e.g. more expressive preferences. Disjunctive Temporal Problems with Preferences (DTPPs) is one of such extensions. DTPPs extend DTPs, i.e. conjunctions of disjunctions of constraints of the form $l \leq x - y \leq u$, where x, y are (real or integer) variables, and l, u are numeric constants, with piece-wise constant preference functions associated to each constraint. The goal is to find an assignment to the variables of the problem that maximizes the sum of the preference values of satisfied disjunctions of constraints (called DTP constraints), where such values are obtained by aggregating the preference functions of the satisfied constraints in it. We consider an utilitarian aggregation of such DTP constraints values, and a "max" semantic for aggregating preference values within DTP constraints: given a (candidate) solution of a DTPP, the preference value of the DTP constraint is defined to

M. Baldoni et al. (Eds.): AI*IA 2013, LNAI 8249, pp. 445–456, 2013.

be the maximum value achieved by any of its satisfied disjuncts (see, e.g. [2]). The actual state-of-the-art approach that considers such aggregation methods is implemented in the DTPP solver MAXILITIS, and is based on an extension of the DTP approach of the solver EPILITIS [3] to deal with piece-wise constant preference functions. Various other approaches have been designed in the literature to deal with DTPPs [4–6, 2], possibly relying on alternative preference aggregation methods (see, e.g. [7, 8]).

In this paper we present an alternative approach that reduces DTPPs to Maximum Satisfiability of a set of Boolean combination of constraints of the form $l \bowtie x - y \bowtie u$, where $\bowtie \in \{<, \leq\}$. At first, we have considered a very natural modeling of the problem where the generated constraints are mutually exclusive, and each is weighted by a preference value: the set is constructed in order to maximize the degree of satisfaction of the DTP constraint. Preliminary experiments report that this solution is impractical. A second solution we propose is, instead, obtained by extending previous work that dealt with constant preference functions only [9], and reduces each DTP constraint to a set of disjunction of constraints, and a non-trivial interplay among their preference values to maximize, as before, the preference value of the DTP constraint. In order to test the effectiveness of our proposal, we have randomly generated DTPPs, following the method originally developed in [7] and then employed in all other papers on DTPPs. In our framework, each problem is then represented as a Satisfiability Modulo Theory (SMT) formula, and the YICES SMT solver, that is able to deal with optimization issue, is employed[1]. An experimental analysis conducted on a wide set of benchmarks, using the same benchmarks setting already employed in past papers, shows that our approach is competitive to, and can be faster than, MAXILITIS.

The rest of the paper is structured as follows. Section 2 introduces preliminaries about DTPs, DTPPs and Maximum Satisfiability. Then, in Section 3 we present our reduction from DTPPs to Maximum Satisfiability of Boolean combination of constraints, while the experimental analysis is presented in Section 4. The paper ends by providing a discussion about the related work in Section 5 and some conclusions in Section 6.

2   Formal Background

Problems involving disjunction of temporal constraints have been introduced in [10], as an extension of the Simple Temporal Problem (STP) [1], which consists of conjunction of different constraints. The problem was referred for the first time as Disjunctive Temporal Problem (DTP) in [11], and is presented in the first subsection. The remaining subsections introduce Maximum Satisfiability of DTPs and DTPPs.

---

[1] YICES showed the best performance in [9] among of number of alternatives, and it is the only SMT solver able to cope with (Partial Weighted) Maximum Satisfiability problems.

2.1 DTP

Let $\mathcal{V}$ be a set of symbols, called *variables*. A *constraint* is an expression of the form $l \bowtie x - y \bowtie u$, where $\bowtie \in \{<, \leq\}$, $x, y \in \mathcal{V}$, and l, u are numeric constants. A *DTP constraint* is a disjunction of constraints having $\bowtie = \leq$ (equivalently seen as a disjunctively intended set of constraints). A *DTP formula*, or simply *formula*, is a conjunction of DTP constraints. A DTP constraint can be either *hard*, i.e. its satisfaction is mandatory, or *soft*, i.e. its satisfaction is not necessary but preferred, and in case of satisfaction it contributes to the generation of high quality solutions according to the aggregation methods employed and defined later. A DTP^A *constraint* is a Boolean combination of constraints.

About the semantics, let the set $\mathcal{D}$ (*domain of interpretation*) be either the set of the real numbers $\mathcal{R}$ or the set of integers $\mathcal{Z}$. An *assignment* is a total function mapping variables to $\mathcal{D}$. Let σ be an assignment and ϕ be a formula composed by hard DTP constraints only. Then, $\sigma \models \phi$ (σ satisfies *a formula ϕ*) is defined as follows

- $\sigma \models l \leq x - y \leq u$ if and only if $l \leq \sigma(x) - \sigma(y) \leq u$;
- $\sigma \models \neg\phi$ if and only if it is not the case that $\sigma \models \phi$;
- $\sigma \models (\wedge_{i=1}^{n}\phi_i)$ if and only if for each $i \in [1, n]$, $\sigma \models \phi_i$; and
- $\sigma \models (\vee_{i=1}^{n}\phi_i)$ if and only if for some $i \in [1, n]$, $\sigma \models \phi_i$.

If $\sigma \models \phi$ then σ is also called a *model* of ϕ. We also say that a formula ϕ is *satisfiable* if and only if there exists a model for ϕ. The DTP is the problem of deciding whether a formula ϕ is satisfiable or not in the given domain of interpretation $\mathcal{D}$. Notice that the satisfiability of a formula depends on $\mathcal{D}$, e.g. the formula

$$x - y > 0 \wedge x - y < 1$$

is satisfiable if $\mathcal{D}$ is $\mathcal{R}$ but unsatisfiable if $\mathcal{D}$ is $\mathcal{Z}$. However, the problems of checking satisfiability in $\mathcal{Z}$ and $\mathcal{R}$ are closely related and will be treated uniformly.

2.2 Max-DTP

Consider now a DTP^A formula ϕ consisting of hard DTP constraints and soft DTP^A constraints. Intuitively, in this case the goal is to find an assignment to the variables in ϕ that satisfies all hard DTP constraints and maximizes the sum of the weights associated to satisfied soft DTP^A constraints. The problem is called Partial Weighted Maximum Satisfiability of DTP^A, and is formally defined as a pair $\langle \phi, w \rangle$, where

1. ϕ is a DTP^A formula consisting of both hard DTP and soft DTP^A constraints, and
2. w is a function that maps DTP^A constraints to positive integer numbers.

More precisely, the goal is to find an assignment σ' for ϕ that satisfies all hard DTP constraints and maximizes the following linear objective function f

$$f = \sum_{d \in \phi, \sigma' \models d} w(d) \tag{1}$$

where d is a soft DTP^A constraint. In the following, for simplicity we will use Max-DTP to refer to the Partial Weighted Maximum Satisfiability problem of mixed DTP and DTP^A constraints as defined above.

2.3   DTPP

DTPP is an extension of DTP, and it is defined as a pair $\langle \phi, w' \rangle$, where

1. ϕ is a DTP formula consisting of both hard and soft DTP constraints, and
2. w' is a (possibly partial) function that maps constraints in soft DTP constraints to piece-wise constant preference functions.

We consider, as before, an utilitarian method for aggregating soft DTP constraints weights: the goal is now to find an assignment σ' for ϕ that (i) satisfies all hard DTP constraints, and (ii) maximizes the sum of weights associated to satisfied soft DTP constraints, i.e. maximizes the linear objective function (1).

It is left to define how weights, corresponding to preference values, are aggregated within soft DTP constraints to "define" their weights $w(d)$ in (1). In our work we consider a prominent semantic for this purpose: the *max* semantic.

Given a constraint $dc := l \leq x - y \leq u$, its preference function $w'(dc)$ is in general defined as:

$$w'(dc) : t \subseteq [l, u] \to [0, R^+]$$

mapping every feasible temporal interval t to a preference value expressing its weight. The *max* semantic [5, 2] defines the weight $w(d)$ of a satisfied soft DTP constraint d as the maximum among the possible preference values of satisfied constraints in d, i.e. given an assignment σ'

$$w(d) := max\{w'(\sigma'(x) - \sigma'(y)) : dc \in d, \sigma' \models dc\}$$

3   Reducing DTPPs to Max-DTPs

As we said before, our main idea is to reduce the problem of solving DTPPs to solving Max-DTPs. Hard DTP constraints remain unchanged in our reduction, while soft DTP constraints need special treatment. Given a soft DTP constraint d, for each constraint dc in d, let L_{dc} be a set of pairs, each pair $\langle DC, v \rangle$ being composed by (i) a set DC of pairs $(\bar{l}, \bar{u})$, representing the end points of intervals, such that $[\bar{l}, \bar{u}] \subseteq [l, u]$, and (ii) the preference value v of the constraints of the type $\bar{l} \bowtie x - y \bowtie \bar{u}$, $\bowtie \in \{\leq, <\}$, extracted from DC, where the variables x, y are obtained from the constraint name. If the preference function is a constant v', L_{dc} is composed by only one pair $\langle\{(l, u)\}, v'\rangle$, i.e. the interval $[l, u]$, representing the constraint $l \leq x - y \leq u$, and its preference value v'.

We need now to "aggregate" the preference values corresponding to different levels of the piece-wise constant functions in the various constraints in order to implement our reduction. The idea is to "merge" the pairs $\langle DC, v\rangle$, representing preference function of constraints, in the same soft DTP constraint; intuitively, this means that, if the candidate solution satisfies at least one of the constraints obtained from DC at preference value v, then a possible preference value for d is v.

More formally, consider aggregating L_{dc_1} and L_{dc_2}, coming from two constraints dc_1 and dc_2 in d, respectively. $L_{dc_1 \vee dc_2} :=$ MERGE(L_{dc_1}, L_{dc_2}) is an operator that

- contains the preference values that are in the preference functions of dc_1 or dc_2; and
- if the preference functions of dc_1 and dc_2 have a common preference value, i.e. L_{dc_1} contains a pair $\langle DC_i, v_i\rangle$, L_{dc_2} contains a pair $\langle DC_j, v_j\rangle$ and $v_i = v_j$, these pairs are merged and $L_{dc_1 \vee dc_2}$ contains a pair $\langle DC_i \cup DC_j, v_i\rangle$.

Moreover, during MERGE pairs $(\bar{l}, \bar{u})$ are attached a subscript, from which we deduce the ordered pair of variables involved in the constraint it represents.

The operator MERGE can be easily generalized to an arbitrary finite number of constraints.

Consider a soft DTP constraint

$$d := dc_1 \vee ... \vee dc_k \tag{2}$$

where $\{dc_1, ..., dc_k\}$ is the set of constraints in d.

The first attempt we considered for our reduction is to express a soft DTP constraint d using soft DTP^A constraints that force the highest preference value associated to satisfied constraints in d to be assigned as weight for d. First, we apply the operator MERGE to all the constraints in d, and related piece-wise constant preference functions, i.e. $L_d :=$ MERGE$(L_{dc_1}, ..., L_{dc_k})$.

Further, consider an ordering on the k pairs in L_d of a dc in d induced by the preference values, i.e. an ordering $\prec$ is which $\langle DC_i, v_i\rangle \prec \langle DC_j, v_j\rangle$ iff $v_i < v_j, 1 \leq i, j \leq k, i \neq j$. For simplicity, from now on we consider the pairs in L_d to be re-ordered according to $\prec$, i.e. DC_1 is the set whose v_1 is maximum among the weights in d, i.e. $v_1 > v_i, 2 \leq i \leq k$, while the set DC_k is such that $v_k < v_i, 1 \leq i \leq k - 1$.

Then, starting from L_d, d and its preference value are expressed by the following $|L_d|$ soft DTP^A constraints: for each $z = 1 ... |L_d|$

$$c_z := \wedge_{i=1}^{z-1} \neg(\vee_{p \in DC_i} dc_p) \wedge (\vee_{p \in DC_z} dc_p), w(d) = w(c_z) = v_z \tag{3}$$

where dc_p is a constraint built from the pair p (we recall that the subscript identifies the variables involved in the constraint, and in which order). The set of constraints is mutually exclusive: considering an assignment, at most one of the constraints in (3) can be satisfied, and the relative value is assigned to d. If a constraint in (3) is satisfied, this is the constraint leading to the maximum value (according to the candidate solution considered).

This is done for each soft DTP constraint in the formula.

Example 1. Consider a soft DTP constraint $dc_1 \lor dc_2$, where $dc_1 : 1 \le x - y \le 10$ and $dc_2 : 5 \le z - q \le 15$. The piece-wise constant preference function associated to dc_1 is

$$f(dc_1) = \begin{cases} 1 & 1 \le x - y \le 3 \\ 2 & 3 < x - y \le 7 \\ 1 & 7 < x - y \le 10 \end{cases} \qquad (4)$$

and can be represented with $L_{dc_1} = \{\langle\{(1,3), (7,10)\}, 1\rangle, \langle\{(3,7)\}, 2\rangle\}$. Regarding dc_2, its preference function is

$$f(dc_2) = \begin{cases} 2 & 5 \le z - q \le 8 \\ 4 & 8 < z - q \le 10 \\ 2 & 10 < z - q \le 15 \end{cases} \qquad (5)$$

represented with $L_{dc_2} = \{\langle\{(5,8), (10,15)\}, 2\rangle, \langle\{(8,10)\}, 3\rangle\}$. We now "merge" L_{dc_1} and L_{dc_2} into $L_{dc_1 \lor dc_2} := \text{MERGE}(L_{dc_1}, L_{dc_2})$ whose result is

$$\{\langle\{(1,3)_1, (7,10)_1\}, 1\rangle, \langle\{(3,7)_1, (5,8)_2, (10,15)_2\}, 2\rangle, \langle\{(8,10)_2\}, 4\rangle\}. \qquad (6)$$

Following (3), the reduction is

$$c_1 : (8 < z - q \le 10), \; w(c_1) = 4$$

$$c_2 : \neg c_1 \land ((3 < x - y \le 7) \lor (5 \le z - q \le 8) \lor (10 < z - q \le 15)), \; w(c_2) = 2$$

$$c_3 : \neg c_1 \land \neg c_2 \land (1 \le x - y \le 3 \lor 7 < x - y \le 10), \; w(c_3) = 1$$

Further note that the preference functions we have considered are characterized by having the left-most sub-interval with both bounds included, while the remaining sub-intervals have only the right bound included: to correctly reproduce the reduction from the set L, we have further assumed that with the subscript we can recognize the left-most sub-interval of each constraint.

This first reduction corresponds to a very natural way of expressing soft DTP constraints; unfortunately, preliminary experiments show that it is inefficient.

A second reduction transforms each soft DTP constraint d to $|L_d|$ soft DTP^A constraints as follows: for each $z = 1 \ldots |L_d|$

$$c'_z := \vee_{i=1}^z \vee_{p \in DC_i} dc_p \qquad (7)$$

The problem is now to define what are the weights associated to each newly defined soft DTP^A constraint, in order to reflect the semantic of our problem. In the previous reduction (2), the constraints occurred positively only once; now there can be many occurrences in the corresponding soft DTP^A constraints in (7) that influence constraints weights adaptation and definition. Our solution starts from the following fact: if the constraint $c'_{|L_d|}$ (i.e. the one that contains all constraints generated with out method) is satisfied, it is safe to consider that it contributes for at least the minimal preference value $v_{|L_d|}$, i.e. the one associated to the set $DC_{|L_d|}$, from which $c'_{|L_d|}$ is constructed. Satisfying the constraint $c'_{|L_d|-1}$ contributes for $v_{|L_d|-1} - v_{|L_d|}$, and given that a constraint c'_z implies all constraints $c'_{z'}$, $z' > z$, these two soft DTP^A constraints together contribute for $v_{|L_d|-1}$. This method is recursively applied up to the set of constraints constructable from DC_1, i.e. c'_1, whose preference value is $v_1 - v_2$ and, given that c'_1 implies all other introduced soft DTP^A constraints, satisfying c'_1 correctly corresponds to assign a weight v_1 to d.

More formally, for each $z = 1 \ldots |L_d|$

$$w(c'_z) = \begin{cases} v_{|L_d|} & z = |L_d| \\ v_z - v_{z+1} & 1 \leq z < |L_d| \end{cases} \tag{8}$$

and, given an assignment σ', $w(d) = \sum_{z \in \{1,\ldots,|L_d|\}, \sigma' \models c'_z} v_z$.

Example 2. Concerning the second reduction, the soft DTP^A constraints that express the constraint d with the preference functions in Example 1 are

$$c'_1 := 8 < z - q \leq 10, \; w(c'_1) = 2$$

$$c'_2 := c'_1 \lor (3 < x - y \leq 7 \lor 5 \leq z - q \leq 8 \lor 10 < z - q \leq 15), \; w(c'_2) = 1$$

$$c'_3 := c'_1 \lor c'_2 \lor (1 \leq x - y \leq 3 \lor 7 < x - y \leq 10), \; w(c'_3) = 1$$

Such reduction works correctly if we consider a single soft DTP constraint. However, considering a formula ϕ, given our reduction, it is possible to have repeated DTP^A constraints in the reduced formula ϕ'. In this case, intuitively, we want each single occurrence in ϕ' to count "separately", given that they take into account different contributions from different soft DTP constraints in ϕ. A solution is to consider a *single occurrence* of the resulting soft DTP^A constraint in ϕ' whose weight is the sum of the weights of the various occurrences. The same applies to the first reduction.

4   Experimental Analysis

We have implemented both reductions, and expressed the resulting formulas as SMT formulas with optimization, then solved with YICES ver. 1.0.38. A preliminary analysis showed that the first reduction is not competitive, thus our

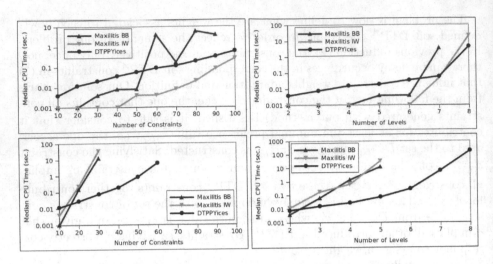

Fig. 1. Results of the evaluated solvers on random DTPPs

experimental analysis compares the performance of our second reduction, called DTPPYICES, with two versions of the MAXILITIS solver, namely MAXILITIS-IW and MAXILITIS-BB. MAXILITIS-IW (IW standing for Iterative Weakening) searches for solutions with a progressively increasing number of violated constraints; MAXILITIS-BB uses a branch and bound approach for reaching the optimal solution. Our experiments aim at comparing the considered solvers on two dimensions, namely the size of the benchmarks and the number of preference levels in the piece-wise constant preference function, as used in past papers on DTPPs, with the same parameter settings. Moreover, we also investigated the performance of the solvers in the case where the preference values of the levels of the piece-wise preference functions are randomly generated. For randomly generating the benchmarks the main parameters considered are: (i) the number k of disjuncts per DTP constraint; (ii) the number n of arithmetic variables; (iii) the number m of DTP constraints; and (iv) number l of levels in the preference functions.[2] For each tuple of values of the parameters, 25 instances have been generated.

The experiments reported in the following ran on PCs equipped with a processor Intel Core i5 running at 3.20 GHz, with 4 GB of RAM, and running GNU Linux Ubuntu 12.04. The timeout for each instance has been set to 300s.

As a first experiment, we randomly generated benchmarks by varying the total amount of constraints, with the following parameters: $k=2$, $m \in \{10, \ldots, 100\}$,

---

[2] The preference functions considered are, as in previous papers, semi-convex piece-wise constant: starting from the lower and upper bounds of the constraints, intervals corresponding to higher preference levels are randomly put within the interval of the immediate lower level, with a reduction factor, up to an highest level. For details see, e.g. [2].

n=0.8 × m, l=5, lower and upper bounds of each constraint taken in $[-50, 100]$. In this setting, the preference value of the i-th levels is i.[3]

The results obtained in the experiment are shown in Figure 1, which is organized as follows. Concerning the left-most plots, in the x axis we show the total amount of constraints, while in the right-most plots the total amount of levels of the piece-wise constant preference function is reported. In the y axis (in log scale), it is shown the related median CPU time (in seconds). MAXILITIS-BB's performance is depicted by blue triangles, MAXILITIS-IW's by using orange upside down triangles, and DTPPYICES performance is denoted by black circles. Plots in the top row have a preference value corresponding to i for the i-th preference level, while plots in the bottom row are related to random DTPPs whose preference values are randomly generated in $\{1, \ldots, 100\}$ (still ensuring to maintain the same shape for preference functions),

Looking at Figure 1, and considering the top-left plot, we can see that the median time of MAXILITIS-BB on benchmarks with 100 constraints runs into timeout. We can also see the up to $m = 80$, MAXILITIS-IW is one order of magnitude of CPU time faster that DTPPYICES, while for $m > 80$ the performance of the solvers are in the same ballpark. Now, considering the same analysis in the case where values of the preference levels are randomly generated, we can see (bottom-left plot) that the picture changes in a noticeable way. Benchmarks are harder than previously: MAXILITIS-BB and MAXILITIS-IW are not able to efficiently cope with benchmarks with $m > 30$. In this case, DTPPYICES is the best solver, and we report that it is able to deal with benchmarks up to $m = 60$.

Detailed results for these benchmarks containing, for each solver and number of constraints, the number of solved instances, and the sum of their solving CPU times, are reported in Table 1.

Our next experiment aims to evaluate the solvers by varying the number of levels in the preference functions, with the following parameters: k=2, n=24, m=30, $l \in \{2, \ldots, 8\}$, lower and upper bounds of each constraint taken in $[-50, 100]$. Top-right and bottom-right plots have the same meaning as before w.r.t. the preference functions. Looking at the top-right plot of Figure 1, we can see that MAXILITIS-IW is the best solver up to $l = 7$, while for $l = 8$, we report that DTPPYICES is faster. Also in this case MAXILITIS-BB does not efficiently deal with the most difficult benchmarks in the suite. Looking now at the plot in the bottom-right, we can see the same picture related to the bottom-left plot: the performance of both versions of MAXILITIS are very similar, while DTPPYICES is the fastest solver: the median CPU time of both MAXILITIS-BB and MAXILITIS-IW runs in timeout for $l > 5$, while DTPPYICES solves all set of benchmarks within the time limit. Along with the previous results, this reveals that MAXILITIS may have specialized techniques to deal with DTPPs whose preference values are of the first type we have analyzed.

Finally, detailed results for each number of levels are reported in Table 2.

---

[3] These benchmarks have been generated using the program provided by Michael D. Moffitt, author of MAXILITIS.

Table 1. Performance of the selected solvers on random DTPPs with different sizes. The first columns ("N") reports the total amount of variables for each pool of DTPPs, while the second one ("M") reports the total amount of constraints. It is followed by three groups of columns, and the label is the solver name. Each group is composed of four columns, reporting the total amount of instances solved within the time limit ("#") and the total CPU time in seconds ("Time") spent, both in the case of fixed preference value corresponding to the level, and randomly generated (groups "Fixed" and "Rand", respectively). In case a solver does not solve any instance, "–" is reported.

N	M	MAXILITIS-BB				MAXILITIS-IW				DTPPYICES			
		Fixed		Rand		Fixed		Rand		Fixed		Rand	
		#	Time	#	Time	#	Time	#	Time	#	Time	#	Time
8	10	25	0.01	25	0.04	25	0.01	25	0.33	25	0.12	25	0.27
16	20	25	0.16	25	16.79	25	0.01	25	62.97	25	0.33	25	0.92
24	30	25	5.75	18	593.82	25	0.02	21	922.82	25	0.55	25	4.78
32	40	24	70.12	3	85.98	25	0.05	9	946.78	25	1.06	25	27.92
40	50	22	27.58	1	108.71	25	0.29	3	334.98	25	1.68	24	278.40
48	60	17	254.24	–	–	25	4.26	–	–	25	2.80	20	225.48
56	70	21	59.28	–	–	25	0.70	–	–	25	4.04	10	900.48
64	80	16	155.92	–	–	25	7.16	–	–	25	5.91	5	110.92
72	90	17	400.46	–	–	25	15.38	–	–	25	9.01	3	225.93
80	100	12	790.15	–	–	25	58.37	–	–	25	12.31	3	144.44

Table 2. Performance of the selected solvers on random DTPPs with different levels. In column "L" we report the total amount of levels, while the rest of the table is organized similarly to Table 1.

L	MAXILITIS-BB				MAXILITIS-IW				DTPPYICES			
	Fixed		Rand		Fixed		Rand		Fixed		Rand	
	#	Time	#	Time	#	Time	#	Time	#	Time	#	Time
2	25	0.01	25	1.70	25	0.01	25	2.86	25	0.10	25	0.20
3	25	0.01	25	7.10	25	0.01	25	47.69	25	0.21	25	0.41
4	25	0.01	21	396.39	25	0.01	25	205.43	25	0.36	25	0.93
5	25	5.81	18	593.82	25	0.02	21	922.82	25	0.55	25	4.78
6	24	33.63	10	679.97	25	4.83	10	614.07	25	1.82	25	22.12
7	21	235.20	2	68.38	23	130.73	2	59.78	25	80.57	21	270.73
8	12	450.63	2	218.46	17	602.64	2	306.20	25	195.52	13	493.60

5   Related Work

MAXILITIS [2, 5], WEIGHTWATCHER [6] and ARIO [4] implement different approaches for solving DTPPs as defined in [7]. MAXILITIS is a direct extension of the DTP solver EPILITIS [3], while WEIGHTWATCHER uses an approach based on Weighted Constraints Satisfaction problems, even if the two methods are similar (as mentioned in, e.g., [6]). ARIO, instead, relies on an approach based on Mixed Logical Linear Programming (MLLP) problems. In our analysis we have used MAXILITIS because the results in, e.g. [2] clearly indicate its superior performance.

About the comparison to MAXILITIS, our solution is easy, yet efficient, and has a number of advantages w.r.t. the approach of MAXILITIS. On the modeling side, it allows to consider (with no modifications) both integer and real variables, while MAXILITIS can deal with integer variables only. Moreover, our implementation provides an unique framework for solving DTPPs, while the techniques proposed in [2] are implemented in two separate versions of MAXILITIS. Finally, our solution is modular, i.e. it is easy to rely on different back-end solvers (or, on a new version of YICES), thus taking advantages on new algorithms and tools for solving our formulas of interest.

6   Conclusions

In this paper we have introduced a general reduction-based approach for solving DTPPs, that reduces these problems to Maximum Satisfiability of DTPs as defined in the paper. An experimental analysis performed with the YICES SMT solver on randomly generated DTPPs shows that our approach is competitive to, and sometimes faster than, the specific implementations of the MAXILITIS solver. The executable of our solver can be found at

http://www.star.dist.unige.it/~marco/DTPPYices/.

Acknowledgment. The authors would like to thank Michael D. Moffitt for providing his solvers and the program for generating random benchmarks, and Bruno Dutertre for his support about YICES.

References

1. Dechter, R., Meiri, I., Pearl, J.: Temporal constraint networks. Artificial Intelligence 49(1-3), 61–95 (1991)
2. Moffitt, M.D.: On the modelling and optimization of preferences in constraint-based temporal reasoning. Artificial Intelligence 175(7-8), 1390–1409 (2011)
3. Tsamardinos, I., Pollack, M.: Efficient solution techniques for disjunctive temporal reasoning problems. Artificial Intelligence 151, 43–89 (2003)
4. Sheini, H.M., Peintner, B., Sakallah, K.A., Pollack, M.E.: On solving soft temporal constraints using SAT techniques. In: van Beek, P. (ed.) CP 2005. LNCS, vol. 3709, pp. 607–621. Springer, Heidelberg (2005)
5. Moffitt, M.D., Pollack, M.E.: Partial constraint satisfaction of disjunctive temporal problems. In: Russell, I., Markov, Z. (eds.) Proc. of the 18th International Conference of the Florida Artificial Intelligence Research Society (FLAIRS 2005), pp. 715–720. AAAI Press (2005)
6. Moffitt, M.D., Pollack, M.E.: Temporal preference optimization as weighted constraint satisfaction. In: Proc. of the 21st National Conference on Artificial Intelligence (AAAI 2006). AAAI Press (2006)
7. Peintner, B., Pollack, M.E.: Low-cost addition of preferences to DTPs and TCSPs. In: McGuinness, D.L., Ferguson, G. (eds.) Proc. of the 19th National Conference on Artificial Intelligence (AAAI 2004), pp. 723–728. AAAI Press/The MIT Press (2004)

8. Peintner, B., Moffitt, M.D., Pollack, M.E.: Solving over-constrained disjunctive temporal problems with preferences. In: Biundo, S., Myers, K.L., Rajan, K. (eds.) Proc. of the 15th International Conference on Automated Planning and Scheduling (ICAPS 2005), pp. 202–211. AAAI (2005)

9. Maratea, M., Pulina, L.: Solving disjunctive temporal problems with preferences using maximum satisfiability. AI Commununications 25(2), 137–156 (2012)

10. Stergiou, K., Koubarakis, M.: Backtracking algorithms for disjunctions of temporal constraints. In: Shrobe, H.E., Mitchell, T.M., Smith, R.G. (eds.) Proc. of the 15th National Conference on Artificial Intelligence (AAAI 1998), pp. 248–253. AAAI Press/The MIT Press (1998)

11. Armando, A., Castellini, C., Giunchiglia, E.: SAT-based procedures for temporal reasoning. In: Biundo, S., Fox, M. (eds.) ECP 1999. LNCS (LNAI), vol. 1809, pp. 97–108. Springer, Heidelberg (2000)

An Intelligent Technique for Forecasting Spatially Correlated Time Series

Sonja Pravilovic[1,2], Annalisa Appice[1], and Donato Malerba[1]

[1] Dipartimento di Informatica, Università degli Studi di Bari Aldo Moro
via Orabona, 4 - 70126 Bari - Italy
[2] Montenegro Business School, Mediterranean University Vaka Djurovica
b.b. Podgorica - Montenegro
{sonja.pravilovic,annalisa.appice,donato.malerba}@uniba.it

Abstract. The analysis of spatial autocorrelation has defined a new paradigm in ecology. Attention to spatial pattern leads to insights that would otherwise overlooked, while ignoring space may lead to false conclusions about ecological relationships. In this paper, we propose an intelligent forecasting technique, which explicitly accounts for the property of spatial autocorrelation when learning linear autoregressive models (ARIMA) of spatial correlated ecologic time series. The forecasting algorithm makes use of an autoregressive statistical technique, which achieves accurate forecasts of future data by taking into account temporal and spatial dimension of ecologic data. It uses a novel spatial-aware inference procedure, which permits to learn the autoregressive model by processing a time series in a neighborhood (spatial lags). Parameters of forecasting models are jointly learned on spatial lags of time series. Experiments with ecologic data investigate the accuracy of the proposed spatial-aware forecasting model with respect to the traditional one.

1 Introduction

In recent years, globalization has significantly accelerated the communication and exchange of experience, but has also increased the amount of data collected as a result of monitoring the spread of economic, social, environmental, atmospheric phenomena. In these circumstances, it is necessary a useful tool for analyzing data that represent the behavior of these phenomena and drawing useful knowledge from these data to predict their future behavior.

Accurate forecasts for the phenomenon behavior can make anticipation of the actions (for example, the prediction of wind speed in a region allows us to define the better strategy to maximize profit in the energy market). In the last two decades, several models of (complex) time series dynamics have been investigated in statistical analysis and machine learning. Forecasting algorithms must determine an appropriate time series model, estimate the parameters and compute the forecasts. They must be robust to unusual time series patterns, and applicable to large numbers of series without user intervention.

M. Baldoni et al. (Eds.): AI*IA 2013, LNAI 8249, pp. 457–468, 2013.

Current time series forecasting methods generally fall into two groups: methods based on statistical concepts and methods based on computational intelligence techniques such as neural networks or genetic algorithms. Statistical forecasting methods are very popular. They include exponential smoothing methods, regression methods and autoregressive integrated moving average (ARIMA) methods. In particular, ARIMA methods are known for the accuracy of forecasts. They give a forecast as a linear function of past data (or the differences of past data) and error values of the time series itself and past data of the explanatory variables. More recently, hybrid methods have started to combine more than one technique. Lee and Tong have described hybrid models which combine genetic programming with an ARIMA model [3]. In any case, the analysis of spatial autocorrelation when learning this kind of models is still overlooked.

Spatial autocorrelation is the correlation among the values of a single variable (i.e. object property) strictly attributable to the relatively close position of objects on a two-dimensional surface. It introduces a deviation from the independent observations assumption of classical statistics [5]. Intuitively, it is a property of random variables taking values, at pairs of locations at certain distance apart, that are more similar (positive autocorrelation) or less similar (negative autocorrelation) than expected for pairs of observations at randomly selected locations [11].

Positive autocorrelation is common in spatial phenomena [6], where it is justified by the first Law of Geography of Tobler [13], according to which "everything is related to everything else, but near things are more related than distant things". This means that by picturing the spatial variation of some observed variable in a map, we may observe regions where the distribution of values is smoothly continuous, with some boundaries possibly marked by sharp discontinuities. Positive spatial autocorrelation suggests that a forecasting model is smoothly continuous in a neighborhood and the inappropriate treatment of data with spatial dependencies, where the spatial autocorrelation is ignored, can obfuscate important insights in the models.

In this paper, we describe an inference procedure, which allows us to obtain a robust and widely applicable intelligent forecasting algorithm. It optimizes the traditional model ARIMA by "jointly" estimating the parameters of a forecasting model over neighboring time series (lags). This new algorithm has been implemented in the function **sARIMA** of the software R. For each spatial lag, we apply all models that are appropriate, optimize the parameters of the model in each case and select the best parameters according to a spatial-aware formulation of the Akaike's Information Criterion (AIC). The point forecasts can be produced by using the best model (with spatial-aware optimized parameters) for as many steps ahead as required.

The paper is organized as follows. In the next Section, we revise basic concepts and background of this work. In Section 3, we describe the function sARIMA in its spatial formulation. In Section 4, we report results of an empirical evaluation with real-world time series data. Finally, some conclusions and future work are drawn in Section 5.

2   Background

In this section, we briefly review the model ARIMA with the spatial extensions
of the model and describe the inference problem that we address in this paper.

2.1   ARIMA Model

ARIMA [1], which is a generalization of the model ARMA [1], is one of the most
popular and powerful forecasting technique.

The autoregressive-moving-average (ARMA) model describes a (weakly) stationary stochastic process Z in terms of two polynomials, one for the auto-
regression and the second for the moving average. Formally,

$$z(t) = c + \epsilon(t) + \sum_{i=1}^{p} \phi(i)L^i z(t) + \sum_{i=1}^{q} \sigma(i)L^i \epsilon(t), \tag{1}$$

where $L^i z(t) = z(t-i)$, p is the auto-regression order, q is the moving average
order, $\phi(i)$ and $\sigma(i)$ are the model coefficients, c is a constant, and the random
variable $\epsilon(\cdot)$ is the white noise.

L is the time lag operator, or backward shift, so that the result of L is a
shift, a translation, of the sequence, backwards. It permits to observe the same
sequence, but from i positions shifted on the left.

After choosing p and q, ARMA models are fitted by least squares regression
to the training time series in order to find the coefficients that minimize the
error term.

ARMA is extended to ARIMA, with the combination of the integrated model
using the differencing operation with order d^1. Formally,

$$(1-L)^d z(t) = c + \epsilon(t) + \sum_{i=1}^{p} \phi(i)L^i(1-L)^d z(t) + \sum_{i=1}^{q} \sigma(i)L^i \epsilon(t), \tag{2}$$

where $1-L$ is the differencing operator.

The differencing operation permits to transform a non-stationary time series
into a stationary one, which is very useful when analyzing real-life time series
data.

The selection of the ARIMA parameters (p, d, q) is not trivial [12]. A good
practice is to look for the smallest p, d and q, which provide an acceptable fit to
the data. Brockwell and Davis [2] (p. 273)] recommend using AICc (AIC with
correction c) for selecting p and q.

By following this idea, Hyndman and Khandakar [8] propose a stepwise algorithm (auto.ARIMA) to determine the parameters of the best ARIMA by
conducting a search over possible models within the order constraints provided.
The function auto.ARIMA is implemented in the software R. d is selected by
using unit-root (KPSS) tests with differenced data and original data. Once d is
selected, p and q are chosen by minimizing the AICc in a step-wise algorithm.

---

[1] The ARIMA model with $d = 0$ is the ARMA model.

2.2   Spatial Extensions of ARIMA

The model ARIMA is fitted to the time series by accounting for the temporal correlations among different data points, without any consideration of the spatial correlations. For geo-physical processes, spatial dimension is another important issue, which cannot be overlooked in real world applications. There are two major methods referring to spatial time series modeling, namely Vector ARIMA [7] and Space-Time ARIMA [9].

Vector ARIMA takes into account the interactions among multiple univariate time series by changing the variable of the univariate formula (see Equation 1) into a multivariate vector.

Space-Time ARIMA is a special case of Vector ARIMA, which emphasizes the spatial dimension in terms of "spatial correlations". Space-Time ARIMA expresses each data point at the time t and the location (x, y) as a linearly weighted combination of data lagged both in space and time.

More recently, a new spatial extension of ARIMA is described in [14], but it is formulated for trajectory data. Trajectory data are provided from mobile sensors: the measured data are the spatial positions of the moving object. This is different from the scenario investigated in this work, where fixed-to-ground sensors measure routinely data for a geo-physical numeric variable.

2.3   Paper Contribution

The innovative contribution of this paper is the definition of a new spatial extension of ARIMA, called sARIMA, which accounts for the spatial correlations of geodata when choosing the parameters of the model. sARIMA uses a variation of the Hyndman and Khandakar algorithm [8], which processes spatial lags of a geo-physical variable. It combines unit root tests on a spatial-expanded time series and minimization of a spatial-aware version of AICc to choose the best parameters for the model ARIMA. Each lag is intended as a single univariate time series geo-located by means of 2-D point coordinates (e.g. latitude and longitude).

Formally, let Z be a geo-physical variable and K be a set of randomly sampled stations, which measure Z at successive points spaced at uniform intervals in time. The station i is one-to-one associated to:

1. a primary lag $z(i)$, which is the sequence of data points, measured at i; and
2. a set of neighboring lags $zN(i)$, which are the sequences of data points measured in a neighborhood sphere of i (iN).

For each station i, the best model geo-located at i is selected according to the spatial-aware corrected Akaike's Information Criterion (AICc*). It is computed on both the primary lag $z(i)$ and the neighboring lags $zN(i)$. This permits to account for spatial autocorrelation without replacing real data with aggregated ones.

Algorithm 1. function sARIMA$(z(i), zN(i)) \mapsto arima(p(i), d(i), q(i))$

Require: $z(i)$: primary lag
Require: $zN(i)$: neighboring lags
Ensure: $arima(p(i), d(i), q(i))$: arima function fitting $z(i)$
1: sort$(zN(i))$
2: **for** $z(j) \in zN(i)$ **do**
3:    $z \leftarrow z \bullet z(j)$
4: **end for**
5: $z \leftarrow z \bullet z(i)$
6: $D \leftarrow$ Canova-Hansen(z)
7: **if** D=0 **then**
8:    $d(i) \leftarrow$ kpss(z)
9: **else**
10:    $d(i) \leftarrow$ kpss$((\text{diff}(z))$
11: **end if**
12: $(p(i), q(i)) \leftarrow stepwiseSearch(z(i), zN(i), d(i))$
13: $arima(p(i), d(i), q(i)) \leftarrow arimaFitting(z(i), p(i), d(i), q(i))$

---

3   sARIMA

The function sARIMA inputs the primary lag $z(i)$ and the set of neighboring lags $zN(i)$ and outputs the model $arima(p(i), d(i), q(i))$, which fits data of the primary lag $z(i)$ with the AR order $p(i)$, the differencing order $d(i)$ and the MA order $q(i)$. The function includes the choice of the best parameter set $(p(i), d(i), q(i))$ based on both $z(i)$ and $zN(i)$.

The algorithm (see Algorithm 1) is three stepped.

1. $d(i)$ is determined by using repeated KPSS tests on the time series obtained by sequencing the neighboring lags and the primary lag in a single time series.
2. $p(i)$ and $q(i)$ are chosen by minimizing the spatial-aware AICc after differencing the data of the lags $d(i)$ times. Similarly to [8], the algorithm uses a stepwise search to traverse the model space of the combinations of p and q.
3. The function $arima$ (see Equation 2) is fitted to $z(i)$ with $p(i)$, $d(i)$ and $q(i)$.

Details of these steps are reported in the followings.

3.1   Choosing d [lines 1-11, Algorithm 1]

The neighborhood $zN(i)$ is sorted in descending order with respect to the distance from i (line 1, Algorithm 1). For each lag pair $z(k), z(j) \in zN(i)$, $z(j)$ precedes $z(k)$ iff $distance(j, i) > distance(k, i)$.

The neighboring lags of $zN(i)$, ordered by distance, and the primary lag $z(i)$ are dealt as consecutive seasons of a spatial-aware seasonal time series z (see Example 1). By following this idea, a multi-lag time series z can be composed by sequencing these lags from the more distant lag to the less distant one from i (lines 2-5, Algorithm 1).

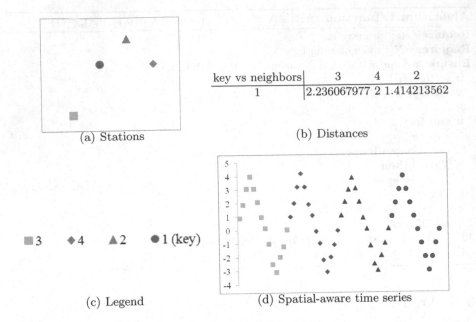

key vs neighbors	3	4	2
1	2.236067977	2	1.414213562

(a) Stations   (b) Distances

(c) Legend   (d) Spatial-aware time series

Fig. 1. Spatial-aware time series generated by sequencing the neighbor time series ordered by descendant distance and the key time series

Example 1. Let us consider the stations reported in Figure 1(a). The circle denotes the key station, while the square, triangle and diamond denote its neighbor stations. These neighbors are ordered according to their Euclidean distances from the key station. Euclidean distances are reported in Figure 1(b). The spatial-aware time series, built by sequencing the time series of these neighbors, is composed of four segment time series, one for each neighbor (see Figure 1(d)). The older segment is the time series of the the more distant neighbor (square station), while the newer segment is the time series of the the less distant neighbor (circle station).

The extended Canova-Hansen test [4] (line 6, Algorithm 1) is used to test whether the seasonal pattern of z changes sufficiently over the neighborhood $N(i)$ to warrant a seasonal unit root, or whether a stable seasonal pattern modeled using fixed dummy variables is more appropriate. By following the interpretation reported in [8], if a stable seasonal pattern is selected (i.e., the null hypothesis is not rejected) then $d(i)$ can be selected on the original data, otherwise $d(i)$ is determined on the seasonally differenced data.

The differencing order $d(i)$ is chosen based on the successive KPSS unit-root tests [10] for the stationarity of the original data or seasonally differencing data (lines 7-11, Algorithm 1). The null hypothesis is that the series is stationary around a deterministic trend. The alternate hypothesis is that the series is difference-stationary. The series is expressed as the sum of deterministic trend, random walk, and stationary error, and the test is the LM test of the hypothesis that the random walk has zero variance.

Initially, $d(i) = 0$. If the test result is significant, we increment $d(i)$ and test the differenced data for a unit root; and so on. We stop this procedure when we obtain the first insignificant result and output $d(i)$.

3.2   Choosing p and q [line 12, Algorithm 1]

The stepwise algorithm to traverse the model space $\theta = (p, q)$ is that described in [8]. It is three-stepped.

Step 1 The best initial model is selected from the following four models:

1. ARIMA(2,d(i),2),
2. ARIMA(0,d(i),0),
3. ARIMA(1,d(i),0),
4. ARIMA(0,d(i),1).

Similarly to [8], the best initial model can be selected via the AICc information criterion (the lower AICc, the best model). The AICc is that defined as follows:

$$AICc(i, \theta) = L^*(z(i), \theta) + \frac{2k(k+1)}{n - k - 1}, \tag{3}$$

where k is the number of parameters in θ; n is the length of the lag $z(i)$ and $L^*(\cdot)$ is the maximum likelihood estimate of θ on the initial states $z(i)$.

Differently from [8], we introduce a spatially weighted formulation of AICc (AICc*), which allows us to jointly estimate the degree of fitness of θ to both the primary lag $z(i)$ and the multiple lags of $zN(i)$, simultaneously. Equation 3 is reformulated as follows:

$$AICc^*(i, \theta) = \frac{L^*(z(i), \theta) + \sum\limits_{z(j) \in zN(i)} w(i, j) L^*(z(j), \theta)}{1 + \sum\limits_{j \in N(i)} w(ij)} + \frac{2k(k+1)}{n - k - 1}, \tag{4}$$

where each $w(j)$ weights neighboring lags. We use the Gaussian kernel to compute these weights:

$$w_{ij} = \begin{cases} e^{(-0.5 d_{ij}^2 / r^2)} & \text{if } d_{ij} \leq r \\ 0 & \text{otherwise} \end{cases}, \tag{5}$$

where r is the radius of the neighborhood sphere and d_{ij} is the Euclidean spatial distance between i and j. The closer j to i, the higher the weight w_{ij}.

The initial model, which minimizes the AICc* amongst all of the starting models, is selected.

Step 2 Variations on the current model are considered. This is done by varying $p(i)$ and/or $q(i)$ from the current model by ± 1. The best model (with lower AIC*) considered so far (either the current model, or one of these variations) becomes the new current model.

Step 3 Repeat Step 2 until no better model (with lower AICc*) can be found.

3.3   Estimating ϕ and σ [line 13, Algorithm 1]

The model with the best estimated parameters $p(i), d(i), q(i)$ is fitted to $z(i)$ by least square regression. The model coefficients ϕ and σ are output. They permit to produce point forecasts for i for as many steps ahead as required.

4   A Case Study

The function sARIMA is implemented in R. In the next subsections, we analyze forecasts produced with sARIMA in an ecological case study.

4.1   Data and Experimental Set-up

We consider data hourly collected by the Texas Commission On Environment Quality (TCEQ) in the time period May 5-19, 2009. Data are obtained from 26 stations (see Figure 2(a)) installed in Texas (http://www.tceq.state.tx.us/).

Stations measure ozone rate, wind speed and temperature. Spatial dimensions are the latitude and longitude of the transmitting stations. We use data in the time period May 5-18, 2009 as training set, data on May 19, 2009 as testing set.

sARIMA is run with the neighborhood radius equal to the maximum distance between nearest neighbor stations, that is,

$$r = \max_{i \in K} \min_{j \in K} (distance(i, j)), \tag{6}$$

where distance($\cdot$) is the Euclidean distance. Table 1 collects the number of neighbors per station and the average correlation between key time series and neighbor time series. Correlation is computed on the training set and averaged on the neighborhood. The higher correlation is observed for the temperature time series. Each station is surrounded by 4.576 neighbor stations in average. Examples of key time series and neighboring time series, which are considered in this study, are shown in Figures 2(b)-2(d).

For the comparative sake, we compare sARIMA to auto.ARIMA. In both cases, we consider the root mean squared error (rmse) to evaluate accuracy of forecasts produced for the testing data of each station.

Table 1. Texas data description: neighborhood size (column 2) and avg correlation (columns 3-5)between key time series and neighboring time series in the training set

station	neighborhood size	Wind Speed	Ozone	Temperature
1	7	0.7980497	0.7385494	0.7893623
2	3	0.8438085	0.5701544	0.9562436
3	8	0.7822543	0.7334237	0.9090337
4	1	0.7617608	0.9041180	0.9880963
5	4	0.7518271	0.8265032	0.9312384
6	8	0.6305772	0.7194442	0.8942110
7	4	0.8032067	0.8386280	0.9298381
8	2	0.6328674	0.8580214	0.9078054
9	5	0.6802551	0.7486119	0.8942556
10	2	0.4619874	0.6471419	0.8564103
11	1	0.7617608	0.9041180	0.9880963
12	3	0.6941623	0.6889025	0.8957058
13	1	0.4549211	0.5382396	0.7546217
14	2	0.3903626	0.3689820	0.7242230
15	5	0.6994974	0.7743656	0.8942556
16	8	0.8060648	0.7522188	0.9174505
17	8	0.7648444	0.5973738	0.8980630
18	4	0.4579509	0.7524598	0.8418243
19	2	0.6103565	0.7903118	0.8385152
20	4	0.7453410	0.4208787	0.8623596
21	7	0.6910777	0.6900899	0.8851737
22	7	0.7868560	0.7190797	0.9254311
23	9	0.6930988	0.4793439	0.8838042
24	7	0.7844528	0.7470145	0.8981438
25	5	0.6174906	0.6914104	0.8296501
26	2	0.8165041	0.7950793	0.9324957
Avg.	4.576	0.689282154	0.703633246	0.885627242
Dev.St	2.544	0.125967968	0.137421848	0.062495225

4.2   Results and Discussion

Table 2 collects the root mean squared error (RMSE) of the forecasts produced for the testing day and the analysis of statistical significance tests (pairwise Wilcoxon signed rank test) comparing squared residuals of the paired test time-series. Both results are collected per station. Table 3 reports the number of stations where sARIMA performs (statistically) better (equal or worse) than auto.ARIMA.

These results show that the inference procedure presented, by accounting for spatial autocorrelation, can, in several cases, optimize the choice of the parameters for the computation of the model ARIMA. The number of stations where sARIMA outperforms or, at worst, performs equally well than auto.ARIMA is always greater than the number of stations where auto.ARIMA outperforms sARIMA.

Table 2. auto.ARIMA (aARIMA) vs sARIMA (RMSE). W (columns 4, 7 and 10) denotes the result of the pairwise Wilcoxon signed rank test comparing auto.ARIMA and sARIMA. (+) means that sARIMA is better than auto.ARIMA (i.e. WT + > WT-), (-) means that auto.ARIMA is better than sARIMA (i.e. WT+ < WT-), (=) means that both algorithms perform equally well (i.e. WT+ = WT-). (++) and (–) denote results in case H_0 (hypothesis of equal performance) is rejected at the 0.05 significance level.

Station	Wind Speed			Temperature			Ozone		
	aARIMA	sARIMA	W	aARIMA	sARIMA	W	aARIMA	sARIMA	W
1	1.82	1.82	=	14.58	8.72	++	23.17	25.54	-
2	4.17	2.71	++	4.03	5.98	–	12.90	15.77	–
3	2.49	3.35	–	17.16	8.56	++	20.30	20.30	=
4	1.68	1.66	+	1.92	2.62	–	15.86	15.86	=
5	2.81	2.77	+	11.54	11.32	+	40.32	21.70	++
6	2.71	3.65	–	4.32	9.31	–	23.55	23.59	-
7	2.05	0.74	++	10.00	9.84	+	31.68	30.50	+
8	3.01	3.01	=	7.78	6.27	++	47.69	21.145	++
9	1.47	1.10	+	4.68	9.57	–	16.64	20.48	-
10	2.22	2.21	+	14.20	6.25	++	23.60	23.67	-
11	2.90	1.64	++	12.99	2.73	++	9.52	9.52	=
12	3.57	3.57	+	4.03	5.75	–	24.83	25.01	-
13	2.53	2.53	=	14.90	14.90	=	24.55	24.55	=
14	3.38	1.87	++	6.80	2.92	++	13.94	12.10	+
15	0.47	0.47	=	11.12	11.26	=	3.79	3.98	-
16	2.41	4.46	–	9.41	7.86	+	29.86	29.86	=
17	3.31	3.85	-	14.51	7.30	++	16.32	16.32	=
18	1.98	2.00	-	5.62	5.52	=	28.20	28.16	+
19	2.31	2.30	+	7.42	7.44	=	33.28	19.49	++
20	2.00	1.99	+	10.43	10.72	=	24.63	24.83	-
21	2.19	2.17	+	6.94	5.39	+	19.20	19.14	+
22	4.16	2.24	++	7.13	10.39	–	25.18	25.60	-
23	3.98	3.99	-	3.99	7.25	–	18.27	23.13	-
24	2.17	2.19	-	9.09	4.52	++	26.71	15.46	++
25	4.01	3.90	+	23.14	12.90	++	15.20	23.90	-
26	3.70	3.22	++	4.80	7.95	–	24.87	25.08	+
Avg	2.676	2.521		9.333	7.823		22.853	20.954	

Table 3. auto.ARIMA vs sARIMA: + (++) is the number of stations where sARIMA (statistically) outperforms auto.ARIMA, - (–) is the number of stations where auto.ARIMA statistically) outperforms sARIMA, (=) is the number of stations where the two functions perform statisitcally equally

Variable	+	++	-	–	=
Wind Speed	9	6	4	3	4
Temperature	4	9	0	8	5
Ozone	5	4	10	1	6

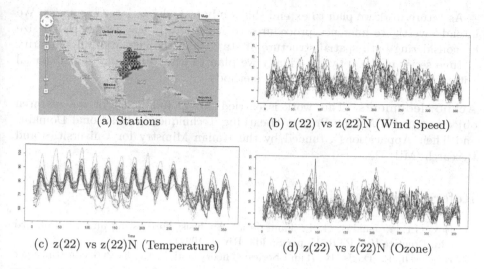

(a) Stations

(b) z(22) vs z(22)N (Wind Speed)

(c) z(22) vs z(22)N (Temperature)

(d) z(22) vs z(22)N (Ozone)

Fig. 2. 2(a) Twenty-six stations, which hourly collected ozone, temperature and wind speed in Texas in the time period May 5-19, 2009. 2(b)-2(d) The time series of wind speed, temperature, ozone measures, which are hourly acquired by both the station 22 (blue colored time series) and the stations in the neighborhood of the station 22

This conclusion is also supported by forecasts produced for both temperature and ozone, where correlation between each key time series and neighbor time series is greater in average (see average statistics reported in Table 1).

Spatial autocorrelation is looked for in a neighborhood sphere. The radius of the sphere is globally determined on the entire network. Future work aims at extending this study by investigating a new mechanism to automatically determine local radius based on the density of stations in a local area and the degree of similarity between time series falling in the candidate neighborhood.

5   Conclusions

In this paper, we have investigated the problem of choosing parameters of a forecasting model for spatially correlated time series data. In particular, we define a new spatial extension of the forecasting model ARIMA. By taking into account spatial autocorrelation of data, we define a function, called sARIMA which jointly infers parameters of a model ARIMA for neighboring time series.

A case study investigates the viability of the presented algorithm in a real-world forecasting application. We consider measurements of ozone, wind speed and temperature hourly collected by twenty six stations in Texas. We compare sARIMA to the state of art function auto.ARIMA. Both are implemented in the software R. Experiments show that sARIMA, in general, outperforms (or performs equally well than) auto.ARIMA by proving that the analysis of spatial autocorrelation can permit a more robust choice of the forecasting parameters in several cases.

As future work we plan to extend this study to multi-variate time series. We intend investigate inference procedures to learn locally the neighborhood size by considering both spatial structure of stations in the network and similarity of time series in the network. Finally we plan to extend the analysis of spatial autocorrelation to hybrid forecasting methods.

Acknowledgments. This work is carried out in fulfillment of the research objectives of PRIN 2009 Project "Learning Techniques in Relational Domains and Their Applications", funded by the Italian Ministry for Universities and Research (MIUR).

References

1. Box, G.E.P., Jenkins, G.M.: Time Series Analysis: Forecasting and Control, 3rd edn. Prentice Hall PTR, Upper Saddle River (1994)
2. Brockwell, P., Davis, R.: Time Series: Theory and Methods, 2nd edn. Springer (2009)
3. Lee, Y.C., Tong, L.: Forecasting time series using a methodology based on autoregressive integrated moving average and genetic programming. Knowledge-Based Systems (24), 66–72 (2011)
4. Canova, F., Hansen, B.: Are seasonal patterns constant over time? a test for seasonal stability. Journal of Business and Economic Statistics (13), 237–252 (1995)
5. Dubin, R.A.: Spatial autocorrelation: A primer. Journal of Housing Economics 7, 304–327 (1998)
6. Goodchild, M.: Spatial autocorrelation. Geo Books (1986)
7. B.H.: Introduction to Multiple Time Series Analysis. Springer (1993)
8. Hyndman, R., Khandakar, Y.: Automatic time series forecasting: The forecast package for r. Journal of Statistical Software 26(3) (2008)
9. Kamarianakis, Y., Prastacos, P.: Space-time modeling of traffic flow. Comput. Geosci. 31(2), 119–133 (2005)
10. Kwiatkowski, D., Phillips, P., Schmidt, P., Shin, Y.: Testing the null hypothesis of stationarity against the alternative of a unit root. Journal of Econometrics (54), 159–178 (1992)
11. Moran, P.: Notes on continuous stochastic phenomena. Biometrika 37(1-2), 17–23 (1950)
12. Sershenfeld, N.A., Weigend, A.S.G.: The future of time series. In: Gershenfeld, A.N., Weigen, A.S. (eds.) Time Series Prediction: Forecasting the Future and Understanding the Past, pp. 1–70 (1993)
13. Tobler, W.: A computer movie simulating urban growth in the Detroit region. Economic Geography 46(2), 234–240 (1970)
14. Yan, Z.: Traj-arima: a spatial-time series model for network-constrained trajectory. In: Geers, D.G., Timpf, S. (eds.) Proceedings of the Second International Workshop on Computational Transportation Science, pp. 11–16. ACM (2010)

Reasoning-Based Techniques for Dealing with Incomplete Business Process Execution Traces

Piergiorgio Bertoli[1], Chiara Di Francescomarino[2],
Mauro Dragoni[2], and Chiara Ghidini[2]

[1] SayService, Trento, Italy
bertoli@sayservice.it
[2] FBK—IRST, Trento, Italy*
{dfmchiara,dragoni,ghidini}@fbk.eu

Abstract. The growing adoption of IT systems to support business activities, and the consequent capability to monitor the actual execution of business processes, has brought to the diffusion of business analysis monitoring (BAM) tools, and of reasoning services standing on top of them. However, in the majority of real settings, due to the different degrees of abstraction and to information hiding, the IT-level monitoring of a process execution may only bring incomplete information concerning the process-level activities and associated artifacts. This may hinder the ability to reason about process instances and executions, and must be coped with. This paper presents a novel reasoning-based approach to recover missing information about process executions, relying on a logical formulation in terms of a satisfiability problem. Ongoing experiments show encouraging results.

1 Introduction

The growing adoption of IT systems to support business activities, and the consequent capability to monitor the actual execution of business processes, has brought to the diffusion of useful business analysis monitoring (BAM) tools, and of reasoning services standing on top of them (see, e.g., Engineering's eBAM, Microsoft's BAM suite in BizTalk, Oracle's BAM, Polymita, IBM's BAM component of WebSphere, just to name a few). These tools and services allow business analysts to observe the evolution of ongoing processes as well as to perform statistical analysis on past executions, and offer the opportunity to identify misalignments between the design and realization of processes, as well as design bottlenecks and hence points of possible improvement.

However, business activity monitoring must deal with a significant difficulty which none of the current approaches, to the best of our knowledge, has tackled: namely, that the observation of process executions may bring, in a great number of cases, only partial information, in terms of which process activities have been executed and what data or artifacts they produced. In fact, the partiality of monitoring information in real settings comes from two undeniable sources. First, the definitions of business processes often integrate activities that are not supported by IT systems, due either to their

---

* This work is supported by "ProMo - A Collaborative Agile Approach to Model and Monitor Service-Based Business Processes", funded by the Operational Programme "Fondo Europeo di Sviluppo Regionale (FESR) 2007-2013 of the Province of Trento, Italy.

M. Baldoni et al. (Eds.): AI*IA 2013, LNAI 8249, pp. 469–480, 2013.
© Springer International Publishing Switzerland 2013

nature (e.g. they consist of human interactions) or to the high level of abstraction of the description, detached from the implementation. Indeed, reflecting this, widely adopted business process modeling formalisms such as BPMN allow for a clear distinction between e.g. IT-supported and human-realized activities. Second, IT systems may expose only partial information about their workings and the data they produce, hence preventing a complete execution tracing.

In this paper, we tackle the problem of reconstructing information about incomplete business process execution traces, proposing and implementing a novel reasoning-based method that leverages on model and domain knowledge. This work brings the following contributions: (i) a novel, reasoning-based method to recover missing information about process executions; (ii) a realization of this approach as an effective prototype that encompasses the adoption and integration of a satisfiability solving engine as part of its architecture. The proposed method is based on the automated derivation of a set of logical rules out of a business process model, consisting of a BPMN diagram enriched with information about the lifecycle of output data objects. For every actual incomplete execution trace, our reasoning approach checks its compliance with the model according to a notion which will be formally given later in the paper, and produces the possible completions for the missing information.

The paper is structured as follows: Section 2 provides an explanatory example which will be used throughout the paper; Section 3 describes technically the problem and our approach to it; and Section 4 and Section 5 present the SAT encoding of our problem and the tool implementing the described approach, respectively. Finally, Section 6 and Section 7 discuss related and future work.

2   Problem and Running Example

The problem that we want to tackle in this paper is how to reconstruct information of incomplete business process execution traces, given some knowledge about the business process model (which describes the behavior of the execution trace) and about its data flow, i.e., how data evolve throughout the process activities. The input to our problem consists of an instance-independent portion, that comprises the business process model and the domain specific (data-flow) knowledge, and an instance-specific portion related to the input trace.

As a simple explanatory example of the problem we want to solve, consider the process described in Figure 1. The process takes inspiration from the procedure for the generation of the Italian fiscal code: the registration module is created (CRM), the personal details of the subject added (APD) and, before adding the fiscal code, according to whether the fiscal code is for a foreigner or for a newborn, the passport/stay permit information ($APPD$) or the parents' data ($APARD$) are added, respectively.

The picture, also reports for each activity the corresponding data-flow, i.e., the *data object* that it shows as output. Data objects are generic data structures, organized into *fields*, and the domain specific knowledge describes how activities affect their content by *creating* them or writing certain fields as their *output*. In detail, in the picture, fields in bold represent fields *created*/filled for the first time by the current activity, while others are fields created by other activities and which can be *shown as output* by the current one.

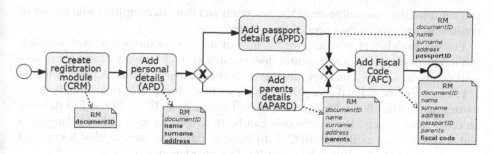

Fig. 1. A simple process for the generation of the Italian fiscal code

Now, suppose, that a run-time monitoring system has been able to trace only some knowledge about the execution of this process, by logging only some activities and the attached documents[1]. For example, let us assume we only have the following partial trace describing the first and the last executed activities and the associated data objects with output fields:

$CRM \{RM.documentID\}$
$AFC \{RM.documentID, RM.name, RM.surname, RM.address, RM.parents, RM.fiscalCode\}$

Exploiting the available knowledge about the process model and the manipulated data, is it possible (and how), to reconstruct the missing information of the partial trace? For instance, in the example, being aware of the process control flow, the fact that the AFC activity has been executed suggests that (i) the APD activity has also been executed; and (ii) either the $APPD$ or the $APARD$ activity has been executed. By looking at the output of the AFC activity, it is possible to understand which among these two alternative paths has been taken. Indeed, the occurrence of the field $RM.parents$ in the AFC activity output suggests that the execution (partially) described by the current trace should have passed through the $APARD$ activity, which is the activity responsible of the creation of the $RM.parents$ field.

While, in this simple example inferring how to fill "gaps" in the trace is quite trivial, in real-world cases, reconstructing missing information is often very complex. In this paper we show how to encode real-world problems in order to be able to automatically reconstruct a partial trace (if it is compliant) or, alternatively, to assess its non-compliance. The output that we expect is hence either (a) the notification that the partial trace is inconsistent with the process model and the domain knowledge, or (b) a set of traces that complete the input partial trace (partially or in full).

3   Approach

To solve the above mentioned problem, our approach is based on a logical formulation of the available knowledge and on the adoption of deduction techniques. In detail, we formally encode the process model and the domain knowledge in terms of logical formulae and we formulate the problem as the problem of deducing all the logical models

---

[1] We assume that, either an activity is not traced at all, or it is completely traced, i.e., it is traced together with all the filled fields it has as output.

that subsume the knowledge about the execution and that are compliant with the set of constraints imposed by the encoded knowledge.

Process models typically describe the flow of the process through activities (which can have associated data) and control flow constructs, such as parallel and alternative split points, as well as synchronization and alternative merge points. A **process model** PM can hence be defined as a pair (FO, C) of flow objects FO (including both activities and control flow objects) and oriented connections $C \subseteq FO \times FO$ defining an order among them. Process models can be described through different languages and notations, such as BPMN, EPC, Petri Nets: in this paper we use BPMN to model processes, however results can be generalized to other languages.

Considering the type of trace we deal with, i.e., traces containing events and associated data, as well as the set of BPMN process elements, we introduce the notion of process model *data flow objects*, i.e., BPMN *Flow Objects* with associated output data (e.g., activities and message events), and we focus our attention on them throughout the paper. Hence, in the specific BPMN case, the set of flow objects (FO) is composed of *data flow objects* (DFO), events without associated data (Ev_{nd}), e.g., escalation events, as well as split (G_S) and merge (G_M) gateways, i.e., $FO = DFO \cup Ev_{nd} \cup G_S \cup G_M$, where both $G_S (= G_{S\_XOR} \cup G_{S\_OR} \cup G_{S\_AND})$ and $G_M (= G_{M\_XOR} \cup G_{M\_OR} \cup G_{M\_AND})$ can be XOR, OR or AND gateways.

Let DFO be a finite set of *data flow object* names and FF a finite set of fields (contained in a data structure)[2] that can be provided as output by a *data flow object*. A domain knowledge K over DFO and FF is a pair $K = (cr, out)$, where $cr : DFO \rightarrow 2^{FF}$ is the function associating to each *data flow object* the fields that it *creates*, while $out : DFO \rightarrow 2^{FF}$ associates to each *data flow object* the fields that it has as *output*.

Let DFO_N be the set of N-instantiations of DFO, $DFO_N = \{dfon_i = (dfo, i) \mid dfo \in DFO \land i = 1, \dots, N\}$; E a finite set of events that can be injectively mapped to the *data flow objects* in DFO_N, via an injective function $\mu_E : E \rightarrow DFO_N \cup \{\perp_E\}$; and D a finite set of data names that can be injectively mapped to fields in FF via an injective function $\mu_D : D \rightarrow FF \cup \{\perp_D\}$. A **trace** T over DFO_N is a totally ordered sequence of events, $T = \langle en_1, \dots, en_Z \rangle$, where each en_i has a number of associated output data, i.e., there exists a function $o_T : E \rightarrow 2^D$, such that $\forall i \in [1, Z]$, $o_T(en_i) = \{d_{i1}, \dots, d_{iX}\}$. Given $e_h, e_k \in DFO$ and $h < k$ denoting that the first occurrence of e_h precedes the first of e_k in T, the total ordering relation $\prec_E$ among events is defined as $en_i = (e_h, p) \prec_E en_j = (e_k, q) \Leftrightarrow (p < q) \lor (p = q \land h < k)$, i.e., either the iteration p at which en_i occurs precedes the iteration q of en_j or, if they occur at the same iteration ($p = q$), en_i occurs before en_j.

Given DFO_N, an **admissible execution path** (AEP) of a process model PM, i.e., an admissible execution behavior of the process model PM, is a totally ordered sequence of executed *data flow objects* $AEP = \langle dfon_1, \dots, dfon_W \rangle$, where the total ordering relationship $\prec_{DFO}$ among executed *data flow objects* is defined as $dfon_i = (dfo_h, p) \prec_{DFO} dfon_j = (dfo_k, q) \Leftrightarrow (p < q) \lor (p = q \land h < k)$ and it respects the semantics defined by the process model PM.

A trace $T = \langle en_1, \dots, en_Z \rangle$ is **compliant** with an admissible path $AEP = \langle dfon_1, \dots, dfon_W \rangle$ if and only if (i) $\forall i \in [1, Z]$, $\mu_E(en_i) \in DFO_{AEP}$, where DFO_{AEP} is

---

[2] For the sake of simplicity a field in FF denotes a field in its specific container data structure.

the subset of DFO_N of symbols occurring in AEP; and (ii) $\forall i, j \in [1, Z], i \neq j$, if $en_i \prec_E en_j$ then $\mu_E(en_i) \prec_{DFO} \mu_E(en_j)$, i.e., the trace ordering does not violate any control flow relationship of the admissible path. A trace $T = \langle en_1, \ldots, en_Z \rangle$ is *compliant* with a process model PM and a domain model $K = (cr, out)$ if and only if (i) PM has at least an admissible path AEP_{PM}, such that T is compliant with AEP_{PM}; and (ii) $\forall i \in [1, Z], \forall d_{ij} \in o_T(en_i), \mu_D(d_{ij}) \in out(\mu_E(en_i))$, i.e., the fields corresponding to the output data of each event in the trace are admissible output fields for the *data flow object* corresponding to the event, according to K.

Given two traces $T_1 = \langle en1_1, \ldots, en1_Z \rangle$ and $T_2 = \langle en2_1, \ldots, en2_W \rangle$, a *partial ordering subsumption relationship* $\sqsubseteq$ exists between them, i.e., $T_1 \sqsubseteq T_2$, if and only if (i) $\forall i \in [1, Z], \exists j \in [1, W]$ such that $en1_i = en2_j$ and $o_{T1}(en1_i) = o_{T2}(en2_j)$; (ii) $\forall i, k \in [1, Z], i \neq k$, if $en1_i \prec_{E_{T1}} en1_k \Rightarrow en2_j \prec_{E_{T2}} en2_g$, where $en1_i = en2_j$ and $en1_k = en2_g$.

Given these definitions, at a high level, our problem can be formulated as follows: "*taking as input a business process model PM, some domain-specific knowledge K and a partially specified execution trace T, produce all the execution traces $\{T_1, \ldots, T_n\}$ that subsume T and are compliant with PM and K*". If no such trace is produced, this denotes that T is not compliant with PM and K. We remark here that, while in the ideal case $n = 1$ and T_1 is a complete trace, even with $n > 1$ and/or incomplete output traces, the partial completions of T are an added value in terms of knowledge on the model behavior.

A similarity can be devised between this problem and problems dealing with incomplete observations such as those faced in planning, model checking or satisfiability checking (SAT), in which, given a reference model, a set of possible model-compliant alternative worlds is generated starting from the partial observations. In particular, it is possible to rephrase our problem as one of satisfiability where, given the model composed by PM and K (in symbols $PM \wedge K$), and a partial assignment T (the partially specified execution trace), one aims at obtaining all (or at least some) *complete* traces consistent with $PM \wedge K$ that supersede T - or to know that no such assignment exist, meaning that T is inconsistent with PM. In symbols, given

$$PM \wedge K \models T \qquad (1)$$

we aim at identifying all complete models of T. It is also clear how our problem relates to SAT-based approaches to reasoning under uncertainty, e.g. in planning [1,2], where for some form of requirement ρ, the problem is recast into a SAT problem $M \models \rho$.

In such a context, to model system dynamics, several SAT-based encodings have been devised. In these cases, logical rules that define the possible system dynamics at each point in time need to be "unfolded", that is instantiated for each step of the execution, to obtain an overall set of constraints that describe how the dynamics of the system can evolve over a given (fixed) number of execution steps. This allows inquiring all possible behaviors over such number of steps, and to detect all the possible strategies, admissible by the model, to finally achieve some facts.

While we also have to deal with the specification of flow dynamics, our situation is however different, since we start from a dynamics of known length, represented as a finite partial trace. As such, our encoding, features a fixed length corresponding to the observed behavior, which is nevertheless guaranteed to cover all the executions implied by the trace. In detail, the SAT input model will contain the instance-independent encoding of the process model and domain knowledge in the form of implication rules related to the flow, and the domain knowledge on data; this will be complemented by the instance-dependent encoding of the trace as assertions of process activities and data logged in the trace file; the discussion of such encoding is reported in the next section.

4   SAT Input Model Encoding

Given a process model defined in the BPMN notation, a specific domain knowledge, and a partially specified execution trace, the purpose of this section is describing how to encode such a knowledge in the form of an input SAT model. In other words, how to obtain the sets of boolean formulae PM, K, and T described in the formula (1).

Given these premises, the model has been encoded by representing all process model *data flow objects* and data object field states as boolean predicates. Logical formulae, in the form of implication rules have then been defined on top of these predicates, to represent the model. In the following we describe the encoding for each of the three different types of knowledge: (i) knowledge related to the process model control flow; (ii) domain knowledge related to stateful data and how they are manipulated by activities; and (iii) instance-specific knowledge related to the execution traces.

4.1   Process Model Encoding

We start with the definition of PM, which encodes, as a set of implication rules, the control flow derived by locally analyzing the process model. The overall idea is that the occurrence of a *data flow object* implies the occurrence of its **preceding** *data flow object*, i.e., the *data flow object* that is connected to the current *data flow object* via sequence flow in BPMN. For example, in the fiscal code running example presented in Section 2, if the APD activity occurs, then the CRM should have been occurred as well. This can be encoded in the form of an implication rule as $APD \rightarrow CRM$. Similarly, the occurrence of a *data flow object* **receiving a message** implies the occurrence of a sender *data flow object*. In other terms, $\forall (fo_1, fo_2) \in C$, such that $fo_1, fo_2 \in DFO_N$, the following rule is generated and added to the process model encoding:

$$fo_2 \rightarrow fo_1 \tag{2}$$

Slightly more complex is the case of split and merge flow points. In this case, indeed, according to the different semantics of the control flow node, different implication rules can be generated. For example, when an activity can be reached from alternative paths, as in the case of AFC, its occurrence implies that at least one of the preceding activities has occurred. For example, if the AFC activity occurs, then the $APPD$ or $APARD$ activities should occur, as well. This can be formulated in terms of implication rules as $AFC \rightarrow APPD \vee APARD$. Moreover, since the two paths are exclusive, either the $APPD$ or the $APARD$ activity can occur, i.e., the occurrence of the first implies the non-occurrence of the second (and viceversa), which can be translated into implication rules as $APPD \rightarrow \neg APARD$. Table 1 reports the translation of the main control flow basic patterns (described in BPMN) into implication rules.

Given the alternative, i.e., XOR and OR, split and merge gateways, respectively, $G_{S\_alt} = G_{S\_XOR} \cup G_{S\_OR}$ and $G_{M\_alt} = G_{M\_XOR} \cup G_{M\_OR}$ and the parallel, i.e., AND, split and merge gateways G_{S_AND} and G_{M_AND}, the encoding function enc is defined as follows:

$$enc(fo) = \begin{cases} fo & fo \in DFO_N \\ \bigvee_{i=1,\ldots,n} enc(fo_i) & fo \in G_{S\_alt} \wedge (fo,fo_i) \in C \ \vee fo \in G_{M\_alt} \wedge (fo_i,fo) \in C \\ \bigwedge_{i=1,\ldots,n} enc(fo_i) & fo \in G_{S\_par} \wedge (fo,fo_i) \in C \ \vee fo \in G_{M\_AND} \wedge (fo,fo_i) \in C \\ TRUE & otherwise \end{cases}$$

For each $(fo_1,fo_2) \in C$, the following implication rule is added to the process model encoding:

$$enc(fo_2) \rightarrow enc(fo_1) \tag{3}$$

Moreover, defined the mutually exclusion function me, as:

$$me(fo) = \bigcup_{\substack{fo_i \\ fo \in G_{S\_XOR} \wedge (fo,fo_i) \in C \ \vee \\ fo \in G_{M\_XOR} \wedge (fo_i,fo) \in C}} enc(fo_i)$$

the set of mutually excluded branches $me(fo)$ is computed for each $fo \in G_{S\_XOR} \cup G_{M\_XOR}$ and, for each $meb_i \in ME_{fo}$, the following rules are added to the process model encoding:

$$meb_i \rightarrow \neg meb_{i+j} \tag{4}$$

with $i \in [1, |me(fo)| - 1]$ and $j \in [1, |me(fo)| - i]$.

Conditions on executed activities can also be encoded adding conditions as further "consequences" of the *data flow object* execution. The last row of Table 1, left, in which conditions related to past executions of *data flow objects* are depicted among square brackets, reports a pattern example and the corresponding implication rules. In other terms, for each sequence flow (fo_1,fo_2) outgoing from a split XOR or OR gateway (i.e., $fo_1 \in G_{S\_XOR} \cup G_{S\_OR}$) with associated *data flow object* condition, $\overline{dfo}$, the following rule is added to the process model encoding:

$$enc(fo_2) \rightarrow \overline{dfo} \tag{5}$$

Finally, cycles represent the most delicate case. Indeed, the encoding is dynamically built depending upon the specific trace under analysis, by unfolding the cycle according to an upper bound value suggested by the trace. Once the cycle and the *data flow objects* it is composed of have been identified in the model, e.g., $cyc = \{cdfo_1, \ldots, cdfo_m\}$ with $cdfo_i \in DFO$, for each trace, the maximum number N of occurrences of events mapping to the same *data flow object* in the cycle is computed. The DFO_N set is computed, a variable for each *data flow object* in the cycle encoded, and the cycle built accordingly. Last row right in Table 1 informally reports the SAT encoding for the cycle in the figure, where N is the maximum number of occurrences of the activities in the cycle in the specific log. In detail, the control flow inside non-overlapping cycles is encoded as usual, instantiating one set of rules for each repetition of the cycle (N or $N - 1$ according to the *data flow object* with the highest occurrence) in the instance-specific input trace. For the first and last *data flow object* of the cycle, instead, two more types of implication rules for each repetition are added: those connecting the first *data flow object* of the ith cycle, $(cdfo_1, i)$, with the last *data flow object* of the $i - 1^{th}$ one, $(cdfo_m, (i - 1))$:

$$(cdfo_1, i) \rightarrow (cdfo_m, (i - 1)) \tag{6}$$

and those connecting the first occurrence of the source *data flow object*, $(cdfo_s, 1)$, with the *data flow objects* preceding the cycle, e.g., $(dfo_i, 1)$ and the last occurrence of the target *data flow object*, $(cdfo_t, N)$, with the first *data flow object* following the cycle, e.g., $(dfo_j, 1)$:

$$(cdfo_s, 1) \rightarrow (dfo_i, 1) \tag{7} \qquad\qquad (dfo_j, 1) \rightarrow (cdfo_t, N) \tag{8}$$

where dfo_i and dfo_j are the first *data flow object* before and after the cycle, respectively. When the cycle has more than one starting or/and ending *data flow object*, the same set of rules is encoded for each of them.

4.2 Domain Knowledge Encoding

Focusing on K, the two types of knowledge related to the relationship among activities and data are exploited for the encoding, i.e., the *creates* (cr) and the *has output* (out) relationships. Being manipulated by activities, a data object field is not an immutable structure, rather, it evolves (i.e., it can also be updated by activities). As such it is a *stateful* field. Here on we denote with F_X a stateful field F in the state (immediately after the execution of a *data flow object*) A such that $F \in out(A)$.

Table 2 reports the encoding for the relationships among activities and data. Note that the encoding provided in the table is restricted to the case in which fields can be created, (updated and) shown as output, but never deleted. While it is a limitation of our current work, this is an assumption which holds in a vast range of cases, whose generalization we leave to future works. The intuition is that if a *data flow object* A has as output a given field F, at least one of the *data flow objects* creating such a field should occur. For instance, in the running example, if, after the execution of the activity AFC, the field $documentID$ is filled, it means that the activity which sets it (CRM) has been executed. We hence encode the corresponding implication as $documentID_{AFC} \rightarrow CRM$. Defined CR_f as the set of *data flow objects* that create the field f, i.e., $CR_f = \{dfon \in DFO_N | f \in cr(dfon)\}$, for each $f \in out(dfon)$ such that $dfon \notin CR_f$, the rule (9) can be added to the encoding of K.

$$f_{dfon} \rightarrow \bigvee_{dfon_i \in CR_f} dfon_i \tag{9} \qquad\qquad \bigvee_{dfon_i \in CR_f} dfon_i \rightarrow f_{dfon} \tag{10}$$

Moreover, if no path exists from the *data flow object* A to none of the activities that create F, then also the inverse relationship occurs, i.e., if F does not occur, it means that none of its creators has occurred. For example, in our running example, since no path exists between AFC and the activities that create the field $passportID$, i.e., $APPD$, the non-occurrence of the field $passportID$ in AFC would imply the non-occurrence of the activity $APPD$. We encode such an implication as $APPD \rightarrow documentID_{AFC}$. In other terms, given $dfon \in DFO_N$ and $f \in out(dfon)$, when no path exists from $dfon$ to none of $dfon_i \in CF_f$, the rule (10) is added to the K encoding.

Table 1. Diagram Translation Rules

BPMN construct	SAT encoding	BPMN construct	SAT encoding
	$B \to A$		$B \to A$
	$B \vee C... \vee Z \to A$ $B \to \neg C$ ... $B \to \neg Z$ ... $C \to \neg Z$ ...		$A \to B \vee C... \vee Z$ $B \to \neg C$ ... $B \to \neg C$ ... $C \to \neg Z$ ...
	$B \vee C... \vee Z \to A$		$A \to B \vee C... \vee Z$
	$B \wedge C... \wedge Z \to A$		$A \to B \wedge C... \wedge Z$
	$B \vee C \to A$ $B \to \neg C$ $B \to D$ $C \to E$		$C \to (A, N)$ $(A, N) \to$ $(A, N-1)$ ... $(A, 2) \to (A, 1)$ $(A, 1) \to B$

4.3   Trace Encoding

Finally, for each trace, the (incomplete) knowledge it contains is also described in terms of partial assignments, in order to obtain T. Table 3 reports the three types of encoding related to the trace T.

Intuitively, each event A in the analyzed process trace has occurred and hence is asserted, i.e., defined as a true predicate: $TRUE \to A$. For instance, in the case of the running example, the activity CRM (occurring in the trace) is encoded as $TRUE \to CRM$. In other terms, for each en_i occurring in trace $T = \langle en_1, \ldots, en_Z \rangle$, the *data flow object* predicate (11) is added to the encoding of T.

$$TRUE \to \mu_E(en_i) \qquad (11) \qquad\qquad TRUE \to \mu_D(d_{ij}) \qquad (12)$$

Table 2. Encoding of K

Domain knowledge	K encoding
A has output F B_1 creates F, ..., B_n creates F	$F_A \rightarrow B_1 \vee \dots \vee B_n$
A has output F B_1 creates F, ..., B_n creates F and no path from A to B_1, ..., no path from A to B_n	$F_A \rightarrow B_1 \vee \dots \vee B_n$ $B_1 \vee \dots \vee B_n \rightarrow F_A$

Table 3. Encoding of T

Trace	T encoding
A	$TRUE \rightarrow A$
$A\{F1, \dots, Fn\}$	$TRUE \rightarrow F1_A$ ... $TRUE \rightarrow Fn_A$
$A\{F1, \dots, Fn\}$ and A has output Fj and $Fj \notin \{F1, \dots, Fn\}$	$TRUE \rightarrow \neg Fj_A$

Similarly, each data object field F occurring in the trace as output of a given *data flow object* A has been actually shown as output of A and hence is added to the rules as a true predicate: $TRUE \rightarrow F_A$. In the running example, for instance, the field shown in output by CRM in the trace is encoded as $TRUE \rightarrow documentID_{CRM}$. More formally, for each $d_{ij} \in o(en_i)$ in trace $T = \langle en_1, \dots, en_Z \rangle$, the corresponding stateful data object field predicate (12) is added to the encoding of T.

Moreover, if, according to K, a *data flow object* A may have as output a given field F but it actually does not in the trace, since we are assuming that an event is either traced completely in the trace or not traced at all, its negation is added to the rules as true predicate: $TRUE \rightarrow \neg F_A$. In case of the running example, for instance, the activity AFC may have as output $passportID$ but the field is not occurring as output of AFC in the trace, hence $TRUE \rightarrow \neg passportID_{AFC}$ should be added to the model. In other terms, for each en_i in trace $T = \langle en_1, \dots, en_Z \rangle$ and for each $f_{ij} \in out(\mu_E(en_i))$, such that $o(en_i)$ does not contain any corresponding d_k in output (i.e., d_k such that $\mu_D(d_k) = f_{ij}$), the following rule is added to the encoding of T:

$$TRUE \rightarrow \neg f_{ij} \tag{13}$$

5   The SATFILLER Tool

The approach has been implemented in the SATFILLER tool. SATFILLER is a Java application taking in input a process model (in the .json format), knowledge on the domain as well as an incomplete trace, and returning either (i) the notification that the input trace is inconsistent with the model or (ii) one or more possible completions for the input trace. SATFILLER is composed of three main modules: a *Model encoder* for encoding the input knowledge into a SAT model, the *SAT-reasoner* and a *Model decoder*, for decoding the output of the reasoner. In detail, SATFILLER uses SAT4J [3], a Java library for solving boolean satisfaction and optimization problems. The *Model encoder* is composed of three submodules for the automatic encoding of the three types of input knowledge (process model, domain knowledge and traces) in the SAT4J format, i.e. the common Conjunctive Normal Form (CNF) Dimacs format. Finally, the *Model decoder* automatically translates the SAT4J output into a (set of) trace(s) completing the incomplete input one.

On-going experiments applying the approach to case studies seem promising. In particular results show that SATFILLER performances in terms of missing information filled are affected not only by process model structure, but also by knowledge on data

---

[3] http://www.sat4j.org

management. Moreover, extra-knowledge gathered from domain experts seem to strongly improve results. Applied to a significant e-government case study, characterized by a process involving about 18 activities, 21 gateways, and 13 different data objects (including in total 75 distinct fields), SATFILLER is able to fill up to the 70% of traces with up to about 63% of missing information.

6 Related Work

The reconstruction of flows of activities of a model given a partial set of information on it can be closely related to several fields of research in which the dynamics of a system are perceived only to a limited extent and hence it is needed to reconstruct missing information. Most of those approaches share the common conceptual view that a model is taken as a reference to construct a set of possible model-compliant "worlds" out of a set of observations that convey limited data. Among qualitative approaches, i.e., approaches relying on equally likely "alternative worlds" rather than probabilistic models, several works have been developed, to model sets of possible worlds in a more effective way than by explicit enumerations, ranging from tree-based representations (BDDs) [3,4], to logical formulae whose satisfaction models implicitly represent the worlds [1,2]. We take this latter choice, which brings to a natural connection of our challenge to one of the core problems in AI: the boolean satisfiability problem.

Remarkably, while satisfiability is known to be NP-complete, several effective implementations are available that, on top of the classic Davis-Putnam [5] algorithm, add optimizations in terms of sub-result caching, search heuristics and so on. Indeed, the wide availability of effective tools for solving SAT is one of the key reasons for the success of SAT-based approaches e.g. in planning or diagnosis.

However our SAT encoding has specific features that differ from those adopted in planning. They stem, in part, from the different view of system dynamics given by the problem: while in the case of planning the task is to build an unknown dynamics (of unknown bounded length), in our case, the task is to complete a given partial dynamics obeying given constraints. This brings to a simplified and effective encoding where an upper-bound limited rule unfolding is considered. Further, the injection of domain knowledge that originates from the relationship among (stateful) data objects has no direct correspondence in the planning approaches, while it is a core part of ours.

The problem of incomplete traces has been faced also in the field of process mining (and in particular of process discovery and conformance), where it still represents one of the challenges [6]. In the field of process conformance, several works have addressed the problem of measuring fitness i.e., the extent to which process models capture the observed behaviour. For example, Rozinat et al. [7] use missing, remaining, produced and consumed tokens for computing fitness. In more recent works [8,9,10], the alignment between event logs and procedural, without [8] and with [9] data, or declarative models [10], is considered. All these works are based on the A* algorithm [11], an algorithm originally proposed to find the path with the lowest cost between two nodes in a directed graph. They all explore the search space of the set of possible moves to find the best one for aligning the log to the model. In detail, Adriansyah et al. [8], focus on procedural models: they align Petri Nets and logs identifying skipped and inserted activities in the process model. De Leoni et al. [9] extend the search space by considering, in the alignment of model and trace, also the data values. In our case, however, both goal and preconditions are slightly different. We assume, indeed, that the model is correct, while logs can either be incomplete or non-conformant at all. In this last case, the goal is not providing a measure of alignment, which can be confused with similar measures of highly incomplete traces, but rather classifying the trace as non-conformant. Moreover, differently from [9], data are not used for weighting a cost function, by looking at their values, but rather their existence is exploited to drive the reconstruction of the complete trace.

7 Conclusion

The paper aims at supporting business analysis activities by tackling the limitations due to the partiality of information often characterizing the business activity monitoring. To this purpose, a novel reasoning method for reconstructing incomplete execution traces, that relies on the formulation of the issue in terms of the satisfiability problem, is presented. First on-going experiment results of the approach, implemented in the SATFILLER tool, look encouraging.

In the future, we plan to validate the proposed approach, as well as extend it considering further data relations (e.g., dependencies among data) and data deletion operations. Moreover, we are interested in exploring the problem of optional execution branches, i.e., how to distinguish whether a branch has not been executed, or it has but all its observations are missing.

References

1. Rintanen, J., Heljanko, K., Niemelä, I.: Planning as satisfiability: parallel plans and algorithms for plan search. Artif. Intell. 170, 1031–1080 (2006)
2. Kautz, H.A., Selman, B.: Planning as satisfiability. In: ECAI, pp. 359–363 (1992)
3. Cimatti, A., Pistore, M., Roveri, M., Traverso, P.: Weak, strong, and strong cyclic planning via symbolic model checking. Artif. Intell. 147, 35–84 (2003)
4. Bertoli, P., Cimatti, A., Roveri, M., Traverso, P.: Planning in nondeterministic domains under partial observability via symbolic model checking. In: Proc. of the 17th Int. Joint Conference on Artificial Intelligence, vol. 1, pp. 473–478 (2001)
5. Davis, M., Putnam, H.: A computing procedure for quantification theory. Journal of the ACM 7, 201–215 (1960)
6. van der Aalst, W.M.P., et al.: Process mining manifesto. In: Daniel, F., Barkaoui, K., Dustdar, S. (eds.) BPM 2011 Workshops, Part I. LNBIP, vol. 99, pp. 169–194. Springer, Heidelberg (2012)
7. Rozinat, A., van der Aalst, W.M.P.: Conformance checking of processes based on monitoring real behavior. Inf. Syst. 33, 64–95 (2008)
8. Adriansyah, A., van Dongen, B.F., van der Aalst, W.M.P.: Conformance checking using cost-based fitness analysis. In: Proc. of EDOC 2011, pp. 55–64 (2011)
9. de Leoni, M., van der Aalst, W.M.P., van Dongen, B.F.: Data- and resource-aware conformance checking of business processes. In: Abramowicz, W., Kriksciuniene, D., Sakalauskas, V. (eds.) BIS 2012. LNBIP, vol. 117, pp. 48–59. Springer, Heidelberg (2012)
10. de Leoni, M., Maggi, F.M., van der Aalst, W.M.P.: Aligning event logs and declarative process models for conformance checking. In: Barros, A., Gal, A., Kindler, E. (eds.) BPM 2012. LNCS, vol. 7481, pp. 82–97. Springer, Heidelberg (2012)
11. Dechter, R., Pearl, J.: Generalized best-first search strategies and the optimality of a*. J. ACM 32, 505–536 (1985)

CDoT: Optimizing MAP Queries on Trees

Roberto Esposito, Daniele P. Radicioni, and Alessia Visconti

Department of Computer Science, University of Torino
{roberto.esposito,daniele.radicioni,alessia.visconti}@unito.it

Abstract. Among the graph structures underlying Probabilistic Graphical Models, trees are valuable tools for modeling several interesting problems, such as linguistic parsing, phylogenetic analysis, and music harmony analysis. In this paper we introduce CDoT, a novel exact algorithm for answering Maximum a Posteriori queries on tree structures. We discuss its properties and study its asymptotic complexity; we also provide an empirical assessment of its performances, showing that the proposed algorithm substantially improves over a dynamic programming based competitor.

1 Introduction

In this paper we introduce CarpeDiem on Trees (CDoT), an algorithm for solving efficiently and exactly Maximum A Posteriori (MAP) queries on tree-structured probabilistic graphical models.

Probabilistic Graphical Models (PGMs) lie at the intersection of probability and graph theory; they sport a rigorous theoretical foundation and provide an abstract language for modeling application domains [1]. PGMs have been successfully applied in fields as diverse as medical diagnosis [2], computer vision [3], and text mining [4]. Informally stated, a PGM is a graph specifying the conditional independence structure of a family of probability distributions. Specifically, vertices are used to represent random variables, and edges are used to model an intervening direct probabilistic interaction between two random variables. The conditional independence structure is the key property used to efficiently handle the otherwise intractable problems of inference and learning.

Among the graph structures underlying PGMs, trees are valuable tools for modeling several interesting problems. They have been successfully applied to solve challenging tasks, such as linguistic parsing [5], phylogenetic analysis [6], and music harmony analysis [7].

Answering MAP queries over a PGM entails finding the assignment to the graph variables that *globally* maximizes the probability of an observation. This differs from optimizing the assignment to each variable in isolation. For instance, in the optical character recognition task, the labeling "*learning*" is probably to be preferred over the labeling "*1earnin9*" –based on known interactions between nearby labels–, even though the observations of the first and last characters might suggest otherwise when considered in isolation. MAP queries are important reasoning tools *per se*, but they are of the utmost importance during learning of

M. Baldoni et al. (Eds.): AI*IA 2013, LNAI 8249, pp. 481–492, 2013.

both the PGM structure and parameters. In fact, in tackling these two tasks, MAP queries are often used as repeatedly called sub-procedures.

The MAP problem is $\mathcal{NP}$-complete in general, but it can be solved in polynomial time on particular graph structures such as chains and trees by variations on the Viterbi algorithm [8]. Several algorithms have been proposed to efficiently solve MAP queries on chains. For instance, CarpeDiem [9] distinguishes between 'local' (associated with vertices) and 'transitional' (associated with edges) information to improve on the quadratic –in the number of labels– complexity of competing approaches. The main intuition therein contained is that local evidence often provides most of the clues needed to optimally solve the problem, while dependencies between variables can be used to discriminate between cases that cannot be differentiated otherwise. Recently, [10] exploited the same intuition to develop an algorithm based on linear programming techniques. In both cases experimental evidence shows impressive improvements with respect to the Viterbi algorithm.

In this paper we introduce CDoT, an algorithm that extends the approach in [9] to the case of tree structured PGMs. The algorithm always returns the optimal answer to the MAP query, often spending only a fraction of the computational resources demanded by competing approaches. We illustrate the algorithm in full details, provide a formal study of the algorithm complexity, and report on an experimentation comparing CDoT with a dynamic programming solution.

2 Formalization

Let us consider a probabilistic graphical model where a tree T encodes conditional independences, and P is the associated probability distribution. Let us assume, without loss of generality that T is a directed tree, and that it has edges oriented from root towards leaves. Each vertex V in T represents a random variable that takes values in the set of labels L_V; in the following, the notation l_V is used to denote a label in L_V. A MAP query over this structure amounts to determining the optimal assignment of labels to variables.[1] An optimal assignment is one that maximizes the cumulative reward of the root R of the tree.

A summary of the notation that we will be using throughout the paper is reported in Table 1.

We consider an execution of the Variable Elimination algorithm using any elimination ordering that eliminates leaves first and then moves toward the root R. The outcome is the evaluation of the expression:

$$\max_{l_R} \left[\sum_{V \in \text{ch}(R)} \underbrace{\max_{l_V} \left[\phi(l_R, l_V) + \sum_{W \in \text{ch}(V)} (\cdots) \right]}_{(a)} \right] \tag{1}$$

---

[1] If more than one optimal assignment exists for the tree, then any one of them is a valid answer to the MAP query.

Table 1. Summary of the adopted notation

Symbol	Description
$|T|$	the number of vertices in T
K	the number of labels per vertex
l_V	a label l in vertex V
λ	a distance from the tree root
$\omega(l_V)$	the cumulative reward up to label l_V
$\tilde{\omega}(l_V)$	the estimated cumulative reward up to label l_V
$\mathcal{L}_\sqsubseteq(V)$	the sorted list of labels for vertex V
$\nu(l_V)$	a local factor
$\tau(l_V \twoheadrightarrow l_W)$	a transition factor
$\mathcal{T}$	the maximal transition weight between any two labels

where $\mathrm{ch}(V)$ is the set of children of V; $\phi(l_V, l_W)$ is a factor that depends only on the labels of variables V and W; and the dots substitute an expression analogous to expression (a). By using standard dynamic programming techniques Expression (1) can be evaluated in $O(|T|K^2)$ where $|T|$ is the number of vertices in T and K is the number of values (i.e., labels) that each variable can assume.

It is always possible to rewrite each $\phi(l_V, l_W)$ as the sum of a transition factor $\tau(l_V \twoheadrightarrow l_W)$ and a local factor $\nu(l_V)$, that is: of a factor $\nu(l_W)$ that only depends on labels of V, and of a factor $\tau(l_V \twoheadrightarrow l_W)$ that accounts for the relationship between the two vertices. In a directed model each factor $\phi(l_V, l_W)$ corresponds to the logarithm of the conditional probability $P(W = l_W | V = l_V)$. The proposed rewriting amounts to factorizing this probability as: $P(W = l_W | V = l_V) = \psi_0(l_W) \cdot \psi_1(l_V, l_W)$ with $\psi_0(l_W) > 0$, then setting $\nu(l_W) = \log(\psi_0(l_W))$ and $\tau(l_V \twoheadrightarrow l_W) = \log(\psi_1(l_V, l_W))$.

Given the above decomposition, we rewrite Expression (1) as:

$$\max_{l_R} \left[\sum_{V \in \mathrm{ch}(R)} \max_{l_V} [\tau(l_R \twoheadrightarrow l_V) + \nu(l_V) + \sum_{W \in \mathrm{ch}(V)} (\cdots)] \right].$$

By defining the cumulative reward $\omega(\cdot)$ as:

$$\omega(l_V) = \nu(l_V) + \sum_{W \in \mathrm{ch}(V)} \max_{l_W} [\tau(l_V \twoheadrightarrow l_W) + \omega(l_W)]$$

the expression can be rewritten as:

$$\max_{l_R} \left[\sum_{V \in \mathrm{ch}(R)} \max_{l_V} [\tau(l_R \twoheadrightarrow l_V) + \omega(l_V)] \right].$$

Finally, by defining $\nu(l_R)$ as 0, Expression (1) can be restated as $\max_{l_R} \omega(l_R)$.

CDoT computes expression $\max_{l_V} \omega(l_V)$ with sub-quadratic complexity. This complexity is achieved when the algorithm succeeds in avoiding the inspection of a number of labels by exploiting an upper bound to the cumulative reward.

Let $\mathcal{T}$ be an upper bound to the maximal transition weight between any two labels, that is: $\mathcal{T} \geq \max_{l_V, l_W} \tau(l_V \to l_W)$. Let $\sqsubseteq$ be a relation such that $l_V \sqsubseteq l'_V$ iff $\nu(l_V) \geq \nu(l'_V)$, and let $\mathcal{L}_\sqsubseteq(V)$ be the list of labels of vertex V, ordered according to $\sqsubseteq$. We say that a label l_V is more promising than label l'_V if $l_V \sqsubseteq l'_V$.

We denote by $\tilde{\omega}(l_V)$ the upper bound for $\omega(l_V)$ defined as:

$$\tilde{\omega}(l_V) = \nu(l_V) + |\text{ch}(V)| \cdot \mathcal{T} + \sum_{W \in \text{ch}(V)} \max_{l_W} \omega(l_W).$$

Two interesting properties of $\tilde{\omega}(l_V)$ are:

$$\forall l_V, l'_V : \nu(l_V) \geq \nu(l'_V) \Leftrightarrow \tilde{\omega}(l_V) \geq \tilde{\omega}(l'_V) \tag{2}$$

$$\forall l_V : \qquad\qquad \tilde{\omega}(l_V) \geq \omega(l_V) \tag{3}$$

Property (2) states that $\tilde{\omega}(l_V)$ is monotone in $\nu(l_V)$. The property follows immediately by noticing that the quantity $|\text{ch}(V)| \cdot \mathcal{T} + \sum_{W \in \text{ch}(V)} \max_{l_W} \omega(l_W)$ for a fixed V does not depend on the actual label l_V. Property (3) states that $\tilde{\omega}(l_V)$ is an upper bound for $\omega(l_V)$. It stems from $a)$ noticing that $\tilde{\omega}(l_V)$ can be obtained by substituting $\mathcal{T}$ in place of $\tau(l_V \to l_W)$ in the definition of $\omega(l_V)$; and $b)$ recalling that by definition $\mathcal{T} \geq \tau(l_V \to l_W)$.

3   The Algorithm

In this section we illustrate the algorithm. We start by introducing the main ideas underlying the optimization strategy, and then we delve into the details of CDoT.

CDoT starts computing $\max_{l_V} \omega(l_V)$ on the vertices that are farther from the tree root and memoizes these results. On a leaf V, there are no children to take into consideration and the label that maximizes the reward is simply the first label in $\mathcal{L}_\sqsubseteq(V)$. If V is not a leaf, we assume to have at hand some procedure to evaluate $\omega(l_W)$ for each label belonging to a child W of V. Let l_V be the first (more promising) label in $\mathcal{L}_\sqsubseteq(V)$ and let us compare it with the next most promising label l'_V. The main insight in CDoT is that it is not always necessary to delve into the inspection of l'_V. In fact, Property (3) implies that it is not necessary to inspect l'_V whenever $\omega(l_V) \geq \tilde{\omega}(l'_V)$. Also, Property (2) implies that if indeed $\omega(l_V) \geq \tilde{\omega}(l'_V)$, then $\omega(l_V)$ is necessarily larger than all remaining labels in $\mathcal{L}_\sqsubseteq(V)$. If $\nu(\cdot)$ is a strong predictor of optimal labels, it is thus likely that the inequality holds and much computation can be saved.

We now consider how to efficiently compute $\omega(l_V)$ through a process similar to that used for $\max_{l_V} \omega(l_V)$. In the forthcoming discussion, we say that we *open* l_V the first time we actually compute its reward $\omega(l_V)$; vice versa, a label l_V is said to be closed if it has never been opened. Whenever we open a label, we memoize its value and assume to have $O(1)$ time access to $\omega(l_V)$ in subsequent calls. The computationally expensive part in the definition of $\omega(l_V)$ is the evaluation of

$\sum_{W \in ch(V)} \max_{l_W} [\tau(l_V \rightarrow l_W) + \omega(l_W)]$. Importantly, each maximization inside the summation can be dealt with independently of the others. Let l^* be the first, most promising, label in $\mathcal{L}_{\sqsubseteq}(W)$ and l_W be the next label in $\mathcal{L}_{\sqsubseteq}(W)$. Remarkably, l^* has to be a previously opened label: it belongs to a vertex already processed (it is farther from the root), and it is the most promising label for that vertex. In contrast, l_W can be either open or closed. If it is open, we can choose between the two labels by comparing $\omega(l_W) + \tau(l_V \rightarrow l_W)$ with $\omega(l^*) + \tau(l_V \rightarrow l^*)$ in $O(1)$. If it is closed, computing the best of the two labels can be expensive. However, we can avoid opening l_W and rule it out as a candidate anyway if

$$\omega(l^*) + \tau(l_V \rightarrow l^*) \geq \widetilde{\omega}(l_W) + \tau(l_V \rightarrow l_W). \tag{4}$$

If additionally it holds $\omega(l^*) + \tau(l_V \rightarrow l^*) \geq \widetilde{\omega}(l_W) + \mathcal{T}$, then *a fortiori* Inequality (4) holds for all subsequent labels in $\mathcal{L}_{\sqsubseteq}(W)$, and l^* is the best possible choice for W.

CDoT is described by Algorithms 1, 2, and 3. Algorithm 1 is the main procedure implementing CDoT. It traverses the input tree starting from the deepest level up to the root. For each level λ (i.e., the set of vertices at depth λ), the algorithm calls the procedure described in Algorithm 2 to determine the best assignment l_V^* for each vertex V in that level. The output of the algorithm is the reward of the best label of the root vertex. Standard dynamic programming techniques can be used to track the assignments to children vertices. The optimal assignment for each vertex can be then reconstructed in linear time by executing a visiting algorithm on the tree.

Algorithm 2 (process_vertex) computes $\max_{l_V} \omega(l_V)$: the best assignment for a given vertex V. It iterates through the labels associated to vertex V computing their reward by means of Algorithm 3 and keeping aside the best label found so far. It returns the best label as soon as it finds that no other label can possibly ameliorate its reward, i.e., as soon as $\omega(l^*) > \widetilde{\omega}(l_V)$ (Algorithm 2, line 4).

Algorithm 3 (open) computes $\omega(l_V)$: the reward for label l_V of vertex V. The algorithm iterates over all children of vertex V integrating the contributions of each one into $\omega(l_V)$ which is initially set to $\nu(l_V)$. In analysing each child label l_W, the algorithm stops as soon as it verifies that current and all forthcoming labels cannot contribute more than the best child label l^* to the final outcome, i.e., it breaks from the loop as soon as the current best reward ω^* for this child is larger than $\widetilde{\omega}(l_W) + \mathcal{T}$ (Algorithm 3, line 7). When this condition is not met, the algorithm has the chance of saving some work anyway: it can exclude that the current label is optimal in $O(1)$ by checking if $\omega^* > \widetilde{\omega}(l_W) + \tau(l_V \rightarrow l_W)$ (Algorithm 3, line 8). Only when also this condition is not met, the algorithm recursively opens the child label.

4 Algorithm Complexity

Let us consider the final step of an execution of CDoT, and assume that for each vertex V, exactly k_V labels have been opened. In this discussion we separately

---

Algorithm 1. (CDoT) Given a tree T, it returns the reward of the best assignment to its vertices

---

input a tree T
1: **for** $\lambda \leftarrow depth(T) \ldots 0$ **do**
2:    **for each** vertex V at level λ **do**
3:        $l_V^* \leftarrow$ process_vertex(V)
4:    **end for**
5: **end for**
6: **return** $\omega\left(l_{root(T)}^*\right)$

Algorithm 2. (process_vertex) Given a vertex V, it computes $\max_{l_V} \omega(l_V)$ and returns the assignment for V that attains the maximum value

---

input a vertex V
1: $l^* \leftarrow$ **dummy label**
2: $\omega^* \leftarrow -\infty$
3: **for all** l_V in list $\mathcal{L}_{\sqsubseteq}(V)$ **do**
4:    **break if** $\omega^* \geq \tilde{\omega}(l_V)$
5:    open(l_V)
6:    **if** $\omega(l_V) > \omega^*$ **then**
7:        $l^* \leftarrow l_V$
8:        $\omega^* \leftarrow \omega(l_V)$
9:    **end if**
10: **end for**
11: **return** l^*

---

consider the time spent to process each vertex of the graph. We define the quantity $\mathbb{T}(V)$ to represent the overall number of steps spent by Algorithms 2 and 3 to process vertex V. Let us define:

$a(l_V)$: the number of steps needed by Algorithm 2 to process label l_V;
$b(l_V)$: the number of steps needed by Algorithm 3 to find the best set of children for label l_V.

We note that $a(l_V)$ does not include the time spent by Algorithm 3 since such time is accounted for by $b(l_V)$. Similarly, $b(l_V)$ does not include neither the time spent by Algorithm 2, nor the time spent by recursive calls to Algorithm 3. In fact, the time spent in recursive calls is taken into account by b values of vertices in previous layers. Then we can compute $\mathbb{T}(V)$ as $\mathbb{T}(V) = \sum_{l_V} a(l_V) + b(l_V)$.

Property 1. $\sum_{l_V} a(l_V)$ is at worst $O(k_V)$.

Proof. Since only k_V labels have been opened at the end of the algorithm, and Algorithm 2 does not do any work on closed labels, the number of steps needed to analyze a vertex V by (the loop in) Algorithm 2 cannot be larger than k_V.

We notice that we are overestimating the cost to analyze each vertex, since k_V is the *overall* number of vertices opened either by Algorithm 2 or by Algorithm 3.

Algorithm 3. (open) Given a label l_V, it computes $\omega(l_V)$

---

input a label l_V

1: $\omega(l_V) \leftarrow \nu(l_V)$
2: **for all** $W \in \text{children}(V)$ **do**
3:    $l^* \leftarrow$ **dummy label**
4:    $\omega^* \leftarrow -\infty$
5:    **for all** l_W in list $\mathcal{L}_{\sqsubseteq}(W)$ **do**
6:      **if** l_W is closed **then**
7:        **break if** $\omega^* > \tilde{\omega}\,(l_W) + \mathcal{T}$
8:        **next if** $\omega^* > \tilde{\omega}\,(l_W) + \tau(l_V \twoheadrightarrow l_W)$
9:        $\text{open}(l_W)$
10:      **end if**
11:      **if** $\omega^* < \omega(l_W) + \tau(l_V \twoheadrightarrow l_W)$ **then**
12:        $l^* \leftarrow l_W$
13:        $\omega^* \leftarrow \omega(l_W) + \tau(l_V \twoheadrightarrow l_W)$
14:      **end if**
15:    **end for**
16:    $\omega(l_V) \leftarrow \omega(l_V) + \omega^*$
17: **end for**

However, this overestimation simplifies the following argument without hindering the result.

Property 2. $\sum_{l_V} b(l_V)$ is $O(k_V \sum_{W \in \text{ch}(V)} k_W)$.

Proof. Since only k_W labels of vertex W have been opened at the end of the algorithm, the two loops in Algorithm 3 iterate altogether at most $\sum_{W \in \text{ch}(V)} k_W$ times. Moreover, since the steps performed by recursive calls are not to be included in $b(l_V)$, all operations are $O(1)$, and the complexity accounted for by $b(l_V)$ is $O(\sum_{W \in \text{ch}(V)} k_W)$. For closed labels $b(l_V)$ is zero since Algorithm 3 would have never been called on such labels. This implies that $\sum_{l_V} b(l_V)$ is at most

$$O\left(\sum_{l_V} \left(I_{[l_V \text{ is open}]} \sum_{W \in \text{ch}(V)} k_W \right) \right) \text{ (where } I_{[x]} \text{ is 1 if } x \text{ is true and 0 otherwise),}$$

thereby resulting in $\sum_{l_V} b(l_V)$ being at most $O\left(k_V \sum_{W \in \text{ch}(V)} k_W \right)$.

Theorem 1. *CDoT has $O(|T|K^2)$ worst case time complexity and $O(|T|K \log K)$ best case time complexity.*

Proof. The complexity of CDoT is $\mathbf{T} = O(|T|K \log K) + \sum_V \mathbb{T}(V)$, where the $O(|T|K \log K)$ term accounts for the time needed to sort the labels in each vertex according to $\sqsubseteq$ and for the time spent by Algorithm 1 to iterate over all the vertices.

By applying Property 1 and Property 2 to the definition of $\mathbb{T}(V)$, we have:

$$\mathbb{T}(V) = \sum_{l_V} a(l_V) + b(l_V) = O(k_V) + O\left(k_V \sum_{W \in \text{ch}(V)} k_W \right)$$

$$= O\left(k_V + k_V \sum_{W \in \text{ch}(V)} k_W \right)$$

which implies $\mathbf{T} = O(|T|K\log K) + \sum_V O\left(k_V + k_V \sum_{W \in \text{ch}(V)} k_W\right)$. By assuming all k_V equal to some constant κ, the above expression reduces to:

$$\mathbf{T} = O(|T|K\log K) + O\left(\sum_V \kappa + \kappa \sum_{W \in \text{ch}(V)} \kappa\right)$$

$$= O(|T|K\log K) + O\left(\kappa(1 + \kappa) \sum_V \sum_{W \in \text{ch}(V)} 1\right)$$

$$= O(|T|K\log K) + O(\kappa^2(|T| - 1))$$

where the last equality holds since the two summations altogether iterate a number of times equal to the number of edges in the tree. The worst case occurs when CDoT opens every label in every vertex. In such case $\kappa = K$ and the above formula reverts to $O(|T|K^2)$. In the best case CDoT opens only one vertex per layer, $\kappa = 1$ and the complexity is $O(|T|K\log K)$.

5 Related Works

Most of the research in solving the MAP problem deals with the much tougher problem of general graphs. In this context, due to the exponential cost of solving this problem exactly, most attempts focused on approximate techniques that trade accuracy for speed. Among approximate methods, we recall Loopy Belief Propagation [11], Linear Programming Relaxations [12], and Branch and Bound methods [13]. All mentioned approaches obtain better time performance at the price of lower accuracy. Another interesting approach is based on min-cuts of a graph built from the PGM [14]: it allows for exact inference on a certain class of binary variables and degrades to approximate inference otherwise.

Exact methods (e.g., variable elimination [15] and belief propagation [16]) basically work by pushing maximization operations to inner levels of the probability expression so to obtain a rewritten, cheaper to compute, expression. Their computational cost depends both on the time needed for rewriting the probability expression and on the time needed for computing it.

A distinguishing feature of our approach is that we assume that the rewritten expression for calculating the probability is given and we provide an algorithm that further improves its computation. To this regards, we are not aware of any competing approach that works in the case of tree-structured PGMs. In contrast, several algorithms have been proposed for the specific case of sequences (e.g., [9,17,10]).

6 Experiments

Before delving into the details of the experimentation we note that while the main discussion focused on directed graphical models, hereafter, as a way to prove the generality of the approach, we report the experiments on undirected trees.

In order to evaluate the performances of the proposed approach, we compare it with the direct evaluation of Expression (1). To these ends, we implemented a Dynamic Programming algorithm (hereafter referred to as the *DP* algorithm) that evaluates the expression. As mentioned, competing approaches deal with the more general case of unconstrained graphs and it would be unfair to compare them with an algorithm specifically built to work on trees. By implementing a dynamic programming version of Expression (1), we are basically comparing CDoT with the Variable Elimination algorithm, but disregarding the time needed to build the formula.

We performed a number of experiments to assess the time performance of the two algorithms by generating tree structures under a number of different problem settings. The parameters defining each experimental setting are:

$|T|$: the number of vertices in T;

K: the number of labels per vertex;

p_λ: the probability of increasing the tree depth. By setting $p_\lambda = 1$ the graph degenerates into a sequence; by setting it to 0, it degenerates into a graph where all vertices are connected to the root;

$\mu_\nu^\uparrow$, $\mu_\nu^\downarrow$: the set of labels in each vertex has been partitioned into two sets $\nu^\uparrow$ and $\nu^\downarrow$. The weights of the labels in the two sets have been drawn randomly from two Normal distributions. The parameters $\mu_\nu^\uparrow$ and $\mu_\nu^\downarrow$ control the mean values of the two distributions: $\mathcal{N}(\mu_\nu^\uparrow, 100)$ and $\mathcal{N}(\mu_\nu^\downarrow, 25)$;

$|\nu^\uparrow|$: the number of labels sampled from $\mathcal{N}(\mu_\nu^\uparrow, 100)$;

μ_τ, σ_τ: the mean and standard deviation for the weights associated with transitions among labels. Specifically, all transition weights have been sampled from the Normal distribution $\mathcal{N}(\mu_\tau, \sigma_\tau^2)$.

To better illustrate the experimentation we define a base experimental setting and derive the other experimental settings by varying some parameters. The defaults used are as follows.

Parameters:	$|T|$	K	p_λ	$\mu_\nu^\uparrow$	$\mu_\nu^\downarrow$	$|\nu^\uparrow|$	$|\nu^\downarrow|$	μ_τ	σ_τ
Values:	100	100	0.4	100	10	10	$K - |\nu^\uparrow|$	10	5

The setting for the base experiment corresponds to a problem involving 100 vertices (99 edges) and 10,000 labels, for a total of 990,000 transitions. By setting $\mu_\nu^\uparrow = 100$, $\mu_\nu^\downarrow = 10$, and $|\nu^\uparrow| = 10$, we define a tree where each vertex has several ($\sim$10%) high-weighted, highly discriminative labels. By setting a $\sigma_\tau = 5$ we define experimental conditions where, on average, the CDoT algorithm is expected to find the optimal solution by inspecting only a fraction of the transitions.

In the experimentation we explore variations to this setting so to compare the performance of the algorithms in different, harder, scenarios.

We generated a total of 100 problem instances for each setting, ran both algorithms on each problem instance and averaged the results.

The algorithms have been implemented in Ruby and run using the standard interpreter (version 1.9.3). All experiments have been executed on computers sporting Intel Xeon dual core CPUs (clock: 2.33GHz, RAM: 8Gb).

7   Results

We compare the performances of CDoT and of the DP algorithm on 36 different settings obtained by varying the following parameters: K, p_λ, $\mu_\nu^\downarrow$, and σ_τ. Results are reported in the following figure:

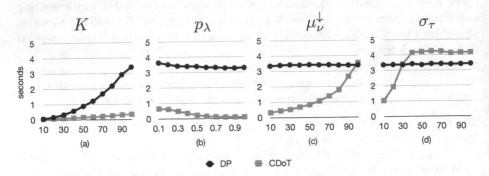

Panel (a) shows the performances of the two algorithms when varying K in $\{10, 20, \ldots, 100\}$. The figure confirms the complexity analysis reported in Section 4 for the best case scenario: the results show an almost linear dependency of CDoT on K. Vice versa, as expected, the DP algorithm shows a non linear (quadratic) dependency on K.

Panel (b) shows that the computational cost of the DP algorithm does not depend on the branching factor of the tree.[2] Interestingly, CDoT performances slightly improve as the branching factor decreases. To explain this behavior let us focus on the bounds at line 4 of Algorithm 2 and at lines 7 and 8 of Algorithm 3. It can be shown that $\omega(l_V) \geq \tilde{\omega}(l_V')$ iff:

$$\nu(l_V) - \nu(l_V') \geq \sum_{W \in \text{ch}(V)} \mathcal{T} - \tau(l_V \twoheadrightarrow l_W^{\rightarrow}) + \omega(l_W^\star) - \omega(l_W^{\rightarrow}) \tag{5}$$

where $l_W^\star$ is the best label assigned to the root of the subtree rooted in W, and $l_W^{\rightarrow}$ is the best label in W for transitions starting from l_V, i.e.: $l_W^{\rightarrow} = \arg\max_{l_W} [\tau(l_V \twoheadrightarrow l_W) + \omega(l_W)]$ and $l_W^\star = \arg\max_{l_W} \omega(l_W)$.
Inequality (5) shows that the bounds are progressively more likely to hold as the number of children of V decreases, thus implying better performances of the CDoT algorithm as the average branching factor decreases.

Panels (c) and (d) show how the performances of the two algorithms vary as the assumptions underlying the CDoT optimization strategy progressively weaken. The CDoT computational cost changes depending on how much the

---

[2] The Figure actually shows a small increase in the performances of the DP algorithm that is not justified by its complexity analysis. Further experiments show that the observed behavior depends on how hashes handling the transitions in the graph are laid out in memory. Importantly, both algorithms share the same data structures, so they are equally affected by this implementation detail.

$\nu(l_V)$ are able to predict the best possible label (i.e., how likely it is that the difference $\nu(l_V) - \nu(l'_V)$ is large), and on how likely high $\tau(l_V \rightarrow l_W)$ are associated with high $\nu(l_V)$.

Panel (c) shows that CDoT is very resilient to fluctuations of the likelihood of $\nu(l_V) - \nu(l'_V)$. Even when the difference is very small (e.g., for $\mu_\nu^\downarrow = 90$) the algorithm still outperforms DP. The same does not hold for σ_τ: panel (d) shows that when σ_τ grows larger than 30, CDoT looses its edge over DP. This is the only case where we observe higher costs in running CDoT instead of DP. Performances are *lower* instead of the *same* due to the cost of sorting the labels to build $\mathcal{L}_\sqsubseteq(V)$ (Algorithm 2 – line 3, and Algorithm 3 – line 5). The actual point where the two curves cross depends on the interaction between the parameter σ_τ and the difference $\mu_\nu^\uparrow - \mu_\nu^\downarrow$. Again, Inequality (5) helps to explain this behavior. In fact, $\mu_\nu^\uparrow - \mu_\nu^\downarrow$ controls the likelihood that the left hand side is large; term $\mathcal{T} - \tau(l_V \rightarrow l_W^\rightarrow)$ tends to get larger values as the standard deviation of τ grows.

As mentioned in Section 1, past evidence on sequences corroborates the hypothesis that real world problems often feature strong evidence associated with vertices, while evidence on transitions are mainly useful to discriminate ambiguous cases. In commenting the above findings, we would like to recall that CDoT is badly affected by high transition variances when it is frequent that high-weighted labels are linked by low-weighted transitions. In fact, it can be argued that in real world problems transition weights mostly corroborate the predictions formulated by considering vertex evidence, rather than contradicting them.

Summarizing, while the worst case complexity of CDoT is the same of DP, in our experiments it outperforms DP in most cases. Furthermore, even in a scenario where evidence on transition often contradicts the one on the vertices, running CDoT instead of DP only requires few additional resources.

8 Concluding Remarks

In this paper we introduced CDoT, a novel algorithm for answering MAP queries on tree structured PGMs. We discussed its properties, specifically the reasons underlying its behavior; and studied its asymptotic complexity, showing that it has $O(|T|K \log K)$ best case complexity and that it is never asymptotically worse than previous approaches. We also provided an empirical assessment of its performances. Experiments bolster the theoretical analysis, showing that the algorithm performances substantially improve over a dynamic programming based competitor.

The provided experimentation is large (totaling 3,600 runs of each algorithm) albeit limited to synthetic data. We acknowledge the importance of an experimentation assessing to what extent real problems match the assumptions underlying the CDoT algorithm. These efforts may indeed shed light on important facets of CDoT behavior, facets that can be useful for deciding when it is worth adopting it. We defer to future work such experimentation.

References

1. Koller, D., Friedman, N.: Probabilistic Graphical Models: Principles and Techniques. MIT Press (2009)
2. Heckerman, D., Horvitz, E., Nathwani, B.: Toward normative expert systems: The Pathfinder project. Knowledge Systems Laboratory, Stanford University (1992)
3. Szeliski, R., Zabih, R., Scharstein, D., Veksler, O., Kolmogorov, V., Agarwala, A., Tappen, M., Rother, C.: A comparative study of energy minimization methods for markov random fields with smoothness-based priors. IEEE Trans. Pattern Anal. Mach. Intell. 30(6), 1068–1080 (2008)
4. Sun, Y., Deng, H., Han, J.: Probabilistic models for text mining. In: Aggarwal, C.C., Zhai, C. (eds.) Mining Text Data, pp. 259–295. Springer (2012)
5. Johnson, M., Griffiths, T., Goldwater, S.: Bayesian inference for PCFGs via Markov Chain Monte Carlo. In: Human Language Technologies 2007, pp. 139–146 (2007)
6. Csűrös, M., Miklós, I.: Streamlining and large ancestral genomes in archaea inferred with a phylogenetic birth-and-death model. Mol. Bio. Evol. 26(9) (2009)
7. Paiement, J.F., Eck, D., Bengio, S.: A Probabilistic Model for Chord Progressions. In: Proc. of the 6th Int. Conf. on Music Information Retrieval, London (2005)
8. Viterbi, A.J.: Error bounds for convolutional codes and an asymptotically optimum decoding algorithm. IEEE Trans. Inf. Theory 13, 260–269 (1967)
9. Esposito, R., Radicioni, D.P.: CarpeDiem: Optimizing the Viterbi Algorithm and Applications to Supervised Sequential Learning. JMLR 10, 1851–1880 (2009)
10. Belanger, D., Passos, A., Riedel, S., McCallum, A.: Speeding up MAP with Column Generation and Block Regularization. In: Proc. of the ICML Workshop on Inferning: Interactions Between Inference and Learning. Omnipress (2012)
11. Murphy, K., Weiss, Y., Jordan, M.: Loopy belief propagation for approximate inference: An empirical study. In: Proc. of the 15th Conf. on Uncertainty in Art. Intell., pp. 467–475 (1999)
12. Wainwright, M., Jaakkola, T., Willsky, A.: Map estimation via agreement on trees: message-passing and linear programming. IEEE Trans. Inf. Theory 51(11), 3697–3717 (2005)
13. Marinescu, R., Kask, K., Dechter, R.: Systematic vs. non-systematic algorithms for solving the mpe task. In: Proc. of the 9th Conf. on Uncertainty in Artificial Intelligence, pp. 394–402. Morgan Kaufmann Publishers Inc. (2002)
14. Boykov, Y., Veksler, O., Zabih, R.: Fast approximate energy minimization via graph cuts. IEEE Trans. Pattern Anal. Mach. Intell. 23(11), 1222–1239 (2001)
15. Zhang, N., Poole, D.: Exploiting causal independence in bayesian network inference. JAIR 5, 301–328 (1996)
16. Weiss, Y., Freeman, W.: On the optimality of solutions of the max-product belief-propagation algorithm in arbitrary graphs. IEEE Trans. Inf. Theory 47(2), 736–744 (2001)
17. Kaji, N., Fujiwara, Y., Yoshinaga, N., Kitsuregawa, M.: Efficient staggered decoding for sequence labeling. In: Proc. of the 48th Meeting of the ACL, pp. 485–494 (2010)

Gesture Recognition for Improved User Experience in a Smart Environment

Salvatore Gaglio, Giuseppe Lo Re, Marco Morana, and Marco Ortolani

DICGIM, University of Palermo – Italy
{salvatore.gaglio,giuseppe.lore,marco.morana,marco.ortolani}@unipa.it

Abstract. Ambient Intelligence (AmI) is a new paradigm that specifically aims at exploiting sensory and context information in order to adapt the environment to the user's preferences; one of its key features is the attempt to consider common devices as an integral part of the system in order to support users in carrying out their everyday life activities without affecting their normal behavior.

Our proposal consists in the definition of a gesture recognition module allowing users to interact as naturally as possible with the actuators available in a smart office, by controlling their operation mode and by querying them about their current state. To this end, readings obtained from a state-of-the-art motion sensor device are classified according to a supervised approach based on a probabilistic support vector machine, and fed into a stochastic syntactic classifier which will interpret them as the basic symbols of a probabilistic gesture language. We will show how this approach is suitable to cope with the intrinsic imprecision in source data, while still providing sufficient expressivity and ease of use.

1 Introduction

Our daily-life activities involve spending an increasingly high amount of time indoor, whether at home or at work, so improving the perceived quality-of-life without intruding into the users' habits is of great importance; this is the purpose of the novel discipline of Ambient Intelligence (AmI) [1], which combines the use of pervasively deployed sensing devices and actuators with advanced techniques from Artificial Intelligence. Due the primary role of the end user, the intrinsic prerequisite of any AmI system (i.e. the presence of "ubiquitous" monitoring tools) must be coupled to the additional requirement of providing the system with efficient functionalities for interaction with the system; moreover, considering the high level of pervasiveness obtainable through currently available devices, the use of equally unobtrusive interfaces is mandatory. An interesting scenario is represented by *smart offices*, where typical applications include energy efficiency control, ambient-assisted living (AAL) and human-building interaction (HBI).

In this work, we build up on our previous experience about a testbed for designing and experimenting with WSN-based AmI applications [2] and we describe a module aimed at assisting users in order to ease their interaction with

M. Baldoni et al. (Eds.): AI*IA 2013, LNAI 8249, pp. 493–504, 2013.

the system itself. In our previous work, we described the deployment of the AmI infrastructure in our department premises; the sensory component is implemented through a Wireless Sensor and Actuator Network (WSAN), whose nodes are equipped with off-the-shelf sensors for measuring such quantities as indoor and outdoor temperature, relative humidity, ambient light exposure and noise level. WSANs extend the functionalities of traditional sensor networks by adding control devices to modify the environment state. The present proposal suggests the use of a motion sensor device as the primary interface between the user and the AmI system. The device we used for sensing motion within the environment is Microsoft Kinect, which, in our vision, plays the dual role of a sensor (since it may be used for some monitoring tasks, i.e. people detection and people counting) and a controller for the actuators. In particular, the latter is performed by training a probabilistic classifier for recognizing some simple hand gestures in order to produce a basic set of commands; since the output of such a classifier is noisy due to the non perfect recognition of the hand shape, we chose to reinforce the classification process by means of a probabilistic syntactic recognizer, realized as a stochastic parser for an ad-hoc gesture language. The use of a grammar for the comprehension of the visual commands is significant as it helps both in dealing with the intrinsically noisy input, and in providing a smoother interface for users.

The paper is organized as follows: after presenting some related work in Section 2, we describe the main modules of the proposed system architecture in Section 3. The assessment of the system in a deployment scenario realized in our department will be discussed in Section 4. Conclusions will follow in Section 5.

2   Related Work

Several works have focused on hand gesture recognition using depth image; a survey is presented in [3], whereas in [4] an approach for hand gesture recognition using Kinect is proposed. The authors preliminary detect the hand by thresholding the depth image provided by the Kinect, then the finger identification task is performed by applying a contour tracing algorithm on the detected hand image. Even if the obtained results are promising, the whole system is based on a number of parameters and thresholds that make the approach unreliable while varying the application scenario. In [5] a novel dissimilarity distance metric based on Earth Mover's Distance (EMD) [6], i.e., Finger-Earth Mover's Distance (FEMD), is presented. The authors showed that FEMD based methods for hand gesture recognition outperform traditional shape matching algorithm both in speed and accuracy.

Besides processing visual input, we are interested in classifying the different hand shapes and effectively translate them into commands for the AmI system. Our system is based on a structural approach to pattern recognition of vectors of features. Most commonly used methods (e.g. nearest-neighbor classifiers, or neural networks) resort to machine-learning techniques for addressing this problem when the features can be represented in a metric space. Our choice, instead, was to work with symbolic input and to rather address a discrete problem by a syntactic pattern recognition method. More specifically, we assume that if user hand gestures are to represent an intuitive language for interacting with the system, then rules of some sort must be used to generate the strings of such language; hence, their inherent structure should not be disregarded, and might turn into a useful support when the system needs to recognize and interpret such strings. This approach is not new and in fact its first proposals date back in time, even though they were referred to different contexts, such as image processing or grammatical inference [7]; recognition based on parsing has been successfully employed also to automatic natural language of handwriting recognition [8].

Such methods have been later expanded also to allow for stochastic grammars, where there are probabilities associated with productions [9], and it has been shown that a grammar may be considered a specification of a prior probability for a class. Error-correcting parsers have been used when random variations occur in an underlying stochastic grammar [10]. Finally probability theory has been applied to languages in order to define the probabilities of each word in a language [11].

3 Providing Smooth Human-System Interaction

Our aim is to provide users of a smart environment with the possibility to interact with the available actuators as naturally as possible, by controlling their operation mode and by querying them about their current state. For instance, they might want to act upon some of the actuators (e.g. air conditioning system, or lighting) by providing a set of consecutive commands resulting in complex configurations, such as turn on the air conditioning system, set the temperature to a certain degree, set the fan speed to a particular value, set the air flow to a specified angle, and so on.

This section describes our proposal for an interface whose use, consistently with AmI philosophy, will not impact on the users' normal behavior; to this aim, we adapt the functionality of a flexible motion sensing input device, namely Microsoft Kinect, to detect simple gestures of the user's hand and to translate them into commands for the actuators. This peripheral has attracted a number of researchers due to the availability of open-source and multi-platform libraries that reduce the cost of developing new algorithms; a survey of the sensor and corresponding libraries is presented in [12,13]. Direct use of the motion sensor is however not viable for our purposes, due to its non negligible intrinsic imprecision; moreover, naively mapping hand gestures into commands would be very awkward for users, and would likely result into them refusing to use the interface altogether.

Our perspective is to consider Kinect as a sensor to transparently gather observations about users' behavior [14], higher-level information may be extracted by such sensed data in order to produce actions for adapting the environment to the users requirements, by acting upon the available actuators. A preliminary version of this approach was presented in [15] with reference to the architecture

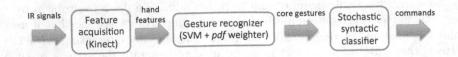

Fig. 1. The main functional blocks of the proposed gesture recognition algorithm

described in [16]. In order to turn the Kinect sensor into an effective human-computer interface, we adopt a probabilistic approach based on recognizing simple hand gestures (i.e., the number of extended fingers) in order to produce a set of commands; however, the output of such a classifier will be affected by noise due to the imperfect recognition of the shape of hand, therefore we chose to reinforce the classification process by means of a grammar. The use of an error-correcting parser for such grammar will add significant improvement to the recognition of visual commands and will make the use of the whole system more natural for the end user.

The overall scheme for our gesture recognition system is shown in Figure 1.

3.1   Capturing and Classifying Hand Gestures

Several vision-based systems have been proposed during the last 40 years for simple gesture detection and recognition. However, the main challenge of any computer vision approach is to obtain satisfactory results not only in a controlled testing environment, but also in complex scenarios with unconstrained lighting conditions, e.g., a home environment or an office. For this reason, image data acquired by multiple devices are usually merged in order to increase the system reliability. In particular, range images, i.e., 2-D images in which each pixel contains the distance between the sensor and a point in the scene, provide very useful information about the elements of the scene, e.g., a moving person, but range sensors used to obtain them are very expensive.

We selected Kinect for our target scenario, thanks to its flexibility, and limited cost; it is equipped with 10 input/output components, as depicted in Figure 2, which make it possible to sense the users and their interactions with the surrounding environment. Its projector shines a grid of infrared dots over the scene, and the embedded IR camera captures the infrared light, then its factory calibration allows to compute the exact position of each projected dot against a surface at a known distance from the camera. Such information is finally used to create depth images of the observed scene, with pixel values representing distances, in order to capture the object position in a three-dimensional space.

An example of hand tracking using Kinect is provided by the OpenNI/NITE packages, whose hand detection algorithm is based on the five gesture detectors listed on the right-hand side of Figure 2, and a hand point listener; however, those APIs are based on a global skeleton detection method, so the hand is defined just as the termination of the arm and no specific information about the hand state (e.g., an open hand vs a fist or the number of extended digits) is

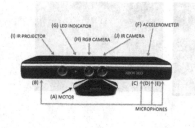

Gesture	Description
wave	hand movement as left-right waving, carried out at least four times in a row.
push	hand movement as pushes towards and away from the sensor.
swipe	hand movement, either up, down, left or right, followed by the hand resting momentarily.
circle	hand that make a full circular motion in either direction.
steady	hand that hasn't moved for some time.

Fig. 2. Kinect components, and the included OpenNI/NITE gesture detectors

provided. For this reason, we only consider this method as the first step of our gesture recognition algorithm, and we add further processing to extract more precise information based on the image representing the area where the hand is located.

The depth images are segmented in order to detect the *hand plane*, i.e. the set of image points that are at the same distance z from the Kinect sensor, where z is the value of the depth image $DI(x, y)$, with x and y indicating the coordinates of the hand as detected by the APIs. Each hand mask is then normalized with respect to scale, and represented as a time-series curve [17]. Such shape representation techniques is also proposed in [5] and is one of the most reliable method for the classification and clustering of generic shapes. A time-series representation allows to capture the appearance of the hand shape in terms of distances and angles between each point along the hand border and the center of mass of the hand region; in particular, as suggested by [5], a time time-series representation can be plotted as a curve where the horizontal axis denotes the angle between each contour vertex, the center point and a reference point along the hand border and the vertical axis denotes the Euclidean distance between the contour vertices and the center point. Figure 3 shows three examples of hand masks, as depth images capture by Kinect (top row); the centroid of the hand region (red cross) is used as center point for the time-series computation, while the lowest point along the hand perimeter (red circle) is used as reference point for computing the angles between the center and the contour points. The time-series curve representation (bottom row) is obtained by plotting the angles in the horizontal axis and the distances on the vertical axis.

The time-series describing the hand shape represents the feature we will analyze in order to discriminate between the set of considered hand gestures; in particular we need to classify each captured hand image according to one of six possible classes, with each hand gesture characterized by the number of extended fingers (from zero in the case of a fist, to five if the hand is open).

The gesture classification has been performed by means of a multi-class Support Vector Machine (SVM) classifier based on a RBF kernel. SVMs are supervised learning models used for binary classification and regression, and multi-class SVMs are usually implemented by combining several binary SVMs according to three main strategies: one-versus-all, one-versus-one, Directed Acyclic

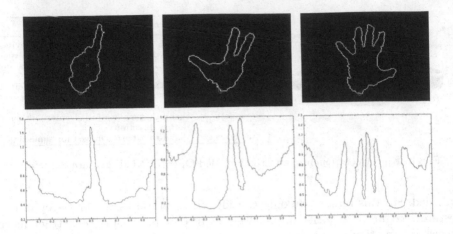

Fig. 3. Gesture representation by means of time-series features

Graphs SVM (DAGSVM). Several studies addressed the issue of evaluating which is the best multi-class SVM method and most works (e.g., [18,19]) claimed that the one-versus-one approach is preferable to other methods. However, what we want to obtain is a membership degree of each detected gesture in the whole set of gestures. Since the output of an SVM is not a probabilistic value, we chose to apply the method proposed by [20] to map the SVM output into a positive class posterior probability by means of a sigmoid function.

3.2   Gesture Language Description

Our Kinect sensor provides an effective way of capturing the input from the user, but although in principle hand gestures could be directly translated into commands, the mere recognition of very basic ones is likely inadequate to cover the broad spectrum of possible instructions with sufficient detail; moreover, a 1-to-1 correspondence between gesture and command would result into an awkward interaction with the system. An additional challenge is represented by the intrinsic imprecision in data obtained through Kinect, which is represented by the probabilistic output of the SVM classifier. In order to obtain an overall satisfying behavior, and to allow users to express a broad set of commands in a natural way starting from elementary and customary gestures, we regard the set of possible commands and queries as strings of a language, such as for instance: "**query** id_1 **status**", or "**set** id_1, id_2, id_3 **start**" for getting the operating status of actuator id_1, or activating actuators id_1, id_2, id_3, respectively.

In this perspective, the language can be precisely defined with the notation borrowed from formal language theory, and hand gestures can be regarded as the symbols of the underlying alphabet of core gestures, assuming we can sample the images acquired by the Kinect with a pre-fixed frequency (i.e., we can identify repetitions of the same symbol); the Kinect sensor only allows to roughly assess the number of extended fingers, with no finer detail; moreover, we will consider an additional symbol representing a separator, corresponding to the case when no gesture is made. The following alphabet will thus constitute the basis for the subsequent discussion:

$$\Sigma = \{\circ, \bullet^1, \bullet^2, \ldots, \bullet^5, \lrcorner\};$$

with $\circ$ indicating the fist (i.e. no recognized finger), $\bullet^n$ indicating that n fingers were detected, and $\lrcorner$ as the separator, when the hand is out of the field of view.

We used this alphabet to code the basic keywords of our language, and we devised a basic grammar capturing a gesture language expressing simple queries and commands to the actuators. So for instance, the proper sequence of gestures by the user (i.e. "$\bullet^1 \lrcorner$") will be understood as the **query** keyword, representing the beginning of the corresponding statement, and similarly for other "visual lexemes".

Even a straightforward language will be able to capture an acceptable range of instructions given by the user in a natural way, and may be easily defined in terms of a formal grammar; its core symbols, however, will inevitably be affected by noise, due to the imprecision in the hand shape classification, and such uncertainty will be expressed by the posterior probabilities attached by the SVM classifier to each instance of the hand shape. In order to interpret instructions for actuators we will thus need to infer the correct gesture pattern to which the incoming symbol sequence belongs.

Following the approach initially described by [21], we chose to regard this problem as a variant of stochastic decoding: we formalized the structure of proper queries and commands by means of context-free grammars expressed in Backus-Naur Form (BNF), and we consequently expect our instructions to conform to one of the pre-defined gesture patterns. In particular, our reference grammars for the language of queries $\mathcal{L}_q$, and the language of commands $\mathcal{L}_c$ are respectively defined as follows:

$S' \rightarrow stat' \mid stat' \; S'$	$S'' \rightarrow stat'' \mid stat'' \; S''$
$stat' \rightarrow$ **query** $idlist$ **status**	$stat'' \rightarrow$ **set** $idlist \; cmd$
$\mid$ **query** $idlist$ **compare** id	$cmd \rightarrow$ **start** $\mid$ **stop**
	$\mid modif$ [**increase** $\mid$ **decrease**]
	$modif \rightarrow$ [**low** $\mid$ **med** $\mid$ **high**]

where the sets of non-terminal symbols, and the start symbol of each grammar are implicitly defined by the productions themselves, and the terminal ones are actually coded in terms of the gesture alphabet[1].

Interpreting gestures may thus be regarded as an instance of a two-class classification problem in which each class of patterns is assumed to be described by the language generated by either of the two grammars. The symbols of alphabet are produced by a noisy sensor, so we must take into account the presence of erroneous symbols; however our noise deformation model will assume that the

---

[1] For the sake of brevity the productions for $idlist$ were not included; unsurprisingly, the list of id's can be described by a regular expression over the available symbols. The regular expression $(\bullet^1 \circ)^+$ was used to define a valid id in our prototypal system.

deformation process does not affect the length of the original string. The only information needed to characterize the process is the probability of corruption for each of the input symbols; more specifically, if $a, b \in \Sigma$ are two terminal symbols (i.e. hand gestures), we take advantage of the training process for the SVM classifier also to compute the conditional probability $r(a|b)$ that a is mistakenly interpreted as b; the probability that an entire string x is deformed to turn into a different y (in other words, that a query may be mistaken for a command, for instance) is given by $p(x|y) = r(a_{1)}|b_1)r(a_{2)}|b_2)...r(a_{n)}|b_n)$.

As pointed out in [22], this kind of stochastic decoding basically corresponds to maximum-likelihood classification of deformed patterns. In our model, the grammars for both $\mathcal{L}_q$ and $\mathcal{L}_c$ are context-free, so each of them may be re-written in Chomsky normal form; this allows us to use the probabilistic version of the traditional Cocke-Younger-Kasami parsing method as described in [21] to reliably recognize gesture patterns and translate them into instructions for the actuators.

At the end of the process, we obtain a reliable translation of a whole sequence of gestures into the corresponding command or query for the actuators. Such approach gives us also the opportunity to exploit the potentialities of the parser used to process the visual language; the underlying structure provided by our formulation in terms of grammars results into a smoother interaction for the user, as well as in greater precision for the overall gesture sequences recognition.

4   Performance Assessment

The proposed method is part of a system aiming for timely and ubiquitous observations of an office environment, namely a department building, in order to fulfill constraints deriving both from the specific user preferences and from considerations on the overall energy consumption.

The system will handle high-level concepts as "air quality", "lighting conditions", "room occupancy level", each one referring to a physical measurement captured by a physical layer. Since the system must be able to learn the user preferences, ad-hoc sensors for capturing the interaction between users and actuators are needed similarly to what is described in [23]. The plan of one office, giving an example of the adopted solutions, is showed in Figure 4.

The sensing infrastructure was realized by means of a WSAN, whose nodes are able to measure temperature, relative humidity, ambient light exposure and noise level. Actuators were available to change the state of the environment by acting on some quantities of interest; in particular the air-conditioning system, the curtain and rolling shutter controllers, and the lighting regulator address this task by modifying the office temperature and lighting conditions. The users' interaction with actuators was captured via the Kinect sensor (Fig. 4-**H**) that was also responsible for detecting the presence and count the number of people on the inside of the office. The Kinect was connected to a miniature fanless PC (i.e., fit-PC2i) with Intel Atom Z530 1.6GHz CPU and Linux Mint OS, that guarantees real-time processing of the observed scene with minimum levels of obtrusiveness and power consumptions (i.e., 6W).

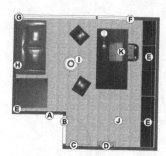

A. RFID Reader
B. Electric Lock
C. Energy Meter
D. Air Conditioning
E. Sensor Node
F. Curtain Controller
G. Rolling Shutter Controller
H. Kinect
I. Lighting Regulator
J. IP Camera
K. Software Sensor

Fig. 4. Monitored office

We evaluated the performance of our system by separately assessing the multi-class SVM classifier and the interpreter for the gesture language. The basic hand gesture classifier was tested under varying lighting conditions and poses showing an acceptable level of robustness, thanks to the fact that the input sensor makes use of depth information in detecting the hand mask, while compensating for the lower quality of the RGB data. Results showed that about 75% of the gestures are correctly classified when the user acts in a range of 1.5 to 3.5 meters from the Kinect; however, greater distances have a negative impact on performance due to the physical limits of the infrared sensor. Moreover we noticed that even when a gesture is misclassified, the correct gesture is the second choice (i.e., rated with the second most probable value) in approximately 70% of the cases. This observation shows that the information provided by the SVM classifier can provide reliable input for the gesture sequence interpreter.

The interpreter functionalities were preliminarily verified by means of a synthetic generator of gestures allowing for the validation of both the alphabet and the grammar we chose. The overall system has been tested by conducting a set of experiments involving 8 different individuals. Each person was positioned in front of Kinect at a distance within the sensor range and was asked to interact with the device by performing a random sequence of 20 gestures chosen from a predefined set of 10 commands, and 10 queries referred to the actuators depicted in Figure 4. Examples of commands and queries, expressed in terms of the grammars described in the previous section, were:

set *light* **low increase**	: increases lighting by a small amount by acting on the dimmer
set *heater* **high increase**	: increases HVAC temperature setpoint by 3°C
set *heater* **stop**	: stops HVAC
query *light, heater* **status**	: gets current status of light and HVAC
query *energymeter*	: gets current value read from energy meter

where *light, heater, energymeter* indicate the id's of the corresponding actuators.

The proposed system was able to correctly classify the input gestures sequences in 88.125% of the cases, corresponding to 141 positives out of 160 inputs;

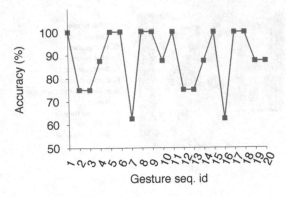

Fig. 5. Accuracy of gesture recognition

Figure 5 shows a plot of the outcome for each trial. While performing translation of the visual commands, the gesture interpreter had to classify the sequences into one of the two classes represented by the languages for queries and commands respectively, thus giving insight about the "structure" of the sentence recognized so far. In a future development, this could be used to provide feedback to the user about the system comprehension of the interaction, resulting in an overall smoother user experience.

5   Conclusions

This work described an approach to the design of a user-friendly interface based on gesture recognition as part of an AmI system; our application scenario is the management of an office environment by means of Kinect, an unobtrusive sensing tool equipped with input/output devices that make it possible to sense the user and their interactions with the surrounding environment. The control of the actuators of the AmI system (e.g., air-conditioning, curtain and rolling shutter) is performed by capturing some simple gestures via the Kinect and recognizing opportunely structured sequences by means of a symbolic probabilistic approach. Besides allowing for a smoother interaction between users and the system, our approach is able to cope with imprecision in basic gesture acquisition thanks to the use of stochastic decoding, which basically corresponds to maximum-likelihood classification of deformed patterns. The system was tested on an actual prototype of a smart office, which we built in our department premises as part of a research project investigating the use of Ambient Intelligence for energy efficiency.

Acknowledgment. This work was partially supported by the SMARTBUILD-INGS project, funded by PO-FESR SICILIA 2007-2013.

References

1. Aarts, E., Marzano, S.: The new everyday: views on ambient intelligence. 010 Publishers (2003)
2. De Paola, A., Gaglio, S., Lo Re, G., Ortolani, M.: Sensor9k: A testbed for designing and experimenting with WSN-based ambient intelligence applications. Pervasive and Mobile Computing 8(3), 448–466 (2012)
3. Suarez, J., Murphy, R.: Hand gesture recognition with depth images: A review. In: 2012 IEEE RO-MAN, pp. 411–417 (September 2012)
4. Li, Y.: Hand gesture recognition using kinect. In: 2012 IEEE 3rd International Conference on Software Engineering and Service Science (ICSESS), pp. 196–199 (June 2012)
5. Ren, Z., Yuan, J., Zhang, Z.: Robust hand gesture recognition based on finger-earth mover's distance with a commodity depth camera. In: Proceedings of the 19th ACM International Conference on Multimedia. MM 2011, pp. 1093–1096. ACM, New York (2011)
6. Rubner, Y., Tomasi, C., Guibas, L.J.: The earth mover's distance as a metric for image retrieval. Int. J. Comput. Vision 40(2), 99–121 (2000)
7. Hopcroft, J.E., Ullman, J.D.: Introduction to Automata Theory, Languages, and Computation, 3/E. Pearson Education (2008)
8. Fu, K.-S., Albus, J.E.: Syntactic pattern recognition and applications, vol. 4. Prentice-Hall, Englewood Cliffs (1982)
9. Liberman, M.Y.: The trend towards statistical models in natural language processing. In: Natural Language and Speech, pp. 1–7. Springer (1991)
10. Lu, S.-Y., Fu, K.-S.: Stochastic error-correcting syntax analysis for recognition of noisy patterns. IEEE Transactions on Computers 100(12), 1268–1276 (1977)
11. Booth, T.L., Thompson, R.A.: Applying probability measures to abstract languages. IEEE Transactions on Computers 100(5), 442–450 (1973)
12. Kean, S., Hall, J., Perry, P.: Meet the Kinect: An Introduction to Programming Natural User Interfaces, 1st edn. Apress, Berkely (2011)
13. Borenstein, G.: Making Things See: 3D Vision With Kinect, Processing, Arduino, and MakerBot. Make: Books. O'Reilly Media, Incorporated (2012)
14. Cottone, P., Lo Re, G., Maida, G., Morana, M.: Motion sensors for activity recognition in an ambient-intelligence scenario. In: 2013 IEEE International Conference on Pervasive Computing and Communications Workshops (PERCOM Workshops), pp. 646–651 (2013)
15. Lo Re, G., Morana, M., Ortolani, M.: Improving user experience via motion sensors in an ambient intelligence scenario. In: PECCS 2013 - Proceedings of the 3rd International Conference on Pervasive Embedded Computing and Communication Systems, Barcelona, Spain, February 19-21, pp. 29–34 (2013)
16. De Paola, A., La Cascia, M., Lo Re, G., Morana, M., Ortolani, M.: User detection through multi-sensor fusion in an ami scenario. In: 2012 15th International Conference on Information Fusion (FUSION), pp. 2502–2509 (July 2012)
17. Keogh, E., Wei, L., Xi, X., Lee, S.-H., Vlachos, M.: Lb keogh supports exact indexing of shapes under rotation invariance with arbitrary representations and distance measures. In: Proceedings of the 32nd International Conference on Very Large Data Bases. VLDB 2006, pp. 882–893. VLDB Endowment (2006)
18. Hsu, C.-W., Lin, C.-J.: A comparison of methods for multiclass support vector machines. IEEE Transactions on Neural Networks 13(2), 415–425 (2002)

19. Duan, K. B., Keerthi, S.S.: Which is the best multiclass SVM method? An empirical study. In: Oza, N.C., Polikar, R., Kittler, J., Roli, F. (eds.) MCS 2005. LNCS, vol. 3541, pp. 278–285. Springer, Heidelberg (2005)
20. Platt, J.C.: Probabilistic outputs for support vector machines and comparisons to regularized likelihood methods. In: Advances in Large Margin Classifiers, pp. 61–74. MIT Press (1999)
21. Fung, L.-W., Fu, K.-S.: Stochastic syntactic decoding for pattern classification. IEEE Transactions on Computers 100(6), 662–667 (1975)
22. Gonzalez, R.C., Thomason, M.G.: Syntactic pattern recognition: An introduction. Addison-Wesley Publishing Company, Reading (1978)
23. Morana, M., De Paola, A., Lo Re, G., Ortolani, M.: An Intelligent System for Energy Efficiency in a Complex of Buildings. In: Proc. of the 2nd IFIP Conference on Sustainable Internet and ICT for Sustainability (2012)

An Efficient Algorithm
for Rank Distance Consensus

Liviu P. Dinu and Radu Tudor Ionescu

Faculty of Mathematics and Computer Science
University of Bucharest, 14 Academiei Street, Bucharest, Romania
ldinu@fmi.unibuc.ro, raducu.ionescu@gmail.com

Abstract. In various research fields a common task is to summarize the information shared by a collection of objects and to find a consensus of them. In many scenarios, the object items for which a consensus needs to be determined are rankings, and the process is called rank aggregation. Common applications are electoral processes, meta-search engines, document classification, selecting documents based on multiple criteria, and many others. This paper is focused on a particular application of such aggregation schemes, that of finding motifs or common patterns in a set of given DNA sequences. Among the conditions that a string should satisfy to be accepted as consensus, are the median string and closest string. These approaches have been intensively studied separately, but only recently, the work of [1] tries to combine both problems: to solve the consensus string problem by minimizing both distance sum and radius.

The aim of this paper is to investigate the consensus string in the rank distance paradigm. Theoretical results show that it is not possible to identify a consensus string via rank distance for three or more strings. Thus, an efficient genetic algorithm is proposed to find the optimal consensus string. To show an application for the studied problem, this work also exhibits a clustering algorithm based on consensus string, that builds a hierarchy of clusters based on distance connectivity. Experiments on DNA comparison are presented to show the efficiency of the proposed genetic algorithm for consensus string. Phylogenetic experiments were also conducted to show the utility of the proposed clustering method. In conclusion, the consensus string is indeed an interesting problem with many practical applications.

Keywords: rank distance, consensus string, genetic algorithm, hierarchical clustering, DNA comparison, DNA sequencing, phylogenetic tree.

1 Introduction

In many important problems of various research fields a common task is to summarize the information shared by a collection of objects and to find a consensus of them. In many scenarios, the object items for which a consensus needs to be determined are rankings. The process of aggregating a set of rankings into a single one is called *rank aggregation*. Common applications of rank aggregation

M. Baldoni et al. (Eds.): AI*IA 2013, LNAI 8249, pp. 505–516, 2013.
© Springer International Publishing Switzerland 2013

schemes are electoral processes, meta-search engines, document classification, selecting documents based on multiple criteria, and many others. Most of these applications are based on an approach that involves combining (or aggregating) classifier schemes. This approach has been intensively studied in recent years [7], various classification schemes [4,5,9] being devised and experimented in different domains.

A particular application of such aggregation schemes is that of finding motifs or common patterns in a set of given DNA sequences. In many important problems in computational biology a common task is to compare a new DNA sequence with sequences that are already well studied and annotated. DNA sequence comparison is ranked in the top of several lists with major open problems in bioinformatics, such as [13]. Sequences that are similar would probably have the same function, or, if two sequences from different organisms are similar, there may be a common ancestor sequence [16]. Typical cases where the problem of finding motifs or common patterns in a set of given DNA sequences occurs are: genetic drugs design with structure similar to a set of existing sequences of RNA, PCR primer design, genetic probe design, antisense drug design, finding unbiased consensus of a protein family, motif finding, etc. All these applications share a task that requires the design of a new DNA or protein sequence that is very similar to (a substring of) each of the given sequences.

A typical combination scheme consists of a set of individual classifiers and a combiner which aggregates the results of the individual classifiers to make the final decision. Among the conditions that a string should satisfy to be accepted as a consensus, the two most important conditions are:

1. Median String: to minimize the sum of distances from all the strings in a given set to the consensus.
2. Closest String: to minimize the maximal distance (or radius) from the strings in a given set to the consensus.

The work of [1] presents a first algorithm which tries to combine both problems: to solve the consensus string problem by minimizing both distance sum and radius. Both problems use a particularly distance, namely Hamming distance. However, various distances usually conduct to different theoretical and empirical results. The aim of this paper is to investigate the consensus string in the rank distance paradigm [4]. Theoretical results show that it is not possible to identify a consensus string via rank distance for three or more strings. Thus, an efficient genetic algorithm is proposed to find the optimal consensus string by adapting the approach proposed by [7]. To show an application for the studied problem, this work also exhibits a clustering algorithm based on consensus string, that builds a hierarchy of clusters based on distance connectivity. Clusters are connected by considering the consensus string, which unifies the approach presented in [5] that considers the median string, and the approach presented in [6] that considers the closest string. Several experiments on DNA comparison are presented to assess the performance of the genetic algorithm for consensus string. Biological experiments are also conducted to show the utility of the proposed clustering method.

The contributions of this paper are:

1. It investigates consensus string under rank distance.
2. It presents an efficient genetic algorithm for consensus string.
3. It gives a hierarchical clustering method for strings based on consensus string.

The paper is organized as follows. Section 2 introduces notation and mathematical preliminaries. Section 3 gives on overview of related work regarding the closest string, the median string and the consensus string, studied under different distance metrics. Section 4 discusses the approach on consensus string problem via rank distance. The genetic algorithm to find and optimal consensus string is given in section 5. The hierarchical clustering algorithm is described in section 6. The experiments using mitochondrial DNA sequences from mammals are presented in section 7. Finally, the conclusion is drawn in section 8.

2   Preliminaries

A ranking is an ordered list that represents the result of applying an ordering criterion to a set of objects. A ranking defines a partial function on $\mathcal{U}$, where for each object $i \in \mathcal{U}$, $\tau(i)$ represents the position of the object i in the ranking τ.

The order of an object $x \in \mathcal{U}$ in a ranking σ of length d is defined by $ord(\sigma, x) = |d + 1 - \sigma(x)|$. By convention, if $x \in \mathcal{U} \setminus \sigma$, then $ord(\sigma, x) = 0$.

Definition 1. *Given two partial rankings σ and τ over the same universe $\mathcal{U}$, the rank distance between them is defined as:*

$$\Delta(\sigma, \tau) = \sum_{x \in \sigma \cup \tau} |ord(\sigma, x) - ord(\tau, x)|.$$

Rank distance is an extension of the Spearman footrule distance [3].

The rank distance is naturally extended to strings. The following observation is immediate: if a string does not contain identical symbols, it can be transformed directly into a ranking (the rank of each symbol is its position in the string). Conversely, each ranking can be viewed as a string, over an alphabet equal to the universe of the objects in the ranking. The next definition formalizes the transformation of strings, that can have multiple occurrences of identical symbols, into rankings.

Definition 2. *Let n be an integer and let $w = a_1 \ldots a_n$ be a finite word of length n over an alphabet Σ. The extension to rankings of w, is defined as $\bar{w} = a_{1,i(1)} \cdots a_{n,i(n)}$, where $i(j) = |a_1 \ldots a_j|_{a_j}$ for all $j = 1, \ldots n$ (i.e. the number of occurrences of a_j in the string $a_1 a_2 \ldots a_j$).*

Rank distance can be extended to arbitrary strings as follows.

Definition 3. *Given $w_1, w_2 \in \Sigma^*$:*

$$\Delta(w_1, w_2) = \Delta(\bar{w}_1, \bar{w}_2).$$

Example 1. Consider the following two strings $x = abcaa$ and $y = baacc$. Then, $\bar{x} = a_1 b_1 c_1 a_2 a_3$ and $\bar{y} = b_1 a_1 a_2 c_1 c_2$. Thus, the rank distance between x and y is the sum of the absolute differences between the orders of the characters in $\bar{x}$ and $\bar{y}$ (for missing characters, the maximum possible offset is considered):

$$\Delta(x, y) = |1 - 2| + |4 - 3| + 5 + |2 - 1| + |3 - 4| + 5 = 14.$$

The transformation of a string into a ranking can be done in linear time (by memorizing for each symbol, in an array, how many times it appears in the string). The computation of the rank distance between two rankings can also be done in linear time in the cardinality of the universe [10]. It is important to note that throughout this paper, the notion of string and the notion of ranking may be used interchangeably.

Let χ_n be the space of all strings of size n over an alphabet Σ and let $P = \{p_1, p_2, \ldots, p_k\}$ be k strings from χ_n. The closest (or centre) string problem (CSP) is to find the center of the sphere of minimum radius that includes all the k strings.

Problem 1. The *closest string problem via rank distance (CSRD)* is to find a minimal integer d (and a corresponding string t of length n) such that the maximum rank distance from t to any string in P is at most d. Let t denote the closest string to P, and d denote the radius. Formally, the goal is to compute:

$$\min_{x \in \chi_n} \max_{i=1..k} \Delta(x, p_i). \tag{1}$$

The median string problem (MSP) is similar to the closest string problem, only that the goal is to minimize the average distance to all the input strings, or equivalently, the sum of distances to all the input strings.

Problem 2. The *median string problem via rank distance (MSRD)* is to find a minimal integer d (and a corresponding string t of length n) such that the average rank distance from t to any string in P is at most d. Let t denote the median string of P. Formally, the goal is to compute:

$$\min_{x \in \chi_n} \operatorname*{avg}_{i=1..k} \Delta(x, p_i). \tag{2}$$

Finally, the consensus string problem is to minimize both the maximum and the average distance to all the input strings.

Problem 3. The *consensus string problem via rank distance* is to find a minimal integer d (and a corresponding string t of length n) such that the sum of the maximum and the average rank distance from t to any string in P is at most d. Let t denote the consensus string of P. Formally, the goal is to compute:

$$\min_{x \in \chi_n} \left(\max_{i=1..k} \Delta(x, p_i) + \operatorname*{avg}_{i=1..k} \Delta(x, p_i) \right). \tag{3}$$

3   Related Work

The closest and the median string problems use a particular distance, and various distance metrics usually conduct to different results. The most studied approach was the one based on edit distance. In [18], it is shown that closest string and median string via edit distance are NP-hard for finite and even binary alphabets. The existence of fast exact algorithms, when the number of input strings is fixed, is investigated in [17]. The first distance measure used for the CSP was the Hamming distance and emerged from a coding theory application [12]. The CSP via Hamming distance is known to be NP-hard [12]. There are recent studies that investigate CSP under Hamming distance with advanced programming techniques such as integer linear programming (ILP) [2]. In recent years, alternative distance measures were also studied. The work of [19] shows that the CSP via swap distance (or Kendall distance) and element duplication distance are also NP-hard.

This paper studies the consensus string problem under rank distance. Rank distance (RD) was introduced in [4] and has applications in biology [7,10], natural language processing, computer science and many other fields. Rank distance can be computed fast and benefits from some features of the edit distance. Theoretically, RD works on strings from any alphabet (finite or infinite). But, in many practical situations the alphabet is of fixed constant size. For example, in computational biology, the DNA and protein alphabets are respectively of size 4 and 20. For some applications, one needs to encode the DNA or protein sequences on a binary alphabet that expresses only a binary property of the molecule, for instance the hydrophoby, which is used in some protocols that identify similar DNA sequences [21]. In [9] it is shown that the CSRD is NP-hard. On the other hand, the median string problem is tractable [8] in the rank distance case. In [7] an approximation for CSRD based on genetic algorithms was proposed.

However, the consensus string problem was only studied recently. The work of [1] presents a first algorithm which tries to combine both problems: to solve the consensus string problem by minimizing both distance sum and radius in the same time. The problem was solved only for the set of three strings in the case of the Hamming distance measure, and, as authors state, the problem is still open for the edit distance or for more than three strings. The consensus problem was also studied for circular strings in [14]. It presents algorithms to find a consensus and an optimal alignment for circular strings by the Hamming distance. This paper investigates the consensus string problem under rank distance.

4   Approach and Discussion

There are a few things to consider for solving the consensus string problem in the case of rank distance. Before going into details, it is important to note that the CSRD and MSRD problems do not have unique solutions. In other words, there are more strings that may satisfy one of the conditions imposed by the two problems. There are three possible directions to follow that could lead to finding the consensus string. The first direction is to obtain all the closest strings, and then search for a median string among the closest strings. The second approach is to obtain all the median strings at first, and then search for a closest string among the median strings. The third and last approach is to directly minimize equation (3). It has been shown that there is at least one string that satisfies both conditions (1) and (2) in the same time, for a set of two strings [11]. However, the consensus problem was not discussed in the case of three or more strings. Theorem 1 shows that it is not always possible to minimize both the radius and the median distance, in the case of three or more strings.

Theorem 1. *A string that minimizes both equations* (1) *and* (2) *does not always exist, in the case of three or more strings.*

Proof. Let $x = 4321$, $y = 4312$ and $z = 2143$ be three strings with letters from $\Sigma = \{1, 2, 3, 4\}$. The closest strings of x, y and z are $CS = \{4213, 4123\}$ at a radius of 4. The median strings are $MS = \{4312, 4312\}$ at a median distance of 10. Observe that $CS \cap MS = \emptyset$. Thus, there is no solution that minimizes both the radius and the median distance. Despite this fact, a consensus string exists. The consensus string that minimizes equation (3) is $c = 4213$.   □

The proof of Theorem 1 points to the fact that a solution for Problem 3 can be found, despite it is sometimes impossible to minimize conditions (1) and (2) in the same time. This leads to the fact that trying to search either for a closest string among median strings, or for a median string among closest strings, might not lead to a solution of Problem 3. The only remaining approach is to find an algorithm to directly minimize equation (3).

Theorem 2. *The consensus problem via rank distance (defined in Problem 3) is NP-hard.*

Proof. Since the closest string problem via rank distance is NP-hard [9], it follows that Problem 3 is also NP-hard.   □

Because Problem 3 is NP-hard, an efficient genetic algorithm is employed to find a close-to-optimal solution for the consensus string problem via rank distance.

5   Genetic Algorithm for Rank Distance Consensus

Genetic algorithms simulate the biological process of natural selection. The algorithm proposed in this work applies this biological process to find a consensus string. It applies a set of operations on a population of chromosomes over a number of generations. The set of operations used are the ones inspired from nature, namely the crossover, the mutation, and the selection.

Algorithm 1. *Genetic algorithm for consensus string*
Input: *A set of strings from an alphabet Σ.*
Initialization: *Generate a random population that represents the first generation.*
Loop: *For a number of generations apply the next operations:*
  1. Apply the crossover according to the probability of having a crossover.
  2. Apply mutations according to the probability of having a mutation.
  3. Sort the chromosomes based on equation (3).
  4. Select the best candidates for the next generation using a density of probability.
Output: *Choose the best individual from the last generation as the optimal consensus string.*

All chromosomes are strings from an alphabet Σ. Each chromosome is a possible candidate for the consensus string. At first, chromosomes are randomly generated. In the case of DNA, chromosomes are actually DNA sequences, with characters from the alphabet $\Sigma = \{A, C, G, T\}$. The crossover operation between two chromosomes is straightforward. First, a random cutting point is generated. The prefixes of the two chromosomes remain in place, while the suffixes of the two chromosomes are swapped. This is the standard crossover operation inspired directly from nature. To apply a mutation to a certain chromosome, one position is randomly chosen. The character found at that position will be changed with another character from the alphabet Σ. Multiple mutations may appear in the same chromosome, although this is very unlikely.

To select the individuals for the next generation from the current one, a density of probability function is used. The new generation is involved in the next iteration of the algorithm. The first step is to sort the individuals using equation (3) so that chromosomes at the top minimize the sum of the maximum and the average rank distance. Then, indexes are generated according to the density of probability function from the top to the bottom of the list of candidates. The indexes close to the top have a greater probability of making it into the next generation. Thus, good candidates for the consensus string will occur more often in the next generation, while bad candidates will occur less often. The density probability function used to select the candidates for the next generation is the normal distribution of mean 0 and variance 0.081 on the interval $[0, 1]$:

$$f(x) = \frac{1}{\sqrt{2 \cdot \pi \cdot 0.0816}} e^{\frac{-x^2}{2 \cdot 0.0816}}$$

The motivation for using this fitness function is based on test results. This fitness functions reduces the number of generations that are required to obtain a close-to-optimal solution. Helped by the crossover and mutation operations, the fitness function has a good generalization capacity: it does not favor certain chromosomes which could narrow the solution space and lead to local minima solutions.

An interesting remark is that the proposed genetic algorithm can also be used to find a consensus substring. This is an important advantage of the algorithm if one wants to a find consensus between strings of different lengths, such as DNA sequences.

6   Hierarchical Clustering Based on Rank Distance Consensus

Many hierarchical clustering techniques are variations on a single algorithm: starting with individual objects as clusters, successively join the two nearest clusters until only one cluster remains. These techniques connect objects to form clusters based on their distance. Such algorithms do not provide a single partitioning of the data set, but an extensive hierarchy of clusters that merge with each other at certain distances. Apart from the choice of a distance function, another decision is needed for the linkage criterion to be used. Popular choices are single-linkage, complete-linkage, or average-linkage. A standard method of hierarchical clustering that uses rank distance is described in [10]. It presents a phylogenetic tree of several mammals comparable to the structure of trees reported by other research studies [15, 20]. But, a hierarchical clustering method designed to deeply integrate rank distance might perform better. The hierarchical clustering method presented here works directly on strings. It is based on rank distance, but instead of a linkage criterion it uses a different approach. More precisely, it determines a centroid string for each cluster and joins clusters based on the rank distance between their centroid strings. In [5] the median string is used as cluster centroid, while in [6] the closest string is used as cluster centroid. The algorithm proposed in this work employs the consensus string as cluster centroid. The algorithm is to be applied on data sets that contain objects represented as strings, such as text, DNA sequences, etc.

Algorithm 2. *Hierarchical clustering based on consensus rank distance*
Initialization: Compute rank distances between all initial strings.
Loop: Repeat until only one cluster remains
  1. Join the nearest two clusters to form a new cluster.
  2. Determine the consensus string of the newly formed cluster.
  3. Compute rank distances from this new consensus string to existing consensus strings.

The analysis of the computational complexity of Algorithm 2 is straightforward. Let m be the number of the input strings. The time required to compute rank distances between the input strings is $O(m^2)$. The algorithm builds a binary tree structure where the leaves are the initial m strings. Thus, it creates $m - 1$ intermediate clusters until only one cluster remains. The most heavy computational step is to determine the consensus string of a cluster which takes $O(n^3)$ time, where n is the string length. Usually n is much greater than m and the algorithm complexity in this case is $O(m \times n^3)$.

7   Experiments

7.1   Data Set

The consensus string is evaluated in the context of two important problems in bioinformatics: the phylogenetic analysis of mammals and the finding of common substrings in a set of given DNA strings. In the phylogenetic experiments presented in this paper, mitochondrial DNA sequence genome of the following 22 mammals, available in the EMBL database (http://www.ebi.ac.uk/embl/), is used: human (*Homo sapiens*, V00662), common chimpanzee (*Pan troglodytes*, D38116), pigmy chimpanzee (*Pan paniscus*, D38113), gorilla (*Gorilla gorilla*, D38114), orangutan (*Pongo pygmaeus*, D38115), sumatran orangutan (*Pongo pygmaeus abelii*, X97707), gibbon (*Hylobates lar*, X99256), horse (*Equus caballus*, X79547), donkey (*Equus asinus*, X97337), Indian rhinoceros (*Rhinoceros unicornis*, X97336), white rhinoceros (*Ceratotherium simum*, Y07726), harbor seal (*Phoca vitulina*, X63726), gray seal (*Halichoerus grypus*, X72004), cat (*Felis catus*, U20753), fin whale (*Balaenoptera physalus*, X61145), blue whale (*Balaenoptera musculus*, X72204), cow (*Bos taurus*, V00654), sheep (*Ovis aries*, AF010406), rat (*Rattus norvegicus*, X14848), mouse (*Mus musculus*, V00711), North American opossum (*Didelphis virginiana*, Z29573), and platypus (*Ornithorhynchus anatinus*, X83427). Additionally, the fat dormouse (*Myoxus glis*, AJ001562) genome is only considered in the DNA comparison experiment. In mammals, each double-stranded circular mtDNA molecule consists of 15,000-17,000 base pairs. DNA from two individuals of the same species differs by only 0.1%. This means, for example, that mtDNA from two different humans differs by less than 20 base pairs. Because this difference cannot affect the study, the experiments are conducted using a single mtDNA sequence for each mammal.

7.2 DNA Comparison

In this section an experiment is performed to show that finding the consensus string for a set of DNA strings is relevant from a biological point of view, thus being an interesting problem for biologists. The genetic algorithm described in section 5 is employed to find the consensus string. Only the rat, the house mouse, the fat dormouse and the cow genomes are used in the experiment. The task is to find the consensus string between the rat and house mouse DNAs, between the rat and fat dormouse DNAs, and between the rat and cow DNAs. The goal of this experiment is to compare the distances associated to the three consensus strings. The cow belongs to the Cetartiodactylae order, while the rat, the house mouse, and the fat dormouse belong to the Rodentia order. Expected results should show that the rat-house mouse distance and the rat-fat dormouse distance are smaller than the rat-cow distance. The same experiment was also conducted in [7] by using three genetic algorithms based on closest string via Hamming distance, edit distance and rank distance, respectively. To compare the results of these algorithms with the results obtained by the genetic algorithm proposed in this work, the same experiment setting is used. The genetic algorithm parameters used to obtain the presented results are: population size - 1800, number of generations - 300, crossover probability - 0.36, mutation probability - 0.1, size of each DNA sequence - 150. The results based on consensus string are presented in Table 1. It shows that the proposed genetic algorithm can efficiently find consensus strings. The reported times are computed by measuring

Table 1. Consensus string results obtained with the genetic algorithm

Consensus Results	rat-house mouse	rat-fat dormouse	rat-cow
Distance	1542	3493	8807
Average time	2.5 seconds	2.6 seconds	2.6 seconds

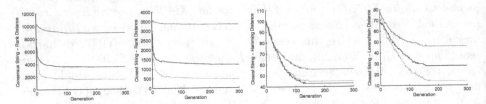

Fig. 1. The distance evolution of the best chromosome at each generation for the Rat-Mouse-Cow experiment. GREEN = rat-house mouse distance, BLUE = rat-fat dormouse, RED = rat-cow distance.

the average time of 10 runs of the genetic algorithm on a computer with Intel Core i7 2.3 GHz processor and 8 GB of RAM memory using a single Core.

Figure 1 shows the distance evolution of the best chromosome at each generation for consensus string via rank distance versus the closest string via rank distance, Hamming distance and Levenshtein distance, respectively. The consensus strings obtained by the genetic algorithm are biologically relevant in that they show a greater similarity between the rat DNA and house mouse DNA and a lesser similarity between the rat DNA and cow DNA. In terms of speed, the proposed method is similar to the other algorithms based on closest string via low-complexity distances (Hamming and rank distance).

7.3   Hierarchical Clustering Experiment

The second experiment is to apply the hierarchical clustering method proposed in section 6 on the 22 DNA sequences. The resulted phylogenetic tree is compared with phylogenetic trees obtained in [5, 6], but also with standard hierarchical clustering methods, such as the one presented in [10]. The phylogenetic tree obtained with Algorithm 2 (using the first 3000 nucleotides) are presented in Figure 2. Analyzing the phylogenetic tree in Figure 2, one can observe that the proposed clustering method is able to separate the Primates, Perissodactylae and Carnivora orders. By considering the Primates cluster, one can observe a really interesting fact: the human is separated from the other Primates. Overall, the phylogenetic tree is relevant from a biological point of view. However, there are two mistakes in the tree. First, the rat is clustered with the sheep instead of the house mouse. Second, the platypus should also be in the same cluster with the opossum and the house mouse. These results are comparable to the phylogenetic trees obtained in [5, 6]. It is interesting to note that the clusters near the root are influenced by the use of closest string, while the clusters near

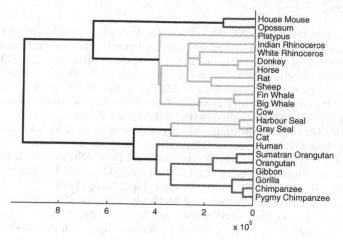

House Mouse
Opossum
Platypus
Indian Rhinoceros
White Rhinoceros
Donkey
Horse
Rat
Sheep
Fin Whale
Big Whale
Cow
Harbour Seal
Gray Seal
Cat
Human
Sumatran Orangutan
Orangutan
Gibbon
Gorilla
Chimpanzee
Pygmy Chimpanzee

8   6   4   2   0

x 10^5

Fig. 2. Phylogenetic tree obtained for 22 mammalian mtDNA sequences using consensus string via rank distance

the leaves are are influenced by the distance itself (in this case, rank distance). Thus, good results are in part an aftermath of rank distance.

By contrast to the results obtained here, in the phylogenetic tree presented in [10] the Perissodactylae and Carnivora orders are mixed together. The standard hierarchical clustering method used in [10] is also unable to join the rat with the house mouse. The phylogenetic tree presented in this work is also comparable to those obtained in other studies [15, 20]. Therefore, the hierarchical clustering method presented in this paper performs better or at least comparable to the state of the art clustering methods.

8   Conclusion

This paper investigated the consensus string problem under the rank distance metric. An efficient genetic algorithm to find a close-to-optimal solution for this problem was described. This approach has the advantage of finding the consensus for any number of strings, being suitable for computational biology applications. The DNA comparison experiment shows that the results of the genetic algorithm are indeed relevant from a biological point of view. Furthermore, phylogenetic experiments show that the tree produced by the hierarchical clustering algorithm that employs consensus string is better or at least comparable with those reported in the literature [5, 10, 15, 20]. The experiments also demonstrate the utility of the two algorithms described in this paper.

Acknowledgments. The authors thank the anonymous reviewers for helpful comments. The contribution of the authors to this paper is equal. This research was supported by a grant of the Romanian National Authority for Scientific Research, CNCS-UEFISCDI, project number PN-II-ID-PCE-2011-3-0959.

References

1. Amir, A., Landau, G.M., Na, J.C., Park, H., Park, K., Sim, J.S.: Consensus optimizing both distance sum and radius. In: Karlgren, J., Tarhio, J., Hyyrö, H. (eds.) SPIRE 2009. LNCS, vol. 5721, pp. 234–242. Springer, Heidelberg (2009)
2. Chimani, M., Woste, M., Bocker, S.: A closer look at the closest string and closest substring problem. In: Proceedings of ALENEX, pp. 13–24 (2011)
3. Diaconis, P., Graham, R.L.: Spearman footrule as a measure of disarray. Journal of Royal Statistical Society. Series B (Methodological) 39(2), 262–268 (1977)
4. Dinu, L.P.: On the classification and aggregation of hierarchies with different constitutive elements. Fundamenta Informaticae 55(1), 39–50 (2003)
5. Dinu, L.P., Ionescu, R.-T.: Clustering Based on Rank Distance with Applications on DNA. In: Huang, T., Zeng, Z., Li, C., Leung, C.S. (eds.) ICONIP 2012, Part V. LNCS, vol. 7667, pp. 722–729. Springer, Heidelberg (2012)
6. Dinu, L.P., Ionescu, R.T.: Clustering Methods Based on Closest String via Rank Distance. In: Proceedings of SYNASC, pp. 207–214 (2012)
7. Dinu, L.P., Ionescu, R.T.: An efficient rank based approach for closest string and closest substring. PLoS ONE 7(6), 37576 (2012)
8. Dinu, L.P., Manea, F.: An efficient approach for the rank aggregation problem. Theoretical Computer Science 359(1-3), 455–461 (2006)
9. Dinu, L.P., Popa, A.: On the closest string via rank distance. In: Kärkkäinen, J., Stoye, J. (eds.) CPM 2012. LNCS, vol. 7354, pp. 413–426. Springer, Heidelberg (2012)
10. Dinu, L.P., Sgarro, A.: A Low-complexity Distance for DNA Strings. Fundamenta Informaticae 73(3), 361–372 (2006)
11. Dinu, L.P., Sgarro, A.: Estimating Similarities in DNA Strings Using the Efficacious Rank Distance Approach, Systems and Computational Biology – Bioinformatics and Computational Modeling. InTech (2011)
12. Frances, M., Litman, A.: On covering problems of codes. Theory of Computing Systems 30(2), 113–119 (1997)
13. Koonin, E.V.: The emerging paradigm and open problems in comparative genomics. Bioinformatics 15, 265–266 (1999)
14. Lee, T., Na, J.C., Park, H., Park, K., Sim, J.S.: Finding consensus and optimal alignment of circular strings. Theoretical Computer Science 468, 92–101 (2013)
15. Li, M., Chen, X., Li, X., Ma, B., Vitanyi, P.M.B.: The similarity metric. IEEE Transactions on Information Theory 50(12), 3250–3264 (2004)
16. Liew, A.W., Yan, H., Yang, M.: Pattern recognition techniques for the emerging field of bioinformatics: A review. Pattern recognition 38(11), 2055–2073 (2005)
17. Nicolas, F., Rivals, E.: Complexities of the centre and median string problems. In: Baeza-Yates, R., Chávez, E., Crochemore, M. (eds.) CPM 2003. LNCS, vol. 2676, pp. 315–327. Springer, Heidelberg (2003)
18. Nicolas, F., Rivals, E.: Hardness results for the center and median string problems under the weighted and unweighted edit distances. Journal of Discrete Algorithms 3, 390–415 (2005)
19. Popov, Y.V.: Multiple genome rearrangement by swaps and by element duplications. Theoretical Computer Science 385(1-3), 115–126 (2007)
20. Reyes, A., Gissi, C., Pesole, G., Catzeflis, F.M., Saccone, C.: Where Do Rodents Fit? Evidence from the Complete Mitochondrial Genome of Sciurus vulgaris. Molecular Biology Evolution 17(6), 979–983 (2000)
21. States, D.J., Agarwal, P.: Compact encoding strategies for dna sequence similarity search. In: Proceedings of the 4th International Conference on Intelligent Systems for Molecular Biology, pp. 211–217 (1996)

Automatic Braille to Black Conversion

Filippo Stanco, Matteo Buffa, and Giovanni Maria Farinella

Dipartimento di Matematica e Informatica
University of Catania, Italy
{fstanco,buffa,gfarinella}@dmi.unict.it
http://iplab.dmi.unict.it

Abstract. The aim of this work is related to the production of inclusive technologies to help people affected by diseases, like the blindness. We present a complete pipeline to convert scanned Braille documents into classic printed text. The tool has been thought as support for assistants (e.g., in the education domain) and parents of blind and partially sighted persons (e.g., children and elderly) for the reading of Braille written documents. The software has been built and tested thanks to the collaboration with experts in the field [1]. Experimental results confirm the effectiveness of the proposed imaging pipeline in terms of conversion accuracy, punctuation, and final page layout.

Keywords: Braille, Optical Character Recognition, Blind, Visually Impaired.

1   Introduction and Motivation

The creation of new technologies to support people with diseases is one of the most important challenges posed by modern society at the research community. Among the diseases that can affect the humans, the visual impairment is increasing. The World Health Organization (WHO) estimated that there were more than 160 millions visually impaired people in the world in 2002, which means more than 2.5% of the total population. Specifically, the study of WHO reported that there were 124 millions of people with low vision capacity and 37 millions were blind [2].

The nowadays society has a high attention for the blind and partially sighted people (e.g., suitable books are available, medications and city maps report information for blind, tactile museums exist, etc.). Industry has produced dedicated hardware (e.g., printers and keyboards) and software (e.g., pdf readers by text to sound transduction, cash machine readers [3], etc.) for the people with low vision ability. Despite these efforts, too few tools have been developed to support assistants and parents of blind and partially sighted people for the reading of documents written for the blind.

The writing system used by blind and partially sighted people is the Braille. It is a tactile writing system invented by Louis Braille in the middle of 19th century.

M. Baldoni et al. (Eds.): AI*IA 2013, LNAI 8249, pp. 517–526, 2013.

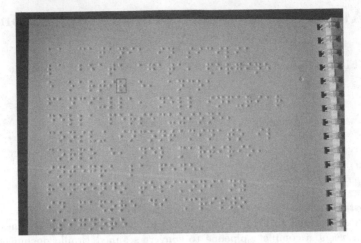

Fig. 1. Example of Braille document. Each character consists of a maximum of six dots arranged in a 3×2 grid (as example see the Braille letter "r" in region indicated by the red bounding box).

Fig. 2. Example of tablet for writing in Braille. On the right the awl used to produce the dots on the paper.

The Braille characters are small rectangular blocks containing tiny palpable dots. Each character consists of a maximum of six dots arranged in a 3×2 grid (Fig. 1).

The number and arrangement of the dots discriminate one character from another (i.e., the different letter of the alphabet and the symbols of punctuation). Some other rules are coded by coupling Braille characters in order to reproduce the classic writing (e.g., an uppercase character is preceded by another Braille symbol with the dots in the first and third row of the first column).

Braille-users can write with a portable braille note-taker like the one shown in Fig. 2. The Braille characters are produced with an awl which is used to impress the dots composing the letters. Braille characters are written from the right to the left in the rear of the Braille page. The reader will scroll the fingertip on the face of the sheet to perceive the dots in relief. The classic printed letters (i.e., the ones written with classic ink) are referred as "black" letters in the slang of blind people.

The present work has been inspired and supported by the Unione Italiana Ciechi e Ipovedenti (Italian Union of the Blind and Partially Sighted) [1] who expressed the need to have technologies useful for non-impaired people (e.g., assistants and parents of blind people) who wish to help blind and partially sighted in the context of education and more in general in other contexts of the public domain. Specifically, in the context of education an automatic Braille to black conversion software is useful to help not-impaired volunteers to interact with blind and visually impaired (e.g., in particular children and elderly) when a Braille document should be read. Indeed, to read a document written for the blind and visually impaired, the Braille system should be known. Despite experts can easily read a Braille document, this is not the case for most of us and more specifically for the parents of blind persons. As indicated by experts of [1], a tool for Braille to black conversion is extremely useful to solve, for instance, disputes between teachers and parents of a blind person. Those disputes usually happen in the context of education when the parents of a blind (who usually do not know the Braille system) do not agree with the outcome of the exams of their children. The problems above have motivated the development of the proposed Braille to black conversion software.

Although other similar works have been proposed in literature [4–8], most of them have been tested on few Braille documents, and no ones seems have been assessed with the support of blind and non blind experts [1] who provided a set of Braille documents considering the different encoding aspects of the Braille writing system. Moreover, the developed software has been properly tested taking into account the different rules encoded in the Braille documents, as well as page layout and punctuation (e.g., the uppercase characters are preceded by a specific Braille code).

The paper is organized as follows: Section 2 details the proposed Braille to black conversion method, whereas in Section 3 the experimental phase and the obtained results are reported. Finally, conclusions and hints for future works are given in Section 4.

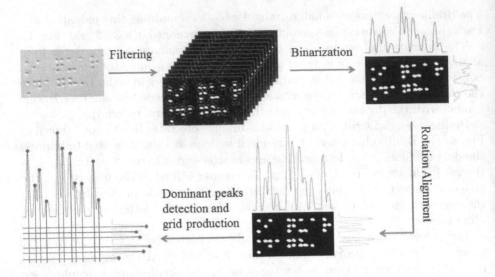

Fig. 3. Overall pipeline related the pre-processing and page layout construction

2   From Braille to Black

The aim of the proposed image processing pipeline is the conversion of a Braille document acquired with a flatbed scanner into text. The overall pipeline is composed by three main steps. As first, a pre-processing procedure is performed to remove rotation variability and to detect the grid layout (i.e., rows and columns) in which the dots composing the Braille characters could be located. Then, a dots detection algorithm is employed and each detected dot is assigned to a location of the grid obtained at the previous step. Finally, starting from the association of the dots to the grid layout and considering the rules of the Braille writing system, the conversion from Braille to black is performed. The three main steps involved in the proposed Braille to black conversion software are detailed in the following sub-sections.

2.1   Preprocessing and Page Layout

In this sub-section we detail the pre-processing step used to remove the rotation variability from the acquired Braille pages. The final aligned image is then processed to find the orthogonal grid layout useful to convert the Braille document into text. The overall pipeline related the pre-processing and page layout production is depicted in Fig. 3.

To the purposes of this step we employ an algorithm based on histogram analysis. The approach is similar to the ones usually employed in bio-medical domain to detect the grid of the spots in Miroarray images [9]. In the proposed algorithm, the scanned image is processed with a bank of rotationally invariant filters and then binarized through a thresholding procedure. Specifically, we filter

the image by employing the Schmid Filter Bank [10] which is composed by 13
isotropic "Gabor-like" filters of the form:

$$S(r, \sigma, \tau) = S_0(\sigma, \tau) + \cos\left(\frac{\pi \tau r}{\sigma}\right) e^{-\frac{r^2}{2\sigma^2}} \qquad (1)$$

This filtering allows the enhancement of the dots in order to exploit a simple
thresholding procedure for binarization purpose. Hence, from the filtered images
we obtain a binary image for each document. The binary image is employed
as input for the rotation alignment and grid layout production (Fig. 3). The
threshold used in our experiments has been fixed through a learning stage taking
in to account the accuracy of detecting dots vs. non-dots pixels on a manually
labeled dataset used as ground truth.

When the binary image is obtained, a procedure to estimate the rotation
of the acquired Braille page is performed. The rotation alignment algorithm
can be formalised as follows. Let M be the binary image obtained with the
aforementioned thresholding procedure, and f a function defined as following:

$$f(M) = max(h_r(M)) + max(h_c(M)) \qquad (2)$$

where $h_r(M)$ and $h_c(M)$ are respectively the distribution related the integral
projections of the rows and columns of the binary image M. Let M_α be the
binary map obtained by rotating M of α radiants. The estimation of the page
rotation angle α^* is defined as:

$$\alpha^* = \operatorname*{argmax}_{\alpha \in \{-\frac{\pi}{32}, \frac{\pi}{32}\}} (f(M_\alpha)) \qquad (3)$$

Both M and the original image I are hence rotated taking into account the estimated rotation angle α^* (Fig. 3). Once the aligned binary image M^* is obtained,
a peak and valley procedure [11] is employed on $h_r(M^*)$ and $h_c(M^*)$ to establish
the dominant peaks corresponding to the horizontal and vertical lines on which
the Braille dots could be located. Finally, the grid layout is produced considering
the rows and columns positions corresponding to the detected dominant peaks
in $h_r(M^*)$ and $h_c(M^*)$ respectively (Fig. 3). This final grid is used as layout to
convert the Braille letters into text as detailed in next sub-sections.

2.2   Dots Detection

The second step of our Braille to black conversion software is devoted to locate
the dots within the Braille page (Fig. 4). To properly detect the dots, the rotationally aligned image is processed with a 3×3 Prewitt filter as first step. This
is done in order to enhance the horizontal discontinuity (Fig. 4b). Then, a Sobel
operator followed by a linear smoothing are employed to highlight the borders of
each dot (Fig. 4c and 4d). The filtered image is hence used as input to detect the
circles (i.e., Braille dots) through the Circle Hough Transform [12] (Fig. 4e and
4f). The position of each detected dot is finally associated to the closest spatial
point of the layout grid obtained in the previous step (see Sub-Section 2.1).

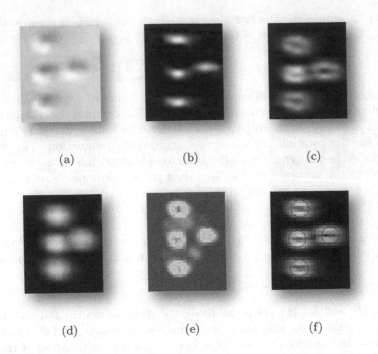

(a)   (b)   (c)

(d)   (e)   (f)

Fig. 4. (a) Original image. (b) Prewitt filter responce. (c) Sobel filtering. (d) Linear smoothing. (e) Circle Hough Transform Map. (f) Detected dots.

In this way, for each point of the grid layout we have a binary digital information (i.e., 1=*dot*, 0=*non-dot*) corresponding to the real Braille dots composing the letters in the document under consideration.

2.3   Document Conversion

At this stage we have a binary map obtained by coupling the grid layout together with the dots locations (see sub-sections 2.1 and 2.2). To convert the binary map into text, we divide the binary information in groups of six binary values by considering matrices with three rows and two columns (Fig. 5). Each matrix correspond to a Braille character. From each 3×2 matrix we obtain a binary code considering the order top to bottom and left to right (Fig. 5). To convert the binary code into text, a binary tree is used. Hence, each six-bits code is given as input to the tree and when the code reach a leaf it is labeled with the corresponding text letter. Note that differently than the ASCII code where consecutive letters or numbers have consecutive eight-bits binary codes (e.g., "a"=01100001 and "b"=01100010), the ordering is not respected in the six-bits codes of the Braille system (e.g, "a"=100000 and "b"=110000). Moreover, we have embedded all the rules related the Braille system in the conversion procedure in order to produce a correct final text document (e.g., numbers are

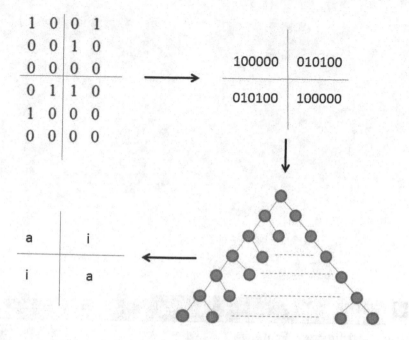

Fig. 5. Document conversion pipeline. Each six-bits binary code is converted to a letter when reach the leaf of the binary tree.

preceded by specific Braille codes, so two character should be considered when a Braille symbol indicating a number emerge). Since this work has been supported by the Unione Italiana Ciechi e Ipovedenti (Italian Union of the Blind and Partially Sighted) [1], both input and output documents used in our experiments are in Italian language. The extension of the proposed method to other languages is straightforward because the Braille alphabet does not change.

3   Experimental Results

A dataset composed by 55 Braille documents coming from journals and books has been used for testing purpose. The dataset contains both geometric and photometric variability (i.e., variability in terms of rotation, page dimensions, color of pages, contrast). The Braille documents have been selected by experts from Unione Italiana Ciechi e Ipovedenti (Italian Union of the Blind and Partially Sighted) by considering the different encoding aspects of the Braille writing system to properly test the proposed imaging pipeline with respect to the different rules embedded in the Braille documents (e.g., the number and position of the Braille dots related to the letter "a" and to the number "1" are identical. An extra Braille symbol, with three dots in the first column and just one dot in the lower row of the second column, precedes the one representing a number). Five random documents have been chosen and each dot within those documents

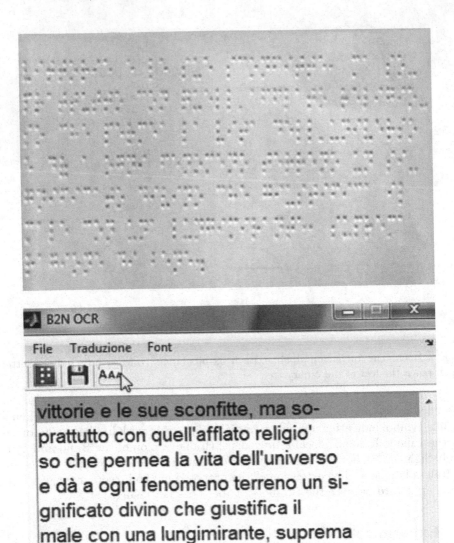

Fig. 6. A Braille document (top) and the related text obtained with the developed software (bottom)

has been manually labeled to have a ground truth for the learning stage. Those labeled documents have been used to learn the parameters involved in the steps related the pre-processing and dot detection (see sub-section 2.1 and 2.2). The remaining 50 Braille documents have been used for testing purposes. The total number of Braille characters in the test set is 40013 with an average of 774 character per document. Each document belonging to the test set has been converted with the developed software. The assessment of the conversion has been done with the help of experts (both blind and not blind) of the Unione Italiana Ciechi e Ipovedenti (Italian Union of the Blind and Partially Sighted) [1], who read each Braille document in order to have a ground truth to be compared with the documents produced by the automatic conversion. The comparison between the document transcription and the text obtained with our software has given an accuracy of 99,90% in terms of characters recognition (i.e., only 42 characters were wrongly recognized among 40013). Most of the unrecognized characters were found in old and very worn Braille documents. Moreover, no errors in terms of page layout and punctuation (i.e., structure and organization of the text) have been found in the converted documents. An example of a part of Braille document given as input to our software with the related conversion (in Italian) is shown in Fig. 6. A video demonstrating the developed software in action can be found at the following URL: `http://iplab.dmi.unict.it/Braille/`.

4   Conclusion and Future Works

We have presented a complete image processing pipeline for Braille to black conversion. The developed software has been assessed with the help of blind and not blind experts in the field on a set of real and representative Braille documents. The experimental results demonstrated good accuracy in terms of letters recognition, punctuation and final page layout. Future works will be devoted to an in-depth comparison of the proposed solution with respect to other approaches in literature (e.g., with respect to punctuation and final page layout). Moreover, the proposed imaging pipeline will be extended in order to be exploited with mobile devices (e.g., smarthpones, tablet) where the constraints related with the limited resources, as well as the unconstrained acquisition in terms of geometric and photometric variability make the task more challenging.

Acknowledgments. The authors would like to thank the Unione Italiana Ciechi e Ipovedenti (Italian Union of the Blind and Partially Sighted) [1] for inspiring, supporting and testing this work.

References

1. Unione Italiana Ciechi e Ipovedenti (Italian Union of the Blind and Partially Sighted), `http://www.uiciechi.it/`
2. Kocur, I., Parajasegaram, R., Pokharel, G.: Global Data on Visual Impairment in the Year 2002. Bulletin of the World Health Organization 82 (2004)

3. Bruna, A.R., Farinella, G.M., Guarnera, C.G., Battiato, S.: Forgery detection and value identification of Euro banknotes. Sensors 13(2), 2515–2529 (2013)
4. Mennens, J., Van Tichelen, L., Francois, G., Engelen, J.J.: Optical recognition of Braille writing using standard equipment. IEEE Transactions on Rehabilitation Engineering 2(4), 207–212 (1994)
5. Ng, C.M., Ng, V., Lau, Y.: Regular feature extraction for recognition of Braille. In: Proceedings of the Third International Conference on Computational Intelligence and Multimedia Applications, pp. 302–306 (1999)
6. Hentzschel, T.W., Blenkhorn, P.: An optical reading system for embossed Braille characters using a twin shadows approach. International Journal of Microcomputer Applications 18(4), 341–354 (1995)
7. Wong, L., Abdulla, W., Hussmann, S.: A software algorithm prototype for optical recognition of embossed Braille. In: Proceedings of the International Conference on Pattern Recognition, vol. 2, pp. 586–589 (2004)
8. Tai, Z., Cheng, S., Verma, P., Zhai, Y.: Braille document recognition using Belief Propagation. Journal of Visual Communication and Image Representation 21(7), 722–730 (2010)
9. Battiato, S., Di Blasi, G., Farinella, G.M., Gallo, G., Guarnera, G.C.: Adaptive Techniques for Microarray Image Analysis with Related Quality Assessment. Journal of Electronic Imaging 16(4), 1–20 (2007)
10. Schmid, C.: Constructing models for content-based image retrieval. In: Proceedings of the IEEE Conference on Computer Vision and Pattern Recognition (2001)
11. Sezgin, M., Sankur, B.: Survey over image thresholding techniques and quantitative performance evaluation. Journal of Electronic Imaging 13(1), 146–168 (2004)
12. Atherton, T.J., Kerbyson, D.J.: Size invariant circle detection. Image and Vision Computing 17(11) (1999)

A Hybrid Neuro–Wavelet Predictor for QoS Control and Stability

Christian Napoli, Giuseppe Pappalardo, and Emiliano Tramontana

Dipartimento di Matematica e Informatica, University of Catania
Viale Andrea Doria 6, 95125 Catania, Italy
{napoli,pappalardo,tramontana}@dmi.unict.it

Abstract. For distributed systems to properly react to peaks of requests, their adaptation activities would benefit from the estimation of the amount of requests. This paper proposes a solution to produce a short-term forecast based on data characterising user behaviour of online services. We use *wavelet analysis*, providing compression and denoising on the observed time series of the amount of past user requests; and a *recurrent neural network* trained with observed data and designed so as to provide well-timed estimations of future requests. The said ensemble has the ability to predict the amount of future user requests with a root mean squared error below 0.06%. Thanks to prediction, advance resource provision can be performed for the duration of a request peak and for just the right amount of resources, hence avoiding over-provisioning and associated costs. Moreover, reliable provision lets users enjoy a level of availability of services unaffected by load variations.

Keywords: Neural networks, wavelet analysis, QoS, adaptive systems.

1 Introduction

General public internet usage has become an essential everyday matter, e.g. widespread use of social–networks and mass communication media, public administration, home banking, etc. Moreover, the request for stable and continuous internet services has reached a pressing priority. Generally, the solutions employed by service providers to adapt server-side resources on-the-fly have been based on *content adaptation* and *differentiated service* strategies [1,9,3,12,13]. However, such strategies could over-deteriorate the service during load peaks. Moreover, for ensuring a minimum level of quality (QoS), even when sudden variations on the number of requests arise, a large number of (over-provisioned) resources is often used, hence incurring into relevant costs and wasted resources for a considerable time interval. Some approaches guarantee a minimum QoS level once the connection has been established (i.e. non-adaptive multimedia services) [11,24], or by using bandwidth adaption algorithms (i.e. adaptive wireless services) [18]. However, such solutions are still liable to an effective loss of QoS for end users, when we consider content quality or availability, and,

M. Baldoni et al. (Eds.): AI*IA 2013, LNAI 8249, pp. 527–538, 2013.

528 C. Napoli, G. Pappalardo, and E. Tramontana

in the worst case, even denial of services (DoS). As an alternative to deterministic scheduling [17] or genetic algorithms [28], neural networks have been used within the area of a multi-objective optimisation problems. In [2] authors provide dynamic admission control while preserving QoS by using hardware-based Hopefield neural networks [33].

When autoregressive moving average models (ARMA), and related generalisations, are used, the underlying assumption, even for the study forecasting host load [10], makes them inappropriate for predictions in non-linear systems exhibiting high levels of variations. Artificial neural networks have been used in several ways to provide an accurate model of the QoS evolution over time. Linear regression models have been applied with the support of neural networks in [16], however without some proper mechanisms, such as time delays or feedback, it is still not possible to dynamically follow the evolution of the extended time series. Machine learning approaches have also been used [25], however such approaches were not designed for on-the-fly adaptation, and are unable to give advantages with respect to user perceived responsiveness [27]. For the above approaches in which the amount of connection requests is unknown, to avoid overloading the server-side, only load balancing and admission control policies have been used. Still when the amount of requests overcomes the available resources, service usability worsening or denial of service cannot be avoided. On the other hand, when more resources can be dynamically allocated, since the amount of the required resources is unknown in advance, it often results in over-provisioning, with negative effects on management and related cost.

This paper investigates the use of Second Generation Wavelet Analysis and Recurrent Neural Network (RNN) to predict over time the amount of connection requests for a service, by using a hybrid wavelet recurrent neural network (WRNN). Recurrent neural networks have been proven powerful enough to gain advantages from the statistical properties of time series. In our experiments, the proposed WRNN has been used to analyse data for the Page view statistics Wikimedia(TM) project, produced by Domas Mituzas and released under Creative Common License[1]. Wavelet analysis has been used in order to reduce data redundancies so as to obtain a representation that can express their intrinsic structure, while the neural networks have been used to have the complexity of non-linear data correlational perform data prediction. Thereby, a relatively accurate forecast of the connection request time series can be achieved even when load peaks arise. The estimated result is fundamental for a management service that performs resource preallocation on demand. The precision of our estimates allows just the right amount of resources to be used.

The rest of this paper is structured as follows. In section 2 the background on wavelet theory is given. Section 3 describes second generation wavelets and how their properties are useful for the proposed forecast. Section 4 provides the proposed neural network configuration. Section 5 reports on the performed experiments and results. Finally, Section 6 draws our conclusions.

---

[1] See dumps.wikimedia.org/other/pagecounts-raw

2   The Basis of Wavelet Theory

This work builds on wavelets and neural networks to model the main characteristics of user behaviour for an online service. Wavelet decomposition is a powerful analysis tool for physical and dynamic phenomena that reduces the data redundancies and yields a compact representation expressing the intrinsic structure of a phenomenon. In fact, the main advantage when using wavelet decomposition is the ability to pack the energy signature of a signal or a time series, and then to express relevant data as a few non-zero coefficients. This characteristic has been proven very useful to optimise the performances of neural networks [14]. Like sine and cosine for Fourier transforms, a wavelet decomposition uses functions, i.e. wavelets, to express a function as a particular expansion of coefficients in the wavelet domain. Once a mother wavelet has been chosen, it is possible, as explained in the following, to create new wavelets by dilates and shifts of the mother wavelet. Such novel generated wavelets, if chosen with certain criteria, eventually form a Riesz basis of the Hilbert space $L^2(\mathbb{R})$ of square integrable functions. Such criteria are at the basis of *wavelet theory* and come from the concept of multiresolution analysis of a signal, also called multiscale approximation. When a dynamic model can be expressed as a time-dependent signal, i.e. described by a function in $L^2(\mathbb{R})$, then it is possible to obtain a multiresolution analysis of such a signal. For the space $L^2(\mathbb{R})$ such an approximation consists in an increasing sequence of closed subspaces which approximate, with a greater amount of details, the space $L^2(\mathbb{R})$, eventually reaching a complete representation of $L^2(\mathbb{R})$ itself. A complete description of multiresolution analysis and the relation with wavelet theory can be found in [20].

One-dimensional decomposition wavelets of order n for a signal $s(t)$ give a new representation of the signal itself in an n-dimensional multiresolution domain of coefficients plus a certain residual coarse representation of the signal in time. For any discrete time step τ then, the corresponding M order wavelet decomposition $\mathbf{W}s(\tau)$ of the signal $s(\tau)$ will be given by the vector

$$\mathbf{W}s(\tau) = [d_1(\tau), d_2(\tau), \ldots, d_M(\tau), a_M(\tau)] \quad \forall \, \tau \in \{\tau_1, \tau_2, \ldots, \tau_N\} \quad (1)$$

where d_1 is the most detailed multiresolution approximation of the series, and d_M the least detailed, and a_M is the residual signal. Such coefficients express some intrinsic time-energy feature of a signal, i.e. features of a time series, while removing redundancies, and offering a well suited representation, which, as described in Section 3, we give as inputs for a neural network.

It is now possible to give a more rigorous definition of a wavelet. Let us take into account a multiresolution decomposition of $L^2(\mathbb{R})$

$$\varnothing \subset V_0 \subset \ldots \subset V_j \subset V_{j+1} \subset \ldots \subset L^2(\mathbb{R})$$

If we call W_j the orthogonal complement V_j, then it is possible to define a wavelet as a function $\psi(x)$ if the set of $\{\psi(x - l) | l \in \mathbb{Z}\}$ is a Riesz basis of W_0 and also meets the following two constraints:

$$\int_{-\infty}^{+\infty} \psi(x)dx = 0 \quad (2)$$

and

$$||\psi(x)||^2 = \int_{-\infty}^{+\infty} \psi(x)\psi^*(x)dx = 1$$

If the wavelet is also an element of V_0 then it exists a sequence $\{g_k\}$ such that

$$\psi(x) = 2\sum_{k\in\mathbb{Z}} g_k\psi(2x - l)$$

then the set of functions $\{\psi_{j,l}|j,l \in \mathbb{Z}\}$ is now a Riesz basis of $L^2(\mathbb{R})$. It follows that a wavelet function can be used to define an Hilbert basis, that is a complete system, for the Hilbert space $L^2(\mathbb{R})$. In this case, the Hilbert basis is constructed as the family of functions $\{\psi_{j,l}|j,l \in \mathbb{Z}\}$ by means of dilation and translation of a mother wavelet function ψ so that $\psi_{j,l} = \sqrt{2^j}\psi(2^j x - l)$. Hence, given a function $f \in L^2(\mathbb{R})$ it is possible to obtain the following decomposition

$$f(x) = \sum_{j,l\in\mathbb{Z}} \langle f|\psi_{j,l}\rangle = \sum_{j,l\in\mathbb{Z}} d_{j,l}\psi_{j,l}(x) \tag{3}$$

where $d_{j,l}$ are called wavelet coefficients of the given function f in the wavelet basis given by the inner product of $\psi_{j,l}$. Likewise, a projection on the space V_j is given by

$$\mathbb{P}_j f(x) = \sum_i \langle f|\varphi_{i,j}\rangle\varphi_{i,j}(x)$$

where $\varphi_{i,j}$ are called dual scaling functions. When the basis wavelet functions coincide with their duals the basis is orthogonal. Choosing a wavelet basis for the multiresolution analysis corresponds to selecting the dilation and shift coefficients. In this way, by performing the decomposition we obtain the $\{d_i|a_M\}$ coefficients sets of (1). From now on we will refer to the described schema as first generation wavelets, whilst Section 3 describes second generation wavelets.

For the present work, we adopted Biorthogonal wavelet decomposition (this wavelet family is described in [20]), for which symmetrical decomposition and exact reconstruction are possible with finite impulse response (FIR) filters [26].

3   Second Generation Wavelets with RNNs

A multiresolution analysis like the one described in Section 2 can be realised by conjugate wavelet filter banks that decompose the signal, and act similarly to a low/high pass filter couple [29]. An advanced solution has been devised to obtain the same decomposition using a lifting and updating procedure. This procedure, named *second generation wavelet*, takes advantage of the properties of multiresolution wavelet analysis starting from a very simple set-up and gradually building up a more complex multiresolution decomposition to have some specific properties. The lifting procedure is made of a space-domain superposition of biorthogonal wavelets developed in [30].

The construction is performed by an iterative procedure called *lifting and prediction*. Lifting consists of splitting the whole time series $x[\tau]$ into two disjoint subsets $x_e[\tau]$ and $x_o[\tau]$, for the even and odd positions, respectively; whereas prediction consists of generating a set of coefficients $d[\tau]$ representing the error of extrapolation of time series $x_o[\tau]$ from series $x_e[\tau]$. Then, an update operation combines the subsets $x_e[\tau]$ and $d[\tau]$ in a subset $a[\tau]$ so that

$$\begin{cases} d[\tau] = x_o[\tau] - \mathbf{P}x_e[\tau] \\ a[\tau] = x_e[\tau] + \mathbf{U}d[\tau] \end{cases} \tag{4}$$

where $\mathbf{P}$ is the prediction operator, and $\mathbf{U}$ is the update operator. Eventually, one cycle for the above procedure creates a complete set of discrete wavelet transforms (DWT) and the relative coefficients $d[\tau]$ and $a[\tau]$. It follows that

$$\begin{cases} x_e[\tau] = a[\tau] - \mathbf{U}d[\tau] \\ x_o[\tau] = d[\tau] + \mathbf{P}x_e[\tau] \end{cases} \tag{5}$$

The said construction yields discrete wavelet filters that preserve only a certain number N of low order polynomials in the time series. Having such low-order polynomials, in turn, makes it possible to apply non-linear predictors without affecting the $a[\tau]$ coefficients, which provide a coarse approximation of the time series itself.

A neural network can be build to perform such a construction, i.e. a neural network would act as an inverse second generation wavelet transform. In [6], a neural network with a rich representation of past outputs like a fully connected recurrent neural network (RNN), known as the Williams-Zipser network or Nonlinear Autoregressive network with eXogenous inputs (NARX) [31], has been proven able to generalise and reproduce the behaviour of $\mathbf{P}$ and $\mathbf{U}$ operators, and to structure itself to behave as an optimal discrete wavelet filter. Moreover, for such a kind of RNNs, when applied to the prediction and modelling of stochastic phenomena, like the considered behaviour of users, which lead to a variable number of access requests in time, real time recurrent learning (RTRL) has been proven to be very effective. A complete description of RTRL algorithm, NARX and RNNs can be found in [32,15].

RTRL has been used to train the RNN and such a trained RNN achieves the ability to perform lifting stages, hence the matching of the time series dynamics at the corresponding wavelet scale. This construction brings the possibility to match non-polynomial and nonlinear signal structures in an optimised straightforward N-dimension means square problem [21]. NARX networks have been proven able to use the intrinsic features of time series in order to predict the following values of the series [23]. One of a class of transfer functions for the RNN has to be chosen to approximate the input-output behaviour in the most appropriate manner. For phenomena having a deterministic dynamic behaviour, the relative time series at a given time point can be modelled as a functional of a certain amount of previous time steps. In such cases, the used model should have some internal memory to store and update context information [19]. This is achieved by feeding the RNN with a delayed version of past data, commonly referred as time delayed inputs [8].

4 Proposed Setup for WRNN Predictor

As stated in Section 3, it would be desired to have a neural network able to perform the wavelet transform as in a recursive lifting procedure. For this, we could use a mother wavelet as transfer function, however mother wavelets lack of some elementary properties needed by a proper transfer function, such as e.g. the absence of local minima and a sufficient graded and scaled response [14]. This leads us to look for a close enough substitute to approximate the properties of a mother wavelet without affecting the functionalities of the network itself. The function classes that more closely approximate a mother waveform have to be found among the Radial Basis Functions (RBFs) that are good enough as transfer functions and partially approximate half of a mother waveform. It is indeed possible to properly scale and shift a couple of RBFs to obtain a mother wavelet. If we define an RBF function as an $f : [-1, 1] \to \mathbb{R}$ then we could dilate and scale it to obtain a new function

$$\tilde{f}(x + 2l) = \begin{cases} +f(2x+1) & x \in [-1, 0) \\ -f(2x-1) & x \in (0, +1] \end{cases} \quad \forall \, l \in \mathbb{Z} \tag{6}$$

With such a definition, starting from the properties of the RBF, it is then possible to verify the following

$$\int_{2h+1}^{2k+1} \tilde{f}(x)dx = 0 \quad \forall \, (h, k) \in \mathbb{Z}^2 : h < k \tag{7}$$

Starting from (6) and (7) it is possible to verify the equations (2) to (3) for the chosen $\tilde{f}$ which we can now call a mother wavelet. The chosen mother wavelet is a composition of two RBF transfer functions that are realised by the proposed neural network to obtain the properties of a wavelet transform. The proposed WRNN has two hidden layers with RBF transfer function.

For this work, the initial dataset was a time series representing access requests coming from users. We call this series $x(\tau)$, where τ is the discrete time step of the data, sampled with intervals of one hour. A biorthogonal wavelet decomposition of the time series has been computed to obtain the correct input set for the WRNN as required by the devised architecture. This decomposition has been achieved by applying the wavelet transform as a recursive couple of conjugate filters (Figure 1) in such a way that the i-esime recursion $\hat{W}_i$ produces, for any time step of the series, a set of coefficients d_i and residuals a_i, and so that

$$\hat{W}_i[a_{i-1}(\tau)] = [d_i(\tau), a_i(\tau)] \quad \forall \, i \in [1, M] \cap \mathbb{N} \tag{8}$$

where we intend $a_0(\tau) = x(\tau)$. The input set can then be represented as an $N \times (M+1)$ matrix of N time steps of a M level wavelet decomposition (see Figure 2), where the τ-esime row represents the τ-esime time step as the decomposition

$$\mathbf{u}(\tau) = [d_1(\tau), d_2(\tau), \dots, d_M(\tau), a_M(\tau)] \tag{9}$$

Each row of this dataset is given as input value to the M input neurons of the proposed WRNN. The properties of this network make it possible, starting from

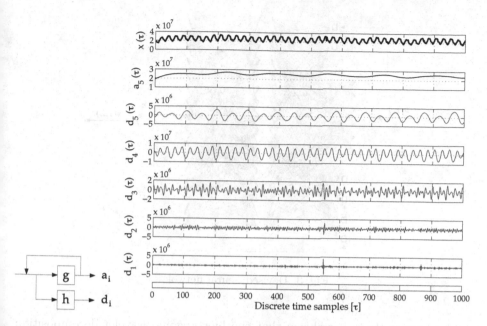

Fig. 1. Filter logic   **Fig. 2.** 5-level biorthogonal 3.7 wavelet decomposition

an input at a time step τ_n, to predict the effective number of access requests at a time step τ_{n+r}. In this way the WRNN acts like a functional

$$\hat{N}[\mathbf{u}(\tau_n)] = x(\tau_{n+r}) \tag{10}$$

where r is the number of time steps of forecast in the future.

5 Experimental Setup

For this work, a 4-level wavelet decomposition has been selected that properly characterises data under analysis. Therefore, the devised WRNN (Figure 3) uses a 5 neuron input layer (one for each level detail coefficient d_i and one for the residual a_5). This WRNN architecture presents two hidden layers with sixteen neurons each and realises a radial basis function as explained in Section 4.

Inputs are given to the WRNN in the following form:

- The wavelet decomposition of the time series $\mathbf{u}(\tau_n)$ for time step τ_n
- The previous delayed decompositions $\mathbf{u}(\tau_{n-1})$ and $\mathbf{u}(\tau_{n-2})$
- The last four delayed outputs $x(\tau_{n+r})$ predicted by the WRNN

Delays and feedback are obtained by using the relative delay lines and operators (D). These feedback lines provide the WRNN with internal memory, hence the modelling abilities for dynamic phenomena.

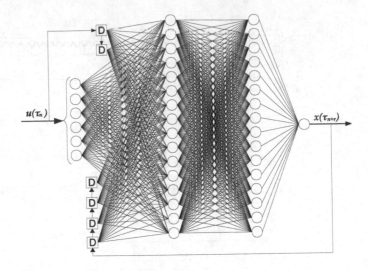

Fig. 3. Devised neural netwok

An accurate study has shown that the biorthogonal wavelet decomposition optimally approximates and denoises the time series under analysis. Such a wavelet family is in good agreement with previous optimal results, obtained by the authors, for the decomposition of other physical phenomena. In fact, such a decomposition splits a phenomenon in a superposition of mutual and concurrent predominant processes with a characteristic time-energy signature. For stochastically-driven processes, such as stellar oscillations [7], renewable energy and systems load [5,4], and for a large category of complex systems [22], wavelet decomposition gives a unique and compact representation of the leading features for a time-variant phenomenon.

For the case study proposed in this paper we have used the raw data of connection requests over time to predict the behaviour of the users of a widely used an internet service, i.e. the one provided by Wikimedia(TM). Raw data were taken from the page-view statistics Wikimedia(TM) project and released by Domas Mituzas under Creative Common License[2]. Original data report the amount of accesses and bytes for the replies that were sampled in time-steps of one hour for each web page accessible in the project. Data were collected for the whole services offered by Wikimedia projects including Wikipedia(r), Wikidictionary(r), Wikibooks(r) and others. Data were gathered and composed by an automatic procedure, obtaining the total requests made to the wikimedia servers for each hour. Therefore, a 2-years long dataset of hourly sampled access requests has been reconstructed. Then, this dataset was decomposed by using a wavelet biorthogonal decomposition identified by the couple of numbers 3.7, which means that are implemented by using FIR filters with 7th order polynomials degree for the decomposition (see also Figure 4 and 5) and 3rd order for the reconstruction.

---

[2] See dumps.wikimedia.org/other/pagecounts-raw

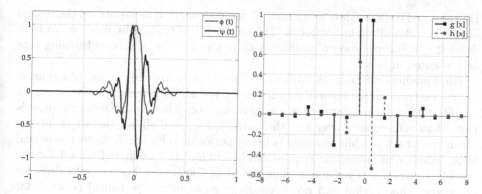

Fig. 4. Biorthogonal wavelet 3.7 decomposition (ψ) and scaling (ϕ) functions

Fig. 5. Biorthogonal wavelet 3.7 high-pass (h) and low-pass (g) filers

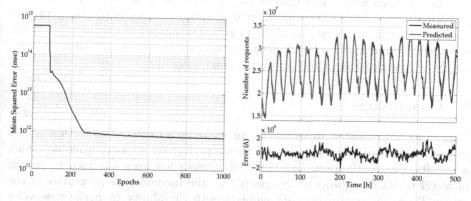

Fig. 6. Error while training the network

Fig. 7. Actual time series and predicted values

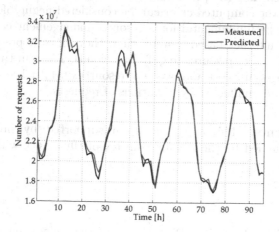

Fig. 8. Top of the curves for actual time series and predicted values, for an arbitrary chosen time-interval

Decomposed data, as shown in Figure 2, were then given as inputs for the neural network as described in Section 4. The network was trained by using a gradient descent back-propagation algorithm with momentum led adaptive learning rate as presented in [15].

For a prediction 6 hours in advance of the time series of the amount of requests, the root mean squared error of prediction for the access requests over time was of $1.3156 \cdot 10^{04}$ requests, which means a relative error of less than 0.6 per thousands (less than six requests over ten thousands). Figure 6 shows the shape of the mean square error while training is being performed. Figure 7 shows the actual time series of the incoming requests in black, and the predicted values for the incoming requests in red for a time period of 500 hours. The actual values for the shown time period had not been given as input to the neural network for training, however have then been used to compare with the predicted values and to compute the error (see the bottom part of Figure 7). A smaller period of time has been shown in Figure 8 to highlight the differences of actual and predicted values. As can be seen, the neural network manages to closely predict even relatively small variations of the trends. The output of the neural network was then given to a resource management service to perform allocation requests in terms of needed bandwidth and virtual machines [12].

6   Conclusions

This paper has provided an ad-hoc architecture for a neural network that is able to predict the amount of incoming requests performed by users when accessing a website. Firstly, the past time series of accesses has been analysed by means of wavelets, which appropriately retain only the fundamental properties of the series. Then, the neural network embeds both the ability to perform wavelet analysis and prediction of future amount of requests. The performed experiments have proven that the provided ensemble is very effective for the desired prediction, since the computed error can be considered negligible.

Estimates can be fundamental for a resource management component, on a server side of an internet based system, since they make it possible to acquire just the right amount of resource (e.g. from a cloud). Then, in turn it is possible to avoid an unnecessary cost and waste of resources, whilst keeping the level of QoS as desired and unaffected by variations of requests.

Acknowledgments. This work has been supported by project PRISMA PON04a2 A/F funded by the Italian Ministry of University within PON 2007-2013 framework.

References

1. Abdelzaher, T.F., Shin, K.G., Bhatti, N.: Performance guarantees for web server end-systems: A control-theoretical approach. IEEE Transactions on Parallel and Distributed Systems 13, 80–96 (2002)

2. Ahn, C.W., Ramakrishna, R.: Qos provisioning dynamic connection-admission control for multimedia wireless networks using a hopfield neural network. Transactions on Vehicular Technology 53(1), 106–117 (2004)
3. Bannò, F., Marletta, D., Pappalardo, G., Tramontana, E.: Tackling consistency issues for runtime updating distributed systems. In: Proceedings of International Symposium on Parallel & Distributed Processing, Workshops and Phd Forum (IPDPSW), pp. 1–8. IEEE (2010), doi:10.1109/IPDPSW.2010.5470863
4. Bonanno, F., Capizzi, G., Gagliano, A., Napoli, C.: Optimal management of various renewable energy sources by a new forecasting method. In: Proceedings of International Symposium on Power Electronics, Electrical Drives, Automation and Motion (SPEEDAM), pp. 934–940. IEEE (2012)
5. Capizzi, G., Bonanno, F., Napoli, C.: A wavelet based prediction of wind and solar energy for long-term simulation of integrated generation systems. In: Proceedings of International Symposium on Power Electronics Electrical Drives Automation and Motion (SPEEDAM), pp. 586–592. IEEE (2010)
6. Capizzi, G., Napoli, C., Bonanno, F.: Innovative second-generation wavelets construction with recurrent neural networks for solar radiation forecasting. Transactions on Neural Networks and Learning Systems 23(11), 1805–1815 (2012)
7. Capizzi, G., Napoli, C., Paternò, L.: An innovative hybrid neuro-wavelet method for reconstruction of missing data in astronomical photometric surveys. In: Rutkowski, L., Korytkowski, M., Scherer, R., Tadeusiewicz, R., Zadeh, L.A., Zurada, J.M. (eds.) ICAISC 2012, Part I. LNCS, vol. 7267, pp. 21–29. Springer, Heidelberg (2012)
8. Connor, J.T., Martin, R.D., Atlas, L.: Recurrent neural networks and robust time series prediction. Transactions on Neural Networks 5(2), 240–254 (1994)
9. Di Stefano, A., Fargetta, M., Pappalardo, G., Tramontana, E.: Supporting resource reservation and allocation for unaware applications in grid systems. Concurrency and Computation: Practice and Experience 18(8), 851–863 (2006), doi:10.1002/cpe.980
10. Dinda, P.A., O'Hallaron, D.R.: Host load prediction using linear models. Cluster Computing 3(4), 265–280 (2000)
11. Epstein, B., Schwartz, M.: Reservation strategies for multi-media traffic in a wireless environment. In: Proceedings of Vehicular Technology Conference, vol. 1, pp. 165–169. IEEE (1995)
12. Giunta, R., Messina, F., Pappalardo, G., Tramontana, E.: Providing qos strategies and cloud-integration to web servers by means of aspects. In: Concurrency and Computation: Practice and Experience (2013), doi:10.1002/cpe.3031
13. Giunta, R., Messina, F., Pappalardo, G., Tramontana, E.: Kaqudai: a dependable web infrastructure made out of existing components. In: Proceedings of Workshops on Enabling Technologies: Infrastructure for Collaborative Enterprises (WETICE). IEEE (2013), doi:10.1109/WETICE.2013.47
14. Gupta, M.M., Jin, L., Homma, N.: Static and dynamic neural networks: from fundamentals to advanced theory. Wiley-IEEE Press (2004)
15. Haykin, S.: Neural networks and learning machines, vol. 3. Prentice Hall New York (2009)
16. Islam, S., Keung, J., Lee, K., Liu, A.: Empirical prediction models for adaptive resource provisioning in the cloud. Future Generation Computer Systems 28(1), 155–162 (2012)
17. Kang, C.G., Kim, Y.J., Hwang, M.J.: Implicit scheduling algorithm for dynamic slot assignment in wireless atm networks. Electronics Letters 34(24), 2309–2311 (1998)

18. Kwon, T., Park, I., Choi, Y., Das, S.: Bandwidth adaption algorithms with multi-objectives for adaptive multimedia services in wireless/mobile networks. In: Proceedings of International Workshop on Wireless Mobile Multimedia, pp. 51–59. ACM (1999)

19. Lapedes, A., Farber, R.: A self-optimizing, nonsymmetrical neural net for content addressable memory and pattern recognition. Physica D: Nonlinear Phenomena 22(1), 247–259 (1986)

20. Mallat, S.: A wavelet tour of signal processing: the sparse way. Academic Press (2009)

21. Mandic, D.P., Chambers, J.: Recurrent neural networks for prediction: Learning algorithms, architectures and stability. John Wiley & Sons, Inc. (2001)

22. Napoli, C., Bonanno, F., Capizzi, G.: Exploiting solar wind time series correlation with magnetospheric response by using an hybrid neuro-wavelet approach. In: Advances in Plasma Astrophysics. Proceedings of the International Astronomical Union, vol. S274, pp. 250–252. Cambridge University Press (2010)

23. Napoli, C., Bonanno, F., Capizzi, G.: An hybrid neuro-wavelet approach for long-term prediction of solar wind. In: Advances in Plasma Astrophysics. Proceedings of the International Astronomical Union, vol. S274, pp. 247–249. Cambridge University Press (2010)

24. Novelli, G., Pappalardo, G., Santoro, C., Tramontana, E.: A grid-based infrastructure to support multimedia content distribution. In: Proceedings of the Workshop on Use of P2P, GRID and Agents for the Development of Content Networks (UPGRADE-CN), pp. 57–64. ACM (2007), doi:10.1145/1272980.1272983

25. Powers, R., Goldszmidt, M., Cohen, I.: Short term performance forecasting in enterprise systems. In: Proceedings of the Eleventh ACM SIGKDD International Conference on Knowledge Discovery in Data Mining, pp. 801–807. ACM (2005)

26. Rabiner, L.R., Gold, B.: Theory and application of digital signal processing, 777 p. Prentice-Hall, Inc., Englewood Cliffs (1975)

27. Schechter, S., Krishnan, M., Smith, M.D.: Using path profiles to predict http requests. Computer Networks and ISDN Systems 30(1), 457–467 (1998)

28. Sherif, M.R., Habib, I.W., Nagshineh, M., Kermani, P.: Adaptive allocation of resources and call admission control for wireless atm using genetic algorithms. Journal on Selected Areas in Communications 18(2), 268–282 (2000)

29. Sweldens, W.: Lifting scheme: A new philosophy in biorthogonal wavelet constructions. In: Proceedings of Symposium on Optical Science, Engineering, and Instrumentation, pp. 68–79. International Society for Optics and Photonics (1995)

30. Sweldens, W.: The lifting scheme: A construction of second generation wavelets. Journal on Mathematical Analysis 29(2), 511–546 (1998)

31. Williams, R.J.: A learning algorithm for continually running fully recurrent neurren neural networks. Neural Computation 1, 270–280 (1989)

32. Williams, R.J., Zipser, D.: Experimental analysis of the real-time recurrent learning algorithm. Connection Science 1(1), 87–111 (1989)

33. Zadeh, M.H., Seyyedi, M.A.: Qos monitoring for web services by time series forecasting. In: Proceedings of International Conference on Computer Science and Information Technology (ICCSIT), vol. 5, pp. 659–663. IEEE (2010)

Author Index